WITHDRAWN BY THE
UNIVERSITY OF MICHIGAN

ELEMENTARY PARTIAL DIFFERENTIAL EQUATIONS with BOUNDARY VALUE PROBLEMS

ELEMENTARY PARTIAL DIFFERENTIAL EQUATIONS with BOUNDARY VALUE PROBLEMS

LARRY C. ANDREWS
University of Central Florida

Academic Press College Division
(Harcourt Brace Jovanovich, Publishers)
Orlando San Diego San Francisco New York
London Toronto Montreal Sydney Tokyo São Paulo

Math
QA
374
.A5241
1986

For Louise, Lori, and Jim

Copyright © 1986 by Academic Press, Inc.
All rights reserved.

No part of this publication may be reproduced or transmitted in any form or by any means, electronic or mechanical, including photocopy, recording, or any information storage and retrieval system, without permission in writing from the publisher.

Academic Press, Inc.
Orlando, Florida 32887

United Kingdom edition published by Academic Press, Inc.
(London) Ltd.
24/28 Oval Road, London NW1 7DX

ISBN: 0-12-059510-9

Library of Congress Catalog Card Number: 85-70364

Printed in the United States of America

math
390-0812
math
6-3-86

CONTENTS

PART II
PARTIAL DIFFERENTIAL EQUATIONS 113

PREFACE

This textbook is an introductory treatment of partial differential equations and boundary value problems. It is intended for students in mathematics, engineering, and physical sciences who are in either their junior or senior year. In some cases it may also serve as an introductory textbook for beginning graduate students. The background necessary to study this book includes the basic calculus sequence and a first course in ordinary differential equations.

The principal objective of the book is to present most of the standard solution techniques used in solving boundary value problems, using a blend of theory and basic physical examples. For instance, a simple physical example is often presented first and used for motivation in developing the accompanying theory. I have tried to present the mathematical theory in a careful manner without the abstract concepts of functional analysis. Most of the major theorems are stated, but some without proof. When a proof is omitted, a reference for the proof is usually cited. While there are numerous physical examples used throughout the text, no special background beyond a basic physics course should be necessary to understand them or to follow the derivations of their governing equations.

The material is organized so that eigenvalue problems and Fourier series are covered before solving partial differential equations by any general techniques. The method of separation of variables, which is the primary tool for solving partial differential equations, is one of the few techniques that works on a wide variety of problems. For this reason it provides a unifying framework in which to solve problems involving the heat, wave, and potential equations. In addition to the method of separation of variables, the methods of Fourier integrals and integral transforms are used where appropriate; the eigenfunction expansion method is used to solve nonhomogeneous equations; and some approximation and numerical techniques are discussed.

To aid the student in studying the material in this book, I have included over 100 worked-out problems and more than 1100 exercises. The exercise sets contain a blend of routine problems, more difficult problems, and some that extend the theory and applications beyond the exposition. More challenging problems, or those used to extend the theory or applications, are marked with a star (⋆). Answers are provided for all odd-numbered problems at the end of the book.

The text is divided into three parts, with the bulk of the material presented in Part II. Part I consists of Chapters 1 and 2, which deal only with ordinary differential equations. The first chapter provides a comprehensive treatment of boundary value problems, including the important study of Sturm–Liouville problems, while the Green's function for both initial value and boundary value problems is presented in Chapter 2. Chapters 3 through 8 make up Part II of the text. The basic concepts of partial differential equations are presented in Chapter 3. Chapter 4 begins with Fourier trigonometric series, moves on to generalized Fourier series, and ends with Fourier integral representations. The integral transforms of Laplace and Fourier are introduced in Chapter 5, and Chapters 6 through 8 provide a systematic coverage of the heat, wave, and potential equations, respectively. Part III contains Chapters 9 and 10, which deal with special functions (primarily Legendre polynomials and Bessel functions) and their role in solving partial differential equations in coordinate systems other than rectangular. As supplementary material, a table of some of the most common Laplace transforms and their inverses is provided in Appendix A, with a similar table of Fourier transforms in Appendix B. Because of the frequent need throughout the text to discuss the solutions of linear systems of equations, a short review of Cramer's rule and its role in developing the theory of simultaneous equations is presented in Appendix C.

Since 1974, a two-course sequence in boundary value problems has been offered at the University of Central Florida to a mix of mathematics, engineering, and physics students. The lecture notes developed for that sequence of courses have finally led to this textbook. However, I have tried to arrange the material so that it is equally well-suited for a more traditional one-semester or one-term course. To help in this regard, I have marked optional sections with an open box (□). These sections can be easily bypassed for a shorter course without any loss of continuity. Naturally, there are other sections and chapters that can be omitted, depending on the instructor. A standard one-semester or one-term course can be developed from material in Chapters 3, 4, 6 through 8, and parts of 1, 5, 9, and 10. In some cases Chapter 1 need not be covered except for a short review on certain selected topics before Chapter 3.

I owe many thanks to all the students who sat through my course over the last ten years while this text was being written. Their patience, understanding, and helpful suggestions are very much appreciated. I wish also to thank my colleagues and friends, Frank L. Salzmann and Robert C. Brigham, who made numerous suggestions while teaching from early drafts of the manuscript. The difficult task of checking and correcting the answers was done by Lisa M. Potchen. I am extremely grateful to Philip Crooke, Vanderbilt University; Edward M. Landesman, University of California, Santa Cruz; Roland DiFranco, University of the Pacific; Herman Gollwitzer, Drexel University; and Monty J. Strauss, Texas Tech University; who served as reviewers and provided many helpful comments. Finally, I wish to express my appreciation to Wesley Lawton, Mathematics Editor, and the entire production staff of Academic Press for their efforts in producing this text.

ELEMENTARY PARTIAL DIFFERENTIAL EQUATIONS with BOUNDARY VALUE PROBLEMS

PART I

ORDINARY DIFFERENTIAL EQUATIONS

CHAPTER 1

BOUNDARY VALUE AND EIGENVALUE PROBLEMS

Most of the theoretical discussions and applications presented in a first course in differential equations (DEs) involve *initial value problems*. This class of problems is characterized by auxiliary conditions all specified at a single value of the independent variable. If the auxiliary conditions are specified at more than one point on the interval of interest, the resulting problem is called a *boundary value problem*. It is primarily this latter class of problems that concern us here. In such problems the auxiliary conditions are usually specified at only two points of the interval, and we refer to these as *two-point boundary value problems*. The DEs associated with boundary value problems are therefore at least second-order, since first-order DEs have only one auxiliary condition and are therefore classified as initial value problems.

In general, initial value problems are "well-behaved" in that they almost always lead to unique solutions. Unfortunately, the same is not true of boundary value problems, and so the general theory of them is inherently more complicated (and more interesting) than that of initial value problems.

We begin our discussion of boundary value problems in Section 1.2 with an introductory application problem involving *steady-state heat conduction* in a thin rod. By deriving the governing DE and describing some possible boundary conditions, we illustrate how boundary value problems can arise naturally in the formulation of a mathematical model of a physical problem. We also use this introductory section for motivation in studying the *general theory* concerning the solutions of boundary value problems in Section 1.3. In sharp contrast with initial value problems, some boundary value problems possess *nontrivial solutions* even when the equation and boundary conditions are all *homogeneous*. This situation is further examined in Section 1.4 in connection with *eigenvalue problems,* a class of problems for which the solution depends upon a parameter λ occuring in the equation. By appropriately selecting λ, an infinite collection of nontrivial solutions (called *eigenfunctions*) can often be found corresponding to an infinite number of values of λ (called *eigenvalues*).

Some additional *physical applications* leading to boundary value problems are discussed in Section 1.5. In particular, we look at the small displacements of an elastic string and an elastic beam, both of which are supporting a distributed load, and related eigenvalue problems including the famous *buckling problem of Euler*.

In Section 1.6 we investigate the general properties shared by the eigenvalues and eigenfunctions belonging to all *regular, periodic,* and *singular Sturm-Liouville systems*. Foremost among these properties is the *orthogonality* of the eigenfunctions, which has far-reaching consequences.

The *method of weighted residuals* is introduced in the final section of the chapter as an *approximation technique* for solving boundary value problems. In addition to handling nonhomogeneous equations, it is a useful method for approximating the eigenvalues and eigenfunctions of various Sturm-Liouville problems.

1.1 INTRODUCTION

A *linear differential equation* (DE) of second-order has the form

$$A_2(x)y'' + A_1(x)y' + A_0(x)y = F(x), \qquad x_1 < x < x_2. \tag{1}$$

The functions $A_0(x)$, $A_1(x)$, and $A_2(x)$ are called the *coefficients* of the DE and are generally assumed to be continuous in the interval (x_1, x_2), and $A_2(x)$ is not identically zero there. In most applications the unknown function y in (1) must satisfy certain restraints or *auxiliary conditions,* the number of which is usually two (equal to the order of the DE). When these conditions are both specified at a single point we call them *initial conditions,* whereas when they are specified at two points (viz., at $x = x_1$ and $x = x_2$), we call them *boundary conditions*. Typical boundary conditions are of the general form

$$\begin{aligned} a_{11}y(x_1) + a_{12}y'(x_1) &= \alpha, \qquad a_{11}^2 + a_{12}^2 \neq 0,\\ a_{21}y(x_2) + a_{22}y'(x_2) &= \beta, \qquad a_{21}^2 + a_{22}^2 \neq 0, \end{aligned} \tag{2}$$

where the a's, α, and β are constants.

The problem of solving (1) subject to the auxiliary conditions (2) is called a *boundary value problem.* Unless the boundary conditions are carefully chosen, a boundary value problem may not have a unique solution, or in fact any solution at all! We say the mathematical problem is *properly-posed* (or *well-posed*) if it satisfies the following requirements:

1. *Existence*—there is at least one solution
2. *Uniqueness*—there is at most one solution
3. *Stability*—the solution depends continuously on the boundary data

The last requirement is important in practice since the boundary data are usually measured, and small errors are naturally introduced in the measurement process. The solution is said to be stable if small errors in these measurements lead, at most, to small changes in the solution. Physical problems associated with boundary value problems often suggest the correct type of boundary conditions under which the problem is properly posed, and for this reason it is helpful to interpret a problem physically, when possible, for guidance in the choice of appropriate boundary conditions. Most boundary conditions arising in practice are specializations of (2), called *unmixed* or *separated boundary conditions,* since the first is specified only at the endpoint $x = x_1$ and the second only at the endpoint $x = x_2$. More general boundary conditions of the form

$$\begin{aligned} a_{11}y(x_1) + a_{12}y'(x_1) + b_{11}y(x_2) + b_{12}y'(x_2) &= \alpha, \\ a_{21}y(x_2) + a_{22}y'(x_2) + b_{21}y(x_1) + b_{22}y'(x_1) &= \beta. \end{aligned} \tag{3}$$

(called *mixed boundary conditions*) are sometimes prescribed, but these are less common than (2). Still other variations occur from time to time.

The DE (1) is said to be *homogeneous* if $F(x) \equiv 0$ in the interval of interest, and similarly, when $\alpha = \beta = 0$ we say the boundary conditions are *homogeneous.** All other specifications of the DE or boundary conditions are called *nonhomogeneous.* In this chapter we will address mostly homogeneous DEs, subject to both homogeneous and nonhomogeneous boundary conditions. An important feature of a linear, homogeneous DE is that if y_1 and y_2 are both solutions of this DE, so is the linear combination $C_1y_1 + C_2y_2$ for any choice of constants C_1 and C_2.

Generally speaking, boundary value problems are closely associated with problems of static equilibrium configurations or steady-state phenomena. One of the oldest applications involving a boundary value problem concerns the buckling of a long, slender rod or column supporting a compressive load. Vertical columns have been used extensively in Greek and Roman structures throughout the centuries. Prior to the time of Euler† in the eighteenth century, columns were designed according to empirical formulas developed by the architects of the early civilizations. Needless to say, by today's standards, such empirical formulas give only a crude estimate of the weight a particular column design can support without collapsing. As a measure of safety such columns were often made much larger than necessary to ensure support of the structure above them. Euler developed the first truly mathematical model that can rather accurately predict the "critical com-

---

* We also say a linear DE or boundary condition is *homogeneous* if whenever y is a solution of the DE or boundary condition, so is Cy for any constant C.

† LEONHARD EULER (1707–1783), considered to be one of the five greatest mathematicians of all time, was born in Basel, Switzerland. Some of his main contributions to mathematics are the addition theorem for elliptic integrals, the analytic treatment of trigonometric functions as numerical ratios, the famous Euler formulas relating sines and cosines to exponentials with imaginary arguments, and the discovery of the gamma function (Section 9.3). He made significant contributions in mechanics, number theory, geometry, fluid dynamics, astronomy, and optics, and he developed much of the theory of differential equations. His collected works fill more than 70 volumes, and although he was blind during the last 17 years of his life, he still continued his work. He died on September 18, 1783, at the age of 76.

pressive load'' that a column can withstand before deformation or ''buckling'' takes place.

The mathematical model developed by Euler is still being used in many designs. It belongs to a particular class of boundary value problems called *eigenvalue problems*. Such a mathematical model is essential today in the design of buildings, aircraft, and marine frameworks, since slender columns play a fundamental role in the plan of these structures. Because the total weight of the structure is often a factor, a column that weighs more than necessary is as unacceptable as one that cannot support its intended load.

1.2 STEADY-STATE HEAT CONDUCTION: AN EXAMPLE

Because boundary value problems are closely associated with physical applications, it may be useful to start our discussion of them by showing how a mathematical model arises out of a given physical problem. In this way we will have an immediate physical interpretation of the terms in the governing DE and of the boundary conditions. Also, this development may help motivate some of the mathematical theory in subsequent sections. For an illustrative example, we choose the steady-state conduction of heat in a cylindrical rod of uniform material. For simplicity, it is assumed that the temperature distribution we seek is the same throughout any cross section of the rod, but may change from cross section to cross section.

Consider a long cylindrical rod or wire coinciding with a portion of the x-axis over $0 \leqslant x \leqslant b$ (see Figure 1.1). To begin, suppose the end of the rod at $x = 0$ is held at constant temperature T_1 and the end at $x = b$ is held at constant temperature T_2. Physical intuition or experience suggests that after a sufficiently long period of time, the temperature inside the rod will no longer change in time, i.e., a *steady-state* temperature distribution will have been reached.* It is this steady-state temperature distribution, which we denote by $y(x)$, that we wish to determine.

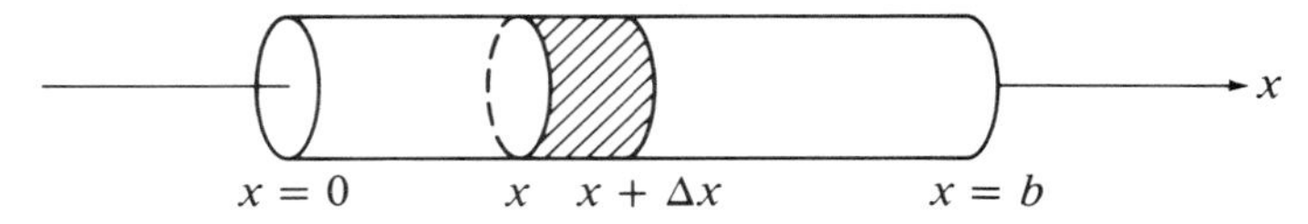

FIGURE 1.1

A cylindrical rod.

The temperatures prescribed at the endpoints of the rod are called *boundary conditions*, and they are mathematically described by

$$y(0) = T_1, \qquad y(b) = T_2. \tag{4}$$

*We assume the medium surrounding the rod is not changing temperature in time, e.g., if the lateral surface is insulated.

At other points of the rod the temperature $y(x)$ will be governed by a certain DE, which we wish to derive. To do so, we will apply the principle of heat balance *(conservation of thermal energy)* to a small slice of the rod between x and $(x + \Delta x)$, as shown in Figure 1.1. Basically, the heat balance requirement is the amount of heat entering the slice at x plus that generated inside the slice, equaling the amount of heat leaving the slice at $(x + \Delta x)$.

Let $Q(x)$ denote the horizontal rate of *heat flow* at the point x, measured in units of heat per unit time per unit area. If A is the cross-sectional area of the rod, then $AQ(x)$ is the rate at which heat enters the slice at x. Similarly, the rate at which heat leaves the slice at $(x + \Delta x)$ is $AQ(x + \Delta x)$. In addition, if heat enters the slice by means other than through the two faces (e.g., through the lateral surface), the rate at which heat is generated in the slice is $A\Delta xF(x)$, where F is the rate of heat generation per unit volume. Combining terms, the assumption of heat balance implies that

$$AQ(x) + A\Delta xF(x) = AQ(x + \Delta x), \tag{5}$$

which can be rewritten as

$$\frac{Q(x + \Delta x) - Q(x)}{\Delta x} = F(x). \tag{6}$$

Observe that the cross-sectional area A drops out. If $Q(x)$ is continuous, then in the limit $\Delta x \to 0$ we find that (6) becomes

$$Q'(x) = F(x), \qquad 0 < x < b. \tag{7}$$

Finally, we need to relate $Q(x)$ and the temperature $y(x)$. From experimental evidence, we know that the heat flow rate $Q(x)$ is proportional to the temperature gradient $y'(x)$.* In symbols,

$$Q(x) = -ky'(x), \tag{8}$$

where k is a positive constant. (The negative sign reflects the fact that heat flows from hotter to cooler regions.) Hence, the substitution of (8) into (7) leads to the governing DE

$$y'' = -\frac{1}{k}F(x), \qquad 0 < x < b. \tag{9}$$

By combining (4) and (9), we obtain the *boundary value problem*

$$y'' = -\frac{1}{k}F(x), \qquad 0 < x < b, \qquad y(0) = T_1, \qquad y(b) = T_2. \tag{10}$$

Other varieties of governing DE and boundary conditions for this heat conduction problem are also possible. For example, if the lateral surface of the rod is insulated and there are no heat sources within the rod, then $F(x) \equiv 0$ and the governing DE (9) reduces to

---

* This is known as *Fourier's law*.

$$y'' = 0, \qquad 0 < x < b. \tag{11}$$

If the lateral surface is not insulated but heat exchanges occur at the lateral surface by convection into the surrounding medium, then the rate of change of heat transfer in the rod is known to be proportional to the difference between the temperature in the rod and the temperature of the surrounding medium, according to *Newton's law of cooling*. Thus, we write

$$F(x) = -h[y(x) - T_0], \tag{12}$$

where h is a positive constant and T_0 is the temperature of the surrounding medium. [Note that heat enters the rod if $y(x) < T_0$, and leaves the rod if $y(x) > T_0$.] In this case the governing DE becomes

$$ky'' - hy = -hT_0, \qquad 0 < x < b. \tag{13}$$

Suppose both ends of the rod are insulated instead of being held at constant temperatures. This means there is no heat loss at the endpoints (i.e., no change in temperature), and we write the boundary conditions as

$$y'(0) = 0, \qquad y'(b) = 0. \tag{14}$$

Finally, if one end of the bar (at $x = b$) permits loss of heat due to radiation into the surrounding medium while the other end is maintained at constant temperature, the boundary conditions take the form

$$y(0) = T_1, \qquad -y'(b) = h[y(b) - T_0], \tag{15}$$

Still other variations of the DE and boundary conditions are possible.

EXAMPLE 1

Find the steady-state temperature distribution in a rod of unit length whose lateral surface is insulated and ends are maintained at temperatures T_1 and T_2.

SOLUTION Because the lateral surface is insulated, the governing DE is (11) and the problem is therefore characterized by

$$y'' = 0, \qquad 0 < x < 1, \qquad y(0) = T_1, \qquad y(1) = T_2.$$

By integrating the DE two successive times, we obtain the general solution

$$y = C_1 + C_2 x.$$

Applying the boundary conditions, we get

$$y(0) = C_1 = T_1,$$

$$y(1) = C_1 + C_2 = T_2,$$

from which we deduce $C_1 = T_1$ and $C_2 = T_2 - T_1$. Hence, the solution we seek is the linear function

$$y = T_1 + (T_2 - T_1)x.$$

Although we have confined our attention here to a simple problem of heat conduction, there are many similarities that carry over to other applications. That is, the same mathematical problem may arise in two or more entirely different situations, and for that reason it is instructive to study the general theory of boundary value problems common to all such applications.

EXERCISES 1.2

In Problems 1–6, determine the steady-state temperature distribution for the given problems.

1. $y'' = 0, \quad y(0) = T_1, \quad y(b) = T_2$
2. $y'' = 0, \quad y'(0) = 0, \quad y(1) = T_2$
3. $y'' = 0, \quad y(0) - y'(0) = T_1, \quad y(1) = 0$
4. $y'' = -T_0, \quad y'(0) = 0, \quad y(1) = 0$
5. $y'' - y = 0, \quad y(0) = T_1, \quad y(b) = T_1$
   Hint: Express the solution in terms of hyperbolic functions $\cosh x = (e^x + e^{-x})/2$ and $\sinh x = (e^x - e^{-x})/2$.
6. $y'' - y = -T_0, \quad y(0) = 0, \quad y(1) = 0$

★7. The temperature in a cooling fin is governed by the boundary value problem

$$y'' = h(y - T_0), \qquad y(0) = T_1, \qquad -y'(b) = h[y(b) - T_0]$$

where h, T_0, and T_1 are constants.
(a) Find the solution $y(x)$.
(b) What solution is obtained in the limit $b \to \infty$?

★8. Let $u(r)$ denote the steady-state temperature distribution in a solid bounded by two concentric spheres of radii $r = 1$ and $r = 2$. If $u = 0$ on the inner surface and $u = T_0$ on the outer surface, where T_0 is constant, find the temperature distribution in the region between the concentric spheres if the governing DE is

$$\frac{d^2}{dr^2}(ru) = 0.$$

★9. Under what conditions will the following problem have a solution?

$$y'' = 0, \qquad y'(0) = T_1, \qquad y'(1) = T_2$$

1.3 BOUNDARY VALUE PROBLEMS

The heat conduction problems discussed in the previous section are specializations of the more general second-order DE

$$A_2(x)y'' + A_1(x)y' + A_0(x)y = F(x), \qquad x_1 < x < x_2 \tag{16}$$

and *unmixed* boundary conditions

$$\begin{aligned} a_{11}y(x_1) + a_{12}y'(x_1) &= \alpha, \qquad (a_{11}^2 + a_{12}^2 \neq 0) \\ a_{21}y(x_2) + a_{22}y'(x_2) &= \beta, \qquad (a_{21}^2 + a_{22}^2 \neq 0). \end{aligned} \tag{17}$$

To discuss the general theory concerning solutions of (16) and (17), it is useful to introduce simplifying notation. First, let us introduce the *differential operator* ($D = d/dx$)

$$M = A_2(x)D^2 + A_1(x)D + A_0(x) \tag{18}$$

for which

$$M[y] \equiv A_2(x)y'' + A_1(x)y' + A_0(x)y. \tag{19}$$

Similarly, we introduce *boundary operators* B_1 and B_2 defined by

$$\begin{aligned} B_1[y] &\equiv a_{11}y(x_1) + a_{12}y'(x_1) \\ B_2[y] &\equiv a_{21}y(x_2) + a_{22}y'(x_2). \end{aligned} \tag{20}$$

In terms of these operators, we can express (16) and (17) more compactly as

$$M[y] = F(x), \qquad x_1 < x < x_2, \qquad B_1[y] = \alpha, \qquad B_2[y] = \beta. \tag{21}$$

Operators M, B_1, and B_2 are examples of what we call *linear operators*. In general, we say an operator M is linear if and only if

$$M[C_1f(x) + C_2g(x)] = C_1M[f(x)] + C_2M[g(x)] \tag{22}$$

where $f(x)$ and $g(x)$ are given functions and C_1 and C_2 are any constants. Verifying that M, B_1, and B_2 all satisfy (22) is left to the exercises (see Problem 17 in Exercises 1.3).

1.3.1 HOMOGENEOUS DEs

When $F(x) \equiv 0$ in (21), we say the DE is *homogeneous* and the problem is then characterized by

$$M[y] = 0, \qquad x_1 < x < x_2, \qquad B_1[y] = \alpha, \qquad B_2[y] = \beta. \tag{23}$$

If y_1 and y_2 are linearly independent solutions of $M[y] = 0$, the general family of solutions is given by

$$y = C_1y_1(x) + C_2y_2(x), \tag{24}$$

where C_1 and C_2 are arbitrary constants. Determining the values of C_1 and C_2 so that (24) also satisfies the boundary conditions in (23) is our remaining task. Imposing the prescribed boundary conditions upon the solution function (24) yields the *system of linear equations*

$$\begin{aligned} C_1B_1[y_1] + C_2B_1[y_2] &= \alpha, \\ C_1B_2[y_1] + C_2B_2[y_2] &= \beta, \end{aligned} \tag{25}$$

from which we deduce the values

$$C_1 = \frac{\begin{vmatrix} \alpha & B_1[y_2] \\ \beta & B_2[y_2] \end{vmatrix}}{\Delta} = \frac{\alpha B_2[y_2] - \beta B_1[y_2]}{\Delta}, \tag{26}$$

$$C_2 = \frac{\begin{vmatrix} B_1[y_1] & \alpha \\ B_2[y_1] & \beta \end{vmatrix}}{\Delta} = \frac{\beta B_1[y_1] - \alpha B_2[y_1]}{\Delta}, \tag{27}$$

(via Cramer's rule*) where Δ is the coefficient determinant defined by

$$\Delta = \begin{vmatrix} B_1[y_1] & B_1[y_2] \\ B_2[y_1] & B_2[y_2] \end{vmatrix} \tag{28}$$

Here we see that the solution of (23) is formally† given by (24), where the constants C_1 and C_2 are explicitly defined by (26) and (27). Owing to the fact that Δ can sometimes be zero, the evaluation of the constants C_1 and C_2 can lead to certain difficulties. For example, the problem (23) generally has *no solution* at all when $\Delta = 0$ unless the numerators of (26) and (27) are also both zero, and then the problem has an *infinite* number of *solutions*.

For instance, suppose $\alpha = \beta = 0$ in (23). The resulting problem is then called a *homogeneous* boundary value problem, and in this case it always has the solution $y = 0$, called the *trivial* solution. If $\Delta \neq 0$, then $C_1 = C_2 = 0$, and $y = 0$ is the only solution of (23). Solutions that are not identically zero are called *nontrivial* solutions. For nontrivial solutions to exist when $\alpha = \beta = 0$, we must also have $\Delta = 0$, which then leads to the indeterminate form 0/0 for both C_1 and C_2. This indeterminate form for both constants suggests that (26) and (27) are equivalent, and one constant, either C_1 or C_2, can be expressed as a multiple of the other through use of either equation. In summary, we have the following theorem.

THEOREM 1.1

If y_1 and y_2 are linearly independent solutions of $M[y] = 0$, then the homogeneous boundary value problem

$$M[y] = 0, \quad x_1 < x < x_2, \quad B_1[y] = 0, \quad B_2[y] = 0$$

has a nontrivial solution if and only if $\Delta = 0$; if $\Delta \neq 0$, the problem has only the trivial solution, $y = 0$.

Let us illustrate Theorem 1.1 with some examples.

EXAMPLE 2

Solve: $y'' + y = 0, \quad 0 < x < \pi, \quad y(0) = 0, \quad y(\pi) = 0.$

SOLUTION The general solution of the DE is

$$y = C_1 \cos x + C_2 \sin x.$$

* See Appendix C for a review of Cramer's rule.

† By *formal* solution we mean one that has not been rigorously verified.

Applying the prescribed boundary conditions, we find

$$y(0) = C_1 \cdot 1 + C_2 \cdot 0 = 0$$

and deduce that $C_1 = 0$, and also

$$y(\pi) = C_2 \cdot 0 = 0$$

which suggests that C_2 is arbitrary. Hence, this problem has an *infinite number of solutions* described by

$$y = C_2 \sin x$$

where C_2 may assume any value.

In most simple problems like Example 2, computing the value of the coefficient determinant Δ is not necessary. This determinant is introduced primarily for theoretical purposes. However, relative to Example 2, we see that

$$\Delta = \begin{vmatrix} \cos(0) & \sin(0) \\ \cos(\pi) & \sin(\pi) \end{vmatrix} = \begin{vmatrix} 1 & 0 \\ -1 & 0 \end{vmatrix} = 0$$

which merely informs us of the existence of a nontrivial solution—it does not aid in the actual determination of the solution.

EXAMPLE 3

Solve: $y'' + y = 0, \qquad 0 < x < \pi/2, \qquad y(0) = 0, \qquad y(\pi/2) = 0.$

SOLUTION As in Example 2, the general solution is

$$y = C_1 \cos x + C_2 \sin x.$$

This time the boundary conditions demand that $C_1 = C_2 = 0$, yielding only the *trivial solution* $y = 0$. In this case

$$\Delta = \begin{vmatrix} \cos(0) & \sin(0) \\ \cos(\pi/2) & \sin(\pi/2) \end{vmatrix} = 1 \neq 0.$$

EXAMPLE 4

Solve: $x^2y'' + xy' + \pi^2 y = 0, \qquad 1 < x < e, \qquad y(1) = 0, \qquad y(e) = 0.$

SOLUTION This time our DE does not have constant coefficients, but is of the type of DE called a *Cauchy-Euler equation*.* The transformation $x = e^t$ yields the constant-coefficient equation

$$\frac{d^2y}{dt^2} + \pi^2 y = 0,$$

---

* See Problem 34 in Exercises 1.3 for more details on the solution of Cauchy-Euler DEs.

whose independent variable is now t. The general solution is

$$y(t) = C_1 \cos \pi t + C_2 \sin \pi t,$$

or in terms of the original variable x, where $t = \log x$, we get

$$y(x) = C_1 \cos(\pi \log x) + C_2 \sin(\pi \log x).*$$

Applying the boundary conditions, we have

$$y(1) = C_1 \cdot 1 + C_2 \cdot 0 = 0,$$

$$y(e) = C_1 \cdot (-1) + C_2 \cdot 0 = 0,$$

and thus conclude that $C_1 = 0$ and C_2 is arbitrary. Hence we obtain the *infinite collection of solutions* ($\Delta = 0$),

$$y = C_2 \sin(\pi \log x).$$

From the linear homogeneous nature of these examples, it follows that if $y = \phi(x)$ is a nontrivial solution of the homogeneous DE and homogeneous boundary conditions, so is $y = C\phi(x)$ for any value of the constant C. That is, corresponding to the nontrivial solution $\phi(x)$ is a family of solutions, each member of which is proportional to $\phi(x)$. For most problems only one such family of nontrivial solutions exists, but situations can arise that lead to two families of solutions corresponding to two linearly independent solutions $\phi(x)$ and $\psi(x)$. For example, both $\cos x$ and $\sin x$ satisfy the homogeneous boundary value problem

$$y'' + y = 0, \qquad y(0) + y(\pi) = 0, \qquad y'(0) + y'(\pi) = 0 \tag{29}$$

and the most general solution is the linear combination

$$y = C_1 \cos x + C_2 \sin x \tag{30}$$

where C_1 and C_2 are both arbitrary constants.

When α and β are not both zero and $\Delta \neq 0$, then (23) will have a *unique nontrivial solution,* since both C_1 and C_2 are uniquely determined by these conditions (and are not both zero). But if $\Delta = 0$ in this case, there will be either *no solution* or *infinitely many solutions*. Again let us illustrate with some examples.

EXAMPLE 5

Solve: $y'' + y = 0, \qquad 0 < x < \pi, \qquad y(0) = 0, \qquad y(\pi) = 2.$

SOLUTION Imposing the prescribed boundary conditions on the general solution

$$y = C_1 \cos x + C_2 \sin x$$

---

* By $\log x$, we mean the *natural logarithm,* also commonly denoted by $\ln x$.

we find

$$y(0) = C_1 = 0,$$

$$y(\pi) = C_2 \sin \pi = 2,$$

which is impossible for any choice of C_2. Hence, because values for both C_1 and C_2 cannot be found, we conclude the problem has *no solution* ($\Delta = 0$).

EXAMPLE 6

Solve: $y'' + y = 0, \quad 0 < x < \pi, \quad y(0) = 1, \quad y(\pi) = -1.$

SOLUTION The boundary conditions applied to the general solution

$$y = C_1 \cos x + C_2 \sin x$$

lead to the relations

$$y(0) = C_1 = 1$$

$$y(\pi) = -C_1 = -1.$$

Hence, we must choose $C_1 = 1$ while C_2 is left arbitrary. In this case we obtain an *infinite number of solutions* given by ($\Delta = 0$)

$$y = \cos x + C_2 \sin x.$$

EXAMPLE 7

Solve: $y'' + y = 0, \quad 0 < x < \pi/2, \quad y(0) = 0, \quad y(\pi/2) = 3.$

SOLUTION Again the general solution is

$$y = C_1 \cos x + C_2 \sin x$$

which, upon imposition of the boundary conditions, leads to

$$y(0) = C_1 = 0$$

$$y(\pi/2) = C_2 = 3.$$

Therefore we obtain the *unique solution* ($\Delta \neq 0$)

$$y = 3 \sin x.$$

Examples 5–7 illustrate that the solution of a boundary value problem depends crucially upon the interval on which it is defined as well as on the values of α and β.

1.3.2 NONHOMOGENEOUS DEs

Based on our knowledge of *nonhomogeneous* DEs from a first course in differential equations, we know that the general solution of

$$M[y] = F(x), \qquad x_1 < x < x_2 \tag{31}$$

is given by the sum of solutions

$$\begin{aligned} y &= y_H(x) + y_P(x) \\ &= C_1 y_1(x) + C_2 y_2(x) + y_P(x), \end{aligned} \tag{32}$$

where $y_H = C_1 y_1(x) + C_2 y_2(x)$* is the general solution of the associated homogeneous DE, $M[y] = 0$, and y_P is any *particular solution* of (31). The particular solution y_P can usually be found by either the method of *undetermined coefficients* or *variation of parameters*.†

By imposing the boundary conditions

$$B_1[y] = \alpha, \qquad B_2[y] = \beta, \tag{33}$$

upon the solution (32), we are led to the system of equations

$$\begin{aligned} C_1 B_1[y_1] + C_2 B_1[y_2] &= \alpha - B_1[y_P] \\ C_1 B_2[y_1] + C_2 B_2[y_2] &= \beta - B_2[y_P] \end{aligned}$$

with solution

$$C_1 = \frac{\begin{vmatrix} \alpha - B_1[y_P] & B_1[y_2] \\ \beta - B_2[y_P] & B_2[y_2] \end{vmatrix}}{\Delta}, \tag{34}$$

$$C_2 = \frac{\begin{vmatrix} B_1[y_1] & \alpha - B_1[y_P] \\ B_2[y_1] & \beta - B_2[y_P] \end{vmatrix}}{\Delta}, \tag{35}$$

where the coefficient determinant Δ is that defined by (28). If $\Delta \neq 0$, then (34) and (35) provide a unique determination of C_1 and C_2, and hence a unique solution of (31) and (33). But if $\Delta = 0$, the problem (31) and (33) will have either no solution or an infinite number of solutions as before. Again, we consider some examples.

EXAMPLE 8

Solve: $y'' + y = x - 1, \quad 0 < x < 1, \quad y(0) = -1, \quad y(1) = 1.$

SOLUTION By inspection, a particular solution is found to be $y_P = x - 1$, and thus the general solution is

$$y = C_1 \cos x + C_2 \sin x + x - 1.$$

---

*The homogeneous solution y_H is often called the *complementary solution* and designated by the symbol y_c in many texts.

†For a review of these techniques, see Sections 4.6 and 4.7 in L. C. Andrews, *Ordinary Differential Equations with Applications*. (Scott, Foresman and Co.: Glenview, Ill.) 1982.

The first boundary condition requires that

$$y(0) = C_1 - 1 = -1,$$

or $C_1 = 0$, and the second condition leads to

$$y(1) = C_2 \sin 1 = 1,$$

from which we find $C_2 = 1/(\sin 1)$. Hence the *unique solution* is $(\Delta \neq 0)$

$$y = \frac{\sin x}{\sin 1} + x - 1.$$

EXAMPLE 9

Solve: $y'' + y = x - 1, \qquad 0 < x < \pi, \qquad y(0) = -1, \qquad y(\pi) = 1.$

SOLUTION From Example 8, the solution satisfying the first boundary condition is

$$y = C_2 \sin x + x - 1.$$

The second boundary condition requires that

$$y(\pi) = \pi - 1 = 1$$

which is impossible. Hence, we conclude the problem has *no solution* $(\Delta = 0)$.

Examples 8 and 9 show once again how critical the prescription of boundary conditions is in order to have a unique solution of the boundary value problem. Our final example illustrates that infinitely many solutions are also possible.

EXAMPLE 10

Solve: $y'' + y = x, \qquad 0 < x < \pi, \qquad y(0) = \pi, \qquad y(\pi) = 0.$

SOLUTION A particular solution is $y_P = x$, and therefore

$$y = C_1 \cos x + C_2 \sin x + x.$$

The boundary conditions lead to the system of equations

$$y(0) = C_1 = \pi,$$

$$y(\pi) = -C_1 + \pi = 0,$$

both of which are satisfied by $C_1 = \pi$ and arbitrary C_2. Thus, we obtain an *infinite number of solutions* given by $(\Delta = 0)$

$$y = \pi \cos x + C_2 \sin x + x.$$

Summarizing the results of this section, we have the following theorem.

THEOREM 1.2

> If y_1 and y_2 are linearly independent solutions of the homogeneous DE, $M[y] = 0$, then the boundary value problem
>
> $$M[y] = F(x), \qquad x_1 < x < x_2, \qquad B_1[y] = \alpha, \qquad B_2[y] = \beta$$
>
> has a unique solution if and only if $\Delta \neq 0$, where
>
> $$\Delta = \begin{vmatrix} B_1[y_1] & B_1[y_2] \\ B_2[y_1] & B_2[y_2] \end{vmatrix}.$$
>
> If $\Delta = 0$, the problem either has no solution or infinitely many solutions.

EXERCISES 1.3

In Problems 1–16, find nontrivial solutions, if they exist. State whether the solution is unique or not.

1. $y'' = 0, \quad y(0) = 0, \quad y(1) = 0$
2. $y'' + y = 0, \quad y(0) = 0, \quad y'(\pi) = 0$
3. $y'' + 9y = 0, \quad y'(0) = 0, \quad y'(\pi) = 0$
4. $y'' + y = 0, \quad y(0) = 0, \quad y(\pi) = 0$
5. $y'' + \pi^2 y = 0, \quad y(0) + y(1) = 0, \quad y'(0) + y'(1) = 0$
6. $y'' - y = 0, \quad y(0) = 0, \quad y(1) = 0$
7. $y'' - 3y' + 2y = 0, \quad y(0) = 0, \quad y(1) = 0$
8. $y'' + y = 0, \quad y(-1) = 0, \quad y(1) = 1$
9. $y'' + y = 0, \quad y(0) = 3, \quad y(\pi) = -2$
10. $y'' + y = 0, \quad y(0) = 3, \quad y(\pi) = -3$
11. $y'' + k^2 y = 0, \quad y(0) = 0, \quad y(1) = 2, \quad (k > 0)$
12. $y'' - y = 0, \quad y(0) = 1, \quad y(1) = -1$
13. $y'' - y = 0, \quad y'(0) + 3y(0) = 0, \quad y'(1) + y(1) = 1$
14. $y'' - 3y' + 2y = 0, \quad y(0) = 1, \quad y(1) = 0$
15. $y'' - 6y' + 9y = 0, \quad y(0) = 0, \quad y(1) = 3$
16. $y'' - y' - 6y = 0, \quad y(0) = 0, \quad y(1) = e^3$
17. Using (22), verify that
    (a) the differential operator ($D = d/dx$)

    $$M = A_2(x)D^2 + A_1(x)D + A_0(x)$$

    is a linear operator.
    (b) the boundary operators B_1 and B_2 defined by (20) are both linear operators.

18. Show that the boundary value problem

$$y'' - k^2 y = 0, \qquad y(0) = 0, \qquad y(1) = 0,$$

cannot have a nontrivial solution for real values of k.

19. For which values of k does the problem

$$y'' + k^2 y = 0, \qquad y(0) = 0, \qquad y(1) = 0, \qquad (k \geqslant 0)$$

have nontrivial solutions? What are the corresponding solutions?

20. Determine the values of b for which the boundary value problem

$$y'' + 4y = 0, \qquad y(0) = 0, \qquad y(b) = 3, \qquad (b > 0)$$

has no solution.

21. Show that the boundary value problem

$$y'' = 0, \qquad y(a) = \alpha, \qquad y(b) = \beta$$

has a unique solution for every choice of a and b ($a \neq b$).

22. Show that the boundary value problem

$$y'' + y = 0, \qquad y(a) = \alpha, \qquad y'(b) = \beta$$

has a unique solution for every choice of a and b such that $a \neq b$, $a - b \neq n\pi/2$, $n = \pm 1, \pm 2, \pm 3, \ldots$.

★23. For which values of a and b (if any) does the boundary value problem

$$y'' + y = 0, \qquad y(a) = 1, \qquad y(b) = 1, \qquad (a < b)$$

have
(a) no solution?
(b) exactly one solution?
(c) more than one solution?

In Problems 24–33, find the solution (if possible). State whether the solution is unique or not.

24. $y'' = -x, \qquad y(0) = 0, \qquad y'(1) = 0$

25. $y'' = x, \qquad y(0) = 1, \qquad y(1) - y'(1) = 0$

26. $y'' = \sin \pi x, \qquad y(0) = 0, \qquad y'(1) = 0$

27. $y'' = 3 \sin \pi x, \qquad y(0) = -2, \qquad y'(1) = 7/\pi$

28. $y'' = 1, \qquad y'(0) = 0, \qquad y'(1) = 1$

29. $y'' + 4y = e^{-x}, \qquad y(0) = -1, \qquad y'(\pi/2) = -e^{-\pi/2}/5$

30. $y'' - y = e^x, \qquad y(0) = 0, \qquad y'(1) = 1$

31. $y'' + 2y' + y = x, \qquad y(0) = -3, \qquad y(1) = -1$

32. $y'' - 6y' + 9y = 9, \qquad y(0) = 1, \qquad y(1) = 3$

33. $y'' - y' - 6y = e^x, \qquad y(0) = 0, \qquad y(1) = 1$

★34. The DE

$$ax^2 y'' + bxy' + cy = 0 \qquad x > 0$$

where a, b, and c are constants, is called a *Cauchy-Euler equation.*

(a) Show that the change of variable $x = e^t$ leads to

$$(D = d/dt) \qquad xy' = Dy, \qquad x^2y'' = D(D-1)y.$$

(b) Using the result of part (a), show that the Cauchy-Euler equation can be transformed into the constant–coefficient equation

$$[aD^2 + (b-a)D + c]y = 0.$$

In Problems 35–40, use the technique of Problem 34 to transform the following Cauchy-Euler equations to constant–coefficient equations, solve the constant–coefficient equation, and transform back to the original variable to obtain the desired solution.

35. $x^2y'' - 5xy' + 25y = 0, \quad y(1) = 0, \quad y(e^\pi) = 0$

★36. $x^2y'' - 5xy' + 5y = 0, \quad y(0) = 0, \quad y(1) = 0$

37. $x^2y'' - xy' + y = 0, \quad y(1) = 1, \quad y(e^2) = 0$

38. $x^2y'' + xy' - y = x, \quad y(1) = 0, \quad y(e) = e/2$

★39. $2x^2y'' + 3xy' - y = 4x^{3/2}, \quad y(0) = 0, \quad y(1) = 0$

40. $xy'' + y' = x, \quad y(1) = 0, \quad y(e) = e^2/4$

★41. *(Abel's formula)** If y_1 and y_2 are linearly independent solutions of the second-order DE

$$y'' + a_1(x)y' + a_0(x)y = 0$$

on some interval where $a_1(x)$ and $a_0(x)$ are continuous, show that the Wronskian† satisfies, for some constant C,

$$W(y_1, y_2)(x) = C \exp\left[-\int a_1(x)\,dx\right].$$

Hint: Show that $dW/dx + a_1(x)W = 0$, and solve this DE for W.

★42. Show that if $y_1(x)$ is a nontrivial solution of

$$y'' + a_1(x)y' + a_0(x)y = 0$$

a second (linearly independent) solution is given by

$$y_2 = y_1(x)\int \frac{\exp\left[-\int a_1(x)\,dx\right]}{y_1^2(x)}\,dx.$$

Hint: Solve the first-order DE $y_1(x)y_2' - y_1'(x)y_2 = W(x)$ for y_2 using Problem 41.

---

*NEILS H. ABEL (1802–1829) was one of six children born into a poor Norwegian family. One of his early accomplishments was proving that the fifth-degree algebraic equation has no radical solution. Stricken with tuberculosis in 1827, he died two years later at the age of 26.

†Recall that the Wronskian is defined by $W(y_1, y_2) = y_1y_2' - y_1'y_2$. It is named after JOZEF M. H. WRONSKI (1778–1853) who studied mathematics in Germany but lived most of his life in France.

1.4 EIGENVALUE PROBLEMS

In the previous section we found that *homogeneous problems,* consisting of homogeneous DEs and homogeneous boundary conditions, can sometimes have nontrivial solutions. Here we wish to examine this special class of problems in more detail.

Frequently the solutions of a DE depend upon a parameter λ which may assume various values during a given discussion. This parameter can appear in the coefficients of the DE, in the boundary conditions, or in both. For a large class of problems of practical significance, the typical equation is of the form

$$M[y] + \lambda y = 0, \qquad x_1 < x < x_2, \tag{36}$$

where $M = A_2(x)D^2 + A_1(x)D + A_0(x)$. In the problems we wish to discuss, we will generally assume that λ appears only in the DE, as in (36), but not in the boundary conditions.

Clearly, the general solution of (36) must depend upon both x and the parameter λ. Thus, if y_1 and y_2 constitute linearly independent solutions of (36), we write the general solution as

$$y = C_1y_1(x, \lambda) + C_2y_2(x, \lambda). \tag{37}$$

Subjecting this solution function to the *homogeneous* boundary conditions

$$B_1[y] = 0, \qquad B_2[y] = 0, \tag{38}$$

leads to a coefficient determinant Δ [given by (28) in Section 1.3] that must also depend upon λ in this situation. Problems of this type arise routinely in solving partial differential equations (Chapters 6–8, 10), but also come up in other applications (e.g., see Sections 1.5.2 and 1.5.4). The basic problem is to determine all values of λ for which $\Delta(\lambda) = 0$; that is, determine all values of λ for which the homogeneous boundary value problem (36) and (38) admits *nontrivial solutions,* and then find the solutions corresponding to these values of λ. These special values of λ are called *eigenvalues,* and the corresponding nontrivial solutions are called *eigenfunctions*. The terms *characteristic values* and *characteristic functions* are also commonly used. The general problem described here is called an *eigenvalue problem,* or *Sturm-Liouville problem* after the two mathematicians who pioneered its investigation.* Special equations (minus boundary conditions) that fall into this classification are the following:

---

*JACQUES C. F. STURM (1803–1855) was a French mathematician of Swiss origin. He is recognized for making the first accurate determination of the velocity of sound in water, his essay on compressible fluids, and his work in differential equations.
JOSEPH LIOUVILLE (1809–1882) provided the first proofs of the existence of transcendental numbers. His other achievements include his research in boundary value problems, the theory of numbers, and differential geometry.

$$y'' + \lambda y = 0, \qquad x_1 < x < x_2 \qquad \text{(Helmholtz equation)}$$

$$(1 - x^2)y'' - 2xy' + \lambda y = 0, \qquad -1 < x < 1 \qquad \text{(Legendre equation)}$$

$$x^2y'' + xy' + (\lambda x^2 - \nu^2)y = 0, \qquad 0 < x < b \qquad \text{(Bessel equation)}$$

$$xy'' + (1 - x)y' + \lambda y = 0, \qquad 0 < x < \infty \qquad \text{(Laguerre equation)}$$

$$y'' - 2xy' + \lambda y = 0, \qquad -\infty < x < \infty \qquad \text{(Hermite equation)}$$

The above DEs are common to many areas of application. For example, Helmholtz's equation appears in vibrating-string problems and in finding the temperature distribution in a rod. Legendre's equation arises in certain problems displaying spherical geometry, Bessel's equation is closely associated with problems involving circular or cylindrical-shaped regions, and the equations of Laguerre and Hermite are conventional in certain quantum mechanics problems.

In applications the eigenvalues and eigenfunctions have important physical interpretations. For example, in vibration problems the eigenvalues are proportional to the squares of the natural frequencies of vibration, while the eigenfunctions provide the natural configuration modes of the system. In buckling problems the eigenvalues are related to the critical loads a column can support before deformation may take place, and the eigenfunctions describe the possible shapes of such deformations. The eigenvalues denote the possible energy states of a system in quantum mechanics problems, and the eigenfunctions are the wave functions. The wave function for a given energy level will yield the electronic distribution for the state corresponding to that energy, as well as other properties. Thus, the interpretations vary widely, as do the areas of application.

EXAMPLE 11

Find the eigenvalues and eigenfunctions of

$$y'' + \lambda y = 0, \qquad 0 < x < 1, \qquad y(0) = 0, \qquad y(1) = 0.$$

SOLUTION We first attempt to find the eigenvalues, which we tacitly assume are real. (More will be said about this assumption very shortly.) Because the solution of the DE may assume different functional forms, depending upon the value of λ, we must consider three separate cases corresponding to these different functional forms.*

CASE I: Let us first assume $\lambda = 0$, which leads to the general solution

$$y = C_1 + C_2x.$$

---

*The auxiliary equation is $m^2 + \lambda = 0$, with roots $m = \pm\sqrt{-\lambda}$. Thus, the three cases correspond to $\lambda = 0$, $\lambda < 0$, and $\lambda > 0$.

The first boundary condition $y(0) = 0$ requires that $C_1 = 0$, and the second condition leads to $C_1 + C_2 = 0$. Hence, $C_1 = C_2 = 0$, so that the only possible solution is the trivial solution $y = 0$. Thus, $\lambda = 0$ is not an eigenvalue of the problem.

CASE II: If λ is negative, it is helpful to set $\lambda = -k^2 < 0$, and then the general solution can be expressed in either the form

$$y = C_1 e^{kx} + C_2 e^{-kx}$$

or

$$y = C_1 \cosh kx + C_2 \sinh kx,$$

where the hyperbolic functions are defined by

$$\cosh z = \tfrac{1}{2}(e^z + e^{-z}), \qquad \sinh z = \tfrac{1}{2}(e^z - e^{-z}).$$

In solving eigenvalue problems on *finite domains,* it is usually preferred to express the general solution in terms of hyperbolic functions rather than exponential functions (although either solution form is acceptable). The reason for this is that certain simplifications take place in determining the constants when using hyperbolic functions; for example, $\cosh(0) = 1$ and $\sinh(0) = 0$. Other basic properties of these functions are taken up in Problem 16 in Exercises 1.4.

By applying the boundary conditions to the above general solution written in terms of hyperbolic functions, we find

$$y(0) = C_1 = 0,$$

$$y(1) = C_2 \sinh k = 0,$$

but since $\sinh k \neq 0$ for $k \neq 0$, we deduce that $C_1 = C_2 = 0$, which leads to only the trivial solution $y = 0$. That is, there are no negative eigenvalues of this problem.

CASE III: Next we set $\lambda = k^2 > 0$, and the general solution is

$$y = C_1 \cos kx + C_2 \sin kx.$$

Application of the boundary conditions leads to $C_1 = 0$ and

$$C_2 \sin k = 0.$$

If we require $C_2 \neq 0$, this last condition can be satisfied only if k is an integral multiple of π, i.e., $k = n\pi$ $(n = 1, 2, 3, \ldots)$.* Hence, the eigenvalues are given by

$$\lambda_n = k_n^2 = n^2\pi^2, \qquad n = 1, 2, 3, \ldots,$$

and the corresponding eigenfunctions are any (nonzero) multiples of

---

* Since $\lambda = k^2$, we can take k positive without loss of generality.

$$y = \phi_n(x) = \sin n\pi x, \qquad n = 1, 2, 3, \ldots$$

(i.e., we set $C_2 = 1$ for convenience).

REMARK: *We could also determine the eigenvalues in Case III by considering the coefficient determinant*

$$\Delta(\lambda) = \begin{vmatrix} 1 & 0 \\ \cos k & \sin k \end{vmatrix} = \sin k,$$

which vanishes, of course, for $k = n\pi$ ($n = 1, 2, 3, \ldots$). However, the actual calculation of $\Delta(\lambda)$ is not necessary in most simple problems, but it may prove useful in more difficult problems.

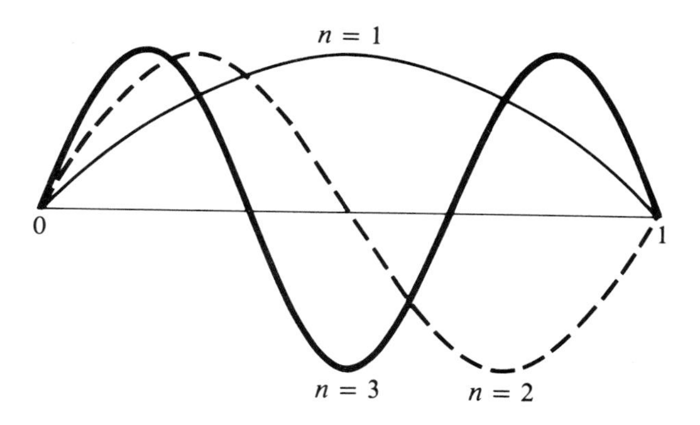

FIGURE 1.2

Eigenfunctions $\phi_n(x) = \sin n\pi x$, $n = 1, 2, 3$, and their zeros on the interval $(0, 1)$.

The first three eigenfunctions of Example 11 are sketched over the interval $[0, 1]$ in Figure 1.2. All eigenfunctions are necessarily zero at the endpoints $x = 0$ and $x = 1$ because of the prescribed boundary conditions. Notice that the first eigenfunction $\phi_1(x) = \sin \pi x$ has no additional zeros on the open interval $(0, 1)$. However, the second eigenfunction $\phi_2(x) = \sin 2\pi x$ has one zero on $(0, 1)$ at $x = 1/2$, and the third eigenfunction $\phi_3(x) = \sin 3\pi x$ has zeros at $x = 1/3$ and $x = 2/3$. In fact, it becomes clear that for any $n \geqslant 1$, the eigenfunction $\phi_n(x) = \sin n\pi x$ has exactly $(n - 1)$ zeros on the open interval $(0, 1)$, which are located at

$$x = \frac{1}{n}, \frac{2}{n}, \frac{3}{n}, \ldots, \frac{(n-1)}{n}.$$

Not only is it true for this special set of eigenfunctions, but a *general property* of eigenfunctions is that the nth eigenfunction $\phi_n(x)$ has exactly $(n - 1)$ zeros on the open interval for which the problem is formulated. This is a remarkable property that also has important physical significance. For instance, in vibration problems the zeros of the eigenfunctions identify the *nodes* (stationary points) for each vibrational mode of the system.

Finally, we point out that, in solving for the eigenvalues in Example 11, we assumed that they were real. The possibility of complex eigenvalues exists, but subsequently we will show that for a certain class of problems (to which Example 11 belongs), only real eigenvalues are possible. Also, because eigenfunctions are determined only to within a multiplicative constant, it is customary in most situations to set the constant to unity.

In some problems the determination of the eigenvalues requires us to find the roots of a transcendental equation. Such equations can often be solved numerically by a simple procedure such as *Newton's method* from calculus. This particular method relies on initial approximations to the roots, the accuracy of which may then be improved upon by an *iterative process*. The number of iterations necessary for a desired accuracy will depend upon the accuracy of the initial estimates of the roots. We will illustrate Newton's method in finding the eigenvalues of Example 12.

EXAMPLE 12

Find the eigenvalues and eigenfunctions belonging to

$$y'' + \lambda y = 0, \qquad 0 < x < 1, \qquad y(0) = 0, \qquad y(1) + y'(1) = 0.$$

SOLUTION Again it is necessary to consider three separate cases.

CASE I: For $\lambda = 0$, the general solution of the DE is

$$y = C_1 + C_2 x.$$

The two boundary conditions demand that $C_1 = 0$ and $C_1 + 2C_2 = 0$, respectively, forcing the trivial solution. Therefore, $\lambda = 0$ is not an eigenvalue.

CASE II: For negative λ, we again set $\lambda = -k^2 < 0$. The general solution is then

$$y = C_1 \cosh kx + C_2 \sinh kx,$$

and the boundary conditions lead to $C_1 = 0$ and

$$C_2(\sinh k + k \cosh k) = 0,$$

or, for $C_2 \neq 0$,

$$k = -\tanh k.$$

To see if there are any solutions of this transcendental equation , we plot the graphs of $u = k$ and $u = -\tanh k$ and look for intersections. From Figure 1.3 it is clear that no intersections of these curves occur for $k > 0$, and hence we conclude that there are no negative eigenvalues.

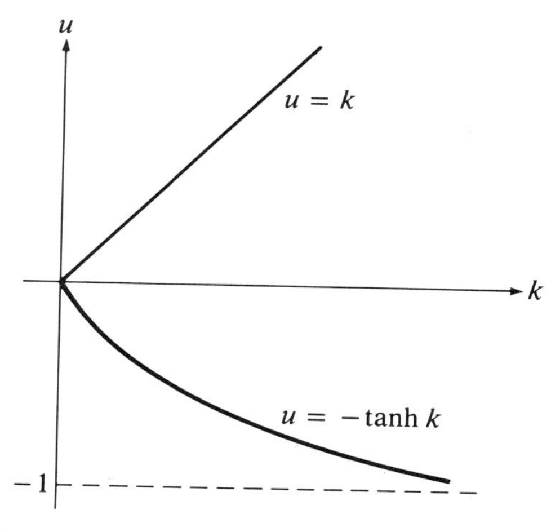

FIGURE 1.3

CASE III: If $\lambda = k^2 > 0$, then the general solution is

$$y = C_1 \cos kx + C_2 \sin kx.$$

The boundary condition at $x = 0$ requires that $C_1 = 0$, and the condition at $x = 1$ leads to

$$C_2(\sin k + k \cos k) = 0,$$

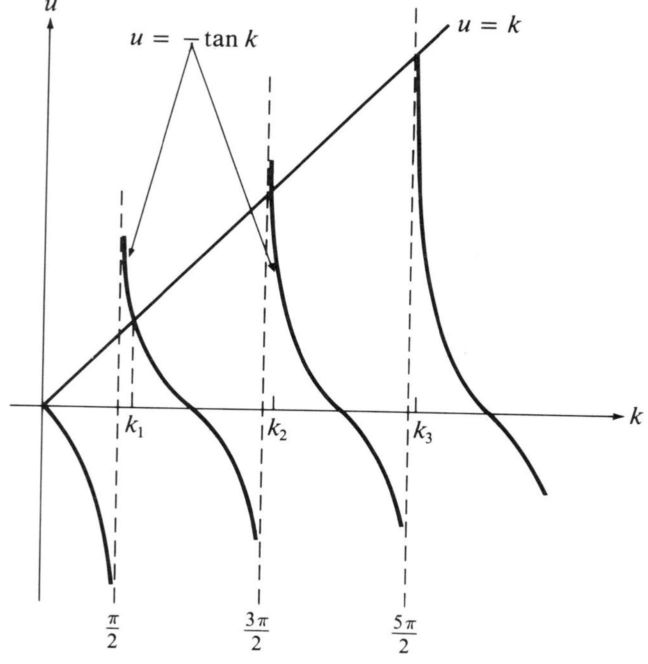

FIGURE 1.4

or, for $C_2 \neq 0$,

$$k = -\tan k.$$

Once again we plot the graphs of $u = k$ and $u = -\tan k$ to see if any intersections occur. Looking at Figure 1.4, we see that there are an infinite number of values k_n $(n = 1, 2, 3, \ldots)$ where intersections occur, but their exact values must be determined numerically. To utilize Newton's method in finding these values, we look for zeros of the function

$$F(x) = x + \tan x.$$

Each value of x for which $F(x) = 0$ can be approximated by the sequence*

$$x_{j+1} = x_j - \frac{F(x_j)}{F'(x_0)}, \qquad j = 0, 1, 2, \ldots,$$

where x_0 is our initial estimate of a root. For instance, from Figure 1.4 we estimate the first zero, denoted by k_1, to be approximately 2. Setting $x_0 = 2$, we find

$$F'(2) = 1 + \sec^2(2) \simeq 6.7744$$

and thus

$$x_{j+1} = x_j - \frac{x_j + \tan x_j}{6.7744}, \qquad j = 0, 1, 2, \ldots.$$

Successively substituting $j = 0, 1, 2, \ldots$ into this last result, we obtain the sequence of approximations

$$x_1 = 2.0273, \qquad x_2 = 2.0286, \qquad x_3 = 2.0287, \qquad \ldots$$

and therefore conclude that $k_1 \simeq 2.0287$. Consequently, $\lambda_1 = k_1^2 \simeq 4.1158$. Using Newton's method for finding additional intersections of the curves in Figure 1.4 will involve more iterations than the first value because of the steepness of the tangent curves near the intersections (see Problem 17 in Exercises 1.4). We do note that after the first few intersections, the remaining intersections of $u = k$ with the graph of $u = -\tan k$ are close to the vertical asymptotes of $-\tan k$. Hence, we deduce that the larger eigenvalues satisfy the approximate relation

$$\lambda_n \simeq \tfrac{1}{4}(2n - 1)^2\pi^2, \qquad n \gg 1.$$

Corresponding to each eigenvalue $\lambda_n = k_n^2$ $(n = 1, 2, 3, \ldots)$ is the eigenfunction

$$\phi_n(x) = \sin k_n x, \qquad n = 1, 2, 3, \ldots$$

(set $C_2 = 1$).

---

*Newton's method is generally given by $x_{j+1} = x_j - F(x_j)/F'(x_j)$, but it is more convenient in many cases to use $x_{j+1} = x_j - F(x_j)/F'(x_0)$ so that $F'(x_j)$ does not have to be evaluated at each step. However, if $F'(x)$ is rapidly changing, then the rate of convergence of the sequence of estimates may be considerably slower by this simpler formula (e.g., see Problem 17 in Exercises 1.4).

The three cases identified in the hunt for eigenvalues in Examples 11 and 12 are directly related to the form of the equation $y'' + \lambda y = 0$. For more general DEs, the cases are usually different. For instance, in solving the general second-order, constant–coefficient DE

$$ay'' + by' + \lambda y = 0, \tag{39}$$

we are led to the auxiliary equation (by setting $y = e^{mx}$)

$$am^2 + bm + \lambda = 0.$$

The roots of this auxiliary equation are given by

$$m = \frac{-b \pm \sqrt{b^2 - 4a\lambda}}{2a},$$

and thus the three cases involved in the search for eigenvalues correspond to

$$\left.\begin{aligned} &b^2 - 4a\lambda = 0, \\ &b^2 - 4a\lambda = k^2 > 0, \\ &b^2 - 4a\lambda = -k^2 < 0 \end{aligned}\right\} \tag{40}$$

(rather than $\lambda = 0$, $\lambda < 0$, and $\lambda > 0$).

Sometimes the eigenvalue problem involves a higher-order DE. The solution procedure is basically the same as that used in solving second-order DEs, as illustrated in our next example.

EXAMPLE 13

Find the eigenvalues and eigenfunctions of

$$y^{(4)} + \lambda y'' = 0, \quad 0 < x < 1, \quad y(0) = y''(0) = 0, \quad y(1) = y'(1) = 0.^*$$

SOLUTION Because we are dealing with a fourth-order DE, four separate boundary conditions are prescribed rather than the customary two associated with second-order equations.

Suppose we start by assuming $\lambda = k^2 > 0$. The resulting DE then possesses the general solution

$$y = C_1 + C_2 x + C_3 \cos kx + C_4 \sin kx,$$

the first two derivatives of which are

$$y' = C_2 + k(-C_3 \sin kx + C_4 \cos kx),$$

$$y'' = -k^2(C_3 \cos kx + C_4 \sin kx).$$

* For higher-order derivatives, we use $y^{(n)} = d^n y/dx^n$ in place of primes.

The two boundary conditions at $x = 0$ yield the equations

$$C_1 + C_3 = 0,$$
$$C_3 = 0,$$

from which we readily deduce $C_1 = C_3 = 0$. The remaining boundary conditions at $x = 1$ lead to the set of equations

$$C_2 + C_4 \sin k = 0,$$
$$C_2 + C_4 k \cos k = 0.$$

This system of equations has a nonzero solution for C_2 and C_4 if and only if the determinant of coefficients satisfies

$$\begin{vmatrix} 1 & \sin k \\ 1 & k \cos k \end{vmatrix} = k \cos k - \sin k = 0.$$

Hence, k must be selected such that $k = \tan k$, which is similar to the situation in Example 12 (see Problem 18 in Exercises 1.4). Once again it can be shown that an infinite number of values k_n satisfy this last relation, giving the set of eigenvalues $\lambda_n = k_n^2$ $(n = 1, 2, 3, \ldots)$.

Observe that when k is one of the values k_n, the two equations above in C_2 and C_4 are then equivalent, and either one can be used to eliminate one of the constants. For example, the first equation gives

$$C_2 = -C_4 \sin k_n,$$

from which we obtain the eigenfunctions (set $C_4 = 1$)

$$\phi_n(x) = \sin k_n x - x \sin k_n, \qquad n = 1, 2, 3, \ldots.$$

In this example it is a routine matter to show that when $\lambda \leq 0$, the DE possesses only the *trivial solution*.* Thus the problem has no additional eigenvalues.

Eigenvalue problems are characterized by both a *homogeneous DE* and *homogeneous boundary conditions*. As we have already observed, such problems always possess the trivial solution ($y = 0$), but it is the possibility of nontrivial solutions that interests us most. The examples illustrated here are rather elementary, but the techniques used apply to more difficult problems. In fact, the eigenvalues and eigenfunctions belonging to certain types of Sturm-Liouville systems have many properties in common, so that solving a simple example of a general class reveals much information characteristic of the whole class. We will discuss these ideas in more detail in Section 1.6.

---

* For $\lambda = -k^2 < 0$, the search for eigenvalues leads to the transcendental equation $k = \tanh k$, which has no solution for $k > 0$.

EXERCISES 1.4

In Problems 1–14, find all real eigenvalues and corresponding eigenfunctions of the given boundary value problem.

1. $y'' + \lambda y = 0, \quad y(0) = 0, \quad y'(1) = 0$
2. $y'' + \lambda y = 0, \quad y'(0) = 0, \quad y(2) = 0$
3. $y'' + \lambda y = 0, \quad y'(0) = 0, \quad y'(\pi) = 0$
4. $y'' + \lambda y = 0, \quad y(0) = 0, \quad y(1) - y'(1) = 0$
5. $y'' + 2y' + (1 - \lambda)y = 0, \quad y(0) = 0, \quad y(1) = 0$
6. $y'' + 4y' + (4 + 9\lambda)y = 0, \quad y(0) = 0, \quad y(2) = 0$
7. $y'' + y' + \lambda y = 0, \quad y(0) = 0, \quad y(1) = 0$
8. $y'' + 2y' + \lambda y = 0, \quad y(0) = 0, \quad y'(2) = 0$
9. $y'' - 3y' + 2\lambda y = 0, \quad y(0) = 0, \quad y(1) = 0$

★10. $y'' + \lambda y = 0, \quad y(0) - y(\pi) = 0, \quad y'(0) - y'(\pi) = 0$

11. $x^2y'' + xy' + \lambda y = 0, \quad y(1) = 0, \quad y(e) = 0$
12. $x^2y'' - xy' + \lambda y = 0, \quad y(1) = 0, \quad y(e) = 0$
13. $\dfrac{d}{dx}(xy') + \dfrac{\lambda}{x} y = 0, \quad y'(1) = 0, \quad y(2) = 0$

★14. $\dfrac{d}{dx}(x^3y') + \lambda xy = 0, \quad y(1) = 0, \quad y'(e^{\pi}) = 0$

★15. Given the eigenvalue problem

$$y'' + \lambda y = 0, \qquad hy(0) + y'(0) = 0, \qquad y(1) = 0,$$

(a) show that if $\lambda = 0$ is an eigenvalue, then $h = 1$.
(b) If $h > 1$, show that there is exactly one negative eigenvalue and that this eigenvalue decreases as h increases.
(c) If $h = 2$, show that the positive eigenvalues for large values of n satisfy the relation

$$\lambda_n \simeq (2n + 1)^2 \frac{\pi^2}{4}, \qquad n \gg 1.$$

(d) If $h < 0$, show that there are no negative eigenvalues.

16. Show that the hyperbolic functions $\cosh x = (e^x + e^{-x})/2$ and $\sinh x = (e^x - e^{-x})/2$ satisfy the relations (see figure)
(a) $\cosh(0) = 1$.
(b) $\sinh(0) = 0$.
(c) $\cosh^2 x - \sinh^2 x = 1$.
(d) $d/dx \cosh x = \sinh x$.
(e) $d/dx \sinh x = \cosh x$.
(f) $\cosh x \simeq \sinh x \simeq e^x/2$ for $x \gg 1$.

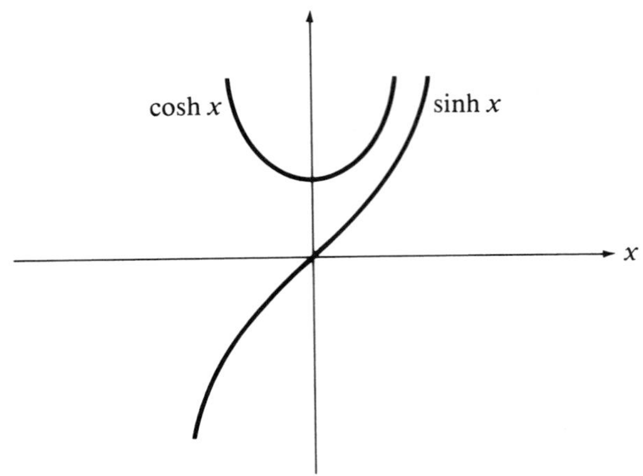

PROBLEM 16

17. Using Newton's method with $x_0 = 5$ and four iterations, approximate the second eigenvalue in Example 12 by employing the formula*
(a) $x_{j+1} = x_j - F(x_j)/F'(x_0), \qquad j = 0, 1, 2, 3.$
(b) $x_{j+1} = x_j - F(x_j)/F'(x_j), \qquad j = 0, 1, 2, 3.$

18. Using Newton's method, approximate the first two eigenvalues in Example 13 to three signficant figures by employing the formula

$$x_{j+1} = x_j - F(x_j)/F'(x_j), \qquad j = 0, 1, 2, \ldots .$$

19. Using Newton's method, approximate the first root of the transcendental equation $\cosh x \cos x + 1 = 0$.

In Problems 20–22, use Newton's method to find an estimate for the first nonzero eigenvalue λ_1. Also determine the form of the eigenfunctions.

20. $y'' + \lambda y = 0, \qquad y(0) - y'(0) = 0, \qquad y'(\pi) = 0$

21. $y'' + \lambda y = 0, \qquad y(0) = 0, \qquad y(\pi) + y'(\pi) = 0$

★22. $y'' + \lambda y = 0, \qquad y(0) - y'(0) = 0, \qquad y(1) + y'(1) = 0$

Problems 23–25 are nonstandard eigenvalue problems because of the way in which λ appears in the DE or boundary condition. Find all real eigenvalues and eigenfunctions in each problem.

★23. $y'' + y' + \lambda(y' + y) = 0, \qquad y'(0) = 0, \qquad y(1) = 0$

★24. $y'' + \lambda y = 0, \qquad y(0) = 0, \qquad y'(1) - \lambda y(1) = 0$

★25. $x^2y'' + \lambda(xy' - y) = 0, \qquad y(1) = 0, \qquad y(2) - y'(2) = 0$

★26. Consider the eigenvalue problem

$$y'' + \lambda y = 0, \qquad y(0) = 0, \qquad y'(1) = 0.$$

* The actual eigenvalue is $\lambda_2 \simeq 24.138$.

Without explicitly solving for them, show that if $\phi_m(x)$ and $\phi_n(x)$ are eigenfunctions corresponding to the eigenvalues λ_m and λ_n, respectively, with $\lambda_m \neq \lambda_n$, then

$$\int_0^1 \phi_m(x)\phi_n(x)\,dx = 0.$$

Hint: Observe that

$$\phi_m'' + \lambda_m\phi_m = 0, \qquad \phi_n'' + \lambda_n\phi_n = 0.$$

Multiply the first of these equations by ϕ_n, the second by ϕ_m, subtract the two equations, and then integrate the result from 0 to 1 using integration by parts and taking into account the boundary conditions.

In Problems 27–30, determine the values of λ for which the given boundary value problem has a *unique solution,* and find the solution in these cases.

27. $y'' + \lambda y = 0, \quad y(0) = 0, \quad y(1) = 1$

28. $y'' + \lambda y = 0, \quad y(-\pi) = 1, \quad y(\pi) = 1$

29. $y'' + 4y' + (4 + 9\lambda)y = 0, \quad y(0) = 1, \quad y(2) = 0$

★30. $y'' + 4y' + \lambda y = 0, \quad y(0) = 0, \quad y'(1) = 2$

In Problems 31–33, find the real eigenvalues and eigenfunctions of the fourth-order DEs by assuming only non-negative eigenvalues (set $\lambda = 0$ and $\lambda = k^4 > 0$).

31. $y^{(4)} - \lambda y = 0, \quad y(0) = y''(0) = 0, \quad y(\pi) = y''(\pi) = 0$

32. $y^{(4)} - \lambda y = 0, \quad y'(0) = y'''(0) = 0, \quad y'(1) = y'''(1) = 0$

★33. $y^{(4)} - \lambda y = 0, \quad y(0) = y'(0) = 0, \quad y(1) = y'(1) = 0$

1.5 PHYSICAL APPLICATIONS

Boundary value problems are associated with determining the equilibrium configuration of some physical system such as a stretched string or a beam supporting an external load, determining the possible buckling modes of a long column under axial loads, finding the steady-state temperature distribution in a long slender rod with prescribed end conditions, and so forth. In this section we will discuss several elementary examples, while others will be taken up in the exercises.

1.5.1 DEFLECTIONS OF AN ELASTIC STRING

To begin, we wish to derive the equation of equilibrium for a string stretched tightly between two supports, located at $x = 0$ and $x = b$, and subjected to a distributed vertical force of intensity $q(x)$ per unit length (see Figure 1.5). It is assumed that the string is under a large tensile force T which remains constant for small displacements y.*

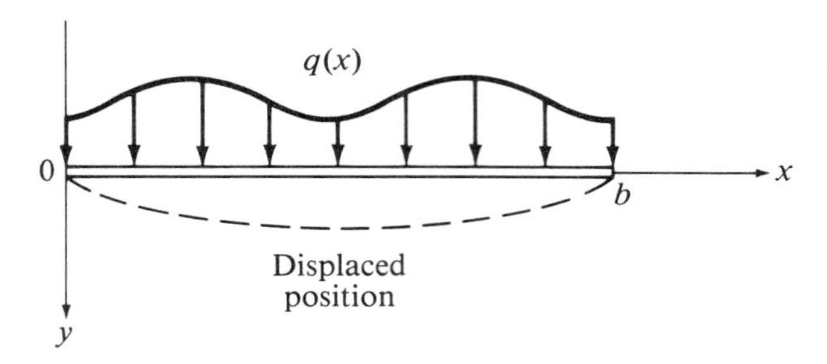

FIGURE 1.5

Elastic string under a load.

Further simplifying assumptions must be made so that the resulting equation of equilibrium does not become too complicated. That is, we wish to approximate *nonlinear behavior* with a *linear model,* a common practice in applications. Our basic assumptions include the following:

1. The mass of the string is constant and the string is perfectly elastic.
2. The tension T on the string is so large that gravitational effects can be ignored.
3. The equilibrium position of the string is a "small" displacement from the x-axis in a vertical plane.

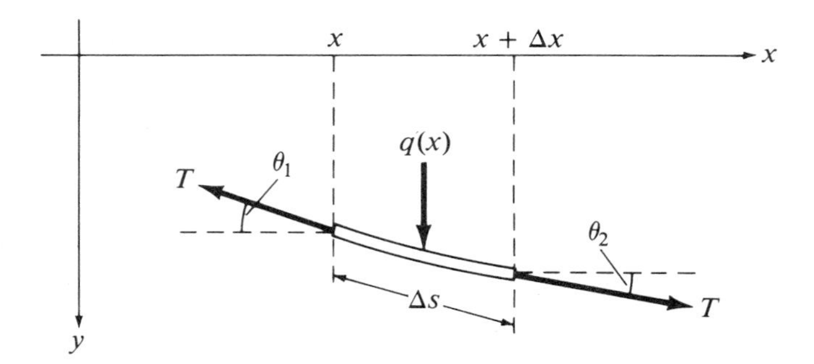

FIGURE 1.6

Small element of the elastic string.

Let us focus our attention on a small element of the string from x to $(x + \Delta x)$, as shown in Figure 1.6. Because $q(x)$ is a distributed load,† the total

* It is customary in such problems to choose the positive y-axis downward.

† That is, a force per unit length.

force acting on the elemental length of string Δs is simply $q(x)\Delta s$ (i.e., the force is essentially constant over the short length Δs). Moreover, for small vertical displacements the slope of the displaced string remains small so that we may use the approximation $\Delta s \simeq \Delta x$ and write the external force over the small element of the string as $q(x)\Delta x$.

For small displacements we only need to consider forces acting in the vertical plane, and equilibrium conditions demand that the sum of all such forces algebraically add to zero. These include the external force and the y-components of the tensile force at x and at $(x + \Delta x)$. Referring to Figure 1.6, the y-component of the tensile force at x is given by

$$-T \sin \theta_1 \simeq -T \tan \theta_1 = -Ty'(x),$$

where we are using the small angle approximation $\sin \theta_1 \simeq \tan \theta_1$ and the fact that $\tan \theta_1$ is simply the slope $y'(x)$ of the string at x. Likewise, at $(x + \Delta x)$ we find the y-component of the tensile force to be

$$T \sin \theta_2 \simeq T \tan \theta_2 = Ty'(x + \Delta x).$$

Hence, summing all vertical forces, we find

$$-Ty'(x) + Ty'(x + \Delta x) + q(x)\Delta x = 0,$$

or, upon division by Δx and T,

$$\left[\frac{y'(x + \Delta x) - y'(x)}{\Delta x}\right] + \frac{1}{T} q(x) = 0. \tag{41}$$

Finally, the limit of (41) as Δx tends to zero leads to the desired equation of equilibrium

$$y'' = -\frac{q(x)}{T}, \qquad 0 < x < b, \tag{42}$$

called the *one-dimensional Poisson equation.**

The string may be attached to various types of supports at the ends $x = 0$ and $x = b$. Most common are the fixed support, which yields zero displacement ($y = 0$) at the endpoint,† and the "free end," which requires the slope to vanish ($y' = 0$) at the endpoint. In order to maintain nearly uniform tension, the free end must be restricted from appreciable movement along the x-axis (i.e., it is not really "free"). This condition could be realized, for example, by considering the free end looped over a frictionless peg that maintains a zero slope on the string at this end while allowing small vertical movements (see Figure 1.7). A third type of boundary condition arises when one end of the string is attached to a yielding support, equivalent to an elastic

---

*Observe that (42) has the same functional form as (12) for the steady-state heat conduction problem. It is common to find the same DE occurring in entirely different applications.

†In some cases, the displacement at the ends may be nonzero (e.g., see Problems 1 and 7 in Exercises 1.5).

spring that exerts a restoring force proportional to its stretch. Such a condition is mathematically described by $Ty' + ky = 0$ at the appropriate endpoint, where k corresponds to the spring constant (see Figure 1.8).

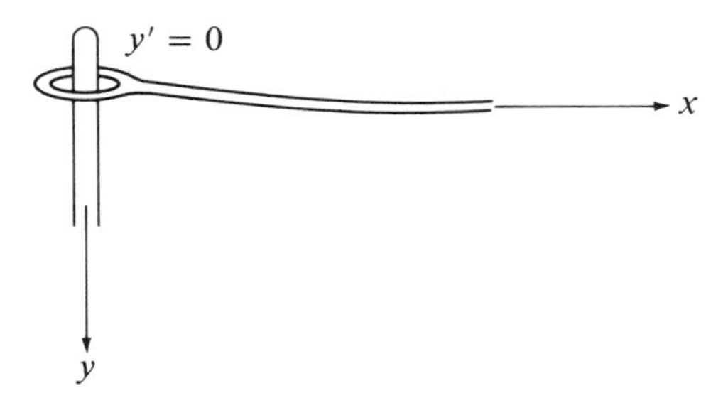

FIGURE 1.7

Free end support.

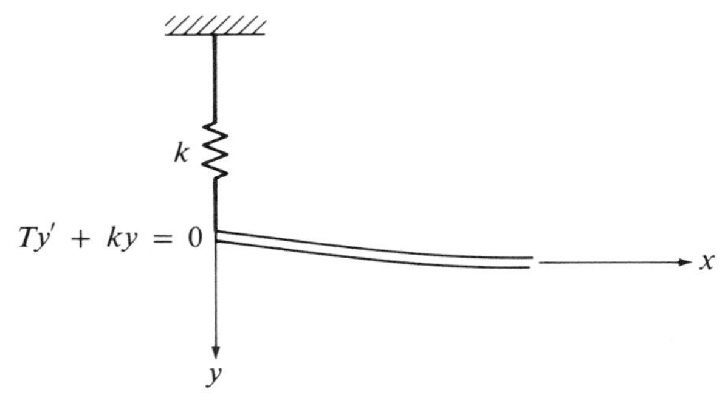

FIGURE 1.8

Yielding support.

EXAMPLE 14

Find the shape of a string stretched tightly between fixed supports (zero displacements) at $x = 0$ and $x = \pi$, and supporting a constant distributed load of intensity q_0 per unit length. Also determine an expression for the maximum displacement of the string and the tension necessary to keep the maximum displacement no more than $0.3q_0$.

SOLUTION The problem is characterized by

$$y'' = -\frac{q_0}{T}, \qquad 0 < x < \pi, \qquad y(0) = 0, \qquad y(\pi) = 0.$$

Two integrations of the DE lead to the general solution

$$y = -\frac{q_0}{2T}x^2 + C_1x + C_2,$$

and the boundary conditions require that $C_1 = q_0\pi/2T$ and $C_2 = 0$. Thus we obtain the parabolic deflection curve

$$y = \frac{q_0}{2T}x(\pi - x).$$

Because the load is symmetrically distributed over the string, the maximum displacement occurs at the midpoint $x = \pi/2$, and is given by*

$$y_{\max} = \frac{q_0\pi^2}{8T}.$$

To keep $y_{\max} \leqslant 0.3q_0$, the tension T must satisfy the relation

$$\frac{q_0\pi^2}{8T} \leqslant 0.3q_0$$

from which we deduce $T \geqslant \pi^2/2.4 \simeq 4.112$.

1.5.2 ROTATING STRING

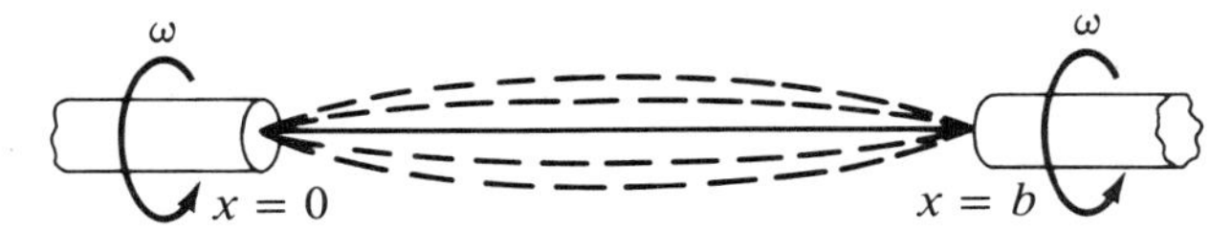

FIGURE 1.9

Rotating string.

Suppose the string discussed in Section 1.5.1 is now connected at $x = 0$ and $x = b$ to the centers of two synchronized rotating shafts whose axes of rotation coincide with the axis of the taut string (see Figure 1.9). If we assume the string has linear density ρ (mass/unit length) and is rotating with uniform angular speed ω, such action generates a distributed inertia force (force/unit length) of magnitude $\rho\omega^2 y$ in a direction transverse to the string.† Under proper conditions, this inertial force displaces the string away from its initial rest configuration. Hence the problem of finding these deflection modes is mathematically equivalent to determining the static equilibrium position of a tightly stretched string subject to the distributed load $q(x) = \rho\omega^2 y$. Making this replacement in (42) and introducing the parameter $\lambda = \rho\omega^2/T$, we can then write (42) in the form of an eigenequation

$$y'' + \lambda y = 0, \qquad 0 < x < b, \tag{43}$$

referred to as the *one-dimensional Helmholtz equation*.

---

* In general, the maximum displacement point is a solution of $y'(x) = 0$.

† The radial acceleration of a point x on the string is $\alpha = v^2/R$, where v is the linear velocity and R is the radius. Since $R = |y|$, and $v = \omega y$, it follows that $\alpha = \omega^2 y$. Multiplied by the mass density ρ, we then obtain the inertia force $\rho\omega^2 y$.

The eigenvalue problem consisting of (43) along with a prescribed set of homogeneous boundary conditions has nontrivial solutions only when λ is one of the eigenvalues. Each eigenvalue λ_n corresponds to a definite value of the angular speed ω_n, called the *critical speed* for that mode of deflection, i.e.,

$$\omega_n = \sqrt{\frac{\lambda_n T}{\rho}}, \qquad n = 1, 2, 3, \ldots .^* \tag{44}$$

For angular speeds smaller than ω_1, the only stable configuration of the string is its undeformed position ($y = 0$) along the x-axis. If the angular speed is increased until $\omega = \omega_1$, the first deflection mode may be obtained. If ω increases further to ω_2, a new configuration mode is possible, and so forth.

For example, when both ends of the string are fixed along the x-axis, the boundary conditions are $y(0) = y(b) = 0$, leading to the set of eigenvalues and eigenfunctions given by

$$\lambda_n = \frac{n^2\pi^2}{b^2}, \qquad \phi_n(x) = C \sin\frac{n\pi x}{b}, \qquad n = 1, 2, 3, \ldots, \tag{45}$$

where C is any (nonzero) constant. The first critical speed and corresponding deflection mode are found to be

$$\omega_1 = \frac{\pi}{b}\sqrt{\frac{T}{\rho}}, \qquad \phi_1(x) = C \sin\frac{\pi x}{b}. \tag{46}$$

Notice that only the shape of the string is determined by $\phi_1(x)$; the amplitude C is apparently arbitrary. This nondetermination of the amplitude is a direct consequence of using a linear model to describe nonlinear behavior. To remove this indeterminacy, we would have to analyze the true behavior of the string more exactly (see Problem 16 in Exercises 1.5). In spite of this particular shortcoming, the linear model used here gives good qualitative information about the rotating string.

□ 1.5.3 DISPLACEMENT CURVE OF AN ELASTIC BEAM

A uniform elastic beam of length b supports a distributed load of intensity $q(x)$ (see Figure 1.10). This load causes the beam to bend or be displaced from its equilibrium configuration along the x-axis, and the curve of its resulting centroidal axis is called the elastic curve, or displacement curve, of the beam. An important problem in strength of materials is to find the equation of this displacement curve. To do so, we will assume that such displace-

*The form of (44) suggests that all eigenvalues λ_n must be positive.

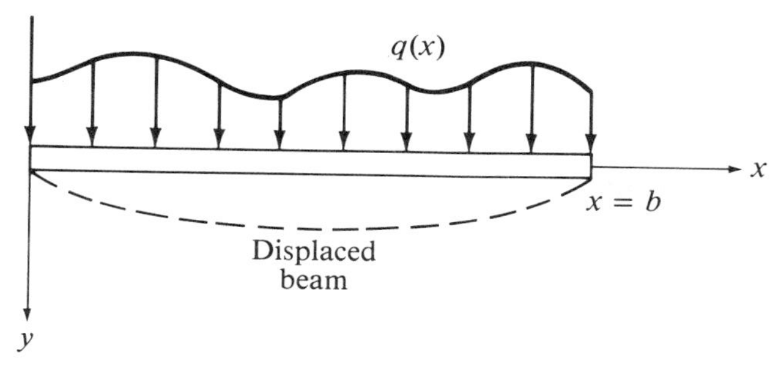

FIGURE 1.10

Elastic beam with load.

ments are small and well within the elastic limit of the beam. Then, by the use of some basic principles of elementary beam theory, we will derive the DE satisfied by the displacement curve.

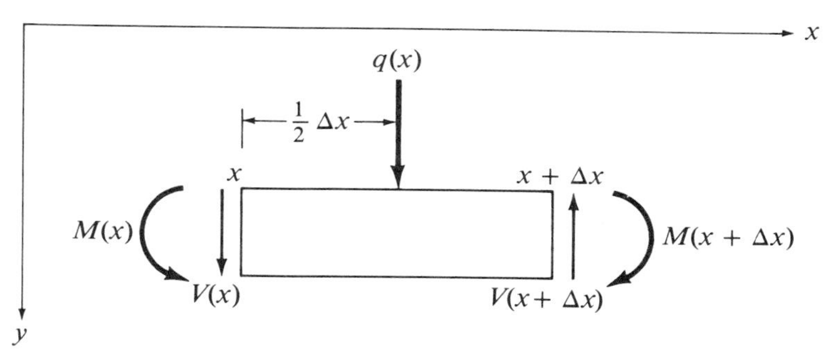

FIGURE 1.11

Small element of the beam.

Let us begin by examining the equilibrium conditions of a small element of the beam over the interval from x to $(x + \Delta x)$ (see Figure 1.11). At each end of the beam element is a shear force V and a bending moment M. Our sign convention is such that V is *positive downward* and M is *positive counterclockwise*. Equilibrium conditions require all forces to sum to zero and all moments taken about one end of the beam to vanish. Summing forces in the vertical direction gives us

$$V(x) + q(x)\Delta x - V(x + \Delta x) = 0, \tag{47}$$

where we are using the approximation that the elemental arc length is $\Delta s \simeq \Delta x$, and also that the force $q(x)$ is approximately constant over this same arc length. Next, taking moments about the end x, we find

$$M(x) - M(x + \Delta x) + V(x + \Delta x)\Delta x - q(x)\Delta x(\tfrac{1}{2}\Delta x) = 0. \tag{48}$$

Rearranging terms in (47) and (48) leads to

$$\frac{V(x+\Delta x)-V(x)}{\Delta x}=q(x),$$

$$\frac{M(x+\Delta x)-M(x)}{\Delta x}=V(x+\Delta x)-\tfrac{1}{2}q(x)\Delta x,$$

and by passing to the limit $\Delta x \to 0$, we obtain the standard equations of elementary beam theory

$$V'(x)=q(x), \qquad M'(x)=V(x). \tag{49}$$

Combining both equations in (49) yields the result

$$M''(x)=q(x), \qquad 0<x<b. \tag{50}$$

At this point we wish to relate the bending moment M to the displacement y. From the elementary theory of small deflections, we have the approximation*

$$M=\frac{EI}{R}=\frac{EIy''}{[1+(y')^2]^{3/2}}\simeq EIy'', \qquad |y'|\ll 1, \tag{51}$$

where E is *Young's modulus*, I is the moment of inertia, and $1/R$ is the curvature of the beam. The term EI can physically be interpreted as a measure of the stiffness of the beam. Thus, writing $M=EIy''$, (50) becomes

$$\frac{d^2}{dx^2}(EIy'')=q(x), \qquad 0<x<b,$$

or, since EI is constant for a uniform beam,

$$y^{(4)}=\frac{q(x)}{EI}, \qquad 0<x<b. \tag{52}$$

Once again we are led to a *nonhomogeneous* DE, but one for which the general solution can be obtained through four integrations of the right-hand side of (52).

Several different types of end conditions are commonly prescribed in the study of beam deformations:

Fixed end: Both displacement and slope must vanish at the *fixed* endpoint; i.e.,

$$y=0, \qquad y'=0. \tag{53}$$

Free end: No moment or shearing force exists at the *free* endpoint; i.e.,

* For example, see S. Timoshenko, *Strength of Materials*, 3rd ed. (Van Nostrand: New York) 1955.

$$EIy'' = 0, \qquad \frac{d}{dx}(EIy'') = 0. \tag{54}$$

Simple support or *hinged end:* No displacement or moment exists at the *simply-supported* endpoint; i.e.,

$$y = 0, \qquad EIy'' = 0. \tag{55}$$

Sliding clamped end: No slope or shear force exists at the *sliding clamped* endpoint; i.e.,

$$y'' = 0, \qquad \frac{d}{dx}(EIy'') = 0 \tag{56}$$

EXAMPLE 15

Determine the displacement curve of a beam that is simply supported at $x = 0$ and $x = 1$, and subject to the distributed load $q(x) = EIx$ (see Figure 1.12). Also find the bending moment and shear force at each point along the beam, and the maximum displacement.

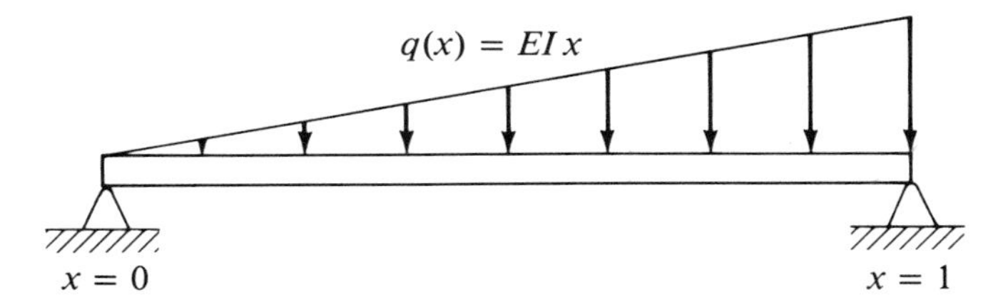

FIGURE 1.12

Beam supporting a load.

SOLUTION The problem is characterized by

$$y^{(4)} = x, \qquad y(0) = y''(0) = 0, \qquad y(1) = y''(1) = 0,$$

where the boundary conditions reflect the fact that the beam is simply supported. By performing four successive integrations of the DE, we obtain the general solution

$$y = \frac{1}{120}x^5 + \frac{1}{6}C_1x^3 + \frac{1}{2}C_2x^2 + C_3x + C_4,$$

where the C's are constants of integration. The two boundary conditions at $x = 0$ require that $C_4 = C_2 = 0$, while the remaining conditions at $x = 1$ lead to $C_1 = -1/6$ and $C_3 = 7/360$. Hence, the displacement curve of the beam is given by

$$y = \frac{x^5}{120} - \frac{x^3}{36} + \frac{7x}{360}.$$

Recalling from (51) that $M(x) = EIy''(x)$, approximately, the bending moment of the beam at any point x along the beam is found to be

$$M(x) = \frac{1}{6} EIx(x^2 - 1).$$

Also, the shear force is related to the bending moment by $V(x) = M'(x)$, so that

$$V(x) = \frac{1}{6} EI(3x^2 - 1).$$

Finally, we note that the maximum displacement of the beam occurs at the point x for which

$$y'(x) = \frac{x^4}{24} - \frac{x^2}{12} + \frac{7}{360} = 0.$$

By defining $F(x) = y'(x)$, we can use Newton's method to approximate this value.* That is, we form the sequence

$$x_{j+1} = x_j - \frac{F(x_j)}{F'(x_0)}, \qquad j = 0, 1, 2, \ldots$$

where $x_0 = 0.5$ is our initial estimate of the point of maximum displacement. Successively substituting $j = 0, 1, 2, \ldots$ into this expression, we find that $x \simeq 0.519$, from which we calculate

$$y_{\max} \simeq 0.0065.$$

□ 1.5.4 BUCKLING OF A LONG COLUMN

Let us now consider a long column or rod of length b that is subjected to an axial compressive force P applied at one end, as shown in Figure 1.13. By

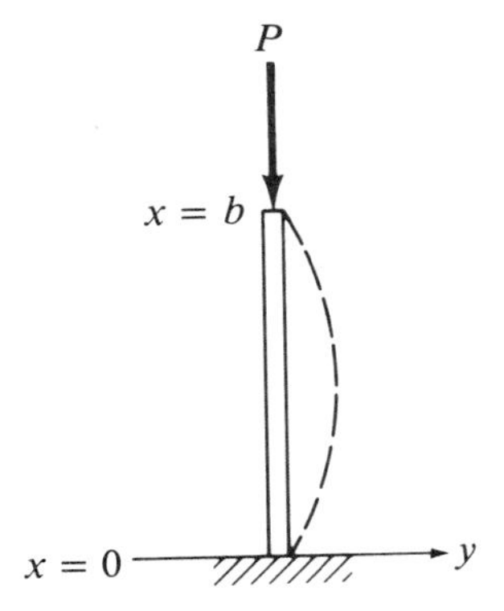

FIGURE 1.13

Long column under axial load P.

---

*This value can also be found in this case by the quadratic formula.

long we mean that the length of the column is much greater than its diameter. We wish to determine the possible modes of lateral deflection from the equilibrium position ($y = 0$) of the column along the x-axis.

This problem was discovered by Euler in the eighteenth century, making it probably the oldest known eigenvalue problem. Slender rods or columns are conventional in many structures such as aircraft, bridges, buildings, and so forth. These columns normally support the weight and loads of the structure above them, and this weight induces a compressive force P on the ends of the column. For sufficiently large values of P, the column will suddenly bow out of its equilibrium state—called a state of *buckling*. Due to Euler's mathematical analysis of columns, engineers were finally able to develop efficient designs on support columns in order to control buckling.

The axial load P affects the column as though a distributed lateral loading of intensity $q(x) = -Py''$ were acting on it. Thus, in determining the possible deformations, we can use the principles of elementary beam theory as applied in Section 1.5.3. The substitution of $q(x) = -Py''$ into Equation (52) leads to the eigenequation

$$y^{(4)} + \lambda y'' = 0, \qquad 0 < x < b, \tag{57}$$

where we have introduced

$$\lambda = \frac{P}{EI}.^{*} \tag{58}$$

Equation (57), along with prescribed homogeneous boundary conditions, constitutes an eigenvalue problem. The allowed values of λ correspond to *critical buckling loads* given by

$$P_n = \lambda_n EI, \qquad n = 1, 2, 3, \ldots. \tag{59}$$

The first critical load P_1 is also known as the *Euler load*. If the axial load P applied to the column is less than the Euler load, the column is stable and will remain in its undeformed equilibrium position. However, when P reaches the value P_1, the theory predicts that the column will assume the shape (under a small disturbance) predicted by the first eigenfunction $\phi_1(x)$, called the *fundamental buckling mode*.

EXAMPLE 16

Determine the possible deflection modes that a simply-supported column of length b will assume, and calculate the largest load the column can support before deformation may take place.

SOLUTION The problem is characterized by

$$y^{(4)} + \lambda y'' = 0, \qquad y(0) = y''(0) = 0, \qquad y(b) = y''(b) = 0,$$

---

* Physical considerations suggest that $\lambda > 0$ for buckling problems.

where we are recalling the boundary conditions (55) for a simply supported column or beam. The general solution of this DE for $\lambda = k^2 > 0$ is

$$y = C_1 + C_2 x + C_3 \cos kx + C_4 \sin kx,$$

and applying the boundary conditions at the end $x = 0$ requires that $C_1 + C_3 = 0$ and $C_3 = 0$. Hence, $C_1 = C_3 = 0$, and the remaining conditions at the end $x = b$ lead to

$$C_2 b + C_4 \sin kb = 0,$$

$$C_4 \sin kb = 0.$$

For nontrivial solutions to exist, we must select $k = n\pi/b$ and set $C_2 = 0$, from which we deduce

$$\lambda_n = \frac{n^2\pi^2}{b^2}, \qquad \phi_n(x) = C_4 \sin \frac{n\pi x}{b}, \qquad n = 1, 2, 3, \ldots.$$

With λ defined as one of the eigenvalues given above, the critical buckling loads become

$$P_n = \frac{n^2\pi^2 EI}{b^2}, \qquad n = 1, 2, 3, \ldots.$$

The largest load the column can support before possible buckling is the Euler load*

$$P_1 = \frac{\pi^2 EI}{b^2},$$

corresponding to the deflection shape

$$\phi_1(x) = C_4 \sin \frac{\pi x}{b}.$$

Once again the linear theory does not provide us with a value of the amplitude C_4 of the deflection.

Although the mathematical theory predicts an infinite number of possible critical loads and deflection modes, only the Euler load and corresponding fundamental buckling mode are actually realized by the column in most situations (if buckling does indeed occur). However, higher-order buckling modes may sometimes be realized, particularly if the column is restrained from movement at some point between its endpoints.

---

*Notice that the critical load P_1 can be increased if the length of the column or beam is decreased. The critical load can also be increased if the moment of inertia I of the beam is increased. The moment of inertia is large for cross-sections that have their masses "far" from the centroid of the beam. It is for this reason that we often find cross-sectional shapes in beams that resemble the letters I and H.

EXERCISES 1.5

In Problems 1–6, determine the displacement curve of a taut string subject to the prescribed external load and boundary conditions. Also find the maximum displacement of the string.

1. $y'' = -x,$

   (a) $y(0) = 0, \quad y(\pi) = \dfrac{\pi}{6}$

   (b) $y'(0) = 0, \quad y(\pi) = 0$

   (c) $y'(0) = 0, \quad y'(\pi) = 0$

2. $y'' = -x^2, \quad y(0) = 0, \quad y'(\pi) = 0$

3. $y'' = -\sin x, \quad y(0) = 0, \quad y(\pi) = 0$

4. $y'' = -x(1 - x), \quad y'(0) = 0, \quad y(1) = 0$

5. $y'' = -(x - \frac{1}{2})^2, \quad y(0) = 0, \quad y(1) = 0$

★6. $y'' = \begin{cases} -1, & 0 < x < 1 \\ 0, & 1 < x < 2, \end{cases} \quad y(0) = 0, \quad y(2) = 0$

★7. A heavy cable between supports at $x = 0$ and $x = 1$ hangs at rest under its own weight, which acts like a uniformly distributed load. The static displacements of the cable therefore satisfy the boundary value problem

$$y'' = -\frac{mg}{T}, \quad y(0) = \alpha, \quad y(1) = \beta,$$

where mg/T is constant.

(a) Show that the cable hangs in a parabolic arc.

(b) If greater accuracy is required in the displacements of the cable, the governing DE should be replaced by the nonlinear DE

$$y'' = -\frac{mg}{T}\sqrt{1 + (y')^2}.$$

By letting $v = y'$, solve the resulting DE by separation of variables and show that the solution satisfying the prescribed boundary conditions when $\alpha = \beta = 0$ is

$$y = \frac{T}{mg}\left\{\cosh\frac{mg}{2T} - \cosh\left[\frac{mg}{T}\left(x - \frac{1}{2}\right)\right]\right\}.$$

(c) Using the polynomial approximation

$$\cosh z \simeq 1 + \frac{z^2}{2!} + \frac{z^4}{4!},$$

estimate the displacement of the cable at $x = 1/2$ by the solution given in (b) and compare with the result obtained by using the solution in (a) when $\alpha = \beta = 0$.

(d) Under the assumption that $mg/T = 1$, compare the numerical values obtained by the solution in (a) when $\alpha = \beta = 0$ and the solution in (b) at the points $x = 0, 0.2, 0.4, 0.6, 0.8, 1$.

8. The transverse displacement of a string under constant tension ($T = 1$), fixed at the end $x = 0$, connected to a spring with spring constant K at $x = 1$, and subject to a transverse force of unity at $x = 1$ and a transverse force $q(x) = x$ per unit length along the string, is characterized by

$$y'' = -x, \qquad y(0) = 0, \qquad y'(1) + Ky(1) = 1.$$

Find the subsequent displacement.

9. When the tension T in a taut string is not constant, show that the linear equation of equilibrium becomes

$$\frac{d}{dx}[T(x)y'] = -q(x).$$

10. Solve the equilibrium equation in Problem 9 when the tension is $T(x) = 1/(1 + x)$ and $q(x) = 1$. Assume the boundary conditions are given by $y(0) = 0$, $y(1) = 0$.

11. Solve the rotating string problem when
(a) one end of the string is fixed and the other is free, i.e.,

$$y(0) = 0, \qquad y'(b) = 0;$$

(b) the physical conditions are as in (a), but

$$y'(0) = 0, \qquad y(b) = 0;$$

(c) both ends are free.

12. Solve the rotating string problem described by

$$y'' + \lambda y = 0, \qquad -a < x < a, \qquad y(-a) = 0, \qquad y(a) = 0.$$

13. If the end $x = b$ of a rotating string is attached to an elastic spring but fixed at the end $x = 0$, show that the eigenvalues must satisfy the transcendental equation

$$-\alpha kb \cos kb = \sin kb,$$

where $k^2 = \lambda$ and the boundary condition at $x = b$ is mathematically described by

$$\alpha by'(b) = -y(b) \qquad (\alpha \text{ constant}).$$

14. For Problem 13,
(a) determine whether $\lambda = 0$ is an eigenvalue.
(b) Find an approximate expression for the larger eigenvalues.

15. When a rotating string is subject to the nonconstant tension $T(x) = x^2$, the eigenequation has the form (see Problem 9)

$$\frac{d}{dx}(x^2y') + \rho\omega^2 y = 0, \qquad 1 < x < e.$$

Assuming boundary conditions $y(1) = y(e) = 0$, find the first critical speed ω_1 and corresponding deflection mode.

★16. Under the assumption that the approximation $\Delta s \simeq \Delta x$ is not valid for the rotating string problem,
(a) show that the governing DE then takes the form

$$\frac{d}{ds}\left(T\frac{dy}{ds}\right) + \rho\omega^2 y = 0, \qquad 0 < x < b.$$

(b) With the added assumption that $T\dfrac{dx}{ds} = T_0 =$ constant, show that the DE in (a) becomes

$$y'' + \lambda y\sqrt{1 + (y')^2} = 0, \qquad 0 < x < b,$$

where $\lambda = \rho\omega^2/T_0$ and $y' = \dfrac{dy}{dx}$.

In Problems 17–21, determine (a) the displacement curve, (b) the bending moment, and (c) the shear force of an elastic beam subject to the prescribed external load and boundary conditions.

17. $y^{(4)} = 1, \quad y(0) = y''(0) = 0, \quad y(1) = y''(1) = 0$

18. $y^{(4)} = x, \quad y(0) = y'(0) = 0, \quad y(1) = y'(1) = 0$

19. $y^{(4)} = x, \quad y(0) = y''(0) = 0, \quad y(1) = y'(1) = 0$

20. $y^{(4)} = x, \quad y(0) = y'(0) = 0, \quad y''(1) = y'''(1) = 0$

21. $y^{(4)} = \sin x, \quad y(0) = y''(0) = 0, \quad y(\pi) = y''(\pi) = 0$

In Problems 22 and 23, use Newton's method (see Example 15 in Section 1.5) to approximate the location of and maximum displacement of the following beams.

22. The beam in Problem 18.

23. The beam in Problem 19.

24. By integrating Equation (57) twice, show that the buckling modes are also solutions of the second-order DE

$$y'' + \lambda y = 0, \qquad \lambda = \frac{P}{EI},$$

when the ends of the column are simply supported, i.e., when

$$y(0) = y''(0) = 0, \qquad y(b) = y''(b) = 0.$$

Using this result,

(a) determine the maximum load P that the rod can support without buckling, assuming EI constant.

(b) If a load $P = \alpha EI$ is applied, where α is a constant, what is the length of the shortest rod for which buckling is likely to occur?

★25. When both ends of a column (of unit length) under an axial compressive force P are fixed, the boundary conditions are

$$y(0) = y'(0) = 0, \qquad y(1) = y'(1) = 0.$$

(a) Show that there are two sets of eigenvalues for this problem which are solutions, respectively, of the equations

$$\sin\frac{k}{2} = 0, \qquad \tan\frac{k}{2} = \frac{k}{2},$$

where $k^2 = \lambda$.

(b) Show that each set of eigenvalues in (a) leads to a different type of deflection mode by finding two sets of eigenfunctions.

26. The deflection modes of a straight shaft of constant density ρ, rotating with uniform angular speed ω about its equilibrium position along the x-axis, are solutions of the eigenequation

$$y^{(4)} - \lambda y = 0, \qquad 0 < x < b,$$

where $\lambda = \rho\omega^2/EI$. Find the characteristic speeds and deflection modes of a shaft of length 4 ft. that is simply supported at each end.

27. Solve Problem 26 for the characteristic speeds when the origin is located at the center of the shaft, i.e., the ends are located at $x = -2$ and $x = 2$.

★28. If the rotating shaft described in Problem 26 is simply-supported at $x = 0$ while the end at $x = b$ is free,
   (a) show that the critical angular speeds are given by

$$\omega_n = k_n^2 \sqrt{\frac{EI}{\rho}}, \qquad n = 1, 2, 3, \ldots,$$

   where k_n $(n = 1, 2, 3, \ldots)$ are the roots of $\tanh bk = \tan bk$.
   (b) Find the deflection modes corresponding to the critical speeds given in (a).
   (c) For the case $b = 1$, use Newton's method (see Example 12 in Section 1.4) to approximate the first critical speed ω_1.

1.6 STURM-LIOUVILLE THEORY

In this section we wish to examine more closely some of the general theory concerning eigenvalue problems, which is one of the deepest and richest theories in all mathematics. For purposes of developing this theory, it is helpful to introduce the notion of a *self-adjoint DE*.

DEFINITION 1.1 A homogeneous, second-order linear DE is said to be in *self-adjoint form* if and only if it has the form

$$p(x)y'' + p'(x)y' + [q(x) + \lambda r(x)]y = 0, \qquad x_1 < x < x_2,$$

or equivalently,

$$\frac{d}{dx}[p(x)y'] + [q(x) + \lambda r(x)]y = 0, \qquad x_1 < x < x_2,$$

where $p(x) > 0$ and $r(x) > 0$ in (x_1, x_2), and $p'(x)$, $q(x)$ and $r(x)$ are all continuous functions in the interval $[x_1, x_2]$.

Despite its appearance, the self-adjoint form is general enough to embrace virtually all second-order linear DEs. For example, the general eigenequation of concern here is of the form

$$A_2(x)y'' + A_1(x)y' + [A_0(x) + \lambda]y = 0 \tag{60}$$

where $A_0(x)$, $A_1(x)$, and $A_2(x)$ are continuous functions in the interval $[x_1, x_2]$, and where $A_2(x) > 0$ on this interval.* This DE is not in self-adjoint form unless $A_1(x) = A_2'(x)$. However, we can easily transform (60) into self-adjoint form by multiplying through by the function $\mu(x) = p(x)/A_2(x)$, producing

$$p(x)y'' + \mu(x)A_1(x)y' + \mu(x)[A_0(x) + \lambda]y = 0. \tag{61}$$

Equation (61) is now in self-adjoint form provided we pick $p(x)$ such that

$$p'(x) = \mu(x)A_1(x) = \frac{p(x)A_1(x)}{A_2(x)}. \tag{62}$$

By solving this first-order DE for $p(x)$, we find†

$$p(x) = \exp\left[\int \frac{A_1(x)}{A_2(x)}\,dx\right]. \tag{63}$$

Also, by further comparison of (61) with the self-adjoint form of Definition 1.1, we identify the functions

$$q(x) = \mu(x)A_0(x) = \frac{p(x)A_0(x)}{A_2(x)} \tag{64}$$

and

$$r(x) = \mu(x) = \frac{p(x)}{A_2(x)}. \tag{65}$$

In addition to providing a "standard form" for all linear, homogeneous second-order equations, the self-adjoint form enjoys certain operational advantages over other forms.

Since much of the subsequent discussion is centered around self-adjoint equations, it is notationally convenient to introduce the special differential operator (where $D = d/dx$)

$$L = D[p(x)D] + q(x), \tag{66}$$

called a *self-adjoint operator.* In terms of L, we can express a homogeneous self-adjoint DE in the compact form

$$L[y] + \lambda r(x)y = 0. \tag{67}$$

In all the remaining sections of the text, we will use the symbol L exclusively to denote a self-adjoint differential operator.

EXAMPLE 17

Put the DE $x^2y'' + xy' + \lambda y = 0$, $x > 0$, in self-adjoint form.

---

* If $A_2(x) < 0$, we simply multiply (60) by -1. Those cases for which $A_2(x) = 0$ at one or both endpoints of the interval of interest are discussed in Section 1.6.4.

† Any (nonzero) multiple of (63) can also be used for $p(x)$.

SOLUTION From (63), we first determine

$$p(x) = \exp\left(\int \frac{1}{x}\,dx\right) = x.$$

Thus, $\mu(x) = p(x)/A_2(x) = 1/x$, and multiplying the original DE through by $\mu(x)$ leads to

$$xy'' + y' + \left(\frac{\lambda}{x}\right)y = 0$$

or equivalently,

$$\frac{d}{dx}(xy') + \left(\frac{\lambda}{x}\right)y = 0.$$

In addition to being self-adjoint, there is another property of the operator L that we desire in much of the following discussion.

DEFINITION 1.2 A self-adjoint operator L is said to be *symmetric* on the interval $[x_1, x_2]$ if and only if

$$\int_{x_1}^{x_2} (uL[v] - vL[u])\,dx = 0$$

for any functions u and v having continuous second-order derivatives on the interval and satisfying the prescribed boundary conditions associated with L.

REMARK: *Eigenvalue problems for which the operator L is* symmetric *are also referred to as* self-adjoint problems *in much of the literature. Thus, an eigenvalue problem can have a self-adjoint operator but not be a self-adjoint problem. Because this terminology can be confusing, we have opted to use the term* symmetric operator *(in place of self-adjoint problem), which has a certain appeal because this same term is used in matrix theory to refer to a special matrix possessing properties similar to those to be discussed.*

Although Definition 1.2 relates the symmetry property to the operator L, we will subsequently find out that this property is closely related to the kind of boundary conditions prescribed along with L. In this regard it is possible that a given operator L is symmetric with one set of boundary conditions but not with a different set, as illustrated in the next example.

EXAMPLE 18

Determine whether the self-adjoint operator $L = D^2$ is symmetric on [0, 1] with respect to the following boundary conditions:

(a) $y(0) = 0, \quad y(1) = 0$

(b) $y(0) - y(1) = 0, \quad y'(1) = 0$

SOLUTION Substituting $L = D^2$ into the integral in Definition 1.2 and using integration by parts, we find

$$\int_0^1 [u(x)v''(x) - v(x)u''(x)]\,dx = u(x)v'(x) - v(x)u'(x)\Big|_0^1$$
$$= [u(1)v'(1) - v(1)u'(1)]$$
$$- [u(0)v'(0) - v(0)u'(0)].$$

For the boundary conditions in (a), it follows that u and v satisfy $u(0) = v(0) = 0$ and $u(1) = v(1) = 0$. Hence, the right-hand side of the above integral vanishes and we conclude that $L = D^2$ is *symmetric* in this case.

In the case of (b), the boundary conditions lead to $u(0) - u(1) = 0$, $v(0) - v(1) = 0$, and $u'(1) = v'(1) = 0$. Based on these relations, the above integral reduces to

$$\int_0^1 [u(x)v''(x) - v(x)u''(x)]\,dx = v(0)u'(0) - u(0)v'(0).$$

Because the right-hand side of this last expression is not necessarily zero, we deduce that $L = D^2$ is *not symmetric* in this case.

LEMMA 1.1

*Lagrange identity.** If $L = D[p(x)D] + q(x)$ is defined on the interval $[x_1, x_2]$, and if u and v are any functions with continuous second-order derivatives on $[x_1, x_2]$, then

$$uL[v] - vL[u] = \frac{d}{dx}[p(x)W(u, v)(x)],$$

where $W(u, v) = uv' - u'v$ is the Wronskian function.

PROOF With $L = D[p(x)D] + q(x)$, we have

$$uL[v] - vL[u] = u\frac{d}{dx}(pv') + uqv - v\frac{d}{dx}(pu') - vqu$$
$$= \frac{d}{dx}(pv'u - pu'v)$$
$$= \frac{d}{dx}[p(x)W(u, v)(x)],$$

which is what we wanted to prove.

*This identity is named after the French mathematician JOSEPH LOUIS LAGRANGE (1736–1813), who is known for his many contributions in mechanics, acoustics, calculus of variations, number theory, algebra, and partial differential equations. He also developed the method of variation of parameters used in solving nonhomogeneous DEs.

Based on Lagrange's identity, we find that

$$\int_{x_1}^{x_2} (uL[v] - vL[u])\, dx = \int_{x_1}^{x_2} \frac{d}{dx} [p(x)W(u, v)(x)]\, dx,$$

from which we deduce

$$\int_{x_1}^{x_2} (uL[v] - vL[u])\, dx = p(x)W(u, v)(x)\Big|_{x_1}^{x_2}. \tag{68}$$

We refer to (68) as *Green's formula,** which is a kind of one-dimensional version of Green's theorem in the plane from advanced calculus. Both Lagrange's identity and Green's formula are fundamental to the general study of boundary value problems.

The following theorem is a simple consequence of Green's formula.

THEOREM 1.3

A self-adjoint operator $L = D[p(x)D] + q(x)$ is a symmetric operator on the interval $[x_1, x_2]$ if and only if

$$p(x)W(u, v)(x)\Big|_{x_1}^{x_2} = 0$$

for any functions u and v that satisfy the prescribed boundary conditions associated with L and have continuous second-order derivatives on the interval $[x_1, x_2]$.

1.6.1 PROPERTIES OF A SYMMETRIC OPERATOR

Once we have established that a given operator L is symmetric, there are several important consequential properties associated with the eigenvalues and eigenfunctions of such an operator. Perhaps the single most important of these properties is the *orthogonality* of the eigenfunctions (see Theorem 1.4). Let us start by defining what we mean by saying two functions are orthogonal.

* Equation (68) is also called the "symmetric form" of Green's formula, which is just one of several similar formulas all referred to as *Green's formula* or *Green's identity*.

DEFINITION 1.3 If f and g are integrable functions on the interval (x_1, x_2), they are said to be *orthogonal* on this interval if and only if

$$\int_{x_1}^{x_2} f(x)g(x)\,dx = 0;$$

they are said to be *orthogonal with respect to a weighting function* $r(x) > 0$ if and only if

$$\int_{x_1}^{x_2} r(x)f(x)g(x)\,dx = 0.$$

REMARK: *The interval of orthogonality in Definition 1.3 may be of infinite extent in some cases, or it may be either open or closed at one end or both ends of the finite interval.*

EXAMPLE 19

Show that $f(x) = \sin 2x$ and $g(x) = \cos x$ are orthogonal on $(-\pi, \pi)$ with weight function $r(x) = 1$.

SOLUTION By using the identity $\sin 2x = 2 \sin x \cos x$, we obtain

$$\begin{aligned}\int_{-\pi}^{\pi} \sin 2x \cos x\,dx &= 2\int_{-\pi}^{\pi} \sin x \cos^2 x\,dx \\ &= -\frac{2}{3}\cos^3 x\Big|_{-\pi}^{\pi} \\ &= 0,\end{aligned}$$

which is what we want to show.

EXAMPLE 20

Show that $f(x) = 1$ and $g(x) = 1 - x$ are orthogonal on $(0, \infty)$ with weight function $r(x) = e^{-x}$.

SOLUTION Using integration by parts, we find

$$\begin{aligned}\int_0^{\infty} e^{-x}(1 - x)\,dx &= -(1 - x)e^{-x}\Big|_0^{\infty} - \int_0^{\infty} e^{-x}\,dx \\ &= 1 - 1 = 0,\end{aligned}$$

from which our desired conclusion follows.

The real significance of orthogonality will not be realized until Chapter 4, where we discuss Fourier series. Please note, however, that our given definition of orthogonality is a generalization of the familiar dot product definition of orthogonality from vector analysis. In fact, much of the development in Chapter 4 concerning orthogonal functions has a vector analog in three-dimensional space.

In the discussion to follow concerning orthogonal functions, it is the orthogonality of the set of eigenfunctions $\{\phi_n(x)\}$ of a symmetric Sturm-Liouville operator that is of primary interest, and for which we have the following important theorem.

THEOREM 1.4

Orthogonality. Let L be a symmetric operator on the interval $[x_1, x_2]$ associated with the eigenequation

$$L[y] + \lambda r(x)y = 0, \qquad x_1 < x < x_2.$$

If λ_n and λ_k are any two distinct eigenvalues of L with corresponding eigenfunctions $\phi_n(x)$ and $\phi_k(x)$, respectively, then $\phi_n(x)$ and $\phi_k(x)$ are orthogonal, i.e.,

$$\int_{x_1}^{x_2} r(x)\phi_n(x)\phi_k(x)\,dx = 0, \qquad n \neq k.$$

PROOF The eigenfunctions $\phi_n(x)$ and $\phi_k(x)$ satisfy the relations

$$L[\phi_n(x)] = -\lambda_n r(x)\phi_n(x),$$

$$L[\phi_k(x)] = -\lambda_k r(x)\phi_k(x).$$

If we multiply the first of these DEs by $\phi_k(x)$ and the second by $\phi_n(x)$, subtract the resulting expressions, and integrate over the interval of interest, we obtain

$$\int_{x_1}^{x_2} \{\phi_k(x)L[\phi_n(x)] - \phi_n(x)L[\phi_k(x)]\}\,dx = (\lambda_k - \lambda_n)\int_{x_1}^{x_2} r(x)\phi_n(x)\phi_k(x)\,dx.$$

Because L is symmetric, it follows from Definition 1.2 that the integral on the left is zero, and hence

$$(\lambda_k - \lambda_n)\int_{x_1}^{x_2} r(x)\phi_n(x)\phi_k(x)\,dx = 0.$$

By hypothesis, $\lambda_n \neq \lambda_k$; thus, we deduce that the integral vanishes and the theorem is proved.

We previously assumed that all eigenvalues are real without showing this to be true. We are now ready to prove our assertion for symmetric operators. The eigenfunctions can also be shown to be real; however, this proof is not presented here.

THEOREM 1.5

> The eigenvalues of a symmetric operator are all real.*

PROOF Suppose there exists some complex eigenvalue λ_k with corresponding eigenfunction $\phi_k(x)$, i.e.,

$$L[\phi_k(x)] + \lambda_k r(x)\phi_k(x) = 0.$$

Since the opertor L is composed of real functions, its complex conjugate $\overline{L}$ equals L. Therefore, by forming the complex conjugate of the above equation, we find

$$\overline{L[\phi_k(x)] + \lambda_k r(x)\phi_k(x)} = L[\overline{\phi_k(x)}] + \overline{\lambda_k} r(x)\overline{\phi_k(x)} = 0.$$

It follows now that $\phi_k(x)$ and $\overline{\phi_k(x)}$ belong to distinct eigenvalues, λ_k and $\overline{\lambda_k}$, respectively, and hence are necessarily orthogonal due to the symmetry of L. This means that

$$\int_{x_1}^{x_2} r(x)\phi_k(x)\overline{\phi_k(x)}\,dx = \int_{x_1}^{x_2} r(x)\,|\,\phi_k(x)\,|^2\,dx = 0,$$

but since the integrand is positive, this integral can never be zero, which leads to a *contradiction*. Our assumption that a complex eigenvalue exists must be false, and the theorem is proved.

Although Theorem 1.5 states that all eigenvalues of a symmetric operator are real, it does not guarantee the existence of any eigenvalues. In more advanced works on DEs, Theorem 1.6 is generally proved, which asserts the existence of an infinite number of eigenvalues starting with a smallest value.†

THEOREM 1.6

> The eigenvalues of a symmetric operator form an infinite sequence ordered in increasing magnitude so that
>
> $$\lambda_1 < \lambda_2 < \cdots < \lambda_n < \cdots,$$
>
> and where $\lambda_n \to \infty$ as $n \to \infty$.

---

* The proof of this theorem can be omitted for the readers not familiar with complex variables.

† For a proof of Theorem 1.6, see R. Courant and D. Hilbert, *Methods of Mathematical Physics*, Vol. 1 (Wiley: New York) 1953, p. 412.

1.6.2 REGULAR STURM-LIOUVILLE SYSTEMS

Most of the eigenvalue problems studied thus far have featured *unmixed* or *separated boundary conditions*. Problems of this type are characterized by

$$\begin{aligned} &L[y] + \lambda r(x)y = 0, \qquad x_1 < x < x_2 \\ &B_1[y] \equiv a_{11}y(x_1) + a_{12}y'(x_1) = 0 \qquad (a_{11}^2 + a_{12}^2 \neq 0), \\ &B_2[y] \equiv a_{21}y(x_2) + a_{22}y'(x_2) = 0 \qquad (a_{21}^2 + a_{22}^2 \neq 0), \end{aligned} \tag{69}$$

where $L = D[p(x)D] + q(x)$. Any eigenvalue problem belonging to this general class is called a *regular Sturm-Liouville system*.

Unmixed homogeneous boundary conditions are of three distinct varieties, which at either endpoint of the interval may assume one of the following forms:

$$\begin{aligned} y &= 0 \\ y' &= 0 \\ hy + y' &= 0 \quad (h \text{ constant}). \end{aligned}$$

These are called, respectively, *boundary conditions of the first, second, and third kinds*. Each kind of boundary condition corresponds to a different type of physical constraint at the endpoint. Some of these physical constraints were discussed in Section 1.2 in connection with steady-state heat conduction in a rod. Other interpretations of these boundary conditions concerning the supports for a taut string undergoing displacements were discussed in Section 1.5.

To show that the operator L associated with a regular Sturm-Liouville system is symmetric, we consider two functions u and v with continuous second derivatives that satisfy the prescribed unmixed boundary conditions in (69). At $x = x_1$, this implies that

$$\begin{aligned} a_{11}u(x_1) + a_{12}u'(x_1) &= 0, \\ a_{11}v(x_1) + a_{12}v'(x_1) &= 0, \end{aligned}$$

but since a_{11} and a_{12} cannot both be zero simultaneously, by definition, it follows that the coefficient determinant vanishes, i.e.,

$$\begin{vmatrix} u(x_1) & u'(x_1) \\ v(x_1) & v'(x_1) \end{vmatrix} = \begin{vmatrix} u(x_1) & v(x_1) \\ u'(x_1) & v'(x_1) \end{vmatrix} = W(u, v)(x_1) = 0.^*$$

By considering the boundary condition at $x = x_2$, we can apply a similar argument to show that $W(u, v)(x_2) = 0$. Thus, we have shown that

$$p(x)W(u, v)(x)\Big|_{x_1}^{x_2} = 0,$$

and by Theorem 1.3 it follows that L is symmetric.

---

* Interchanging the rows and columns of a determinant does not alter its value.

Because a regular Sturm-Liouville system has a symmetric operator, the eigenvalues and eigenfunctions of such a system have the important properties stated in Theorems 1.4, 1.5, and 1.6. In addition, we have the following result for these systems.

THEOREM 1.7

The eigenvalues of a regular Sturm-Liouville system are simple, i.e., only one eigenfunction belongs to each eigenvalue.

PROOF Assume $\phi_n(x)$ and $\phi_k(x)$ are distinct eigenfunctions belonging to the same eigenvalue λ_n. At $x = x_1$, each eigenfunction must satisfy the prescribed boundary condition so that

$$a_{11}\phi_n(x_1) + a_{12}\phi_n'(x_1) = 0,$$

$$a_{11}\phi_k(x_1) + a_{12}\phi_k'(x_1) = 0.$$

Since a_{11} and a_{12} cannot both be zero, we claim that the coefficient determinant must vanish, and thus (as above)

$$W(\phi_n, \phi_k)(x) = 0.$$

If the Wronskian of two solutions of a DE vanishes at one point on the solution interval, it must be identically zero on this interval.* Hence, $\phi_n(x)$ and $\phi_k(x)$ are proportional (linearly dependent) and therefore represent the same eigenfunction.

1.6.3 PERIODIC STURM-LIOUVILLE SYSTEMS

The second major class of eigenvalue problems that we consider are those of the form

$$\begin{aligned} &L[y] + \lambda r(x)y = 0, \qquad x_1 < x < x_2 \\ &y(x_1) = y(x_2), \qquad y'(x_1) = y'(x_2), \end{aligned} \tag{70}$$

where $L = D[p(x)D] + q(x)$ and where

$$p(x_1) = p(x_2).$$

Such problems are called *periodic Sturm-Liouville systems*.† We leave it to the reader this time to show that the operator L is symmetric for this class of problems (see Problem 7 in Exercises 1.6).

---

* See Theorem 4.4 in L. C. Andrews, *Ordinary Differential Equations with Applications* (Scott, Foresman: Glenview, Ill.) 1982. Also see Problem 41 in Exercises 1.3.

† Periodic Sturm-Liouville systems are a particular member of a larger class of Sturm-Liouville systems featuring *mixed* boundary conditions, the general forms of which are given by Equation (3) in Section 1.1.

A prime distinction of periodic Sturm-Liouville systems is that the eigenvalues are not necessarily simple, as illustrated in the next example.

EXAMPLE 21

At a particular instant of time, the transverse deflections of a vibrating circular membrane at a fixed distance from the center of the membrane are related to solutions of the eigenequation

$$y'' + \lambda y = 0, \qquad -\pi < \theta < \pi,$$

where $y = y(\theta)$ is a function of the polar angle θ. To insure that the deflections remain single-valued, and thus conform to physical considerations, the periodic boundary conditions

$$y(-\pi) = y(\pi), \qquad y'(-\pi) = y'(\pi),$$

are imposed. We wish to determine the possible modes of deflection.

SOLUTION Here $p(\theta) = 1$, so that $p(-\pi) = p(\pi)$. When $\lambda = 0$, the general solution of the DE is

$$y = C_1 + C_2\theta.$$

In order to accommodate the periodic boundary conditions, we must satisfy

$$y(-\pi) - y(\pi) = -2C_2\pi = 0,$$

$$y'(-\pi) - y'(\pi) = C_2 - C_2 = 0,$$

which implies that $C_2 = 0$ while C_1 is left arbitrary. Thus,

$$\lambda_0 = 0, \qquad \phi_0(\theta) = 1$$

constitute an eigenvalue–eigenfunction pair. For $\lambda = k^2 > 0$, the general solution takes the familiar form

$$y = C_1 \cos k\theta + C_2 \sin k\theta.$$

This time the boundary conditions lead to (upon simplification)

$$C_1 \sin k\pi = 0, \qquad C_2 \sin k\pi = 0.$$

Hence, for $k = n$ ($n = 1, 2, 3, \ldots$), we see that both C_1 and C_2 remain arbitrary. We conclude, therefore, that to each eigenvalue

$$\lambda_n = n^2, \qquad n = 1, 2, 3, \ldots$$

there corresponds *two* eigenfunctions, which in the general case we may write as

$$\phi_n(\theta) = C_1 \cos n\theta + C_2 \sin n\theta, \qquad n = 1, 2, 3, \ldots$$

and

$$\psi_n(\theta) = C_3 \cos n\theta + C_4 \sin n\theta, \qquad n = 1, 2, 3, \ldots.$$

where C_1, C_2, C_3 and C_4 are constants such that $\phi_n(\theta)$ and $\psi_n(\theta)$ are linearly independent.

Verifying that no negative eigenvalues exist is left to the exercises.

In situations like that in Example 21 we sometimes select the simplest combination of linearly independent eigenfunctions. For instance, by choosing $C_2 = C_3 = 0$ (and $C_1 = C_4 = 1$) in the result of Example 21, we obtain the linearly independent eigenfunctions

$$\phi_n(\theta) = \cos n\theta, \qquad n = 1, 2, 3, \ldots$$

$$\psi_n(\theta) = \sin n\theta, \qquad n = 1, 2, 3, \ldots.$$

Moreover, by making this choice of eigenfunctions we observe that

$$\int_{-\pi}^{\pi} \cos n\theta \sin n\theta \, d\theta = 0, \qquad n = 1, 2, 3, \ldots,$$

a result that is easy to verify by simple integration methods. Hence, not only are $\cos n\theta$ and $\sin n\theta$ the simplest combination of linearly independent eigenfunctions in Example 21, but they are also *orthogonal*.

In general, if two linearly independent eigenfunctions $\phi_n(x)$ and $\psi_n(x)$ can be found corresponding to a single eigenvalue λ_n, it is always possible to find linear combinations of $\phi_n(x)$ and $\psi_n(x)$ that are also orthogonal on the specified interval. A general procedure for obtaining orthogonal eigenfunctions in such cases is discussed in Problem 25 in Exercises 1.6. As a final comment here concerning the multiplicity of eigenvalues, we note that for second-order DEs no more than *two* linearly independent eigenfunctions can exist for a single eigenvalue.

1.6.4 SINGULAR STURM-LIOUVILLE SYSTEMS

Many of the most interesting and most common Sturm-Liouville systems in practice are classified as *singular* as defined below. These singularities change the general nature of the system, especially in the form of boundary conditions necessary to ensure that the operator L is symmetric, i.e., ensure that the resulting eigenfunctions are mutually orthogonal.

DEFINITION 1.4 A Sturm-Liouville system is said to be *singular* if one or more of the following events occur on the interval $[x_1, x_2]$:

(a) $p(x_1) = 0$ and/or $p(x_2) = 0$;*

(b) $p(x)$, $q(x)$, or $r(x)$ becomes infinite at $x = x_1$ or $x = x_2$ (or both); and

(c) either x_1 or x_2 (or both) are infinite.

---

*This situation corresponds to $A_2(x) = 0$ in Equation (60).

Some DEs that fall into this category include the following:

$$\frac{d}{dx}[(1-x^2)y'] + \lambda y = 0, \qquad -1 < x < 1 \qquad \text{(Legendre equation)}$$

$$\frac{d}{dx}(xy') - \frac{\nu^2}{x}y + \lambda xy = 0, \qquad 0 < x < b \qquad \text{(Bessel equation)}$$

$$\frac{d}{dx}(xe^{-x}y') + \lambda e^{-x}y = 0, \qquad 0 < x < \infty \qquad \text{(Laguerre equation)}$$

$$\frac{d}{dx}(e^{-x^2}y') + \lambda e^{-x^2}y = 0, \qquad -\infty < x < \infty \qquad \text{(Hermite equation)}$$

Legendre's equation is singular since $p(x) = 1 - x^2$ is zero at both endpoints $x = \pm 1$. For Bessel's equation, we find $p(x) = x$ vanishes at $x = 0$, and for $\nu \neq 0$, the function $q(x) = -\nu^2/x$ becomes infinite at $x = 0$. Laguerre's equation is singular at $x = 0$ since $p(x) = xe^{-x}$ goes to zero, but also one endpoint of the interval is infinite.* In the case of Hermite's equation, the singularity is due entirely to the infinite endpoints of the interval, where also $p(x) = e^{-x^2} \to 0$ as $|x| \to \infty$.

To ensure that a singular Sturm-Liouville system has a symmetric operator, we require

$$\int_{x_1}^{x_2} (uL[v] - vL[u])\,dx = p(x)W(u, v)(x)\Big|_{x_1}^{x_2} = 0, \tag{71}$$

(from Theorem 1.3) where u and v are any continuous, twice differentiable functions satisfying the prescribed boundary conditions of the Sturm-Liouville system. For example, if there is a singularity at $x = x_1$, we impose boundary conditions such that

$$\lim_{x\to x_1^+} p(x)W(u, v)(x) = 0\dagger \tag{72}$$

$$p(x_2)W(u, v)(x_2) = 0. \tag{73}$$

When the singularity arises specifically from $p(x_1) = 0$, then (72) is satisfied, for instance, by prescribing the condition

$$y(x),\ y'(x) \quad \text{finite as} \quad x \to x_1^+, \tag{74}$$

and (73) is safisfied by prescribing any condition of the form

$$a_{21}y(x_2) + a_{22}y'(x_2) = 0. \tag{75}$$

Of course, conditions other than (74) could be prescribed in order to satisfy (72), but any other condition is normally too restrictive in practice to allow for a nontrivial solution of the problem. Hence, we select this weaker modified type of boundary condition in most instances.

---

*Note also that $p(x) = xe^{-x} \to 0$ as $x \to \infty$.

†By $x \to x_1^+$, we mean a right-hand limit. Similarly, we denote a left-hand limit by $x \to x_2^-$.

The case of a singularity at $x = x_2$ is similarly treated. For example, if it should happen that $p(x_2) = 0$, we will require that $y(x)$ and $y'(x)$ remain finite as $x \to x_2^-$. Finally, if $p(x_1) = p(x_2) = 0$, we impose the boundary conditions

$$y(x),\ y'(x) \quad \text{finite as} \quad x \to x_1^+ \quad \text{and} \quad x \to x_2^-. \tag{76}$$

Singularities arise quite often in practice when a boundary point of the problem is not truly a physical boundary. For example, inside a circular domain of radius b we restrict the radial variable r such that $0 \leq r \leq b$. The point $r = 0$ is not a physical boundary of the circular domain, and yet it is clearly a mathematical boundary. In such cases we generally have to impose the condition that the solution of the problem remain bounded at $r = 0$.

Sometimes the singular eigenvalue problem does not involve a symmetric operator. This is the case, for example, when we consider the temperature distribution in a long rod that is mathematically modeled as infinite in extent.* In these cases we do not expect the eigenfunctions (if any) and eigenvalues to satisfy the conditions of Theorems 1.4–1.6. Consider the following example.

EXAMPLE 22

Find the eigenvalues and eigenfunctions of

$$y'' + \lambda y = 0, \qquad 0 < x < \infty, \qquad y(0) = 0, \qquad y, y' \quad \text{finite as} \quad x \to \infty.$$

SOLUTION For $\lambda = k^2 > 0$, the general solution is

$$y = C_1 \cos kx + C_2 \sin kx.$$

The first boundary condition requires that $C_1 = 0$, and thus

$$y = C_2 \sin kx.$$

Both $y = C_2 \sin kx$ and $y' = kC_2 \cos kx$ remain finite as $x \to \infty$, and so we conclude that $\lambda = k^2$ is an eigenvalue for *any* real (positive) number k. The corresponding eigenfunction is

$$\phi(x) = \sin kx.$$

We leave it to the reader to show that there are no further eigenvalues for $\lambda \leq 0$.

Example 22 illustrates that nonsymmetric operators can lead to results quite different from those associated with symmetric operators. For example, the eigenvalues are not discrete as is the case for symmetric operators, but form what we call a *continuum* of eigenvalues. The operator is nonsymmetric in this example because the boundary conditions are not appropriate for (71) to be satisfied. That is, since $p(x) = 1$ for all x, it does not vanish in the limit $x \to \infty$.

---

*This problem is discussed in Section 6.4.

EXERCISES 1.6

In Problems 1–6, put each equation in self-adjoint form.

1. $xy'' + \lambda y = 0, \qquad x > 0$
2. $y'' - y' + \lambda y = 0$
3. $xy'' + (1 - x)y' + \lambda y = 0, \qquad x > 0$
4. $(1 - x^2)y'' - 2xy' + \lambda y = 0, \qquad -1 < x < 1$
5. $x^2y'' + xy' + (\lambda x^2 - n^2)y = 0, \qquad x > 0$
6. $x^2(x^2 + 1)y'' + 2x^3y' + \lambda y = 0, \qquad x > 0$
7. Verify that a periodic Sturm-Liouville system has a symmetric operator.
8. Given that $r(x) = 1$, verify that the following functions form an orthogonal set on the interval $(0, \pi)$:
   (a) $\phi_n(x) = \sin nx, \qquad n = 1, 2, 3, \ldots.$
   (b) $\phi_n(x) = \cos nx, \qquad n = 0, 1, 2, \ldots.$
9. An orthogonal set of functions $\{\phi_n(x)\}$ satisfying the relation

$$\int_{x_1}^{x_2} [\phi_n(x)]^2\, dx = 1, \qquad n = 1, 2, 3, \ldots$$

   is said to be *normalized* on the given interval. Determine the constants C_n so that
   (a) $\phi_n(x) = C_n \sin nx, \qquad n = 1, 2, 3, \ldots$, becomes normalized on $(0, \pi)$.
   (b) $\phi_n(x) = C_n \cos nx, \qquad n = 0, 1, 2, \ldots$, becomes normalized on $(0, \pi)$.
10. Let $p_n(x)$, $n = 0, 1, 2, \ldots$, be an infinite set of polynomials satisfying the three conditions
    (a) $p_n(x)$ is of degree n,
    (b) $p_n(1) = 1, \qquad n = 0, 1, 2, \ldots$, and
    (c) $\displaystyle\int_{-1}^{1} p_n(x)p_k(x)\, dx = 0, \qquad n \neq k.$

    Determine the first *three* members of this set.
    Hint: Set $p_0(x) = a_1$, $p_1(x) = b_1 + b_2x$, and $p_2(x) = c_1 + c_2x + c_3x^2$, and solve for the unknown constants.
11. Repeat Problem 10 for the set of polynomials satisfying the conditions
    (a) $p_n(x)$ is of degree n,
    (b) $\displaystyle\int_0^{\infty} e^{-x}p_n(x)p_k(x)\, dx = 0, \qquad n \neq k$, and
    (c) $\displaystyle\int_0^{\infty} e^{-x}[p_n(x)]^2\, dx = 1.$

★12. If $L = D^4$, show that the Lagrange identity takes the form

$$uL[v] - vL[u] = \frac{d}{dx}(uv''' - u'''v - u'v'' + u''v').$$

Use this result to show that if $\phi_n(x)$ and $\phi_k(x)$ are eigenfunctions of the fourth-order eigenvalue problem

$$y^{(4)} - \lambda y = 0, \qquad y(0) = y''(0) = 0, \qquad y(1) = y''(1) = 0,$$

corresponding to distinct eigenvalues λ_n and λ_k, then $\phi_n(x)$ and $\phi_k(x)$ are orthogonal on $(0, 1)$.

Note: A fourth-order differential operator L is called self-adjoint if it has the form $L = D^2[s(x)D^2] + D[p(x)D] + q(x)$.

In Problems 13–15, show directly (without Theorem 1.4) that the eigenfunctions of the given Sturm-Liouville problem are orthogonal without explicitly solving for them, and state the orthogonality condition. (See Problem 26 in Exercises 1.4.)

13. $y'' + \lambda(1 + x)y = 0, \qquad y(0) = 0, \qquad y'(1) = 0$
14. $x^2y'' + \lambda y = 0, \qquad y(1) = 0, \qquad y'(2) = 0$
15. $xy'' + y' + \lambda xy = 0, \qquad y'(0) = 0, \qquad y(1) = 0$
16. If y_1 and y_2 are solutions of $L[y] = 0$, where L is self-adjoint, show that for some constant C

$$p(x)W(y_1, y_2)(x) = C.$$

Hint: See Problem 41 in Exercises 1.3.

In Problems 17–19, find the eigenvalues and eigenfunctions of the regular Sturm-Liouville system.

17. $y'' - 3y' + \lambda y = 0, \qquad y'(0) = 0, \qquad y'(\pi) = 0$
18. $x^2y'' + 3xy' + \lambda y = 0, \qquad y(1) = 0, \qquad y(5) = 0$

★19. $\dfrac{d}{dx}[(x^2 + 1)y'] + \dfrac{\lambda}{x^2 + 1}y = 0, \qquad y(0) = 0, \qquad y(1) = 0$

Hint: Let $x = \tan\theta$.

20. Show that $\lambda = 0$ is an eigenvalue of the regular Sturm-Liouville system

$$\frac{d}{dx}[p(x)y'] + \lambda r(x)y = 0, \qquad y'(0) = 0, \qquad y'(1) = 0.$$

In Problems 21–23, find the eigenvalues and eigenfunctions of the periodic Sturm-Liouville system.

21. $y'' + \lambda y = 0, \qquad y(-1) = y(1), \qquad y'(-1) = y'(1)$
22. $y'' + \lambda y = 0, \qquad y(0) = y(2\pi), \qquad y'(0) = y'(2\pi)$
23. $y'' + \lambda y = 0, \qquad y(0) = y(1), \qquad y'(0) = y'(1)$
24. Show that there are no negative eigenvalues in Example 21.
25. Let $\phi_n(x)$ and $\psi_n(x)$ be linearly independent eigenfunctions on the interval $x_1 < x < x_2$ corresponding to the same eigenvalue λ_n. Given the linear combinations

$$f_n(x) = \phi_n(x), \qquad g_n(x) = \psi_n(x) - \alpha\phi_n(x),$$

find a value of α so that

$$\int_{x_1}^{x_2} r(x)f_n(x)g_n(x)\,dx = 0.$$

26. Show that $\lambda = 0$ is an eigenvalue of multiplicity two for

$$y'' + \lambda y = 0, \qquad y'(0) = y'(1), \qquad y'(0) + y(0) = y(1).$$

Find an orthogonal pair of eigenfunctions belonging to $\lambda = 0$.
Hint: See Problem 25.

In Problems 27 and 28, find the eigenvalues and eigenfunctions of the singular Sturm-Liouville system.

★27. $x^2y'' + xy' + \lambda y = 0, \qquad y, y' \text{ finite as } x \to 0^+, \qquad y(1) = 0$

★28. $y'' + \lambda y = 0, \qquad y'(0) = 0, \qquad y, y' \text{ finite as } x \to \infty$

In Problems 29 and 30, determine the proper types of boundary conditions necessary to ensure the orthogonality of the resulting eigenfunctions.

29. $\dfrac{d}{dx}[(1 - x^2)y'] + \lambda y = 0, \qquad -1 < x < 1$

30. $\dfrac{d}{dx}(xy') + \lambda xy = 0, \qquad 0 < x < 1$

★31. Given the eigenequation

$$\frac{d}{dx}[p(x)y'] + [q(x) + \lambda r(x)]y = 0,$$

(a) multiply by $y(x)$ and integrate by parts to show that

$$-\int_{x_1}^{x_2} [p(x)(y')^2 - q(x)y^2]\,dx + \lambda \int_{x_1}^{x_2} r(x)y^2\,dx + p(x)yy'\Big|_{x_1}^{x_2} = 0.$$

(b) If the boundary conditions are such that

$$p(x)yy'\Big|_{x_1}^{x_2} = 0,$$

and $q(x) < 0$, show that the eigenvalues are strictly positive.

★32. If a Sturm-Liouville system has prescribed boundary conditions such that the eigenfunctions satisfy the inequalities

$$y(x_1)y'(x_1) \geqslant 0, \qquad y(x_2)y'(x_2) \leqslant 0,$$

and if $q(x) \leqslant 0$, show that all eigenvalues are strictly positive unless $q(x) \equiv 0$, in which case they are non-negative.
Hint: Use Problem 31(a).

□ 1.7 APPROXIMATION METHODS: WEIGHTED RESIDUALS

Perhaps it is correct to say that most mathematical problems encountered in practice are either difficult or impossible to solve exactly by known analytical techniques. Even in cases where exact solutions are explicitly obtained, they frequently are too cumbersome or complex for interpretation and numerical evaluation. Therefore, it becomes either necessary or convenient to employ an approximation or numerical method that will yield accurate numerical estimates of the true solution of the problem. Historically, it is for this reason that approximation and numerical techniques have received considerable attention in the applied areas of mathematics.

Analytical "solutions" generally provide more qualitative information and possibly a better understanding of the physical problem being studied than numerical "solutions," which consist of a set of numerical values. Because of this, approximation methods are often preferred in practice to numerical techniques, even though the numerical solution may be more accurate. The most widely used approximation methods for boundary value problems have been unified into one standardized technique called the *method of weighted residuals*. Essentially, the technique is based upon a procedure of selecting a finite series of "standard functions," frequently simple polynomials, with unspecified coefficients. This finite series represents a trial solution or approximate solution of the DE, but is selected in such a way that it satisfies exactly the prescribed boundary conditions. It is then forced to satisfy other physical or mathematical characteristics of the problem by any one of several methods, which are all based upon minimizing the "error of the fit."

1.7.1 NONHOMOGENEOUS EQUATIONS

To begin, let us consider the *nonhomogeneous* boundary value problem

$$L[y] = f(x), \qquad x_1 < x < x_2, \qquad B_1[y] = \alpha, \qquad B_2[y] = \beta, \tag{77}$$

where L is a self-adjoint operator* and B_1 and B_2 are unmixed boundary operators. We start by constructing the *linear trial solution*

$$\psi = g_0(x) + \sum_{j=1}^{N} c_j g_j(x), \tag{78}$$

which approximates the true solution y of (77). The c's in (78) are unknown constants, and the $g_j(x)$, $j = 1, 2, 3, \ldots, N$ are linearly independent functions chosen so that they satisfy the *homogeneous* boundary conditions

$$B_1[g_j] = 0, \qquad B_2[g_j] = 0, \qquad j = 1, 2, 3, \ldots, N. \tag{79}$$

The remaining function $g_0(x)$ is selected as any suitable function for which

$$B_1[g_0] = \alpha, \qquad B_2[g_0] = \beta. \tag{80}$$

Thus, by our choice of functions $g_0(x)$, $g_j(x)$, $j = 1, 2, 3, \ldots, N$, we see that the trial solution ψ has been chosen to satisfy the prescribed boundary conditions exactly for all choices of the unknown c's.

REMARK: *If* $\alpha = \beta = 0$, *we choose* $g_0(x) \equiv 0$.

---

*Greater accuracy is achieved in some approximation methods when the DE is first put into self-adjoint form.

Other than the conditions cited, choosing the g's is to a large extent arbitrary. In physically motivated problems the general nature of the solution is often known in advance and can be used in selecting these functions. Accuracy in the technique is enhanced in general by taking larger values of N.

In determining the c's, we first define the *error of the fit* or *residual*

$$E_N(x) = L[\psi] - f(x), \qquad x_1 < x < x_2. \tag{81}$$

The object is to minimize $E_N(x)$ in some meaningful fashion. We will discuss two ways in which this is accomplished—the *collocation method* and *Galerkin's method.*

REMARK: *Observe that if ψ were picked as the exact solution y, then $E_N(x) = 0$ for all x in the interval $x_1 < x < x_2$.*

Collocation Method: The collocation method is the simplest to apply. Here we simply set $E_N(x) = 0$ at N points distributed throughout the open interval $x_1 < x < x_2$, i.e., set

$$E_N(x) = 0 \quad \text{at} \quad x = X_1, X_2, \ldots, X_N. \tag{82}$$

This action leads to N simultaneous (nonhomogeneous) equations in c_1, c_2, $c_3, \ldots, c_N$. Solving these N equations for the c's yields our solution given by (78). The location of the N points is frequently determined by intuition, physical considerations, or practical considerations such as computational simplicity. In many cases we simply choose them so that they are evenly distributed throughout the interval.*

A disadvantage of the collocation method is that the approximate solution ψ may vary greatly with the location of the collocation points. This problem is somewhat alleviated by taking N "sufficiently large."

Galerkin's Method: The method developed by B. G. Galerkin in 1915 is perhaps the most widely used of the approximation methods. The general procedure is similar to the collocation method, but here we choose the N constants $c_1, c_2, c_3, \ldots, c_N$ in such a way that

$$\int_{x_1}^{x_2} E_N(x) g_k(x)\, dx = 0, \qquad k = 1, 2, 3, \ldots, N. \tag{83}$$

Hence, (83) will also lead to N simultaneous equations from which the c's may be computed.

REMARK: *The collocation method and Galerkin's method are specific examples of* weighted residual methods. *In general, weighted residual methods rely on setting the weighted integral of the residual to zero, i.e.,*

---

*The N points should be chosen so that none of the trial functions $g_j(x)$ has a zero at any of the collocation points.

$$\int_{x_1}^{x_2} E_N(x) w_k(x)\, dx = 0, \qquad k = 1, 2, 3, \ldots, N$$

for some set of weight functions $\{w_k(x)\}$. *For the collocation method the weight functions are*

$$w_k(x) = \delta(x - x_k),$$

where δ *is the* Dirac delta function *(see Section 2.3), and in the case of Galerkin's method the weight functions* $w_k(x)$ *are chosen to be the trial functions* $g_k(x)$. *Another weighted residual method that is commonly used is the* least squares method *for which*

$$w_k(x) = \frac{\partial E_N(x)}{\partial c_k}.*$$

EXAMPLE 23

Using the set of linearly independent polynomials

$$g_j(x) = x^j(1 - x), \qquad j = 1, 2, 3, \ldots, N,$$

find an approximate solution of the boundary value problem

$$y'' + y = x - 1, \qquad y(0) = -1, \qquad y(1) = 1.$$

SOLUTION We first observe that each $g_j(x)$ satisfies the *homogeneous* boundary conditions

$$g_j(0) = 0, \qquad g_j(1) = 0, \qquad j = 1, 2, 3, \ldots, N.$$

A function $g_0(x)$ satisfying the prescribed *nonhomogeneous* boundary conditions is given by $g_0(x) = 2x - 1$. Other choices for $g_0(x)$ are also possible, such as $g_0(x) = -\cos \pi x$ (see Problem 1 in Exercises 1.7). Using our first choice, the general trial solution takes the form

$$\psi = 2x - 1 + \sum_{j=1}^{N} c_j x^j (1 - x).$$

To begin, let us take the case $N = 1$. Thus,

$$\psi = 2x - 1 + c_1 x(1 - x),$$

leading to the residual

$$E_1(x) = \psi'' + \psi - x + 1 = x + c_1(x - x^2 - 2).$$

For the collocation method we will set $E_1 = 0$ at $x = 1/2$, i.e.,

$$\frac{1}{2} + c_1\left(\frac{1}{2} - \frac{1}{4} - 2\right) = 0.$$

* For a deeper discussion of the weighted residual methods, see B. A. Finlayson, *The Method of Weighted Residuals and Variational Principles* (Academic Press: New York) 1972.

Solving, we find

$$c_1 = \frac{2}{7} \simeq 0.286$$

and thus

$$\psi = 2x - 1 + 0.286x(1 - x).$$

The Galerkin method requires that

$$\int_0^1 E_1(x)x(1 - x)\,dx = 0,$$

or

$$\int_0^1 [x^2 - x^3 + c_1(x^4 - 2x^3 + 3x^2 - 2x)]\,dx = 0.$$

Evaluating the integral, we get

$$\frac{1}{12} - \frac{3}{10}c_1 = 0,$$

from which we determine

$$c_1 = \frac{5}{18} \simeq 0.278.$$

Therefore, our approximate solution is

$$\psi = 2x - 1 + 0.278x(1 - x).$$

If we retain two terms in the trial solution, then

$$\psi = 2x - 1 + c_1x(1 - x) + c_2x^2(1 - x),$$

and in this case

$$E_2(x) = x + c_1(x - x^2 - 2) + c_2(x^2 - x^3 + 2 - 6x).$$

Collocating at $x = 1/3$ and $x = 2/3$, we obtain the set of equations

$$\frac{16}{9}c_1 - \frac{2}{27}c_2 = \frac{1}{3},$$

$$\frac{16}{9}c_1 + \frac{5}{27}c_2 = \frac{2}{3},$$

with solution $c_1 \simeq 0.195$ and $c_2 \simeq 0.173$; hence, in this case

$$\psi = 2x - 1 + 0.195x(1 - x) + 0.173x^2(1 - x).$$

TABLE 1.1 Approximate Solutions of $y'' + y = x - 1$, $y(0) = -1$, $y(1) = 1$, by the Method of Weighted Residuals

x	COLLOCATION (ONE TERM)	GALERKIN (ONE TERM)	COLLOCATION (TWO TERMS)	EXACT VALUE
0	−1	−1	−1	−1
0.1	−0.774	−0.775	−0.781	−0.781
0.2	−0.554	−0.556	−0.563	−0.564
0.3	−0.340	−0.342	−0.348	−0.349
0.4	−0.131	−0.133	−0.137	−0.137
0.5	0.072	0.070	0.070	0.070
0.6	0.269	0.267	0.272	0.271
0.7	0.460	0.458	0.466	0.466
0.8	0.646	0.644	0.653	0.653
0.9	0.826	0.825	0.832	0.831
1.0	1	1	1	1

All three approximate solutions obtained in Example 23 are compared in Table 1.1 with the exact solution (see Example 8)

$$y = \frac{\sin x}{\sin 1} + x - 1.$$

EXAMPLE 24

Find an approximate solution to the string problem of Example 14 in Section 1.5.1 using $g_j(x) = \sin jx$, $j = 1, 2, 3, \ldots, N$. For the sake of simplicity, assume $q_0/T = 1$.

SOLUTION The string problem is characterized by

$$y'' = -1, \qquad y(0) = 0, \qquad y(\pi) = 0.$$

Because the prescribed boundary conditions are homogeneous, we can choose $g_0(x) \equiv 0$, and thus our trial solution is

$$\psi = \sum_{j=1}^{N} c_j \sin jx.$$

For illustrative purposes we will select $N = 2$ so that the trial solution takes the form

$$\psi = c_1 \sin x + c_2 \sin 2x.$$

The corresponding residual is

$$E_2(x) = \psi'' + 1 = -c_1 \sin x - 4c_2 \sin 2x + 1.$$

We will set $E_2(x) = 0$ at $x = \pi/3$ and $x = 2\pi/3$, which leads to

$$-\frac{\sqrt{3}}{2} c_1 - 2\sqrt{3}\, c_2 = -1,$$

$$-\frac{\sqrt{3}}{2} c_1 + 2\sqrt{3}\, c_2 = -1.$$

Solving for c_1 and c_2 yields

$$c_1 = \frac{2}{\sqrt{3}} \simeq 1.155, \qquad c_2 = 0,$$

so that for the collocation method

$$\psi = 1.155 \sin x.$$

For Galerkin's method, we require that

$$\int_0^{\pi} E_2(x) \sin x \, dx = 0, \qquad \int_0^{\pi} E_2(x) \sin 2x \, dx = 0,$$

which leads to

$$-\frac{\pi}{2} c_1 = -2,$$

$$-2\pi c_2 = 0,$$

or

$$c_1 = \frac{4}{\pi} \simeq 1.273, \qquad c_2 = 0.$$

Thus, for the Galerkin method our solution is

$$\psi = 1.273 \sin x.$$

TABLE 1.2 Approximate solutions of $y'' = -1$, $y(0) = 0$, $y(\pi) = 0$, by the Method of Weighted Residuals with $N = 2$.

x	COLLOCATION	GALERKIN	EXACT VALUE
0	0	0	0
$\pi/8$	0.442	0.487	0.540
$\pi/4$	0.817	0.900	0.925
$\pi/2$	1.155	1.273	1.234
$3\pi/4$	0.817	0.900	0.925
$7\pi/8$	0.442	0.487	0.540
π	0	0	0

The approximate solutions found in Example 24 are compared with the exact solution $y = x(\pi - x)/2$ in Table 1.2.

Neither the collocation method nor the Galerkin method led to a very accurate solution in Example 24, even though we started with a two-term trial solution. The reason for this is that the second term of the trial solution was rejected by both methods ($c_2 = 0$), and thus we essentially had only a one-term trial solution, which is not generally expected to give very accurate results. To understand why the second term did not contribute to the solution, we note from the solution values listed in Table 1.2 that the solution is symmetric about the midpoint $x = \pi/2$, a fact that can be deduced from the formulation of the problem. A trial solution should be chosen so that each function $g_j(x)$ reflects this symmetry property. Note that the function $g_1(x) = \sin x$ is symmetric about $x = \pi/2$ but $g_2(x) = \sin 2x$ is *not,* and hence the reason that the methods required $c_2 = 0$. Only those $g_j(x) = \sin jx$ with *odd* index j have the proper symmetry, and so in Example 24 a better two-term trial solution is

$$\psi = c_1 \sin x + c_3 \sin 3x.$$

(See Problem 2 in Exercises 1.7.) However, in using this trial solution we must select the collocation points other than $x = \pi/3$ and $x = 2\pi/3$ to avoid the zeros of $\sin 3x$. For example, we might choose $x = \pi/4$ and $x = \pi/2$.

EXAMPLE 25

Consider a uniform beam of length b that is resting on an elastic foundation with constant modulus K (lb/ft/displacement). If the beam is supporting a uniformly distributed load q_0 and is simply supported at both ends (see Figure 1.14), the problem is characterized by

$$EIy^{(4)} + Ky = q_0, \qquad 0 < x < b$$

$$y(0) = y''(0) = 0, \qquad y(b) = y''(b) = 0.$$

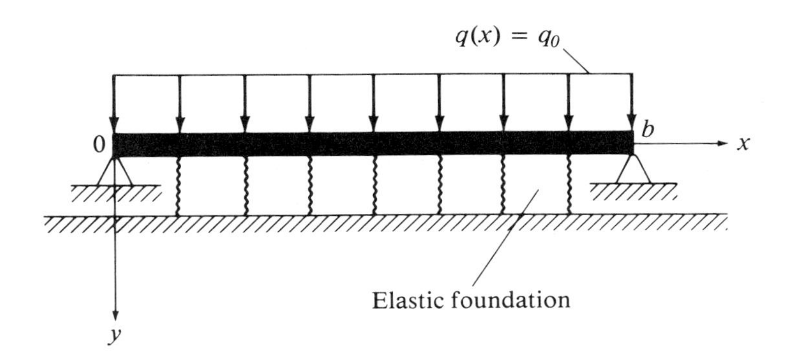

FIGURE 1.14

Beam resting on elastic foundation.

Use the collocation method with two terms to determine an approximate solution for the displacement of the beam.

SOLUTION In this case it may be best to first make the equation *nondimensional* by introducing the new parameters

$$z = \frac{x}{b}, \qquad w(z) = \frac{EI}{q_0 b^4}\, y(z), \qquad k^4 = \frac{Kb^4}{EI}.$$

The problem now assumes the form

$$w^{(4)} + k^4 w = 1, \qquad w(0) = w''(0) = 0, \qquad w(1) = w''(1) = 0,$$

where differentiation is with respect to z.

For a trial solution we take

$$\psi(z) = c_1 \sin \pi z + c_3 \sin 3\pi z,$$

where we are recognizing the fact that the displacements are symmetric about $z = 1/2$. Calculating the residual, we obtain

$$E_2(z) = w^{(4)} + k^4 w - 1 = (\pi^4 + k^4)c_1 \sin \pi z + (81\pi^4 + k^4)c_3 \sin 3\pi z - 1,$$

and by setting $E_2(z) = 0$ at $z = 1/4$ and $z = 1/2$, we are led to the set of equations

$$(\pi^4 + k^4)c_1 + (81\pi^4 + k^4)c_3 = \sqrt{2},$$

$$(\pi^4 + k^4)c_1 - (81\pi^4 + k^4)c_3 = 1,$$

the simultaneous solution of which is

$$c_1 = \frac{\sqrt{2} + 1}{2(\pi^4 + k^4)}, \qquad c_3 = \frac{\sqrt{2} - 1}{2(81\pi^4 + k^4)}.$$

In terms of the original parameters, our solution takes the form

$$\psi(x) = \frac{q_0 b^4}{EI}\left[c_1 \sin \frac{\pi x}{b} + c_3 \sin \frac{3\pi x}{b}\right],$$

where c_1 and c_2 are given above.

For a fixed value of N, the collocation method is generally less accurate than the Galerkin method. The main advantage of the collocation method is ease of implementation, but the Galerkin method is generally preferred when using only one-term or two-term approximations because of its greater accuracy.

1.7.2 EIGENVALUE PROBLEMS

The weighted residual methods are also useful techniques for approximating the eigenvalues and eigenfunctions of Sturm-Liouville systems. For second-order DEs, we can always put the DE in the self-adjoint form

$$L[y] + \lambda r(x)y = 0, \qquad x_1 < x < x_2, \tag{84}$$

and define the *residual* by

$$E_N(x) = L[\psi] + \Lambda r(x)\psi, \tag{85}$$

where Λ approximates the eigenvalues of the given Sturm-Liouville system and

$$\psi = \sum_{j=1}^{N} c_j g_j(x) \tag{86}$$

approximates the eigenfunctions. Because the boundary conditions for eigenvalue problems are homogeneous, it is not necessary to include the term $g_0(x)$ in the trial solution (86). We select the trial functions $g_j(x)$ by primarily the same criteria listed in the last section.

Basically, the technique from this point is essentially the same as before, except that now the system of equations in the unknown c's will be *homogeneous*. Hence, in order to obtain nontrivial solutions for the c's we must require the coefficient determinant of the c's to vanish (see Cramer's rule in Appendix C). Doing so leads to what is called the *secular equation*, an Nth degree polynomial in Λ. The roots of this polynomial furnish approximations to the first N eigenvalues of the given Sturm-Liouville system.*

EXAMPLE 26

Approximate the first two eigenvalues and eigenfunctions of

$$y'' + \lambda y = 0, \qquad y(0) = 0, \qquad y(1) = 0,$$

using the trial solution $\psi = c_1 x(1 - x) + c_2 x^2(1 - x)$.

SOLUTION The residual is

$$E_2(x) = \psi'' + \Lambda\psi = -[2 - \Lambda(x - x^2)]c_1 + [2 - 6x + \Lambda(x^2 - x^3)]c_2,$$

and by setting $E_2(x) = 0$ at $x = 1/3$ and $x = 2/3$, we obtain the system of equations (for the collocation method)

$$\left(2 - \frac{2}{9}\Lambda\right)c_1 - \frac{2}{27}\Lambda c_2 = 0,$$

$$\left(2 - \frac{2}{9}\Lambda\right)c_1 + \left(2 - \frac{4}{27}\Lambda\right)c_2 = 0.$$

For nonzero values of c_1 and c_2 we require that

$$\begin{vmatrix} 2 - \frac{2}{9}\Lambda & -\frac{2}{27}\Lambda \\ 2 - \frac{2}{9}\Lambda & 2 - \frac{4}{27}\Lambda \end{vmatrix} = \frac{4}{243}\Lambda^2 - \frac{16}{27}\Lambda + 4 = 0,$$

*The method is not uniformly accurate for all eigenvalues. Greater error is always observed in the larger eigenvalues for any N.

from which we calculate

$$\Lambda_1 = 9, \qquad \Lambda_2 = 27.$$

By substituting $\Lambda_1 = 9$ into the above system of equations involving c_1 and c_2, we find that $c_2 = 0$ while c_1 is arbitrary. Hence, the first eigenfunction is approximated by

$$\psi_1 = x(1 - x)$$

(for $c_1 = 1$). Similarly, substituting $\Lambda_2 = 27$ into the same system of equations leads to

$$-4c_1 - 2c_2 = 0,$$

from which we conclude that $c_2 = -2c_1$, leaving c_1 arbitrary, and hence, setting $c_1 = 1$

$$\begin{aligned}\psi_2 &= x(1 - x) - 2x^2(1 - x)\\ &= x(1 - 3x + 2x^2).\end{aligned}$$

Using the Galerkin method, we set

$$\int_0^1 E_2(x)x(1 - x)\,dx = 0, \qquad \int_0^1 E_2(x)x^2(1 - x)\,dx = 0,$$

which yields the system of equations

$$\begin{aligned}\left(\frac{1}{3} - \frac{1}{30}\Lambda\right)c_1 + \left(\frac{1}{6} - \frac{1}{60}\Lambda\right)c_2 = 0,\\ \left(\frac{1}{6} - \frac{1}{60}\Lambda\right)c_1 + \left(\frac{2}{15} - \frac{1}{105}\Lambda\right)c_2 = 0.\end{aligned}$$

By equating the coefficient determinant to zero, this time we obtain a secular equation with solutions

$$\Lambda_1 = 10, \qquad \Lambda_2 = 42.$$

These eigenvalues lead to the same eigenfunction approximations found above by the collocation method.

For the sake of comparison, we recall that the exact eigenvalues for this problem were previously found to be

$$\lambda_1 = \pi^2 \simeq 9.870, \qquad \lambda_2 = 4\pi^2 \simeq 39.478$$

(Example 11 in Section 1.4). Here we see that the Galerkin method has once again provided better approximations than the collocation method. Of course, greater precision can be achieved in both approximation schemes if additional terms are included in the trial solution.

EXAMPLE 27

Using a one-term trial solution and the Galerkin method, approximate the largest load P that a column of length b can support before buckling may take place. The problem is characterized by

$$EIy^{(4)} + Py'' = 0, \qquad 0 < x < b$$

$$y(0) = y''(0) = 0, \qquad y(b) = y''(b) = 0$$

(see Example 16 in Section 1.5.4).

SOLUTION To use a one-term polynomial approximation, we require that the trial solution satisfy (we set $c_1 = 1$ for convenience)

$$\psi'' = x(x - b)$$

so that $\psi''(0) = \psi''(b) = 0$. Therefore, following two integrations we obtain

$$\psi = \frac{1}{12}x^4 - \frac{b}{6}x^3 + C_1x + C_2,$$

and by setting $\psi(0) = 0$ and $\psi(b) = 0$, we find that $C_1 = b^3/12$ and $C_2 = 0$. We thus have produced the trial solution

$$\psi = \frac{1}{12}x^4 - \frac{b}{6}x^3 + \frac{b^3}{12}x,$$

which satisfies all four boundary conditions. Writing $\Lambda = P/EI$, we find the residual to be

$$E_1(x) = 2 + \Lambda x(x - b),$$

and by setting

$$\int_0^b E_1(x)\left(\frac{1}{12}x^4 - \frac{b}{6}x^3 + \frac{b^3}{12}x\right) dx = 0$$

we obtain

$$\Lambda_1 = \frac{168}{17b^2} \simeq \frac{9.882}{b^2}.$$

The largest load is then

$$P \simeq 9.882\frac{EI}{b^2}$$

(the Euler load is proportional to the first eigenvalue), which is very close to the exact value

$$P = \pi^2\frac{EI}{b^2} \simeq 9.870\frac{EI}{b^2}.$$

On the other hand, the collocation method with $E_1(x) = 0$ at $x = b/2$ would lead to the less accurate result $P \simeq 8EI/b^2$.

EXERCISES 1.7

In Problems 1–6, use the given trial solution function and the collocation method to find an approximate solution.

1. $y'' + y = x - 1, \quad y(0) = -1, \quad y(1) = 1: \quad \psi = -\cos \pi x + c_1 x(1 - x)$

2. $y'' = -1, \quad y(0) = 0, \quad y(\pi) = 0: \ \psi = c_1 \sin x + c_3 \sin 3x$
   Construct a table of numerical values like Table 1.2 and compare with the exact solution. [Set $E_2(x) = 0$ at $x = \pi/4$ and $x = \pi/2$.]

3. $y^{(4)} + y = x, \quad y(0) = y''(0) = 0, \quad y(1) = y''(1) = 0:$
   $\psi = c_1(x^4/12 - x^3/6 + x/12)$

4. $\dfrac{d}{dx}[(1 + x)^{-1}y'] = 1, \quad y(0) = 0, \quad y(1) = 0: \quad \psi = c_1 x(1 - x)$

5. $2x^2y'' + 3xy' - y = 4x^{3/2}, \quad y(0) = 0, \quad y(1) = 0:$
   $\psi = c_1 x(1 - x) + c_2 x^2(1 - x)$

★6. $\dfrac{d}{dx}[(1 + ay)y'] = 0, \quad y(0) = 0, \quad y(1) = 10: \ \psi = 10x + c_1 x(1 - x)$
   (a) Assume $a = 0.1$ and compare your answer with the exact solution at $x = 0.2, 0.4, 0.6, 0.8$.
   (b) Repeat (a) when $a = 1$. Comment on the accuracy of the approximate solutions as the parameter a increases.
   Note: The solution y represents the steady-state temperature across a slab with a temperature dependent conductivity $k(y) = 1 + ay$.

In Problems 7–12, use the given trial solution function and the Galerkin method to find an approximate solution.

7. Problem 1

8. Problem 2

9. Problem 3

10. Problem 4

11. Problem 5

★12. Problem 6

13. If the uniform beam in Example 25 is subject to the distributed load $q(x) = q_0(b - x)$, the problem is characterized by

$$EIy^{(4)} + Ky = q_0(b - x), \qquad 0 < x < b$$

$$y(0) = y''(0) = 0, \qquad y(b) = y''(b) = 0.$$

   Using the trial solution

$$\psi = c_1 \sin \pi x/b + c_2 \sin 2\pi x/b,$$

   find an approximation of the displacement y
   (a) by the collocation method;
   (b) by the Galerkin method.

In Problems 14–17, use the given trial solution and the collocation method to find an approximation to the first eigenvalue.

14. $y'' + \lambda y = 0, \quad y'(0) = 0, \quad y(1) = 0\colon \quad \psi = c_1(1 - x^2)$

15. $y'' + \lambda y = 0, \quad y(0) = 0, \quad y'(1) = 0\colon \quad \psi = c_1 x(1 - x/2)$

16. $x^2 y'' + 2xy' + \lambda y = 0, \quad y(1) = 0, \quad y(3) = 0\colon \quad \psi = c_1(x - 1)(3 - x)$

17. $\dfrac{d}{dx}(xy') + \lambda xy = 0, \quad y(0)$ finite, $\quad y(1) = 0\colon \quad \psi = c_1(1 - x^2)$

In Problems 18–21, use the given trial solution and the Galerkin method to find an approximation to the first eigenvalue.

18. Problem 14

19. Problem 15

20. Problem 16

21. Problem 17

22. By Galerkin's method, approximate the first two eigenvalues of

$$y'' + \lambda xy = 0, \qquad y(0) = 0, \qquad y(1) = 0,$$

(a) using trial solution $\psi = c_1 x(1 - x) + c_2 x^2(1 - x)$;
(b) using trial solution $\psi = c_1 \sin \pi x + c_2 \sin 2\pi x$.

23. By the collocation method, approximate the first two eigenvalues and eigenfunctions of

$$xy'' + \lambda y = 0, \qquad y(0) = 0, \qquad y(1) = 0,$$

(a) using trial solution $\psi = c_1 x(1 - x) + c_2 x^2(1 - x)$;
(b) using trial solution $\psi = c_1 \sin \pi x + c_2 \sin 2\pi x$.

24. Using the trial solution $\psi = c_1(2x^2 - 3x) + c_3(x^3 - 2x)$, approximate the first two eigenvalues and eigenfunctions of

$$y'' + \lambda y = 0, \qquad y(0) = 0, \qquad y(1) + y'(1) = 0$$

(a) by the collocation method;
(b) by the Galerkin method.

*25. The anharmonic oscillator problem in quantum mechanics is characterized by

$$\frac{H^2}{2m} y'' - \frac{1}{2} m\omega^2 x^2 y - \epsilon x^4 y + \lambda y = 0, \qquad -\infty < x < \infty$$

$$\lim_{x \to \pm\infty} y = 0,$$

where all parameters are constant. The eigenvalues λ_n represent the energy levels possible for the system, the ground state being determined by the smallest eigenvalue λ_0. Using the trial solution $\psi = c_1 \exp(-m\omega x^2/2H)$ and Galerkin's method, approximate the ground state energy and show that for $\epsilon = 0$ it reduces to the well-known result $\lambda_0 = \omega H/2$.

★26. Show that Galerkin's method for the special Sturm-Liouville problem

$$L[y] + \lambda r(x)y = 0, \qquad y(x_1) = 0, \qquad y(x_2) = 0,$$

leads to the system of equations

$$\sum_{j=1}^{N} (a_{jk} - \Lambda b_{jk})c_j = 0, \qquad k = 1, 2, \ldots, N$$

where

$$a_{jk} = \int_{x_1}^{x_2} [p(x)g_j'(x)g_k'(x) - q(x)g_j(x)g_k(x)]\, dx$$

and

$$b_{jk} = \int_{x_1}^{x_2} r(x)g_j(x)g_k(x)\, dx.$$

27. Use the results of Problem 26 to solve Example 26 with $N = 2$.

28. Use the results of Problem 26 to solve Problem 22.

CHAPTER 2

THE METHOD OF GREEN'S FUNCTIONS

Historically the use of a Green's function to solve DEs grew out of a study of a special partial differential equation and boundary condition called the Dirichlet problem (see Chapter 8). Later it was discovered that a similar function could also be used in the analysis of ordinary differential equations featuring nonhomogeneities. Although some of these functions were originally given different names for different problems, today we collectively refer to them all as *Green's functions*.

In Section 2.2 we show how the solution of a second-order initial value problem is always the superposition of a function y_H, which physically represents the response of a system to initial conditions alone, and a function y_P, which can be interpreted as the response of a system at rest that is later subjected to an external *forcing function*. In constructing y_P by the method of variation of parameters, we are led to the notion of a *one-sided Green's function*. Anticipating the physical interpretation of the one-sided Green's function, we discuss the concept of an *impulse function* in Section 2.3. A more general definition of a Green's function is presented here and the basic properties associated with this more general function are also examined.

The method of Green's functions is extended in Section 2.4 to *boundary value problems*. Both the physical interpretation of the Green's function for boundary value problems and the conditions under which the Green's function fails to exist are examined. It is this failure to always exist that most clearly distinguishes the Green's function associated with boundary value problems from the Green's function associated with initial value problems, which always exists when the coefficients of the DE, in *normal form,* are continuous.

2.1 INTRODUCTION

When a physical system is subject to some external disturbance, a nonhomogeneity arises in the mathematical formulation of the problem, either in the DE or in the auxiliary conditions, or both. When the DE is nonhomogeneous, a *particular solution* of the equation can be found by applying either the method of undetermined coefficients or the variation of parameters technique. In general, however, such techniques lead to a particular solution that has no special physical significance. While this is perfectly acceptable from a purely mathematical point of view, it does not provide the kind of information about, or insight into, the problem that is possible by constructing the particular solution in a more specific manner.

It is the purpose of this chapter to introduce a systematic procedure for solving nonhomogeneous DEs that not only has a special physical significance, but lends itself as well to the development of general solution formulas. The method of attack is attributed to George Green,* whose name occurs so abundantly throughout classical analysis. It is a powerful tool based upon the construction of a specific function known as the *Green's function*. This function measures the response of a system due to a point source somewhere on the fundamental domain, and all other solutions due to different source terms are found to be superpositions of the Green's function, also called the *fundamental solution* in this context.

2.1.1 A PHYSICAL EXAMPLE

To help motivate our discussion of the Green's function, let us solve a simple introductory example. Consider the motion of a particle of mass m in a resistive medium under the influence of an external force $F(t)$. If the resistive force is proportional to the velocity v of the particle, then the equation of motion is

$$m\frac{dv}{dt} = F(t) - cv, \tag{1}$$

(via Newton's second law of motion) where c is a positive constant whose value is determined by the nature of the resistive medium. Assuming the velocity of the particle at time $t = 0$ is v_0, the velocity at all later times is then a solution of the *initial value problem*

$$m\frac{dv}{dt} + cv = F(t), \qquad t > 0, \qquad v(0) = v_0. \tag{2}$$

The solution of (2) is readily found to be†

$$v(t) = e^{-ct/m}\left(v_0 + \frac{1}{m}\int_0^t e^{c\tau/m}F(\tau)\,d\tau\right). \tag{3}$$

The advantage of this solution formula is that it provides a solution of the problem for any initial velocity v_0 and external force $F(t)$. Not only that, but it is composed of parts that each have important physical interpretations. For instance, when $F(t) \equiv 0$, the solution reduces to

$$v(t) = v_0 e^{-ct/m},$$

---

*GEORGE GREEN (1793–1841) gained recognition for his important works concerning the reflection and refraction of sound and light waves. He also extended the work of Poisson in the theory of electricity and magnetism.

†For example, see Section 2.5.2 in L. C. Andrews, *Ordinary Differential Equations with Applications* (Scott, Foresman and Co.: Glenview, Ill.) 1982.

whereas when the initial velocity is zero, i.e., $v_0 = 0$, (3) becomes

$$v(t) = \frac{1}{m} e^{-ct/m} \int_0^t e^{c\tau/m} F(\tau)\, d\tau.$$

Thus, by rewriting (3) as the sum

$$v(t) = \underbrace{v_0 e^{-ct/m}}_{v_H(t)} + \underbrace{\frac{1}{m} e^{-ct/m} \int_0^t e^{c\tau/m} F(\tau)\, d\tau}_{v_P(t)}, \tag{4}$$

we see that $v_H(t)$ represents the "free velocity" of the particle in the absence of an external force, and $v_P(t)$ describes the motion of a particle that is initially at rest and then subject at time $t = 0$ to the external force $F(t)$.

Because of the physical interpretations derived from (4), and the ease of using this solution formula for different values of v_0 and different forcing functions $F(t)$, it is desirable to obtain similar solution formulas for initial value problems involving DEs of higher order. Our pursuit of such formulas is what leads us, in subsequent sections, to the notion of a Green's function.

2.2 INITIAL VALUE PROBLEMS

The conventional way we have been taught to solve DEs with prescribed auxiliary conditions is to first find a general solution, and then impose the auxiliary conditions to find numerical values for the arbitrary constants. The main disadvantage in this approach is that it usually gives very little insight as to how the system being studied responds to each input parameter independently, and yet such information is often vital in understanding the fundamental characteristics of the system.

In order to obtain a solution whose form readily provides important information about the physical system, we will attempt to develop a solution formula for higher order DEs that is comparable to (4). We will initially address only second-order DEs, since these are the most common DEs in practice, and generalize the results to equations of order n in the exercises.

Initial value problems involving second-order DEs have the general form*

$$\begin{aligned} &A_2(t)y'' + A_1(t)y' + A_0(t)y = F(t), \qquad t > t_0 \\ &y(t_0) = k_0, \qquad y'(t_0) = k_1. \end{aligned} \tag{5}$$

For purposes of developing the following theory, it is helpful to put the DE in *normal form*

---

*Since initial value problems generally involve time, we will designate the independent variable by t, rather than by x.

$$y'' + a_1(t)y' + a_0(t)y = f(t), \tag{6}$$

obtained from (5) by division by $A_2(t)$. Hence, $a_1(t) = A_1(t)/A_2(t)$, $a_0(t) = A_0(t)/A_2(t)$, and $f(t) = F(t)/A_2(t)$. We also find it convenient to introduce the *normal differential operator*

$$N = D^2 + a_1(t)D + a_0(t), \tag{7}$$

where $D = d/dt$, so that (5) can be expressed in the more compact form

$$N[y] = f(t), \qquad t > t_0, \qquad y(t_0) = k_0, \qquad y'(t_0) = k_1. \tag{8}$$

Closely associated with (8) are the problems described by

$$N[y] = 0, \qquad y(t_0) = k_0, \qquad y'(t_0) = k_1 \tag{9}$$

and

$$N[y] = f(t), \qquad y(t_0) = 0, \qquad y'(t_0) = 0. \tag{10}$$

In (9) we note that the DE is homogeneous while the initial conditions are nonhomogeneous, whereas in (10) it is the DE that is nonhomogeneous and the initial conditions homogeneous. The solution of (9), which we will denote by y_H, physically represents the response of the system described by (8) entirely due to the initial conditions in the absence of external disturbances. On the other hand, the solution of (10), denoted by y_P, represents the response of the same system, which is at rest until time $t = t_0$, at which time it is subject to the external input $f(t)$. In this regard the function $f(t)$ is often referred to as a *forcing function*. Problem 26 in Exercises 2.2 shows that the sum of solutions of (9) and (10) is the solution of (8).

The homogeneous DE, $N[y] = 0$, in (9) has the general solution

$$y_H = C_1y_1(t) + C_2y_2(t), \tag{11}$$

where y_1 and y_2 are linearly independent solutions. Hence, their Wronskian is nonzero, i.e.,

$$W(y_1, y_2)(t) = y_1(t)y_2'(t) - y_1'(t)y_2(t) \neq 0, \qquad t \geqslant t_0. \tag{12}$$

Imposing the prescribed initial conditions on y_H, we get the simultaneous equations

$$C_1y_1(t_0) + C_2y_2(t_0) = k_0,$$

$$C_1y_1'(t_0) + C_2y_2'(t_0) = k_1,$$

the solution of which leads to the *unique* determination

$$C_1 = \frac{k_0y_2'(t_0) - k_1y_2(t_0)}{W(y_1, y_2)(t_0)}, \tag{13}$$

$$C_2 = \frac{k_1y_1(t_0) - k_0y_1'(t_0)}{W(y_1, y_2)(t_0)}. \tag{14}$$

Notice that when $k_0 = k_1 = 0$, we get $C_1 = C_2 = 0$, so that necessarily $y_H = 0$. The physical implication of this result is that a system which is initially at rest and not subject to any external disturbance must remain at rest.

THEOREM 2.1

> If $N = D^2 + a_1(t)D + a_0(t)$, where $a_0(t)$ and $a_1(t)$ are continuous functions for $t \geqslant t_0$, then the homogeneous initial value problem
>
> $$N[y] = 0, \qquad t > t_0, \qquad y(t_0) = 0, \qquad y'(t_0) = 0,$$
>
> has only the trivial solution $y = 0$.

2.2.1 ONE-SIDED GREEN'S FUNCTION

Now that we have solved (9), we concentrate on solving the remaining problem (10), which we renumber as

$$N[y] = f(t), \qquad y(t_0) = 0, \qquad y'(t_0) = 0. \tag{15}$$

To solve (15) we use the method of *variation of parameters*. This begins with the assumption that there is a solution of (15) of the form

$$y_P = u(t)y_1(t) + v(t)y_2(t), \tag{16}$$

where $y_1(t)$ and $y_2(t)$ are the linearly independent solutions of the associated homogeneous DE, and $u(t)$ and $v(t)$ are two functions to be determined.

By differentiating (16), we obtain

$$y_P' = u(t)y_1'(t) + v(t)y_2'(t) + u'(t)y_1(t) + v'(t)y_2(t).$$

We now impose the assumption

$$u'(t)y_1(t) + v'(t)y_2(t) = 0, \tag{17}$$

and therefore conclude that

$$y_P' = u(t)y_1'(t) + v(t)y_2'(t). \tag{18}$$

Another differentiation of (18) yields

$$y_P'' = u(t)y_1''(t) + v(t)y_2''(t) + u'(t)y_1'(t) + v'(t)y_2'(t). \tag{19}$$

Next, the substitution of (16), (18), and (19) into the DE in (15) leads to

$$\begin{aligned} N[y_P] &= y_P'' + a_1(t)y_P' + a_0(t)y_P \\ &= u(t)\underbrace{[y_1'' + a_1(t)y_1' + a_0(t)y_1]}_{\text{zero}} + v(t)\underbrace{[y_2'' + a_1(t)y_2' + a_0(t)y_2]}_{\text{zero}} \\ &\quad + u'(t)y_1'(t) + v'(t)y_2'(t). \end{aligned} \tag{20}$$

The bracketed expressions above are both zero because of the assumption that y_1 and y_2 are both solutions of the associated homogeneous DE. Thus, if y_P is indeed a solution of the DE in (15), then $N[y_P] = f(t)$, and we deduce from (20) that

$$u'(t)y_1'(t) + v'(t)y_2'(t) = f(t). \tag{21}$$

Using Cramer's rule, the simultaneous solution of (17) and (21) gives us

$$u'(t) = -\frac{y_2(t)f(t)}{W(y_1, y_2)(t)}, \qquad v'(t) = \frac{y_1(t)f(t)}{W(y_1, y_2)(t)}, \tag{22}$$

where $W(y_1, y_2)(t)$ is the nonvanishing Wronskian [recall Equation (12)]. Finally, in order to satisfy the prescribed homogeneous initial conditions in (15), we wish to take definite integrals of (22) in finding the functions $u(t)$ and $v(t)$. The proper choice, which we will verify below, turns out to be

$$u(t) = -\int_{t_0}^{t} \frac{y_2(\tau)f(\tau)}{W(y_1, y_2)(\tau)}\,d\tau, \qquad v(t) = \int_{t_0}^{t} \frac{y_1(\tau)f(\tau)}{W(y_1, y_2)(\tau)}\,d\tau. \tag{23}$$

The substitution of these results into (16) yields, upon combining integrals,

$$y_P = \int_{t_0}^{t} \frac{y_1(\tau)y_2(t) - y_1(t)y_2(\tau)}{W(y_1, y_2)(\tau)} f(\tau)\,d\tau. \tag{24}$$

However, by defining the function

$$g_1(t, \tau) = \frac{y_1(\tau)y_2(t) - y_1(t)y_2(\tau)}{W(y_1, y_2)(\tau)} = \frac{\begin{vmatrix} y_1(\tau) & y_2(\tau) \\ y_1(t) & y_2(t) \end{vmatrix}}{W(y_1, y_2)(\tau)}, \tag{25}$$

we can express our particular solution in the form

$$y_P = \int_{t_0}^{t} g_1(t, \tau)f(\tau)\,d\tau. \tag{26}$$

What remains now is to verify that this choice of y_P actually satisfies the homogeneous initial conditions in (15). First, when $t = t_0$ we get the immediate result

$$y_P(t_0) = \int_{t_0}^{t_0} g_1(t_0, \tau)f(\tau)\,d\tau = 0. \tag{27}$$

Next, using the Leibniz formula (from calculus)

$$\frac{d}{dt}\int_{t_0}^{t} F(x, t)\,dx = \int_{t_0}^{t} \frac{\partial F}{\partial t}(x, t)\,dx + F(t, t),$$

we find

$$y_P'(t_0) = \int_{t_0}^{t_0} \frac{\partial g_1}{\partial t}(t_0, \tau)f(\tau)\,d\tau + g_1(t_0, t_0)f(t_0) = 0, \tag{28}$$

where we recognize that $g_1(t_0, t_0) = 0$ by definition. Hence, the initial conditions are satisfied.

The function $g_1(t, \tau)$ is called the *one-sided Green's function*. Its construction depends only upon knowledge of the homogeneous solutions $y_1(t)$ and $y_2(t)$. And since these solutions always exist for a normal operator N with continuous coefficients, it follows that the one-sided Green's function always *exists* for such operators, and moreover, is *unique*.

REMARK: *For proper identification of the function f in (26), it is important that the nonhomogeneous DE (15) is in* normal form.

EXAMPLE 1

Use the method of Green's function to solve

$$y'' + y = \sin t, \qquad t > 0, \qquad y(0) = 1, \qquad y'(0) = -1.$$

SOLUTION We first rewrite the problem as two problems:

$$y'' + y = 0, \qquad y(0) = 1, \qquad y'(0) = -1$$

and

$$y'' + y = \sin t, \qquad y(0) = 0, \qquad y'(0) = 0.$$

In the first case the general solution is

$$y_H = C_1 \cos t + C_2 \sin t,$$

and imposing the initial conditions, we find

$$y_H(0) = C_1 + C_2 \cdot 0 = 1,$$

$$y_H'(0) = -C_1 \cdot 0 + C_2 = -1.$$

Hence, $C_1 = 1$ and $C_2 = -1$, so that

$$y_H = \cos t - \sin t.$$

For the second problem we observe that $W(\cos t, \sin t) = 1$, and consequently

$$g_1(t, \tau) = \begin{vmatrix} \cos \tau & \sin \tau \\ \cos t & \sin t \end{vmatrix} = \cos \tau \sin t - \cos t \sin \tau.$$

Using (26), we have

$$\begin{aligned} y_P &= \int_0^t g_1(t, \tau) \sin \tau \, d\tau \\ &= \sin t \int_0^t \cos \tau \sin \tau \, d\tau - \cos t \int_0^t \sin^2 \tau \, d\tau \\ &= \frac{1}{2}(\sin t - t \cos t), \end{aligned}$$

and by combining y_H and y_P, we deduce that

$$\begin{aligned} y &= \cos t - \sin t + \frac{1}{2}(\sin t - t\cos t) \\ &= \left(1 - \frac{1}{2}t\right)\cos t - \frac{1}{2}\sin t. \end{aligned}$$

EXAMPLE 2

Use the method of Green's function to solve

$$t^2y'' - 3ty' + 3y = 2t^4e^t, \qquad t > 1, \qquad y(1) = 0, \qquad y'(1) = 2.$$

SOLUTION The two problems we need to solve are

$$t^2y'' - 3ty' + 3y = 0, \qquad y(1) = 0, \qquad y'(1) = 2,$$

$$y'' - \frac{3}{t}y' + \frac{3}{t^2}y = 2t^2e^t, \qquad y(1) = 0, \qquad y'(1) = 0,$$

where we have divided the DE in the second problem by t^2 to put it in normal form. We recognize the DE as a Cauchy-Euler equation,* whose homogeneous solution is

$$y_H = C_1t + C_2t^3.$$

Imposing the initial conditions in the first problem above, we find that $C_2 = -C_1 = 1$, and hence

$$y_H = t^3 - t.$$

The Wronskian of $y_1 = t$ and $y_2 = t^3$ is $W(t, t^3) = 2t^3$. Therefore the one-sided Green's function is

$$g_1(t, \tau) = \frac{1}{2\tau^3}\begin{vmatrix} \tau & \tau^3 \\ t & t^3 \end{vmatrix} = \frac{\tau t^3 - t\tau^3}{2\tau^3}.$$

From the normal form of the DE we recognize $f(t) = 2t^2e^t$, and thus the solution of the second problem above is

$$\begin{aligned} y_P &= \int_1^t g_1(t, \tau)(2\tau^2e^\tau)\,d\tau \\ &= t^3\int_1^t e^\tau\,d\tau - t\int_1^t \tau^2e^\tau\,d\tau \\ &= (t - t^3)e + 2t(t - 1)e^t. \end{aligned}$$

By writing $y = y_H + y_P$, we get the intended solution

$$y = (t^3 - t)(1 - e) + 2t(t - 1)e^t.$$

* For solution techniques of Cauchy-Euler equations, see Problem 34 in Exercises 1.3.

TABLE 2.1 Table of One-Sided Green's Functions

	OPERATOR	$g_1(t, \tau)$
1.	D^2	$t - \tau$
2.	$D^n, \quad n = 2, 3, 4, \ldots$	$\dfrac{(t-\tau)^{n-1}}{(n-1)!}$
3.	$D^2 + b^2$	$\dfrac{1}{b} \sin b(t-\tau)$
4.	$D^2 - b^2$	$\dfrac{1}{b} \sinh b(t-\tau)$
5.	$(D-a)(D-b), \quad a \neq b$	$\dfrac{1}{a-b}[e^{a(t-\tau)} - e^{b(t-\tau)}]$
6.	$(D-a)^2$	$(t-\tau)e^{a(t-\tau)}$
7.	$(D-a)^n, \quad n = 2, 3, 4, \ldots$	$\dfrac{(t-\tau)^{n-1}}{(n-1)!} e^{a(t-\tau)}$
8.	$D^2 - 2aD + a^2 + b^2$	$\dfrac{1}{b} e^{a(t-\tau)} \sin b(t-\tau)$
9.	$D^2 - 2aD + a^2 - b^2$	$\dfrac{1}{b} e^{a(t-\tau)} \sinh b(t-\tau)$
10.	$t^2D^2 + tD - b^2$	$\dfrac{\tau}{2b}\left[\left(\dfrac{t}{\tau}\right)^b - \left(\dfrac{\tau}{t}\right)^b\right]$

Construction of the one-sided Green's function using (25) is mostly mechanical. In solving initial value problems by this technique we find that the same DE appears many times, but with different forcing functions and initial conditions. Since the one-sided Green's function is not dependent upon the forcing function or initial conditions, all initial value problems with the same operator N have the same one-sided Green's function. For this reason it is convenient to tabulate some of the most common differential operators and their corresponding one-sided Green's function. A few such entries are provided in Table 2.1.

EXERCISES 2.2

In Problems 1–10, determine the one-sided Green's function for the given operator ($D = d/dt$).

1. D^2
2. $D^2 - 5$
3. $D^2 + 5$
4. $D^2 + 4D + 4$
5. $4D^2 - 8D + 5$
6. $D^2 - D - 2$
7. $t^2D^2 + tD - 16$
8. $t^2D^2 - tD + 1$

★9. $D[(1 - t^2)D]$

★10. $tD^2 - (1 + 2t^2)D$

In Problems 11–25, use the one-sided Green's function to solve the given initial value problem.

11. $y'' - y = 1, \quad y(0) = 0, \quad y'(0) = 1$

12. $y'' + y = e^{t-1}, \quad y(1) = 0, \quad y'(1) = 0$

13. $y'' - 3y' - 4y = e^{-t}, \quad y(2) = 3, \quad y'(2) = 0$

14. $y'' + y = 2 \csc t \cot t, \quad y(\pi/2) = 1, \quad y'(\pi/2) = 1$

15. $y'' + y' - 2y = -4, \quad y(0) = 2, \quad y'(0) = 3$

16. $2y'' + y' - y = t + 1, \quad y(0) = 1, \quad y'(0) = 0$

17. $y'' + 4y' + 6y = 1 + e^{-t}, \quad y(0) = 1, \quad y'(0) = -4$

18. $y'' + y' + 2y = \sin 2t - 2 \cos 2t, \quad y(0) = 0, \quad y'(0) = 0$

19. $y'' + 2y' + y = 3te^{-t}, \quad y(0) = 4, \quad y'(0) = 2$

20. $y'' - 6y' + 9y = t^2e^{3t}, \quad y(0) = 2, \quad y'(0) = 6$

21. $t^2y'' + 7ty' + 5y = t, \quad y(1) = 0, \quad y'(1) = 0$

22. $t^2y'' - 5ty' + 8y = 2t^3, \quad y(1) = 1, \quad y'(1) = 0$

23. $t^2y'' - 6y = \log t, \quad y(1) = 1/6, \quad y'(1) = -1/6$

24. $t^2y'' + ty' + 4y = \sin(\log t), \quad y(1) = 1, \quad y'(1) = 0$

25. $4t^2y'' + 8ty' + y = 3t^{-1/2} \log t^2, \quad y(1) = 0, \quad y'(1) = 1$

26. If y_H and y_P satisfy, respectively, the initial value problems

$$N[y] = 0, \quad y(t_0) = k_0, \quad y'(t_0) = k_1$$

and

$$N[y] = f(t), \quad y(t_0) = 0, \quad y'(t_0) = 0,$$

verify that the sum $y = y_H + y_P$ is a solution of

$$N[y] = f(t), \quad y(t_0) = k_0, \quad y'(t_0) = k_1.$$

27. For all $\tau \geqslant t_0$, show that
(a) $g_1(\tau, \tau) = 0$.
(b) $g_1(t, \tau)$, along with its first and second derivatives with respect to t, is continuous.

28. Show that, for a fixed value of τ, the function $y(t) = g_1(t, \tau)$ is a solution of the initial value problem

$$N[y] = 0, \quad y(\tau) = 0, \quad y'(\tau) = 1.$$

Hint: Use the result of Problem 27.

29. Given the first-order initial value problem,

$$y' + a_0(t)y = f(t), \quad y(t_0) = k_0,$$

show that it has the solution

$$y = \frac{k_0 y_1(t)}{y_1(t_0)} + \int_{t_0}^{t} g_1(t, \tau) f(\tau)\, d\tau,$$

where y_1 is a solution of the associated homogeneous DE and $g_1(t, \tau) = y_1(t)/y_1(\tau)$.

★30. Use the result of Problem 29 to show that*

$$y = \left[\frac{1}{2}\sqrt{\pi}\,\mathrm{erf}(t) + 3\right]e^{t^2}$$

is the solution of the initial value problem

$$y' - 2ty = 1, \qquad y(0) = 3.$$

31. The small motions $y(t)$ of an undamped spring-mass system are governed by the initial value problem

$$my'' + ky = F(t), \qquad y(0) = y_0, \qquad y'(0) = v_0,$$

where m is the mass, k is the spring constant, and $F(t)$ is an external (driving) force. Show that the one-sided Green's function associated with this system is

$$g_1(t, \tau) = \frac{1}{\omega_0}\sin[\omega_0(t - \tau)], \qquad \omega_0 = \sqrt{k/m}.$$

32. Using the one-sided Green's function given in Problem 31, find the response of the spring-mass system given that the system is initially at rest (i.e., $y_0 = v_0 = 0$) and then subject to the driving force
(a) $F(t) = P$ (constant)
(b) $F(t) = P\cos\omega t, \quad \omega \neq \omega_0$.
(c) $F(t) = P\cos\omega_0 t$.

33. When resistive forces are taken into account for the spring-mass system in Problem 31, the motions are called *damped*. In such cases the DE is modified to read

$$my'' + cy' + ky = F(t),$$

where c is a positive constant. Show that the one-sided Green's function for the following cases of damping is given by

(a) *underdamped* ($c^2 < 4mk$):

$$g_1(t, \tau) = \frac{1}{\mu}e^{-c(t-\tau)/2m}\sin[\mu(t - \tau)],$$

where $\mu = (4mk - c^2)^{1/2}/2m$.

(b) *critically damped* ($c^2 = 4mk$):

$$g_1(t, \tau) = (t - \tau)e^{-c(t-\tau)/2m}.$$

(c) *overdamped* ($c^2 > 4mk$):

$$g_1(t, \tau) = \frac{1}{\alpha}e^{-c(t-\tau)/2m}\sinh[\alpha(t - \tau)],$$

where $\alpha = (c^2 - 4mk)^{1/2}/2m$.

---

* $\mathrm{erf}(t)$ is the *error function* defined by

$$\mathrm{erf}(t) = \frac{2}{\sqrt{\pi}}\int_0^t e^{-\tau^2}\,d\tau.$$

34. When the motions of a spring–mass system are damped, that part of the particular solution y_P that does not vanish in the limit as $t \to \infty$ is called the *steady-state solution*. Use the one-sided Green's function given in Problem 33(a) to find the steady-state solution of the spring-mass system when the driving force is

(a) $F(t) = P$ (constant). (b) $F(t) = P \cos \omega t$.

(c) $F(t) = \begin{cases} 1, & 0 < t < 1 \\ 0, & t > 1. \end{cases}$ (d) $F(t) = \begin{cases} \sin t, & 0 < t < \pi \\ 0, & t > \pi. \end{cases}$

★35. Consider the third-order initial value problem

$$y''' + a_2(t)y'' + a_1(t)y' + a_0(t)y = f(t),$$

$$y(t_0) = k_0, \qquad y'(t_0) = k_1, \qquad y''(t_0) = k_2.$$

(a) Show that the one-sided Green's function is defined by

$$g_1(t, \tau) = \frac{\begin{vmatrix} y_1(\tau) & y_2(\tau) & y_3(\tau) \\ y_1'(\tau) & y_2'(\tau) & y_3'(\tau) \\ y_1(t) & y_2(t) & y_3(t) \end{vmatrix}}{W(y_1, y_2, y_3)(\tau)},$$

where y_1, y_2, y_3 are linearly independent solutions of the associated homogeneous DE.

(b) Establish the solution formula

$$y = \int_{t_0}^{t} g_1(t, \tau)f(\tau)\, d\tau + C_1y_1(t) + C_2y_2(t) + C_3y_3(t),$$

and derive determinant expressions for C_1, C_2, and C_3 similar to those in (13) and (14).

(c) Generalize the results in (a) and (b) to nth-order problems.

In Problems 36–41, use the result of Problem 35 to construct the one-sided Green's function for the given operator.

36. D^n $(n = 2, 3, 4, \ldots)$

37. $D^2(D^2 - 1)$

38. $D(D^2 + 4)$

★39. $D^3 + \frac{5}{2}D^2 - \frac{3}{2}$

40. $D^4 - 1$

★41. $D^3 - 6D^2 + 11D - 6$

In Problems 42 and 43, use the one-sided Green's function to find the solution (see Problem 35).

★42. $y''' + y = (t - 1)e^{t-1}, \quad y(1) = 0, \quad y'(1) = 0, \quad y''(1) = 1$

★43. $y''' - y'' + 4y' - 4y = 1, \quad y(0) = 0, \quad y'(0) = 1, \quad y''(0) = -1$

2.3 IMPULSE FUNCTIONS

In certain applications it is convenient to introduce the concept of an impulse, which is the result of a sudden excitation administered to a system, such as a sharp blow or voltage surge. Let us imagine that the sudden excitation, which we will denote by $d_a(t)$, has a nonzero value over the short interval of time $a - \epsilon < t < a + \epsilon$, but is otherwise zero. The total *impulse* (force times duration) imparted to the system is thus defined by

$$I = \int_{-\infty}^{\infty} d_a(t)\,dt = \int_{a-\epsilon}^{a+\epsilon} d_a(t)\,dt \qquad (\epsilon > 0). \tag{29}$$

The value of I is a measure of the strength of the sudden excitation.

In order to provide a mathematical model of the function $d_a(t)$, it is convenient to think of it as having a constant value over the interval $a - \epsilon \leqslant t \leqslant a + \epsilon$ (see Figure 2.1). Furthermore, we wish to choose this constant value in such a way that the total impulse given by (29) is unity. Hence, we write

$$d_a(t) = \begin{cases} \dfrac{1}{2\epsilon}, & a - \epsilon \leqslant t \leqslant a + \epsilon \\ 0, & \text{otherwise.} \end{cases} \tag{30}$$

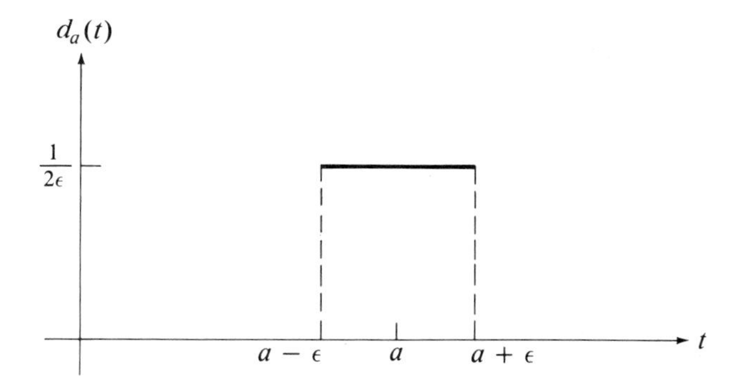

FIGURE 2.1

Impulse function.

Now let us idealize the function $d_a(t)$ by requiring it to act over shorter and shorter intervals of time by allowing $\epsilon \to 0$. Although the interval about $t = a$ is shrinking to zero, we still want $I = 1$, i.e.,

$$\lim_{\epsilon\to 0} I = \lim_{\epsilon\to 0} \int_{-\infty}^{\infty} d_a(t)\,dt = 1. \tag{31}$$

We can use the result of this limit process to define an "idealized" *unit impulse function,* $\delta(t - a)$, which has the property of imparting a unit impulse to the system at time $t = a$ but being zero for all other values of t. The defining properties of this function are therefore

$$\begin{aligned} &\delta(t - a) = 0, \qquad t \neq a \\ &\int_{-\infty}^{\infty} \delta(t - a)\,dt = 1. \end{aligned} \tag{32}$$

By a similar kind of limit process, it is possible to define the integral of a product of the unit impulse function and any continuous and bounded function f; that is,

$$\int_{-\infty}^{\infty} \delta(t-a)f(t)\,dt = \lim_{\epsilon\to 0}\int_{-\infty}^{\infty} d_a(t)f(t)\,dt$$
$$= \lim_{\epsilon\to 0}\frac{1}{2\epsilon}\int_{a-\epsilon}^{a+\epsilon} f(t)\,dt.$$

Recalling that

$$\int_a^b f(t)\,dt = f(\xi)(b-a), \qquad a<\xi<b, \tag{33}$$

which is the *mean value theorem* of the integral calculus, we find that

$$\int_{-\infty}^{\infty} \delta(t-a)f(t)\,dt = \lim_{\epsilon\to 0}\frac{1}{2\epsilon}\cdot f(\xi)\cdot 2\epsilon$$

for some ξ in the interval $a-\epsilon<\xi<a+\epsilon$. Consequently, in the limit we see that $\xi\to a$, and deduce the property*

$$\int_{-\infty}^{\infty} \delta(t-a)f(t)\,dt = f(a). \tag{34}$$

Obviously the "function" $\delta(t-a)$, also known as the *Dirac delta function,*† is not a function in the usual sense of the word. It has significance only as part of an integrand. It is an example of what is commonly called a *generalized function*. In dealing with this function, it is usually best to avoid the idea of assigning "functional values" and instead refer to its integral property (34), even though it has no meaning as an ordinary integral. Following more rigorous lines, generalized functions can be defined as a limit of an infinite sequence of well-behaved functions (see Problems 12 and 13 in Exercises 2.3)

EXAMPLE 3

Solve the initial value problem

$$y''+y=\delta(t-\pi), \qquad t>0, \qquad y(0)=0, \qquad y'(0)=0.$$

SOLUTION Since the initial conditions are homogeneous, we only need consider the particular solution; i.e., $y=y_P$. The one-sided Green's function for the operator $N=D^2+1$ was shown in Example 1 to be

* Note that $\xi=\xi(\epsilon)$, so that $\lim_{\epsilon\to 0} f[\xi(\epsilon)] = f[\lim_{\epsilon\to 0}\xi(\epsilon)] = f(a)$.

† Named after PAUL A. M. DIRAC (1902–1984), who was awarded the Nobel Prize (with E. SCHRÖDINGER) in 1933 for his work in quantum mechanics.

$$g_1(t, \tau) = \sin t \cos \tau - \cos t \sin \tau = \sin(t - \tau).$$

Thus it follows that

$$y = \int_0^t \sin(t - \tau)\delta(\tau - \pi)\, d\tau$$

$$= \begin{cases} 0, & t < \pi \\ \sin(t - \pi), & t \geq \pi \end{cases}$$

by use of (34).

We can interpret this solution as the response of some system that remains at rest until time $t = \pi$, when it is subjected to a unit impulse. After time $t = \pi$, the response of the system follows that of a simple sinusoid with frequency equal to that of the natural frequency of the system (see Figure 2.2).

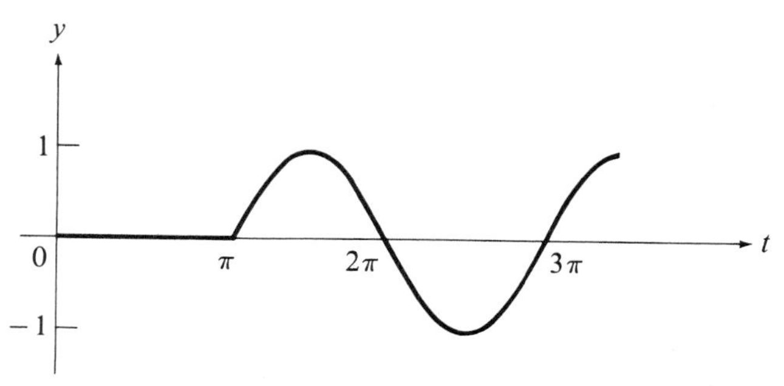

FIGURE 2.2

□ 2.3.1 PHYSICAL INTERPRETATION AND A MORE GENERAL GREEN'S FUNCTION

The unit impulse function defined in the previous section is also useful for providing us with a physical interpretation of the one-sided Green's function. To see this, let us consider the initial value problem

$$N[y] = \delta(t - a), \qquad y(t_0) = 0, \qquad y'(t_0) = 0, \tag{35}$$

for which the initial conditions are homogeneous. The solution of (35) is formally given by

$$y = \int_{t_0}^t g_1(t, \tau)\delta(\tau - a)\, d\tau = \begin{cases} 0, & t_0 \leq t < a \\ g_1(t, a), & t \geq a, \end{cases} \tag{36}$$

where we are using the integral property (34) of the impulse function and where $g_1(t, \tau)$ is the one-sided Green's function associated with the operator N. Thus the function $g_1(t, a)$ represents the response of the system described by (35) for $t > a$, which was formerly at rest and then subjected to a unit disturbance (impulse) at time $t = a$.

Based on the above interpretation of $g_1(t, \tau)$, let us introduce the more general function

$$g(t, \tau) = \begin{cases} 0, & t_0 \leq t < \tau \\ g_1(t, \tau), & \tau \leq t < \infty, \end{cases} \tag{37}$$

where we imagine t varying for a fixed value of τ. In solving DEs by means of the Green's function, however, it is t that is taken to be fixed while τ is allowed to vary. Hence, we rewrite (37) as

$$g(t, \tau) = \begin{cases} g_1(t, \tau), & t_0 \leq \tau \leq t \\ 0, & t < \tau < \infty, \end{cases} \tag{38}$$

so that we can express the response of (35) to a general input function f as

$$y = \int_{t_0}^{\infty} g(t, \tau) f(\tau)\, d\tau, \tag{39}$$

where the integration now takes place over all $\tau \geq t_0$. The function $g(t, \tau)$ can be interpreted as the response of the system for *all* time t due to a unit impulse delivered at time $t = \tau$. We will refer to $g(t, \tau)$ as simply the *Green's function,* or *influence function* as it is sometimes called.*

Our primary purpose for introducing a more general function $g(t, \tau)$ is that the properties of this function, given in Definition 2.1, can be used for motivational purposes in constructing a similar function in the solution of *boundary value problems* in the next section.

For a fixed value of τ, we see that $g(t, \tau)$ must necessarily satisfy the DE, $N[g] = \delta(t - \tau)$, where N is the differential operator associated with the construction of $g_1(t, \tau)$. The Green's function must also satisfy the homogeneous initial conditions $g(t_0, \tau) = (\partial g/\partial t)(t_0, \tau) = 0$ for any t_0 less than or equal to the fixed value of τ. It is a continuous function for all values of t, since $g_1(\tau, \tau) = 0$ and $g_1(t, \tau)$ is itself a continuous function of t by definition. However, there is a jump discontinuity of unit magnitude in the first derivative of $g(t, \tau)$ at $t = \tau$. That is, for $t < \tau$, the derivative is clearly zero, while

$$\left.\frac{\partial g}{\partial t}\right|_{t=\tau^+} = \frac{y_1(\tau)y_2'(\tau) - y_1'(\tau)y_2(\tau)}{W(y_1, y_2)(\tau)} = 1.\dagger \tag{40}$$

The jump discontinuity in the derivative turns out to be an essential feature of the Green's function.

In summary, we have the following definition.

DEFINITION 2.1 The *Green's function* $g(t, \tau)$ associated with the nonhomogeneous initial value problem

$$N[y] = f(t), \qquad t > t_0, \qquad y(t_0) = k_0, \qquad y'(t_0) = k_1,$$

---

* The *influence function* in some textbooks is defined as a constant multiple of the function $g(t, \tau)$ defined by (38).

† By $t = \tau^+$, we mean a right-hand limit.

where $N = D^2 + a_1(t)D + a_0(t)$, $D = d/dt$, is a function satisfying the following conditions ($\tau > t_0$):*

(a) $N[g] = \delta(t - \tau)$ (τ fixed)

(b) $g(t_0, \tau) = 0, \qquad \dfrac{\partial g}{\partial t}(t_0, \tau) = 0$

(c) $g(\tau^+, \tau) = g(\tau^-, \tau)$ (i.e., g is continuous at $t = \tau$)

(d) $\left.\dfrac{\partial g}{\partial t}\right|_{t=\tau^-}^{t=\tau^+} = 1$

EXERCISES 2.3

In Problems 1–6, use the one-sided Green's function to find the solution of the given initial value problem.

1. $y'' + 2y' + 2y = \delta(t - \pi), \quad y(0) = 1, \quad y'(0) = 0$
2. $y'' + y = \delta(t - \pi) - \delta(t - 2\pi), \quad y(0) = 0, \quad y'(0) = 1$
3. $y'' + y = \delta(t - \pi) + 3\cos 2t, \quad y(0) = 0, \quad y'(0) = 0$
4. $y'' - y = 2\delta(t - 1), \quad y(0) = 1, \quad y'(0) = -1$
5. $y'' + y = 5\delta(t - \pi/2)\sin t, \quad y(0) = 0, \quad y'(0) = 2$

★6. $y''' + y = e^t + \delta(t - 1), \quad y(0) = 1, \quad y'(0) = 1, \quad y''(0) = 2$

7. If $a > 0$, show that

$$\int_{-\infty}^{\infty} \delta(at)f(t)\,dt = \frac{1}{a}f(0).$$

★8. Using integration by parts, show formally that

(a) $\displaystyle\int_{-\infty}^{\infty} \delta'(t)f(t)\,dt = -f'(0).$

(b) $\displaystyle\int_{-\infty}^{\infty} \delta^{(n)}(t)f(t)\,dt = (-1)^n f^{(n)}(0), \qquad n = 1, 2, 3, \ldots.$

★9. Show formally that $h'(t) = \delta(t)$, where $h(t)$ is the *Heaviside unit function* defined by

$$h(t) = \begin{cases} 0, & t < 0 \\ 1, & t \geq 0. \end{cases}$$

Hint: Use integration by parts to show that

$$\int_{-\infty}^{\infty} h'(t)f(t)\,dt = f(0).$$

---

*Conditions (c) and (d) are merely consequences of conditions (a) and (b).

★10. Show that $\delta(t - x) = \delta(x - t)$.

★11. Show formally that

$$f(t)\delta(t - a) = f(a)\delta(t - a),$$

and use this result to deduce that $t\delta(t) = 0$.

Hint: Show that $\int_{-\infty}^{\infty} g(t)[f(t)\delta(t - a)]\,dt = \int_{-\infty}^{\infty} g(t)[f(a)\delta(t - a)]\,dt.$

★12. Consider the sequence of rectangle functions defined by ($n = 1, 2, 3, \ldots$)

$$\psi_n(t) = \begin{cases} \dfrac{n}{2}, & |t| < \dfrac{1}{n} \\ 0, & |t| > \dfrac{1}{n}. \end{cases}$$

(a) Show that for each n the area enclosed by the rectangle is unity and deduce that

$$\lim_{n\to\infty} \int_{-\infty}^{\infty} \psi_n(t)\,dt = f(0).$$

(b) More generally, if f is any function continuous at $t = 0$ and everywhere bounded, show that

$$\lim_{n\to\infty} \int_{-\infty}^{\infty} \psi_n(t)f(t)\,dt = 1.$$

★13. Any sequence of continuous and differentiable functions $\psi_1(t), \psi_2(t), \ldots, \psi_n(t), \ldots$ satisfying the conditions of Problem 12(a), (b), is called a *delta sequence*. Show that the following sequences satisfy the condition of Problem 12(a):

(a) $\psi_n(t) = \dfrac{n}{\pi(1 + n^2t^2)}, \qquad n = 1, 2, 3, \ldots.$

(b) $\psi_n(t) = \dfrac{n}{\sqrt{\pi}} e^{-n^2t^2}, \qquad n = 1, 2, 3, \ldots.$

Hint: $\int_{-\infty}^{\infty} e^{-x^2}\,dx = \sqrt{\pi}$

14. Develop the Green's function $g(t, \tau)$ for the operator $N = D^2 + b^2$ and show that it satisfies conditions (b), (c), and (d) of Definition 2.1.

15. Repeat Problem 14 for $N = D^2 - 4D + 5$.

2.4 BOUNDARY VALUE PROBLEMS

When the prescribed auxiliary conditions are in the form of *boundary conditions,* we find that it is convenient once again to develop the particular solution y_P of a nonhomogeneous DE in terms of a Green's function as we did in Section 2.2 for initial value problems. Although the Green's function for boundary value problems is defined similar to that for initial value problems, a major distinction is that the Green's function for boundary value problems does not always exist.

2.4.1 GREEN'S FUNCTION

The general boundary value problem that concerns us is characterized by

$$\begin{aligned} M[y] &= F(x), \qquad x_1 < x < x_2, \\ B_1[y] &\equiv a_{11}y(x_1) + a_{12}y'(x_1) = \alpha, \\ B_2[y] &\equiv a_{21}y(x_2) + a_{22}y'(x_2) = \beta, \end{aligned} \tag{41}$$

where the differential operator M is defined by

$$M = A_2(x)D^2 + A_1(x)D + A_0(x),$$

where $D = d/dx$. For purposes of developing a Green's function for this problem, it is usually preferred to first put the DE in self-adjoint form.* Recalling Definition 1.1, we obtain the self-adjoint form of $M[y] = F(x)$ by simply multiplying the DE by the function $\mu(x) = p(x)/A_2(x)$, where

$$p(x) = \exp\left[\int \frac{A_1(x)}{A_2(x)}\,dx\right]. \tag{42}$$

Doing so, the problem described by (41) now assumes the form

$$L[y] = f(x), \qquad x_1 < x < x_2, \qquad B_1[y] = \alpha, \qquad B_2[y] = \beta, \tag{43}$$

where $f(x) = p(x)F(x)/A_2(x)$, $q(x) = p(x)A_0(x)/A_2(x)$, and

$$L = D[p(x)D] + q(x). \tag{44}$$

Building upon our experience with initial value problems, we assume the solution of (43) has the general form $y = y_H + y_P$, where y_H satisfies the boundary value problem

$$L[y] = 0, \qquad B_1[y] = \alpha, \qquad B_2[y] = \beta, \tag{45}$$

and y_P is a solution of

$$L[y] = f(x), \qquad B_1[y] = 0, \qquad B_2[y] = 0. \tag{46}$$

Both y_H and y_P have physical interpretations. For example, if we imagine (43) to describe the steady-state temperature distribution in a thin rod, then y_H is the equilibrium solution of the rod whose ends exchange heat energy with the surrounding medium in some fashion, but which is free of heat sources. The solution y_P is the temperature distribution in the rod due to a heat source

---

*For constructing a Green's function for DEs *not* in self-adjoint form, see Problem 46 in Exercises 2.4.

proportional to $f(x)$, and where the ends of the rod are in a surrounding medium at temperature 0° C.

The general solution of (45) is

$$y_H = C_1 y_1(x) + C_2 y_2(x), \tag{47}$$

where y_1 and y_2 are linearly independent solutions of the homogeneous DE. The constants C_1 and C_2 are then determined by imposing the nonhomogeneous boundary conditions (see Section 1.3.1). Let us now consider the problem described by (46).

We begin by assuming the solution of (46) can be expressed in the integral form (by analogy with initial value problems)

$$y_p = -\int_{x_1}^{x_2} g(x, s)f(s)\, ds, \tag{48}$$

where $g(x, s)$ is the Green's function we wish to define. (The minus sign in (48) is chosen so that $g(x, s)$ will have the proper physical interpretation. We will discuss this more in Section 2.4.2.) If we formally apply the differential operator L to both sides of (48), assuming commutativity of L with integration, we find that

$$L[y_P] = L\left[-\int_{x_1}^{x_2} g(x, s)f(s)\, ds\right] = -\int_{x_1}^{x_2} L[g]f(s)\, ds.$$

We now argue that if y_P is indeed a solution of the DE in (46), then the right-hand side of this last expression must equal $f(x)$. This will happen provided

$$L[g] = -\delta(x - s), \qquad x_1 < x < x_2, \tag{49}$$

where $\delta(x - s)$ is the Dirac delta function (see Section 2.3).

In order to uniquely determine the function $g(x, s)$, we must find conditions other than (49) that contribute to its definition. Specifically, let us now impose the homogeneous boundary conditions in (46) on the solution (48), which leads to

$$B_1[y_P] = -\int_{x_1}^{x_2} B_1[g]f(s)\, ds = 0,$$

$$B_2[y_p] = -\int_{x_1}^{x_2} B_2[g]f(s)\, ds = 0.$$

Because f can be almost any function, these relations are possible only if

$$B_1[g] = 0, \qquad B_2[g] = 0. \tag{50}$$

Consequently, we have shown that the Green's function we are seeking is a solution of the boundary value problem

$$L[g] = -\delta(x - s), \quad x_1 < x < x_2, \quad B_1[g] = 0, \quad B_2[g] = 0, \tag{51}$$

where s is fixed and $x_1 < s < x_2$. Although this problem is quite similar to the problem described by (46), the forcing function in (51) is a delta function rather than an arbitrary function f. This means that solving the problem for g will be somewhat simpler than solving the corresponding problem for y, and once the Green's function has been found for a particular operator L and set of boundary conditions, it may be used for solving (46) any number of times where only the function f changes from problem to problem. It is this feature of the Green's function, coupled with its physical interpretation, that makes it most useful in applications.

The presence of the delta function in (51) suggests that the behavior of $g(x, s)$ in the vicinity of $x = s$ is somewhat peculiar. To investigate this behavior, we start with (48) to obtain

$$y_P(s^+) - y_P(s^-) = -\int_{x_1}^{x_2} [g(s^+, s) - g(s^-, s)]f(s)\, ds.^*$$

Because the solution of a DE must be a continuous function, the left-hand side of the above expression vanishes and, since f is arbitrary, we deduce that

$$g(s^+, s) = g(s^-, s), \tag{52}$$

which implies that $g(x, s)$ is continuous at $x = s$. Going one step further, we now wish to investigate the behavior of the derivative of $g(x, s)$. Here we formally integrate both sides of (49) with respect to x from $x = s^-$ to $x = s^+$, and find that

$$\left[p(x)\frac{\partial g}{\partial x}\right]\Bigg|_{x=s^-}^{x=s^+} + \int_{s^-}^{s^+} q(x)g(x, s)\, dx = -\int_{s^-}^{s^+} \delta(x - s)\, dx.$$

From the continuity of both $q(x)$ and $g(x, s)$ at $x = s$, it follows that the integral on the left-hand side of this expression is zero. Also, using the integral property of the delta function and the fact that $p(x)$ is continuous and nonzero on $[x_1, x_2]$, this last expression reduces to

$$\frac{\partial g}{\partial x}\Bigg|_{x=s^-}^{x=s^+} = -\frac{1}{p(s)}, \tag{53}$$

where we have divided by $p(s)$. This result asserts that at $x = s$, the derivative of $g(x, s)$ has a jump discontinuity of magnitude $1/p(s)$.

The nonrigorous manipulations just performed serve to motivate the following definition of the Green's function for boundary value problems.

DEFINITION 2.2 The Green's function $g(x, s)$ associated with the boundary value problem

$$L[y] = f(x), \qquad x_1 < x < x_2, \qquad B_1[y] = \alpha, \qquad B_2[y] = \beta,$$

---

* We are using the standard notation $y(s^+) = \lim_{\epsilon\to 0^+} y(s + \epsilon)$ and $y(s^-) = \lim_{\epsilon\to 0^+} y(s - \epsilon)$, which are right-hand and left-hand limits.

where $L = D[p(x)D] + q(x)$, is a function satisfying the following conditions $(x_1 < s < x_2)$:*

(a) $L[g] = -\delta(x - s), \qquad x_1 < x < x_2 \quad (s \text{ fixed})$

(b) $B_1[g] = 0, \qquad B_2[g] = 0$

(c) $g(s^+, s) = g(s^-, s)$

(d) $\left.\dfrac{\partial g}{\partial x}\right|_{x=s^-}^{x=s^+} = -\dfrac{1}{p(s)}$

Based on Definition 2.2, an explicit formula for the Green's function can now be constructed. We first observe from condition (a) that if either $x < s$ or $x > s$, then $L[g] = 0$ from definition of the delta function. Next, if z_1 and z_2 are solutions of the homogeneous DE, $L[g] = 0$, selected in such a way that

$$B_1[z_1] = 0, \qquad B_2[z_2] = 0, \tag{54}$$

then it follows from conditions (a) and (b) of Definition 2.2 that the Green's function has the form

$$g(x, s) = \begin{cases} u(s)z_1(x), & x < s \\ v(s)z_2(x), & x > s, \end{cases} \tag{55}$$

where u and v are functions to be determined. Imposing conditions (c) and (d) from Definition 2.2 on (55), it follows that the unknown functions u and v must be chosen such that

$$\begin{aligned} v(s)z_2(s) - u(s)z_1(s) &= 0, \\ v(s)z_2'(s) - u(s)z_1'(s) &= -\frac{1}{p(s)}. \end{aligned} \tag{56}$$

The simultaneous solution of (56) yields

$$\begin{aligned} u(s) &= -\frac{z_2(s)}{p(s)W(z_1, z_2)(s)}, \\ v(s) &= -\frac{z_1(s)}{p(s)W(z_1, z_2)(s)}, \end{aligned} \tag{57}$$

where $W(z_1, z_2) = z_1 z_2' - z_1' z_2$ is the Wronskian function. It is a curious fact that if y_1 and y_2 are any solutions of the same homogeneous DE, then $p(x)W(y_1, y_2)(x)$ is a constant (see Problem 16 in Exercises 1.6 and Problem 32 in Exercises 2.4). Thus, since z_1 and z_2 are two such solutions, we write

$$p(s)W(z_1, z_2)(s) = p(x)W(z_1, z_2)(x) = C, \tag{58}$$

and the Green's function in (55) takes the form

---

*Conditions (c) and (d) are consequences of (a) and (b).

$$g(x, s) = \begin{cases} -z_1(x)z_2(s)/C, & x_1 < x < s \\ -z_1(s)z_2(x)/C, & s \leqslant x < x_2, \end{cases} \tag{59}$$

or equivalently*

$$g(x, s) = \begin{cases} -z_1(s)z_2(x)/C, & x_1 < s \leqslant x \\ -z_1(x)z_2(s)/C, & x < s < x_2. \end{cases} \tag{60}$$

Because C is constant, it follows from inspection of (60) that the Green's function is symmetrical in x and s, i.e., $g(x, s) = g(s, x)$. This property is sometimes quite useful in the construction of Green's function by other methods (e.g., numerical techniques, etc.).

EXAMPLE 4

Construct the Green's function associated with the boundary value problem

$$y'' + k^2y = f(x), \qquad 0 < x < 1, \qquad y(0) = \alpha, \qquad y(1) = \beta, \qquad k \neq 0.$$

SOLUTION Linearly independent solutions of the associated homogeneous DE are given by $y_1 = \cos kx$ and $y_2 = \sin kx$. In order to construct solutions z_1 and z_2, we assume they are some linear combinations of y_1 and y_2. For instance, writing

$$z_1 = c_{11} \cos kx + c_{12} \sin kx$$

and applying the first boundary condition, $z_1(0) = 0$, suggests the choice $c_{11} = 0$ and $c_{12} = 1$, leading to

$$z_1 = \sin kx.$$

Similarly, by writing

$$z_2 = c_{21} \cos kx + c_{22} \sin kx$$

and imposing the second boundary condition, $z_2(1) = 0$, we find

$$c_{21} \cos k + c_{22} \sin k = 0.$$

We wish to choose c_{21} and c_{22} so that this last relation is satisfied, but not with c_{21} and c_{22} both zero. Among other possible choices, we will select $c_{21} = \sin k$ and $c_{22} = -\cos k$, so that

$$z_2 = \sin k \cos kx - \cos k \sin kx = \sin k(1 - x).$$

What remains is the determination of the constant C given by (58). We first observe that the operator $L = D^2 + k^2$ is self-adjoint, where $p(x) = 1$. Hence, it follows that

---

* Equation (59) is a convenient representation of the Green's function when we think of s as fixed and x the variable, and (60) is the form we want when x is fixed and s is the variable. For example, in y_P, given by (48), it is x that is assumed fixed under the integral while s varies.

$$C = p(x)W(z_1, z_2)(x) = \begin{vmatrix} \sin kx & \sin k(1-x) \\ k\cos kx & -k\cos k(1-x) \end{vmatrix},$$

from which we deduce

$$C = -k \sin k.$$

Because the Green's function involves division by C, we must now restrict $k \neq n\pi$ ($n = 1, 2, 3, \ldots$) to avoid division by zero. Under this condition, we get

$$g(x, s) = \begin{cases} \dfrac{\sin ks \sin k(1-x)}{k \sin k}, & 0 < s \leqslant x \\[2ex] \dfrac{\sin kx \sin k(1-s)}{k \sin k}, & x < s < 1. \end{cases}$$

If we allow $k = n\pi$ ($n = 1, 2, 3, \ldots$), then $C = 0$ and the Green's function defined by (60) cannot be constructed. Also, in this case, we find that

$$z_1 = z_2 = \sin n\pi x,$$

which means that the related homogeneous problem

$$y'' + n^2\pi^2 y = 0, \qquad y(0) = 0, \qquad y(1) = 0$$

has a nontrivial solution.*

Example 4 illustrates an important fact— that the Green's function for boundary value problems does not always exist! That is, if $C = 0$, then $g(x, s)$ cannot be constructed by (60). To clarify this point, let us consider the general nonhomogeneous problem

$$L[y] = f(x), \qquad x_1 < x < x_2, \qquad B_1[y] = \alpha, \qquad B_2[y] = \beta. \tag{61}$$

If the associated homogeneous problem

$$L[y] = 0, \qquad B_1[y] = 0, \qquad B_2[y] = 0 \tag{62}$$

has a nontrivial solution, then z_1 and z_2 are linearly dependent and necessarily $W(z_1, z_2) = 0$, or $C = 0$. Consequently, the construction of the Green's function is doomed to fail, and moreover, under these circumstances the problem described by (61) generally has no solution. On the other hand, if (62) has only the trivial solution, then it can be shown that z_1 and z_2 are always linearly independent. In this case $C \neq 0$, and the Green's function can be uniquely constructed by (60).

Although we have already discussed the solution of (45) in a general sense, it is perhaps worth pointing out that instead of representing y_H in terms

---

* See Theorem 2.2.

of y_1 and y_2 [see (47)], it may be more convenient in some situations to use the new linearly independent solutions z_1 and z_2. In this case, we write

$$y_H = C_1 z_1(x) + C_2 z_2(x). \tag{63}$$

This particular combination of solutions provides the simplest determination of the constants C_1 and C_2. For example, imposing the nonhomogeneous boundary conditions in (45) on the solution (63), we obtain

$$C_1 = \frac{\beta}{B_2[z_1]}, \qquad C_2 = \frac{\alpha}{B_1[z_2]}. \tag{64}$$

The complete solution of (41) [or (43) or (61)] is then

$$\begin{aligned} y &= y_H + y_P \\ &= C_1 z_1(x) + C_2 z_2(x) - \int_{x_1}^{x_2} g(x, s) f(s)\, ds, \end{aligned} \tag{65}$$

where C_1 and C_2 are given by (64).

In summary, we state the following theorem.

THEOREM 2.2

> Let $p(x)$, $q(x)$, and $f(x)$ be continuous functions on the interval $[x_1, x_2]$, with $p(x) \neq 0$. Then, the boundary value problem
>
> $$L[y] = f(x), \qquad x_1 < x < x_2, \qquad B_1[y] = \alpha, \qquad B_2[y] = \beta,$$
>
> where $L = D[p(x)D] + q(x)$, either has a unique solution given by (65), or else the associated homogeneous problem
>
> $$L[y] = 0, \qquad B_1[y] = 0, \qquad B_2[y] = 0.$$
>
> has a nontrivial solution.

Theorem 2.2 is widely known as the *Fredholm alternative*. Formulated as an eigenvalue problem, it could be rephrased as saying that either the nonhomogeneous problem has a unique solution or else the associated homogeneous problem has an eigenfunction corresponding to a zero eigenvalue.

Let us illustrate the Green's function method with some examples.

EXAMPLE 5

Use the method of Green's function to solve

$$y'' + y = \sin x, \qquad 0 < x < \frac{\pi}{2}, \qquad y(0) = 1, \qquad y\left(\frac{\pi}{2}\right) = -1.$$

SOLUTION Following the approach we used in solving initial value problems, we decompose our problem into the following problems:

$$y'' + y = 0, \qquad y(0) = 1, \qquad y\left(\frac{\pi}{2}\right) = -1$$

$$y'' + y = \sin x, \qquad y(0) = 0, \qquad y\left(\frac{\pi}{2}\right) = 0.$$

The first problem above has the general solution

$$y_H = C_1 \cos x + C_1 \sin x,$$

which, subject to the prescribed nonhomogeneous boundary conditions, leads to $C_1 = -C_2 = 1$. Hence,

$$y_H = \cos x - \sin x.$$

Linearly independent solutions of the homogeneous DE, each of which satisfies one homogeneous boundary condition, are readily found to be

$$z_1 = \sin x, \qquad z_2 = \cos x.$$

Since $p(x) = 1$ and

$$W(z_1, z_2)(x) = \begin{vmatrix} \sin x & \cos x \\ \cos x & -\sin x \end{vmatrix} = -1,$$

we have $C = p(x)W(z_1, z_2)(x) = -1$. The Green's function is, accordingly,

$$g(x, s) = \begin{cases} \sin s \cos x, & 0 < s \leqslant x \\ \sin x \cos s, & x < s < \pi/2. \end{cases}$$

The second problem shown above is already in self-adjoint form. Therefore, $f(x) = \sin x$ and the solution is

$$\begin{aligned} y_P &= -\int_0^{\pi/2} g(x, s) \sin s \, ds \\ &= -\cos x \int_0^x \sin^2 s \, ds - \sin x \int_x^{\pi/2} \cos s \sin s \, ds, \end{aligned}$$

which simplifies to

$$y_P = -\frac{1}{2} x \cos x.$$

Combining results, we have

$$y = y_H + y_P = \left(1 - \frac{x}{2}\right) \cos x - \sin x.$$

EXAMPLE 6

Use the method of Green's function to solve

$$2x^2y'' + 3xy' - y = 4x^{3/2}, \qquad 0 < x < 1, \qquad y(0) = 0, \qquad y(1) = 0.$$

SOLUTION The function y_H is a solution of the homogeneous problem

$$2x^2y'' + 3xy' - y = 0, \qquad y(0) = 0, \qquad y(1) = 0.$$

This is a Cauchy-Euler equation whose general solution is

$$y_H = C_1x^{1/2} + C_2x^{-1}.$$

In order to satisfy the first boundary condition $y(0) = 0$, we must set $C_2 = 0$, while C_1 is arbitrary. However, the second boundary condition requires that $y(1) = C_1 = 0$, and thus we obtain the trivial solution

$$y_H = 0.$$

This is precisely what we need to have happen if the original problem is to have a unique solution.

In solving for y_P, we must first put the DE in self-adjoint form so that we may properly identify $p(x)$ and $f(x)$. Recalling Equation (42), we find

$$p(x) = \exp\left[\int \left(\frac{3x}{2x^2}\right) dx\right] = \exp\left(\frac{3}{2}\int \frac{dx}{x}\right) = x^{3/2},$$

and so dividing the DE by $2x^2$ and multiplying by $p(x) = x^{3/2}$, we obtain the self-adjoint form

$$\frac{d}{dx}(x^{3/2}y') - \frac{1}{2}x^{-1/2}y = 2x, \qquad y(0) = 0, \qquad y(1) = 0.$$

Linear combinations of $x^{1/2}$ and x^{-1} suitable for constructing the Green's function are found to be

$$z_1 = x^{1/2}, \qquad z_2 = x^{1/2} - x^{-1}.$$

An easy calculation shows that

$$C = p(x)W(z_1, z_2)(x) = x^{3/2}\left(\frac{3}{2}x^{-3/2}\right) = \frac{3}{2},$$

and hence we deduce that

$$g(x, s) = \begin{cases} -\dfrac{2}{3}s^{1/2}(x^{1/2} - x^{-1}), & 0 < s \leq x \\[2ex] -\dfrac{2}{3}x^{1/2}(s^{1/2} - s^{-1}), & x < s < 1. \end{cases}$$

The solution we seek is therefore (here $y = y_P$, since $y_H = 0$)

$$\begin{aligned} y &= -\int_0^1 g(x, s)(2s)\, ds \\ &= \frac{2}{3}(x^{1/2} - x^{-1})\int_0^x s^{1/2}(2s)\, ds + \frac{2}{3}x^{1/2}\int_x^1 (s^{1/2} - s^{-1})(2s)\, ds, \end{aligned}$$

which reduces to

$$y = \frac{4}{5} x^{1/2}(x - 1).$$

2.4.2 PHYSICAL INTERPRETATION OF THE GREEN'S FUNCTION

To aid us in finding a physical interpretation of the Green's function, we wish to consider an example of the string deflection problem of Section 1.5.1, which is characterized by

$$y'' = -\frac{q(x)}{T}, \qquad 0 < x < b, \qquad y(0) = 0, \qquad y(b) = 0. \tag{66}$$

Here T is the (constant) tension in the string, $q(x)$ is a distributed load, and the boundary conditions suggest that the ends of the string are fixed on the x-axis. Linearly independent solutions of $y'' = 0$, each of which satisfies one boundary condition, are

$$z_1 = x, \qquad z_2 = b - x, \tag{67}$$

and the corresponding Green's function is

$$g(x, s) = \begin{cases} \dfrac{s}{b}(b - x), & 0 < s \leq x \\[2ex] \dfrac{x}{b}(b - s), & x < s < b. \end{cases} \tag{68}$$

For any loading $q(x)$, the displacement of the string is then determined by*

$$y = \frac{1}{T} \int_0^b g(x, s) q(s)\, ds. \tag{69}$$

For example, if $q(x) = q_0$ (uniform load) and $b = \pi$, we obtain from (69) the familiar result (see Example 14 in Section 1.5.1)

$$y = \frac{q_0}{2T} x(\pi - x), \qquad 0 < x < \pi. \tag{70}$$

Suppose the load term $q(x)$ in (66) is now a single force P concentrated at the point $x = c$, $0 < c < b$. Mathematically, we write

$$q(x) = P\delta(x - c), \tag{71}$$

where $\delta(x - c)$ is the delta function, and in this case the response of the string is

---

* Here we find that $y_H = 0$.

$$y = \frac{P}{T}\int_0^b g(x, s)\delta(s - c)\, ds = \frac{P}{T}\, g(x, c),$$

or

$$y = \begin{cases} \dfrac{Px}{bT}(b - c), & 0 < x \leqslant c \\ \dfrac{Pc}{bT}(b - x), & c < x < b. \end{cases} \tag{72}$$

Consequently, the Green's function $g(x, c)$, to within a positive multiplicative constant, is actually the shape of the string in response to the force concentrated at $x = c$ (see Figure 2.3). Also, from the symmetry of the Green's function, it follows that the response of the string at the point c is the same as that at $(b - c)$ when the same force is now applied at $(b - c)$.*

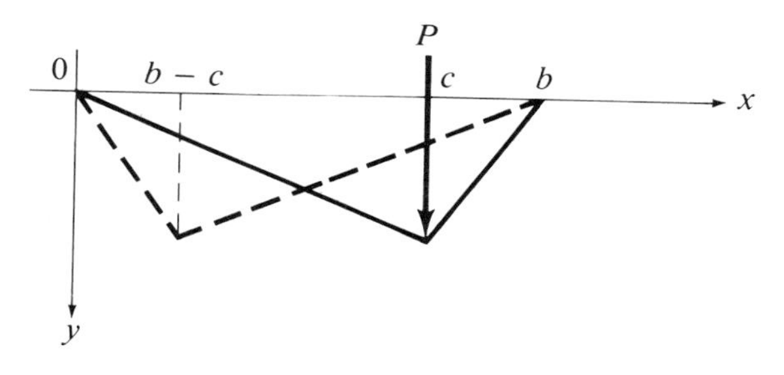

FIGURE 2.3

In general, we say that the Green's function is the response of a system due to a unit disturbance, or unit input, somewhere on the fundamental domain. For this reason, it is also called a *fundamental solution* of the boundary value problem, and all other solutions are simply *superpositions* of this fundamental solution.

The physical interpretation of the Green's function lends itself to empirical construction of this function, which then appears in the form of a matrix of numerical values. For example, in studying small deflections of an elastic string, or possibly an elastic beam, a number of points $c_1, c_2, \ldots, c_k$ may first be selected along the axis of the string. By applying unit loads successively at these k points, and in each case measuring the deflections at several other points, generally k points also, a matrix of numerical values describing the Green's function can be obtained. In this fashion, the deflections of the string can be determined in a numerical sense for any other loading (see Problem 44 in Exercises 2.4).

*This particular property is often called *Maxwell reciprocity*.

EXERCISES 2.4

In Problems 1–6, put the DE in self-adjoint form.

1. $xy'' + y = x^2, \quad x > 0$
2. $y'' - y' + y = 3e^x$
3. $x^2y'' - xy' + y = xe^{-x}, \quad x > 0$
4. $x^2y'' - 4xy' - 6y = x^7, \quad x > 0$
5. $xy'' + y' + xy = 4x^2e^{-2x}, \quad x > 0$
6. $y'' - (\tan x)y' + y = 5x, \quad -\pi/2 < x < \pi/2$

In Problems 7–15, construct the Green's function, if possible, for the specified operator and boundary conditions.

7. $L = D^2$;
   (a) $y(a) = 0, \quad y(b) = 0$ (b) $y(0) = 0, \quad y'(1) = 0$
   (c) $y'(0) = 0, \quad y(1) = 0$ (d) $y'(0) = 0, \quad y'(1) = 0$
8. $L = D^2 + 1$;
   (a) $y(0) = 0, \quad y(1) = 0$ (b) $y(0) = 0, \quad y'(1) = 0$
   (c) $y'(0) = 0, \quad y(1) = 0$ (d) $y'(0) = 0, \quad y'(1) = 0$
9. $L = D^2 - 1$;
   (a) $y(0) = 0, \quad y(1) = 0$ (b) $y(0) + y'(0) = 0, \quad y(1) = 0$
10. $L = D(e^{3x}D) + 2e^{3x}; \quad y(0) = 0, \quad y(1) = 0$
11. $L = D(x^{-4}D) - 6x^{-6}; \quad y(0) = 0, \quad y(1) = 0$
12. $L = D(x^2D) + \dfrac{1}{4}; \quad y(1) = 0, \quad y(e) = 0$
13. $L = D^2 - \dfrac{1}{x^2}; \quad y(1) = 0, \quad y(e) = 0$
14. $L = D(xD) + \dfrac{1}{x}; \quad y(1) = 0, \quad y(e) = 0$
15. $L = x^2D^2 + 2xD + 1; \quad y(1) = 0, \quad y(e^{\pi/\sqrt{3}}) = 0$
16. For which values of λ will the Green's function for the boundary value problem
$$y'' + \lambda y = f(x), \quad y(0) = 0, \quad y(a) = 0$$
not exist?

In Problems 17–28, find the solution, if possible, by the method of Green's function.

17. $y'' = -x, \quad y(0) = 0, \quad y'(1) = 0$
18. $y'' = x, \quad y(0) = 1, \quad y(1) - 2y'(1) = 0$
19. $y'' = \sin \pi x, \quad y(0) = 0, \quad y'(1) = 0$
20. $y'' = 3 \sin \pi x, \quad y(0) = -2, \quad y'(1) = 7/\pi$

21. $y'' + y = 1, \quad y'(0) = 0, \quad y'(1) = 0$

22. $y'' + 4y = e^{-x}, \quad y(0) = -1, \quad y'(\pi/2) = e^{-\pi/2}/5$

23. $y'' - y = x^2, \quad y(0) = 0, \quad y(1) = 0$

24. $y'' - y = e^x, \quad y(0) = 0, \quad y'(1) = 1$

25. $y'' + 4y = 1 - \delta(x - 1), \quad y(0) = 1, \quad y(2) = 0$

26. $y'' + 2y' + y = x, \quad y(0) = -3, \quad y(1) = -1$

★27. $x^2y'' - 3xy' + 3y = 2x^4e^x, \quad y(0) = 0, \quad y(1) = 0$

28. $xy'' + y' = x, \quad y(1) = 0, \quad y(e) = e^2/4$

29. Prove that the self-adjoint operator $L = D[p(x)D] + q(x)$ is a linear operator, i.e., for any constants C_1 and C_2, that

$$L[C_1y_1 + C_2y_2] = C_1L[y_1] + C_2L[y_2].$$

30. If y_P satisfies the boundary value problem

$$L[y] = f(x), \quad B_1[y] = 0, \quad B_2[y] = 0,$$

and y_H satisfies

$$L[y] = 0, \quad B_1[y] = \alpha, \quad B_2[y] = \beta,$$

show that the sum $y = y_P + y_H$ is a solution of

$$L[y] = f(x), \quad B_1[y] = \alpha, \quad B_2[y] = \beta.$$

Hint: Use Problem 29.

★31. Explain why the Green's function (60) is unique even though z_1 and z_2 are not unique. Proof?

32. Show that $p(x)W(z_1, z_2)(x) = C$ (constant), where z_1 and z_2 are solutions of $L[y] = 0$ satisfying $B_1[z_1] = 0$, $B_2[z_2] = 0$.
Hint: Show that

$$\frac{d}{dx}[p(x)W(z_1, z_2)(x)] = 0.$$

33. Consider the boundary value problem

$$y'' = -1, \quad y'(0) = 0, \quad y'(1) = -1.$$

(a) Can the Green's function (60) be constructed?
(b) Does the problem have a solution? If so, find it.

34. Show that the Green's function associated with the boundary value problem

$$y'' + \pi^2y = \pi^2x, \quad y(0) = 1, \quad y(1) = 0$$

does not exist, but that the problem still has solutions given by

$$y = \cos \pi x + C \sin \pi x + x \quad (C \text{ arbitrary}).$$

35. Show that, while Problem 34 has solutions, the subsidiary problems (see Problem 30) for y_P and y_H fail to have solutions. Does this contradict the result of Problem 30?

36. Verify conditions (c) and (d) of Definition 2.2 for the functions

$$p(x) = 1 - x^2, \qquad g(x, s) = \begin{cases} \dfrac{1}{2} \log \dfrac{1+s}{1-s}, & 0 < s \leqslant x \\[2ex] \dfrac{1}{2} \log \dfrac{1+x}{1-x}, & x < s < 1. \end{cases}$$

37. Given that $g(x, s) = g(s, x)$ and the following information, determine the function $p(x)$ and the Green's function for $x > s$:

(a) $g(x, s) = -\log s, \quad x < s$ (b) $g(x, s) = \dfrac{x}{2s}(1 - s^2), \quad x < s$

38. The Green's function for a given self-adjoint operator is

$$g(x, s) = \begin{cases} \dfrac{1}{2} xs^3 - \dfrac{5}{2} xs + x, & 0 < s \leqslant x \\[2ex] \dfrac{1}{2} x^3 s - \dfrac{5}{2} sx + s, & x < s < 1. \end{cases}$$

(a) Determine $p(x)$. (b) Determine $q(x)$.

★39. Given the boundary value problem

$$\frac{d}{dx}[(1 - x^2)y'] = f(x), \qquad y(0) = 0, \qquad y(1) \text{ finite},$$

(a) show that the Green's function is

$$g(x, s) = \begin{cases} \dfrac{1}{2} \log \dfrac{1+s}{1-s}, & 0 < s \leqslant x \\[2ex] \dfrac{1}{2} \log \dfrac{1+x}{1-x}, & x < s < 1. \end{cases}$$

Hint: Solutions z_1 and z_2 do not have to vanish at singular points, but must remain finite.

(b) Solve the problem using the given Green's function when $f(x) = 1$.
(c) Solve the problem using the given Green's function when $f(x) = x$.

★40. By direct construction, show that the Green's function associated with the singular boundary value problem

$$y'' - k^2 y = f(x), \qquad -\infty < x < \infty, \qquad y \text{ finite as } |x| \to \infty,$$

is given by

$$g(x, s) = \frac{1}{2k} e^{-k|x-s|}, \qquad k > 0.$$

★41. By direct construction, show that the Green's function associated with the singular boundary value problem

$$y'' - k^2 y = f(x), \qquad 0 < x < \infty, \qquad y(0) = 0, \qquad y \text{ finite as } x \to \infty,$$

is given by $(k > 0)$

$$g(x, s) = \begin{cases} \dfrac{1}{k} e^{-kx} \sinh ks, & 0 < s \leqslant x \\ \dfrac{1}{k} e^{-ks} \sinh kx, & x < s < \infty. \end{cases}$$

42. Determine the deflection curve of a string of unit length governed by

$$y'' = -\frac{q(x)}{T}, \qquad y(0) = 0, \qquad y(1) = 0,$$

when the load $q(x)$ is
(a) equally concentrated loads P at $x = 1/4$ and $x = 3/4$.
(b) a concentrated load of intensity P at $x = 1/4$, a concentrated load of intensity $-2P$ at $x = 1/2$, and a concentrated load of intensity $3P$ at $x = 3/4$.

43. Referring to Problem 42, find the deflection curve when the string is supporting the distributed load $q(x) = x$ and a concentrated load of intensity P at

(a) $x = 1/2$. (b) $x = 1/4$.

★44. Under laboratory conditions, unit loads are applied successively to an elastic string of unit length ($0 \leqslant x \leqslant 1$) at the points $c_1 = 1/4$, $c_2 = 1/2$, $c_3 = 3/4$, and the deflections at all three points are measured and tabulated each time. By so doing, the following matrix of numerical values is generated for the Green's function,

$$g(x, s) \equiv [g_{jk}] = \begin{pmatrix} 0.19 & 0.13 & 0.06 \\ 0.13 & 0.25 & 0.13 \\ 0.06 & 0.13 & 0.19 \end{pmatrix}.$$

If we denote the deflection due to an external load $q(x)$ by $y_j = y(c_j)$, and assume the string is fixed at $x = 0$ and $x = 1$, it can be shown that

$$y_j = \frac{1}{4T} \sum_{k=1}^{3} g_{jk} q_k, \qquad j = 1, 2, 3,$$

where $q_k = q(c_k)$.
(a) Determine y_j (for $j = 1, 2, 3$) when $q(x) = x$. Also, solve the problem exactly and compare the two solutions at these points. Assume $T = 1$.
(b) Assuming the deflections of the string at other points can be reasonably approximated by straight line segments connecting the known deflection points at $x = 0, 1/4, 1/2, 3/4, 1$, compute $y(1/8)$, $y(3/8)$, $y(5/8)$ and $y(7/8)$, and compare these values with the exact values at these points (see figure).

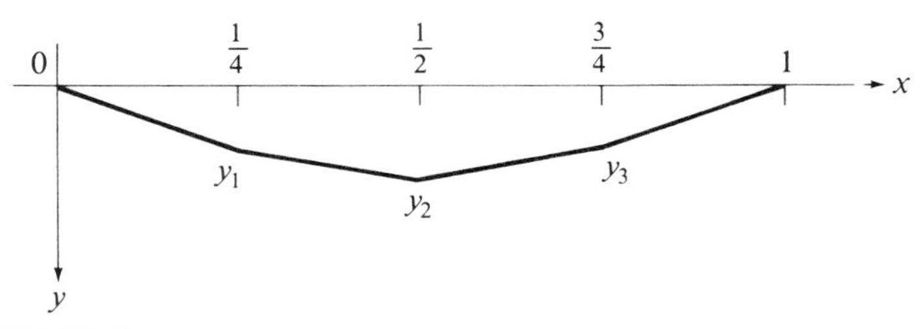

PROBLEM 44

★45. By directly solving the boundary value problem

$$y^{(4)} = -\delta(x - s), \qquad y(0) = y'(0) = 0, \qquad y(1) = y'(1) = 0,$$

(a) show that its solution yields the Green's function

$$g(x, s) = \begin{cases} \frac{1}{6} s^2(1 - x)^2(2xs + s - 3x), & 0 < s \leq x \\ \frac{1}{6} x^2(1 - s)^2(2sx + x - 3s), & x < s < 1. \end{cases}$$

(b) Show that g, g_x, and g_{xx} are all continuous at $x = s$, but that g_{xxx} has a jump discontinuity at $x = s$.

★46. Consider the nonhomogeneous DE

$$A_2(x)y'' + A_1(x)y' + A_0(x)y = F(x), \qquad x_1 < x < x_2,$$

where $A_2(x) \neq 0$ for $x_1 \leq x \leq x_2$.

(a) If $y_1(x)$ and $y_2(x)$ are two linearly independent solutions of the associated homogeneous DE, show that the method of variation of parameters (Section 2.2) leads to

$$y_P = -\int_{x_1}^{x} \frac{y_1(x)y_2(s)F(s)}{W(y_1, y_2)(s)A_2(s)}\, ds - \int_{x}^{x_2} \frac{y_1(s)y_2(x)F(s)}{W(y_1, y_2)(s)A_2(s)}\, ds.$$

(b) By defining the function

$$K(x, s) = \begin{cases} \dfrac{y_1(x)y_2(s)}{W(y_1, y_2)(s)A_2(s)}, & x_1 < s \leq x \\ \dfrac{y_1(s)y_2(x)}{W(y_1, y_2)(s)A_2(s)}, & x < s < x_2 \end{cases}$$

show that the solution in part (a) can be written as

$$y_P = -\int_{x_1}^{x_2} K(x, s)F(s)\, ds.$$

(c) For the special case when $A_2(x) = p(x)$, $A_1(x) = p'(x)$, $y_1(x) = z_2(x)$, and $y_2(x) = z_1(x)$, show that the function $K(x, s)$ defined in part (b) is the Green's function $g(x, s)$. [NOTE: The function $K(x, s)$ can also be used as a type of Green's function, although in general it is not symmetrical. Furthermore, the solution y_P in (b) will not satisfy homogeneous boundary conditions unless the solutions $y_1(x)$ and $y_2(x)$ are chosen to be $z_2(x)$ and $z_1(x)$, respectively.]

BIBLIOGRAPHY

Courant, R., and D. Hilbert. *Methods of Mathematical Physics*. Vol. 1. New York: Interscience, 1953.

Finlayson, B. A. *The Method of Weighted Residuals and Variational Principles*. New York: Academic Press, 1972.

Friedmann, B. *Principles and Techniques of Applied Mathematics*. New York: John Wiley & Sons, 1957.

Greenberg, M. D. *Applications of Green's Function in Science and Engineering*. Englewood Cliffs, N.J.: Prentice-Hall, 1961.

———. *Foundations of Applied Mathematics*. Englewood Cliffs, N.J.: Prentice-Hall, 1978.

Hildebrand, F. B. *Advanced Calculus for Applications*. 2nd ed. Englewood Cliffs, N.J.: Prentice-Hall, 1976.

Kreider, D. L., D. R. Ostberg, R. G. Kuller, and F. W. Perkins. *An Introduction to Linear Analysis*. Reading, Mass.: Addison-Wesley, 1966.

Pipes, L. A. *Applied Mathematics for Engineers and Physics*. 2nd ed. New York: McGraw-Hill, 1958.

Rabenstein, A. L. *Introduction to Ordinary Differential Equations*. New York: Academic Press, 1972.

PART II

PARTIAL DIFFERENTIAL EQUATIONS

CHAPTER 3

AN INTRODUCTION TO PARTIAL DIFFERENTIAL EQUATIONS

In the analysis of certain mechanical and electrical systems, we are led to *ordinary DEs* with time as the independent variable. Some typical examples include spring-mass systems and electrical circuits. In these cases we say the system contains "lumped parameters," such as the masses of a mechanical system or the elements of an electric circuit concentrated at a single point. While lumping parameters in this way is satisfactory for some problems, it is not adequate for others. For example, in determining the deflections of an elastic string set in motion (e.g., by plucking it), we find it necessary to assume that the mass of the string is continuously distributed over some interval of the x-axis. The deflections of the string in this instance will then be a function of both the x-coordinate and time t, and the governing DE a *partial differential equation.* In general, partial differential equations are prominent in those physical and geometrical problems involving functions that depend upon more than one independent variable. The most commonly occurring partial differential equations in practice are those of the second order. Consequently, our treatment of partial differential equations will be limited primarily to second-order equations.

We classify second-order and linear partial differential equations in two independent variables in Section 3.2 by the same classification scheme used for conics, viz., *elliptic, hyperbolic,* or *parabolic*. For the special sub-class of *constant–coefficient equations,* we find that *general solution formulas* are readily obtained for all three classes of equations. One of the rare examples where such general solutions can be used to develop *particular solutions* of the partial differential equation satisfying certain auxiliary conditions is presented here, and leads to the classical *d'Alembert solution of the wave equation*.

The *equations of mathematical physics* are introduced in Section 3.3. Important specializations of these equations that are later discussed in detail are the *heat equation* (governing diffusion processes), *wave equation* (governing vibrational phenomena), and *potential equation* (governing steady-state processes). We also discuss the important solution technique called *separation of variables,* which is used extensively throughout the remainder of the text.

3.1 INTRODUCTION

The boundary value problems studied up to now have all been formulated as ordinary differential equations (ODEs). Historically, however, the theory of boundary value problems grew out of a study of certain partial differential equations (PDEs) encountered in mathematical physics. Thus an investigation into the theory of such equations seems appropriate if only for historical reasons. The importance of PDEs, of course, lies beyond any historical significance since these equations are fundamental to almost every facet of classical physics and engineering. Also, a study of boundary value problems involving PDEs can serve to unite various topics like eigenvalue problems (discussed in Chapter 1) and Fourier series (to be discussed in Chapter 4), bringing them into sharper perspective as well as extending the theory of these topics to more complex problems.

When partial derivatives are required in the mathematical formulation of some physical phenomenon, the resulting equation is called a *partial differential equation*. The variables involved may be time and/or one or more spatial coordinates. It is convenient to indicate partial derivatives by writing independent variables as subscripts. Thus, we will write

$$u_x \quad \text{for} \quad \frac{\partial u}{\partial x}, \qquad u_{xx} \quad \text{for} \quad \frac{\partial^2 u}{\partial x^2}, \qquad u_{xy} \quad \text{for} \quad \frac{\partial^2 u}{\partial y\, \partial x},$$

and so on. It is generally assumed that u satisfies conditions so that $u_{xy} = u_{yx}$.

PDEs are classified as to order, linearity, and homogeneity in much the same way as ODEs. For example, the *order* of a PDE is the order of the partial derivative of highest order appearing in the equation. The equation is *linear* if the unknown function and all its derivatives are of the first degree algebraically. We say a linear PDE is *homogeneous* if whenever u is a solution of the PDE, so is Cu for any constant C.

By *solution* of a PDE we mean a function u that has all partial derivatives occurring in the PDE that, when substituted into the equation, reduces it to an identity for all independent variables. For example, let us consider the simple first-order PDE

$$u_x = 3x^2 + 7y^5, \tag{1}$$

where x and y are independent variables and u is the unknown. Holding y fixed and integrating (1) with respect to x, we find

$$u(x, y) = x^3 + 7xy^5 + F(y), \tag{2}$$

where F is *any* differentiable function of y. Thus, it follows that

$$u(x, y) = x^3 + 7xy^5 + e^y,$$

$$u(x, y) = x^3 + 7xy^5 - \sin y,$$

$$u(x, y) = x^3 + 7xy^5 + 2y^{3/2},$$

and so on, are all solution functions of (1). Likewise, it is easy to verify that

$$u(x, y) = x + 2y,$$
$$u(x, y) = \cos(x + 2y),$$
$$u(x, y) = 3e^x e^{2y} = 3e^{x+2y},$$

each satisfy the equation

$$2u_x - u_y = 0. \tag{3}$$

A *general solution* of a PDE is a collection of all the solutions of the equation. For instance, (2) is a general solution of (1) and

$$u(x, y) = G(x + 2y) \tag{4}$$

is a general solution of (3), where G is any differentiable function of $x + 2y$. Here we find one of the most fundamental differences between the general solution of an ODE and that of a PDE—the general solution of an ODE contains *arbitrary constants,* whereas that of a PDE involves *arbitrary functions.*

In practice one is seeking a *particular solution* of the PDE that satisfies certain *auxiliary conditions* arising out of a given physical situation. Most of the time it is impossible, or at least difficult, to find the needed particular solution by specializing the general solution of the PDE as is done for ODEs. This is so because it is very difficult to specialize arbitrary functions except in special situations. Therefore, it is usually best in solving PDEs to find a suitable set of particular solutions, each of which satisfies some of the prescribed auxiliary conditions, and then combine these particular solutions in some fashion so that the resulting solution function satisfies all the prescribed conditions.*

3.1.1 PROPERLY-POSED PROBLEMS

The problems consisting of solving a PDE subject to certain auxiliary conditions in the form of boundary and/or initial conditions are called *initial-boundary value problems,* or more simply, *boundary value problems.* In the study of these problems, three basic questions arise of chief importance:

1. Does a solution *exist?*
2. Is the solution *unique?*
3. Is the solution *stable?*

Questions concerning existence and uniqueness are standard in any study of DEs. The third question above deals with the problem of whether the solution

---

*This solution technique is called *separation of variables*.

depends continuously upon the prescribed data (both boundary and initial conditions). That is, do small changes in the prescribed data produce only small changes in the values of the solution function at each point? This question is of great concern in applications wherein the auxiliary data are determined most often by measurements and hence are only approximate, not exact as we assume in theoretical discussions. We would certainly hope that small errors in these measurements would produce only small errors in the solution function. A boundary value problem possessing a unique, stable solution is called *properly-posed* or *well-posed.* The reader is cautioned that a problem with too many prescribed boundary and/or initial conditions is an overspecified problem and may not have a solution, and a problem that has too few prescribed conditions does not have a unique solution.

Much work has been carried out over the years to determine the types of auxiliary conditions that must be prescribed so that a given boundary value problem is properly-posed, but such an analysis here would be too lengthy for our purposes. We will point out, however, that boundary and initial conditions that frequently arise in the description of physical phenomena fall mainly into four categories:*

1. *Cauchy Conditions*—Values of the unknown function u and possibly its derivative u_t, where $0 \leqslant t < \infty$, are prescribed on the "boundary" $t = 0$. Initial conditions fall into this category.
2. *Dirichlet Conditions*—The unknown function u is specified at each point on the boundary of the region of interest.
3. *Neumann Conditions*—Values of the normal derivative of the unknown function u are prescribed at each point on the boundary of the region of interest.
4. *Robin Conditions*—Values of the sum of the unknown function u and its normal derivative are prescribed at each point on the boundary of the region of interest.

A typical example illustrating some of these prescribed conditions is given by

$$u_{xx} = u_{tt}, \qquad 0 < x < p, \qquad t > 0$$
$$u(0, t) = T_1, \qquad u_x(p, t) = 0, \qquad t > 0$$
$$u(x, 0) = f(x), \qquad u_t(x, 0) = g(x), \qquad 0 < x < p.$$

A Dirichlet condition is prescribed at the boundary point $x = 0$, a Neumann condition occurs at the boundary point $x = p$, and Cauchy conditions are prescribed at $t = 0$.

---

* Dirichlet, Neumann, and Robin conditions are also called, respectively, boundary conditions of the first, second, and third kind.

3.2 CLASSIFICATION OF SECOND-ORDER PDES IN TWO VARIABLES

Linear, homogeneous PDEs of the second order in two independent variables have the form

$$A(x, y)u_{xx} + B(x, y)u_{xy} + C(x, y)u_{yy} + D(x, y)u_x + E(x, y)u_y + H(x, y)u = 0, \tag{5}$$

where A, B, C, D, E, and H are the coefficients of the equation. In developing the general theory of such equations, it is helpful to classify them as *hyperbolic, parabolic,* or *elliptic* according to the scheme used in studying conic sections:

$$\begin{aligned} &B^2 - 4AC > 0 \quad &\text{(hyperbolic)},\\ &B^2 - 4AC = 0 \quad &\text{(parabolic)},\\ &B^2 - 4AC < 0 \quad &\text{(elliptic)}. \end{aligned} \tag{6}$$

Equations belonging to the same classification exhibit many features in common, and thus by solving a particular member of any class we are exposed to certain fundamental characteristics shared by all members of that class. The classification scheme is also important in determining the method of attack when numerical techniques are used for finding solutions.

The proper types of auxiliary conditions to be prescribed for (5), which lead to a properly-posed problem, depend upon the class of the PDE. For most cases involving an elliptic equation, the unknown function (or possibly its derivative) is specified around the entire boundary enclosing the region of interest, i.e., a Dirichlet or Neumann condition. However, this is not necessarily the case for a parabolic or hyperbolic equation since one of the independent variables, such as time, is usually unbounded. In these latter two cases we usually specify Cauchy conditions in combination with a Dirichlet or Neumann condition. Thus, the type of auxiliary conditions leading to a properly-posed problem for one kind of PDE is often not correct for another kind. Furthermore, the classification of an equation can vary from point to point when the coefficients A, B, and C are functions of x and y. For instance, the PDE

$$u_{xx} - xu_{yy} + u = 0,$$

where $B^2 - 4AC = 4x$, is of the hyperbolic type in the half-plane $x > 0$, elliptic for $x < 0$, and parabolic along the y-axis.*

* For a more detailed discussion of the general classification scheme, see H. F. Weinberger, *Partial Differential Equations,* Chapter II (Blaisdell: New York), 1965, or R. Dennemeyer, *Introduction to Partial Differential Equations and Boundary Value Problems,* Chapter 2 (McGraw-Hill: New York), 1968.

3.2.1 CONSTANT–COEFFICIENT EQUATIONS

An important subclass of (5) are those equations of the form

$$Au_{xx} + Bu_{xy} + Cu_{yy} = 0, \tag{7}$$

where A, B, and C are constants. For such equations, general solutions can always be found.

To find a general solution of (7), it is best to put the equation in a more suitable form. This we do by the linear transformation of coordinates*

$$\begin{aligned} r &= ax + by, \\ s &= cx + dy, \end{aligned} \qquad (ad - bc \neq 0) \tag{8}$$

where a, b, c, and d are constants yet to be determined. From the chain rule, we find

$$\begin{aligned} u_x &= u_r r_x + u_s s_x = au_r + cu_s, \\ u_y &= u_r r_y + u_s s_y = bu_r + du_s, \end{aligned}$$

and

$$\begin{aligned} u_{xx} &= r_x^2 u_{rr} + 2r_x s_x u_{rs} + s_x^2 u_{ss} + r_{xx} u_r + s_{xx} u_s \\ &= a^2 u_{rr} + 2acu_{rs} + c^2 u_{ss}, \\ u_{xy} &= abu_{rr} + (ad + bc)u_{rs} + cdu_{ss}, \\ u_{yy} &= b^2 u_{rr} + 2bdu_{rs} + d^2 u_{ss}. \end{aligned}$$

The substitution of these expressions into (7), followed by some rearranging of terms, leads to the transformed PDE

$$(Aa^2 + Bab + Cb^2)u_{rr} + (Ac^2 + Bcd + Cd^2)u_{ss} + [2Aac + B(ad + bc) + 2Cbd]u_{rs} = 0. \tag{9}$$

By suitable choices of a, b, c, and d, we can make the first two terms of (9) vanish. If $A \neq 0$, then among other possible choices we might select these constants such that $b = d = 1$, and such that a and c are the roots m_1 and m_2 of the quadratic equation

$$Am^2 + Bm + C = 0. \tag{10}$$

For example, if we choose

$$a = m_1 = \frac{-B + \sqrt{B^2 - 4AC}}{2A} \tag{11}$$

---

*The transformation (8) used here is analogous to the rotation of axes that we do for the purpose of obtaining a more convenient coordinate representation in studying conic sections.

and

$$c = m_2 = \frac{-B - \sqrt{B^2 - 4AC}}{2A}, \tag{12}$$

then (9) reduces to

$$[2Am_1m_2 + B(m_1 + m_2) + 2C]u_{rs} = 0,$$

or, since $m_1 + m_2 = -B/A$ and $m_1m_2 = C/A$, this last equation becomes

$$\frac{1}{A}(4AC - B^2)u_{rs} = 0. \tag{13}$$

Here the analogy with conic sections, which depends upon whether the discriminant $B^2 - 4AC$ is positive, negative, or zero, becomes clearer.

From (6), we see that both the hyperbolic and elliptic cases require that $B^2 - 4AC \neq 0$, and thus (13) reduces to simply

$$u_{rs} = 0, \tag{14}$$

which has the general solution

$$u(r, s) = F(r) + G(s). \tag{15}$$

Consequently, since the equations of transformation (8) are now specifically described by

$$\begin{aligned} r &= m_1x + y, \\ s &= m_2x + y, \end{aligned} \tag{16}$$

where m_1 and m_2 are given by (11) and (12), respectively, the general solution (15) as a function of x and y can be written as

$$u(x, y) = F(m_1x + y) + G(m_2x + y). \tag{17}$$

When the discriminant is positive (hyperbolic case), the roots m_1 and m_2 are real and distinct, whereas the roots are complex conjugates when the discriminant is negative (elliptic case), i.e., $m_2 = \overline{m}_1$.

In the parabolic case the discriminant is zero and (13) reduces to the identity $0 = 0$. Here we find that $m_1 = m_2 = m$, and so our transformation (8) degenerates since $r = s$, or equivalently, $ad - bc = 0$. In this case we look for a new transformation involving one of the variables, say s, which is a linear combination of x and y but not proportional to r. For example, if we choose the new linear transformation

$$\begin{aligned} r &= mx + y, \\ s &= x, \end{aligned} \tag{18}$$

then $a = m$, $b = 1$, $c = 1$, and $d = 0$, and (9) reduces to

$$u_{ss} = 0 \tag{19}$$

$(A \neq 0)$ with general solution

$$u(r, s) = F(r) + sG(r). \tag{20}$$

In terms of x and y [see (18)] this solution takes the form

$$u(x, y) = F(mx + y) + xG(mx + y). \tag{21}$$

REMARK: *If $A = 0$, then the above results can still be used provided we interchange the roles of x and y, and the roles of A and C.*

EXAMPLE 1

Find the general solution of $u_{xx} + u_{yy} = 0$.

SOLUTION This PDE is the well-known *potential equation,* or *Laplace's equation.* We first observe that $B^2 - 4AC = -4$, which classifies the PDE as elliptic. To find the general solution, we must find the roots of the quadratic equation (10), which in this case is

$$m^2 + 1 = 0.$$

Hence, we see that $m_1 = i$ and $m_2 = -i$, and thus the general solution is

$$u(x, y) = F(y + ix) + G(y - ix).$$

REMARK: *Although the solution in Example 1 is represented in terms of complex functions, the solutions of the potential equation obtained in practice are usually real. This is accomplished by taking certain linear combinations of complex solutions, analogous to the situation in solving linear ODEs with constant coefficients when the roots of the auxiliary equation are complex.*

EXAMPLE 2

Find the general solution of $u_{xx} - 2u_{xy} + u_{yy} = 0$.

SOLUTION This time the PDE is of the *parabolic* type since $B^2 - 4AC = 0$. The related quadratic equation is

$$m^2 - 2m + 1 = 0,$$

with roots $m_1 = m_2 = 1$. Using the general form (21) leads to the solution

$$u(x, y) = F(x + y) + xG(x + y).$$

3.2.2 D'ALEMBERT'S SOLUTION OF THE WAVE EQUATION

In this section we wish to consider one of the rare examples of a PDE subject to certain auxiliary conditions that can be solved by first finding the general solution and then imposing the auxiliary conditions. The method was introduced in 1747 by d'Alembert,* six years before Bernoulli† provided his series solution by separating the variables. At first d'Alembert thought he had developed a technique that could be applied to a large class of equations, but he soon realized this was not so.

The *one-dimensional wave equation* (see also Chapter 7)

$$u_{xx} = c^{-2}u_{tt} \qquad (c \text{ constant}) \tag{22}$$

is one of the simplest examples of a PDE occurring in practice. It is the governing equation for small deflections of an elastic string that is stretched tightly and given some initial deflection and velocity. For now, let us suppose the string is infinitely long ($-\infty < x < \infty$) and that the initial deflection and velocity of the string are specified, respectively, by the Cauchy conditions

$$u(x, 0) = f(x), \qquad u_t(x, 0) = g(x). \tag{23}$$

Writing (22) in the form $u_{tt} - c^2u_{xx} = 0$, we note that the equation is hyperbolic since $B^2 - 4AC = 4c^2 > 0$. Also, $m_1 = -m_2 = c$, providing us with the general solution

$$u(x, t) = F(x + ct) + G(x - ct), \tag{24}$$

where F and G are any two everywhere defined, differentiable functions. The initial conditions (23) imposed upon this general solution lead to the relations

$$\begin{aligned} u(x, 0) &= F(x) + G(x) = f(x), \\ u_t(x, 0) &= c[F'(x) - G'(x)] = g(x). \end{aligned} \tag{25}$$

If we differentiate the first equation in (25) with respect to x and solve simultaneously with the remaining equation, we find

$$F'(x) = \frac{1}{2}\left[f'(x) + \frac{1}{c}\,g(x)\right],$$

$$G'(x) = \frac{1}{2}\left[f'(x) - \frac{1}{c}\,g(x)\right],$$

which upon integration yields

---

*JEAN LE ROND D'ALEMBERT (1717–1783) was a French mathematician well known for his work in mechanics and the fundamental theorem of algebra, which incidentally is still called d'Alembert's theorem in France.. He was first educated in law and then studied medicine before he resolved to devote all his time to mathematics. He wrote on a variety of topics including planetary perturbation, music, and philosophy.

†DANIEL BERNOULLI (1700–1782) was the son of Johann from the famous family of Swiss mathematicians. He became a close friend of Euler, who was a student of Johann Bernoulli. Daniel and his brother Nicolaus were both appointed by Catherine I of Russia to the Academy of Sciences in St. Petersburg. He made contributions in fluid mechanics (Bernoulli's principle) and probability theory, and he solved the oscillating chain problem by an infinite series that we now call *Bessel's function*.

$$F(x) = \frac{1}{2}\left[f(x) + \frac{1}{c}\int_0^x g(z)\,dz\right] + C_1,$$
$$G(x) = \frac{1}{2}\left[f(x) - \frac{1}{c}\int_0^x g(z)\,dz\right] + C_2, \tag{26}$$

where C_1 and C_2 are constants of integration. Because we permit the arguments in F and G to extend over all real numbers, the results (26) can be generalized to the set of expressions

$$F(x + ct) = \frac{1}{2}\left[f(x + ct) + \frac{1}{c}\int_0^{x+ct} g(z)\,dz\right] + C_1,$$

$$G(x - ct) = \frac{1}{2}\left[f(x - ct) - \frac{1}{c}\int_0^{x-ct} g(z)\,dz\right] + C_2,$$

which added together lead us to *d'Alembert's solution*

$$u(x, t) = \frac{1}{2}\left[f(x + ct) + f(x - ct) + \frac{1}{c}\int_{x-ct}^{x+ct} g(z)\,dz\right]. \tag{27}$$

The constant $(C_1 + C_2)$ has been set to zero in order to satisfy the initial condition $u(x, 0) = f(x)$.

Suppose that $g(x) \equiv 0$. Then the functions $f(x + ct)$ and $f(x - ct)$ that are left in the solution (27) have interesting physical interpretations as wave phenomena. The function $f(x + ct)$, plotted as a function of x alone, is exactly the same in shape as the initial deflection $f(x)$, but with every point on it displaced a distance ct to the left of the corresponding point in $f(x)$. Thus the function $f(x + ct)$ represents a wave of displacement traveling to the left with velocity c.* In the same manner we can interpret $f(x - ct)$ as a displacement wave traveling with velocity c to the right (see Figure 3.1). The solution $u(x, t) = [f(x + ct) + f(x - ct)]/2$ is then simply the superposition of these two *traveling waves*.

It is interesting to note that at a given point (x_0, t_0) of the xt–plane, d'Alembert's solution (27) shows that the solution value $u(x_0, t_0)$ depends only upon the values of $f(x)$ and $g(x)$ along the segment of the x-axis between $x_0 - ct_0$ and $x_0 + ct_0$ (see Figure 3.2). That is, $f(x)$ and $g(x)$ can be arbitrarily modified outside this interval without changing the value $u(x_0, t_0)$. For this reason the interval $[x_0 - ct_0, x_0 + ct_0]$ is called the *interval of dependency* of the point (x_0, t_0). The two lines of Figure 3.2 connecting the point (x_0, t_0) with the endpoints of the interval of dependency are called the *characteristics* of the wave equation. The equations of these characteristics are given by

$$x - ct = x_0 - ct_0, \qquad x + ct = x_0 + ct_0,$$

*Consequently, for the wave equation we say a disturbance or wave *propagates with finite speed.* It turns out that this is a common characteristic of *hyperbolic* equations, but is not the case for parabolic equations such as the heat equation (see Section 6.2.5).

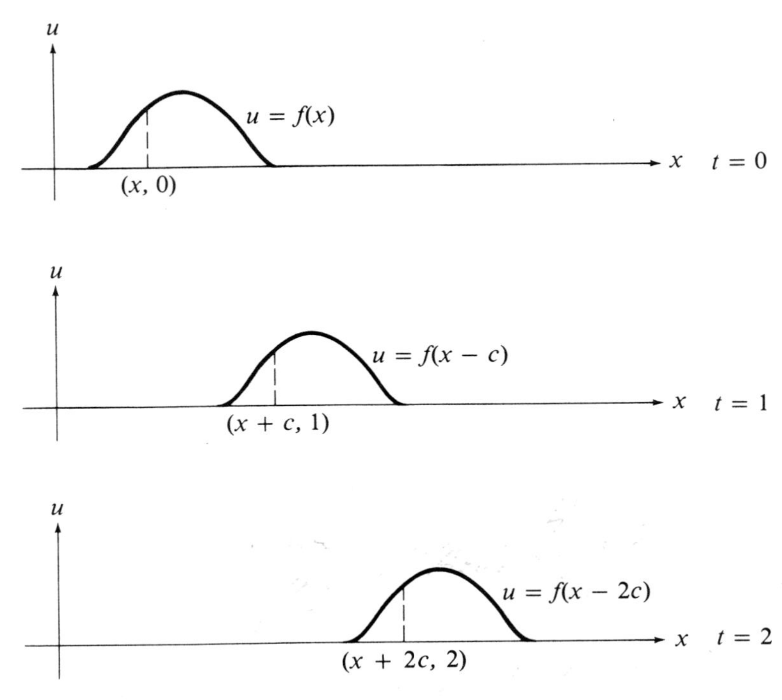

FIGURE 3.1

Traveling wave moving to the right.

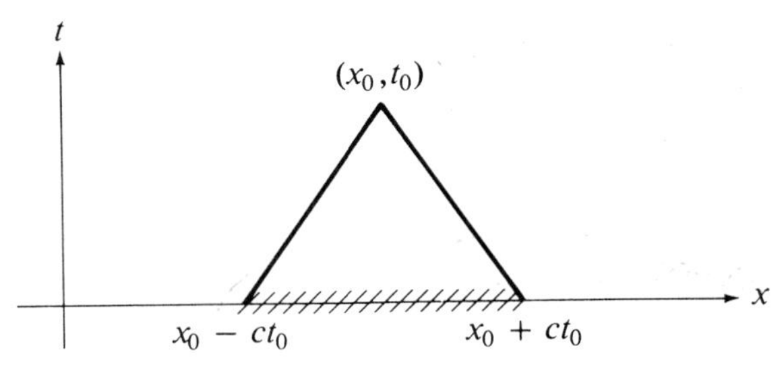

FIGURE 3.2

Interval of dependency and domain of determinacy.

and they are significant in that they form the boundaries of the triangular region in Figure 3.2, which is known as the *domain of determinacy* of the interval $[x_0 - ct_0, x_0 + ct_0]$. The values of $u(x, t)$ are uniquely determined by d'Alembert's solution everywhere inside this triangular region by the values of $f(x)$ and $g(x)$ prescribed along the interval of dependency. For instance, if $f(x)$ and $g(x)$ are identically zero on the interval of dependency, then $u(x, t)$ is also zero everywhere inside the triangular region. Consequently, two solutions of (22) and (23) whose initial values coincide over the interval of dependency will coincide in value everywhere inside the domain of determinacy (but may differ outside this triangular region).

EXERCISES 3.2

In Problems 1–8, classify the PDE as hyperbolic, parabolic, or elliptic and find its general solution.

1. $u_{xx} - 3u_{xy} + 2u_{yy} = 0$
2. $4u_{xx} - 7u_{xy} + 3u_{yy} = 0$
3. $u_{xx} + a^2u_{yy} = 0 \quad (a \neq 0)$
4. $a^2u_{xx} + 2au_{xy} + u_{yy} = 0 \quad (a \neq 0)$
5. $4u_{tt} - 12u_{xt} + 9u_{xx} = 0$
6. $2u_{xt} + 3u_{tt} = 0$
7. $u_{xx} + 2u_{xy} + 5u_{yy} = 0$
8. $8u_{xx} - 2u_{xy} - 3u_{yy} = 0$

In Problems 9–14, determine for which values of x and y the PDE is hyperbolic, parabolic, and elliptic.

9. $u_{xx} - xu_{yy} = 0$
10. $u_{xx} - 2xu_{xy} + yu_{yy} = 0$
11. $u_{xx} + 2xu_{xy} + (1 - y^2)u_{yy} = 0$
12. $xu_{xx} + xu_{xy} + yu_{yy} = 0$
13. $(1 + y^2)u_{xx} + (1 + x^2)u_{yy} = 0$
14. $u_{xx} + xu_{xy} + yu_{yy} - xyu_y = 0$
15. Show that the transformation

$$r = x + y, \qquad s = -5x + y,$$

reduces the PDE

$$u_{xx} + 4u_{xy} - 5u_{yy} + 6u_x + 3u_y - 9u = 0$$

to the *canonical form*

$$u_{rs} = \frac{1}{4}(u_r - 3u_s - u).$$

16. Show that the transformation

$$r = 3x + y, \qquad s = y.$$

reduces the PDE

$$u_{xx} - 6u_{xy} + 9u_{yy} + 2u_x + 3u_y - u = 0$$

to the *canonical form*

$$u_{ss} = -u_r - \frac{1}{3}u_s + \frac{1}{9}u.$$

17. Show that the transformation

$$r = -x + y, \qquad s = 2x,$$

reduces the PDE

$$u_{xx} + 2u_{xy} + 5u_{yy} + u_x - 2u_y - 3u = 0$$

to the *canonical form*

$$u_{rr} + u_{ss} = \frac{1}{4}(u_r - 2u_s + 3u).$$

18. If u depends only upon the radial distance r from the origin and upon time t, the wave equation can be written in the form

$$\frac{\partial^2}{\partial r^2}(ru) = c^{-2}\frac{\partial^2}{\partial t^2}(ru).$$

Show that the general solution of this PDE can be expressed as

$$u(r, t) = \frac{1}{r}F(r + ct) + \frac{1}{r}G(r - ct).$$

19. A uniform infinite string is given the initial displacement

$$u(x, 0) = f(x) = \frac{1}{1 + 2x^2}$$

and released from rest, i.e., $g(x) = 0$.
(a) Determine its subsequent displacements.
(b) Plot the displacement curves corresponding to $ct = 0, 1/2, 1$.

20. A uniform infinite string is given the initial displacement

$$u(x, 0) = f(x) = \begin{cases} 1 - |x|, & |x| < 1 \\ 0, & |x| \geq 1 \end{cases}$$

and released from rest. Determine its subsequent displacement.

★21. A uniform string occupying the domain $0 \leq x < \infty$ begins its motion with initial displacement $f(x)$ and initial velocity $g(x)$. Show that its motion can be found as the motion of the right-half of a doubly infinite string provided that the initial displacement $f(-x)$ and initial velocity $g(-x)$ for the negative extension of the string satisfy the relation

$$f(x) + f(-x) = -\frac{1}{c}\int_{-x}^{x} g(z)\,dz.$$

★22. If a semi-infinite string begins its motion with initial displacement and velocity given by

$$f(x) = \frac{\sin 2x}{c} \quad \text{and} \quad g(x) = 2,$$

and if the negative extension of the string is imagined to have the initial displacement $f(-x) = 0$, find the proper initial velocity for the negative portion of the string.
Hint: Use the result of Problem 21.

3.3 EQUATIONS OF MATHEMATICAL PHYSICS

The most frequently encountered PDEs in practice are members of the classical *equations of mathematical physics*. The majority of these can be obtained by suitably specializing the form

$$\nabla^2 u + \alpha u = \beta u_{tt} + \gamma u_t - F, \tag{28}$$

where α, β, and γ are certain specified physical constants and F is a specified function of position (and possibly time). The differential operator ∇^2 is called the *Laplacian operator*, and the quantity $\nabla^2 u$ is called simply the *Laplacian*.

The Laplacian is a measure of the difference between the value of u at a point and the average value of u in a small neighborhood of the point. Since this difference influences the further space–time evolution of the unknown function u in problems of diffusion processes, wave propagation, and potential theory, we find that the Laplacian is fundamental to most of the equations of mathematical physics. In rectangular coordinates the Laplacian takes the form

$$\nabla^2 u = u_{xx} + u_{yy} + u_{zz}. \tag{29}$$

It is often necessary to consider coordinate systems other than rectangular—the most advantageous in a particular problem generally being dictated by the shape of the region of interest. For such problems we must find expressions comparable to (29) for the Laplacian in these other coordinate systems.

Specific equations arising out of (28) include the following:

$$\nabla^2 u = a^{-2} u_t \qquad \text{(heat equation)}$$

$$\nabla^2 u = c^{-2} u_{tt} \qquad \text{(wave equation)}$$

$$\nabla^2 u = 0 \qquad \text{(potential equation)}$$

$$\nabla^2 u + k^2 u = 0 \qquad \text{(Helmholtz equation)}$$

$$\nabla^2 u = -F \qquad \text{(Poisson equation)}$$

These PDEs play an important role in many diverse areas of application. Of course, other PDEs occur in certain applications that are not specializations of (28), but we will not give them separate treatment. In fact, no effort will be made to present a general theory of PDEs. We limit our discussion to the basic equations of diffusion processes (parabolic equations), wave propagation (hyperbolic equations), and potential theory (elliptic equations). A detailed study of these equations, however, reveals much information about similar but more general PDEs.

3.3.1 SEPARATION OF VARIABLES

The general method of attack that is most useful in solving PDEs of the type (28) is the so-called *separation of variables technique,* or *Bernoulli product method.* This method should not be confused with a similarly named procedure for solving ODEs. For purposes of illustrating the basic idea used in this solution technique, we will apply it to the *one-dimensional wave equation*

$$u_{xx} = c^{-2} u_{tt}, \tag{30}$$

where c^2 is a constant.

We seek a solution of (30) that is a product of a function of x alone by a function of t alone, i.e., we assume solutions exist of the form

$$u(x, t) = X(x)W(t). \tag{31}$$

Because each factor depends on only one variable, we find

$$u_{xx} = X''(x)W(t), \qquad u_{tt} = X(x)W''(t).$$

(For notational ease, we will use primes throughout to indicate differentiation with respect to whatever variable is in parentheses.) The substitution of these expressions for u_{xx} and u_{tt} into (30) and subsequent division by the product $X(x)W(t)$ leads to

$$\frac{X''(x)}{X(x)} = \frac{W''(t)}{c^2 W(t)}. \tag{32}$$

In (32) we have "separated the variables" since the left-hand side contains only functions of x and the right-hand side only functions of t. Because the left-hand side is independent of t, the same is true of the right-hand side. Similarly, the right-hand side does not depend on x and thus the left-hand side cannot. If (32) is independent of x and t, both sides can be equated to a common constant, $-\lambda$; that is,

$$\frac{X''(x)}{X(x)} = -\lambda, \qquad \frac{W''(t)}{c^2 W(t)} = -\lambda.$$

The negative sign of the constant is not required but merely conventional.* These last two relations lead to separate ODEs for the unknown factors $X(x)$ and $W(t)$, which are

$$X''(x) + \lambda X(x) = 0, \tag{33}$$

$$W''(t) + \lambda c^2 W(t) = 0. \tag{34}$$

Both (33) and (34) can be readily solved by techniques discussed in earlier chapters. However, the solutions of these DEs will take on various forms depending upon the value of λ (which we will assume is real). If we set $\lambda = k^2 > 0$, there follows

$$X(x) = C_1 \cos kx + C_2 \sin kx \tag{35}$$

and

$$W(t) = C_3 \cos kct + C_4 \sin kct. \tag{36}$$

Hence, since $u(x, t) = X(x)W(t)$, we deduce that

$$u(x, t) = (C_1 \cos kx + C_2 \sin kx)(C_3 \cos kct + C_4 \sin kct), \tag{37}$$

where C_1, C_2, C_3, and C_4 are all arbitrary constants. Likewise, if $\lambda = -k^2 < 0$, then

$$u(x, t) = (C_1 \cosh kx + C_2 \sinh kx)(C_3 \cosh kct + C_4 \sinh kct), \tag{38}$$

and finally if $\lambda = 0$, we have

$$u(x, t) = (C_1 + C_2 x)(C_3 + C_4 t). \tag{39}$$

Direct verification that (37), (38), and (39) all satisfy the wave equation (30) is a simple matter and left to the reader.

---

* We are not losing any generality by writing the separation constant as $-\lambda$, since any real constant written this way can still be positive, negative, or zero.

Although we have produced several solution forms of (30), viz., (37)–(39), the solutions that we seek in practice must also satisfy certain prescribed boundary and/or initial conditions that identify the unspecified constants. The general procedure for finding such solutions involves the concept of a *Fourier series,* which we discuss in the following chapter. Therefore, we will delay any further discussion of the method of separation of variables until we are familiar with Fourier series.

Finally, we should mention that the technique of separating the variables is not suitable for all PDEs. In fact, it works for only a very few types of equations. For example, if the PDE has variable coefficients, it is often not possible to separate the variables. In some cases we find the PDE may be separable in one coordinate system but not in another, and the coordinate system is usually dictated by the shape of the boundary of the region in which we are seeking a solution. And then, even if the PDE itself is separable in a particular coordinate system, it is of no avail unless the prescribed boundary conditions are homogeneous and specified along the coordinate curves, such as $x =$ constant and $y =$ constant in rectangular coordinates. In spite of all the shortcomings listed above, the separation of variables method is still one of the most powerful and elegant techniques available. And although the field of application of this technique is narrow, some of the most important PDEs that arise in practice fall into this field. In Chapters 6–8 we will use separation of variables extensively in solving problems of heat conduction, wave propagation, and potential theory.

EXAMPLE 3

Use separation of variables to solve*

$$u_x + 2u_y = 0, \qquad u(0, y) = 3e^{-2y}.$$

SOLUTION By writing $u(x, y) = X(x)Y(y)$ and substituting this product form into the PDE, we obtain

$$X'(x)Y(y) + 2X(x)Y'(y) = 0,$$

which can also be expressed in the form

$$\frac{X'(x)}{2X(x)} = -\frac{Y'(y)}{Y(y)}.$$

Here we have separated the variables, and by equating each side of the equation to the constant λ, we get the ODEs

$$X'(x) - 2\lambda X(x) = 0,$$

$$Y'(y) + \lambda Y(y) = 0.$$

*This is an example of a boundary value problem that can be solved by separation of variables without the need for Fourier series.

These equations have solutions given, respectively, by

$$X(x) = C_1 e^{2\lambda x} \quad \text{and} \quad Y(y) = C_2 e^{-\lambda y},$$

and thus

$$u(x, y) = X(x)Y(y) = Ae^{\lambda(2x-y)},$$

where we write $A = C_1C_2$.

If we now use the auxiliary condition, we find

$$u(0, y) = 3e^{-2y} = Ae^{-\lambda y},$$

and hence deduce that $A = 3$ and $\lambda = 2$. Our solution is therefore given by

$$u(x, y) = 3e^{2(2x-y)}.$$

□ 3.3.2 SOLUTIONS BY ODE METHODS

Most PDEs must be solved by a general solution technique, such as separation of variables or a transform method, or in some cases by a numerical procedure. Occasionally, however, the PDE of interest is simple enough that its form may suggest a method of solution. This was the case in solving Equation (1), which is an example of a class of PDEs that can be solved by methods of ODEs, using one independent variable at a time.

EXAMPLE 4

Find a solution of the boundary value problem

$$u_{xy} - u_y = 5, \qquad u_y(0, y) = 3y^2, \qquad u(x, 0) = 0.$$

SOLUTION By writing the PDE as

$$\frac{\partial}{\partial y}(u_x - u) = 5,$$

we can hold x fixed and integrate with respect to y to obtain

$$u_x - u = 5y + F(x),$$

where F is an arbitrary, differentiable function of x. For fixed y, this last equation is a first-order linear DE, whose general solution is readily found to be

$$\begin{aligned} u(x, y) &= e^x G(y) - 5y + e^x \int e^{-x} F(x)\, dx \\ &= e^x G(y) - 5y + H(x), \end{aligned}$$

where we define $H(x) = e^x \int e^{-x} F(x)\, dx$.

Now imposing the first auxiliary condition, we have

$$u_y(0, y) = G'(y) - 5 = 3y^2,$$

from which we deduce

$$G(y) = y^3 + 5y + C,$$

where C is a constant. Hence,

$$u(x, y) = e^x(y^3 + 5y + C) - 5y + H(x),$$

and the second auxiliary condition leads to

$$u(x, 0) = Ce^x + H(x) = 0,$$

or $H(x) = -Ce^x$. Our solution now takes the form

$$u(x, y) = e^x(y^3 + 5y) - 5y.$$

EXERCISES 3.3

1. By setting $C_1 = C_2 = C_3 = C_4 = 1$, show that the solution (37) can be expressed in the general form

$$u(x, t) = F(x + ct) + G(x - ct).$$

In Problems 2 and 3, find all solution forms by using the method of separation of variables.

2. $u_{xx} = u_t$

3. $u_{xx} + u_{yy} = 0$

4. Show that

$$a^2u_{xx} = u_{tt} + 2bu_t, \qquad (a, b \text{ constants})$$

has solutions of the form

$$u(x, t) = (C_1 \cos kx + C_2 \sin kx)W(t),$$

and determine the possible forms for $W(t)$.

5. Show that

$$u_{rr} + \frac{2}{r}u_r = u_{tt}$$

has solutions of the form

$$u(r, t) = \frac{v(r)}{r}\cos nt, \qquad n = 0, 1, 2, \ldots.$$

Find a DE that $v(r)$ must satisfy and find its general solution.

★6. Given the PDE

$$u_{xx} + u_{yy} + u_{zz} = 0,$$

assume a product solution form $u(x, y, z) = V(x, y)Z(z)$, and
(a) find the DEs satisfied by $V(x, y)$ and $Z(z)$.
(b) Let $V(x, y) = X(x)Y(y)$ and separate the variables once again using a new separation constant. What DEs do $X(x)$ and $Y(y)$ satisfy?

In Problems 7–12, find solutions by the method of separation of variables.

7. $u_x = u_y, \quad u(0, y) = 2e^{3y}$

8. $u_x + u = u_y, \quad u(x, 0) = 4e^{-3x}$

9. $u_{xx} = u_{tt}, \quad u(0, t) = 0, \quad u(\pi, t) = 0, \quad u(x, 0) = \sin 3x, \quad u_t(x, 0) = 0$

10. $x^2 u_{xy} + 3y^2 u = 0, \quad u(x, 0) = e^{1/x}$

11. $u_{xx} = \dfrac{2}{k} u_t + u, \quad u(1, t) = 0, \quad u_x(0, t) = -be^{-kt}, \quad (k, b \text{ constants})$

12. $u_{xx} + u_{tt} + 6u_x + 9u = 0, \quad u(0, t) = 0, \quad u_{xt}(0, t) = 6 \cos 3t$
Hint: The solution is periodic in t.

In Problems 13–17, use ODE methods to find the solution satisfying the prescribed auxiliary conditions.

13. $u_x = \sin y, \quad u(0, y) = 0$

14. $u_{yy} = x^2 \cos y, \quad u(x, 0) = 0, \quad u(x, \pi/2) = 0$

15. $u_{xy} = 4xy + e^x, \quad u_y(0, y) = y, \quad u(x, 0) = 2$

16. $u_{xy} + 4u_x = 2x, \quad u(0, y) = 1, \quad u_x(x, 0) = 0$

17. $u_{xy} = u_x + 2, \quad u(0, y) = 0, \quad u_x(x, 0) = x^2$

CHAPTER 4

FOURIER SERIES AND FOURIER INTEGRALS

One of the most important problems in mathematical analysis is the determination of various representations of a given function. A particular representation of a function often enables us to deduce properties of that function that are not as readily ascertained by a different representation. Power series are especially useful in this regard, but here we wish to extend our notion of infinite series to include those involving sines and cosines, called *Fourier series*.

In the first three sections of this chapter we consider Fourier series of *periodic functions* having fundamental period 2π. Euler's formulas for the *Fourier coefficients* are formally derived and the convergence of the series is discussed. When the function involved is either *even* or *odd,* the full Fourier trigonometric series reduces to either a *cosine series* for an even function or a *sine series* for an odd function. The rather lengthy proof of the (pointwise) *convergence theorem* is presented in Section 4.4.

The theory is extended in Section 4.5 to include functions with more *general periods* and to certain *nonperiodic functions* defined over *finite intervals*. In Section 4.6 we introduce the notion of a *generalized Fourier series,* which includes the Fourier trigonometric series as a special case. We then use these series to solve certain types of nonhomogeneous boundary value problems and also to construct what is called a *bilinear formula* of the Green's function.

In Section 4.7 the *Fourier integral theorem* is formally derived by considering the limit of a Fourier series representation as the period of the represented function tends to infinity. The Fourier integral theorem is the basis for developing the *Fourier integral representation* of a nonperiodic function. The general theory concerning Fourier integral representations closely parallels that of Fourier series.

4.1 INTRODUCTION

The concept of an infinite series dates back as far as the ancient Greeks such as Archimedes (287–212 B.C.), who summed a geometric series in order to compute the area under a parabolic arc. In the eighteenth century, power series expansions for functions like e^x, $\sin x$, and $\arctan x$ were first published by the Scottish mathematician C. Maclaurin (1698–1746), and British mathematician B. Taylor (1685–1731) generalized this work by providing power series expansions about some point other than $x = 0$.

By the middle of the eighteenth century it became important to study the possibility of representing a given function by infinite series other than power series. D. Bernoulli showed that the mathematical conditions imposed by physical considerations in solving the vibrating-string problem were formally satisfied by functions represented as infinite series involving sinusoidal functions. In the early 1800s, the French physicist J. Fourier* came across similar representations and announced in his work on heat conduction that an "arbitrary function" could be expanded in a series of sinusoidal functions. Some of Fourier's work lacked rigor, but nevertheless he provided the first real impetus to the subject now bearing his name. Fourier analysis, or harmonic analysis as it is now often called, has turned out to be tremendously important in virtually all areas of pure and applied mathematics and the physical sciences. It is one of the best examples of a mathematics tool that was invented to solve a specific problem and has turned out to be an important tool for solving many other problems.

4.2 FOURIER SERIES OF PERIODIC FUNCTIONS

A function f is called *periodic* if there exists a constant $T > 0$ for which $f(x + T) = f(x)$ for all x. The smallest value of T for which the property holds is called the *fundamental period*, or simply, the *period* (see Figure 4.1). It follows that if $f(x + T) = f(x)$, then also

$$f(x \pm T) = f(x \pm 2T) = f(x \pm 3T) = \dots = f(x).$$

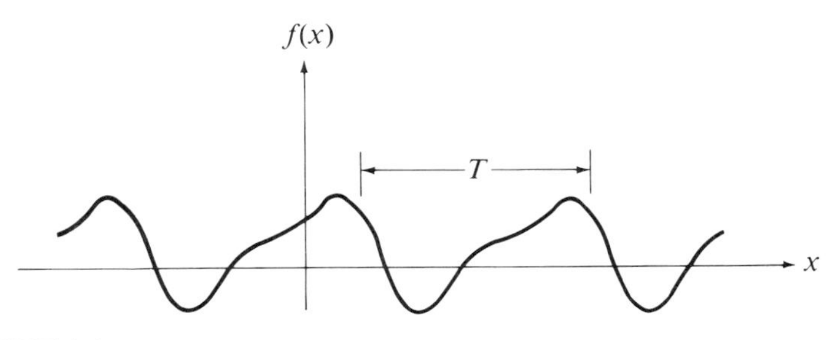

FIGURE 4.1

A periodic function.

*JEAN BAPTISTE JOSEPH FOURIER (1768–1830) is known mainly for his work on the representation of functions by trigonometric series in his studies on the theory of heat conduction. His basic papers, presented to the Academy of Sciences in Paris in 1807 and 1811, were criticized by the referees (most strongly by Lagrange) for a lack of rigor and consequently were not then published. However, when publishing the classic *Théorie analytique de la Chaleur* in 1822, he also incorporated his earlier work almost without change. He died in Paris, leaving his final publication to be finished by Navier.

Periodic functions appear in a wide variety of physical problems, such as those concerning vibrating springs and membranes, planetary motion, a swinging pendulum, and musical sounds, to name a few. The modern theory of light is based on wave mechanics, which features certain periodic characteristics. Many of these phenomena involve periodic functions of a complicated nature, so in order to better understand such functions it is desirable to express them in terms of a set of simple periodic functions. Doing so has the effect of decomposing a periodic phenomenon into its simple harmonic components.

Perhaps the simplest set of periodic functions is given by

$$1; \quad \cos x, \ \sin x; \quad \cos 2x, \ \sin 2x; \quad \cos 3x, \ \sin 3x; \ \ldots;$$

which all have period 2π in common.* If f is any other periodic function with period 2π, it may seem reasonable to look for a representation of f in terms of the above simple sinusoidal functions. The series arising in this connection will be of the form

$$f(x) = A_0 + \sum_{n=1}^{\infty} (a_n \cos nx + b_n \sin nx), \tag{1}$$

where $A_0, a_1, a_2, \ldots, b_1, b_2, \ldots,$ are constants. Such a series is called a *Fourier (trigonometric) series.*

REMARK: *Because the validity of (1) has not been established, it is customary in many textbooks to replace the equal sign* $=$ *with the symbol* $\sim$ *until conditions of convergence have been discussed. Nonetheless, we will continue to use the equal sign here with the understanding that its appearance does not imply any type of convergence.*

To obtain expressions for evaluating the constants in (1) it is easiest to assume that (1) is a convergent series and proceed in a purely formal fashion. That is the approach we use here. Our method depends upon the evaluation of certain definite integrals involving the sines and cosines appearing in (1). First, if n and k are any nonzero integers, it can be shown that

$$\int_{-\pi}^{\pi} \cos nx \, dx = \int_{-\pi}^{\pi} \sin nx \, dx = \int_{-\pi}^{\pi} \sin nx \cos kx \, dx = 0, \tag{2}$$

and also

$$\int_{-\pi}^{\pi} \cos nx \cos kx \, dx = \int_{-\pi}^{\pi} \sin nx \sin kx \, dx = \begin{cases} 0, & k \neq n \\ \pi, & k = n. \end{cases} \tag{3}$$

These integral formulas can be derived directly through simple integration techniques and are left to the Exercises (see Problem 2 in Exercises 4.2).

The integral relations (2) suggest that integrating (1) from $-\pi$ to π will

*Any constant is a periodic function. See Problem 1 in Exercises 4.2.

greatly simplify the right-hand side. To do this, we must tacitly assume that termwise integrtation is justified.* Proceeding in this fashion, we obtain

$$\int_{-\pi}^{\pi} f(x)\,dx = A_0\int_{-\pi}^{\pi} dx + \sum_{n=1}^{\infty}\left(a_n\int_{-\pi}^{\pi} \cos nx\,dx + b_n\int_{-\pi}^{\pi} \sin nx\,dx\right).$$

In view of the integral relations (2), we see that each term of the series integrates to zero, and from the remaining nonzero integrals, we find

$$A_0 = \frac{1}{2\pi}\int_{-\pi}^{\pi} f(x)\,dx. \tag{4}$$

This identifies the constant A_0 as the *average value* of $f(x)$ over the interval $[-\pi, \pi]$. Next, we multiply (1) by $\cos kx$ and integrate termwise to obtain

$$\int_{-\pi}^{\pi} f(x)\cos kx\,dx = A_0\int_{-\pi}^{\pi} \cos kx\,dx + \sum_{n=1}^{\infty}\left(a_n\int_{-\pi}^{\pi} \cos nx \cos kx\,dx\right.$$

$$\left. + b_n\int_{-\pi}^{\pi} \sin nx \cos kx\,dx\right).$$

(In the printed equations above, the integrals $\int_{-\pi}^{\pi} \cos nx\,dx$, $\int_{-\pi}^{\pi} \sin nx\,dx$, $\int_{-\pi}^{\pi} \cos kx\,dx$ and $\int_{-\pi}^{\pi} \sin nx \cos kx\,dx$ are each struck through with an arrow marked 0, and $\int_{-\pi}^{\pi} \cos nx \cos kx\,dx$ with an arrow marked $0\ (n \neq k)$.)

Because of (2) and (3), all terms integrate to zero except for the coefficient of a_n corresponding to $n = k$, and here we get

$$\int_{-\pi}^{\pi} f(x)\cos kx\,dx = a_k\int_{-\pi}^{\pi} \cos^2 kx\,dx = \pi a_k,$$

or

$$a_k = \frac{1}{\pi}\int_{-\pi}^{\pi} f(x)\cos kx\,dx, \qquad k = 1, 2, 3, \ldots. \tag{5}$$

In the same fashion, if we multiply the series (1) by $\sin kx$ and integrate the result termwise, we generate the final formula

$$b_k = \frac{1}{\pi}\int_{-\pi}^{\pi} f(x)\sin kx\,dx, \qquad k = 1, 2, 3, \ldots. \tag{6}$$

Constants defined by (4)–(6) are known as *Fourier coefficients* (also called *Euler's formulas*). It is customary in the literature to set $A_0 = a_0/2$ so that we can write the above formulas more compactly as

$$a_n = \frac{1}{\pi}\int_{-\pi}^{\pi} f(x)\cos nx\,dx, \qquad n = 0, 1, 2, \ldots \tag{7}$$

(now changing the index back to n) and

---

*The assumption of termwise integration does not necessarily follow for all convergent series. In the case of Fourier series, however, we can always justify such a procedure (see Theorem 4.5), even in some cases when the series (1) diverges.

$$b_n = \frac{1}{\pi}\int_{-\pi}^{\pi} f(x)\sin nx\,dx, \qquad n = 1, 2, 3, \ldots. \tag{8}$$

Thus, (1) now takes the form

$$f(x) = \frac{1}{2}a_0 + \sum_{n=1}^{\infty}(a_n \cos nx + b_n \sin nx). \tag{9}$$

Writing the constant term as $a_0/2$ does not aid in its computation, only in the compactness of the formula (7). In general we find that a_0 must be evaluated separately from the rest of the a's.

REMARK: *Note that formulas (7) and (8) for the Fourier coefficients depend only upon values of f in the interval $[-\pi, \pi]$. If (9) is indeed a convergent series in this interval, it follows from the periodicity of the trigonometric functions that the series converges for all x, and thus it defines the periodic function f for all values of x.*

EXAMPLE 1

Find the Fourier series of the function $f(x) = |x|$, $-\pi \leq x < \pi$, $f(x + 2\pi) = f(x)$. The graph of this function is the triangular wave shown in Figure 4.2.

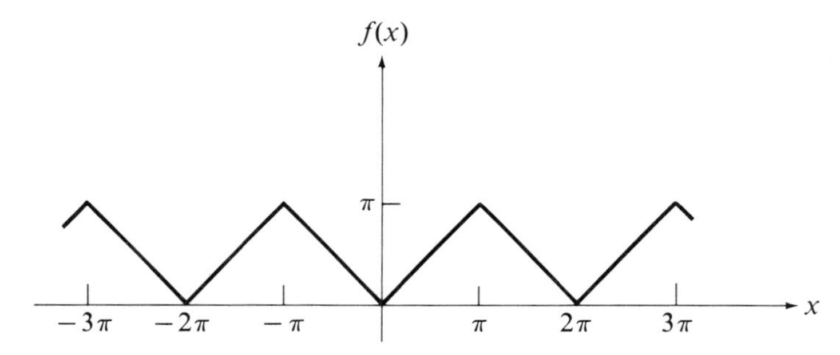

FIGURE 4.2

Triangular wave.

SOLUTION The substitution of $f(x) = |x|$ into (7) and (8) leads to

$$a_0 = \frac{1}{\pi}\int_{-\pi}^{\pi}|x|\,dx = -\frac{1}{\pi}\int_{-\pi}^{0} x\,dx + \frac{1}{\pi}\int_{0}^{\pi} x\,dx = \pi,$$

$$a_n = -\frac{1}{\pi}\int_{-\pi}^{0} x\cos nx\,dx + \frac{1}{\pi}\int_{0}^{\pi} x\cos nx\,dx = \frac{2}{\pi n^2}[\cos n\pi - 1],$$

$n = 1, 2, 3, \ldots$, and

$$b_n = -\frac{1}{\pi}\int_{-\pi}^{0} x\sin nx\,dx + \frac{1}{\pi}\int_{0}^{\pi} x\sin nx\,dx = 0, \qquad n = 1, 2, 3, \ldots.$$

Since $\cos n\pi = (-1)^n$, $n = 1, 2, 3, \ldots$, we can write

$$a_n = \frac{2}{\pi n^2}[(-1)^n - 1] = \begin{cases} -\dfrac{4}{\pi n^2}, & n = 1, 3, 5, \ldots, \\ 0, & n = 2, 4, 6, \ldots \end{cases}$$

and thus the Fourier series becomes

$$f(x) = \frac{\pi}{2} - \frac{4}{\pi}\sum_{\substack{n=1 \\ (\text{odd})}}^{\infty} \frac{\cos nx}{n^2}.$$

By replacing the index n with new index $(2n - 1)$, we can also express this Fourier series in the form

$$f(x) = \frac{\pi}{2} - \frac{4}{\pi}\sum_{n=1}^{\infty} \frac{\cos(2n-1)x}{(2n-1)^2}.$$

EXAMPLE 2

Find the Fourier series of the periodic function f that is defined by

$$f(x) = \begin{cases} 0, & -\pi \leqslant x < 0 \\ x, & 0 \leqslant x < \pi, \end{cases} \qquad f(x + 2\pi) = f(x).$$

(See Figure 4.3.)

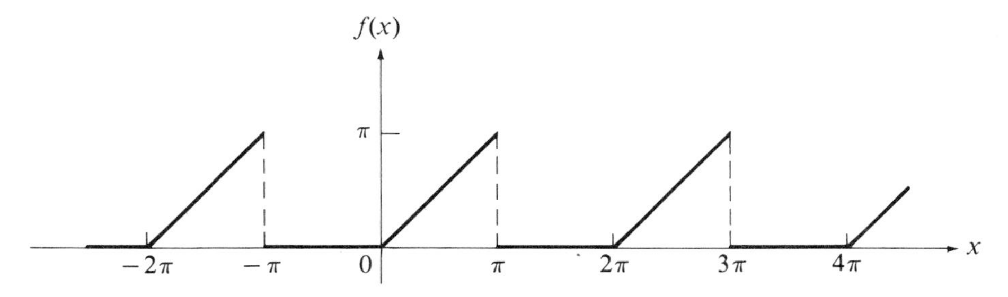

FIGURE 4.3

SOLUTION The Fourier coefficients are given by

$$a_0 = \frac{1}{\pi}\int_{-\pi}^{\pi} f(x)\,dx = \frac{1}{\pi}\int_0^{\pi} x\,dx = \frac{\pi}{2},$$

$$a_n = \frac{1}{\pi}\int_0^{\pi} x\cos nx\,dx = \begin{cases} -\dfrac{2}{\pi n^2}, & n = 1, 3, 5, \ldots, \\ 0, & n = 2, 4, 6, \ldots \end{cases}$$

and

$$b_n = \frac{1}{\pi}\int_0^{\pi} x\sin nx\,dx = \frac{(-1)^{n-1}}{n}, \qquad n = 1, 2, 3, \ldots.$$

Substituting these results into the series (9), we obtain

$$f(x) = \frac{\pi}{4} - \frac{2}{\pi}\left(\cos x + \frac{\cos 3x}{3^2} + \frac{\cos 5x}{5^2} + \cdots\right)$$
$$+ \left(\sin x - \frac{\sin 2x}{2} + \frac{\sin 3x}{3} - \cdots\right),$$

or

$$f(x) = \frac{\pi}{4} - \sum_{n=1}^{\infty}\left[\frac{2}{\pi}\frac{\cos(2n-1)x}{(2n-1)^2} + \frac{(-1)^n}{n}\sin nx\right].$$

Our treatment of Fourier series thus far has been purely formal. For example, our derivation of the Fourier coefficients a_n and b_n has involved some assumptions regarding termwise integration of an infinite series. However, if the function f is integrable on the interval $[-\pi, \pi]$, then the Fourier coefficients can be defined outright by (7) and (8), and the resulting series (9) can be studied on its own merits in regards to its convergence and validity as a series representation of the function f that generates the series. We henceforth wish to adopt this point of view in our further discussion of Fourier series.

4.2.1 COSINE AND SINE SERIES

If the function f we wish to express in a Fourier series is either even or odd, some simplification takes place in the calculation of the Fourier coefficients. Suppose f is defined either on the entire x-axis or some finite interval so that $x = 0$ is the midpoint of the interval. We say that f is an *even function* if it is true that

$$f(-x) = f(x)$$

for every x in the interval. If, on the other hand, it is true that

$$f(-x) = -f(x)$$

for every x in the interval, we say that f is an *odd function*. See Figure 4.4 for illustrations of an even and an odd function.

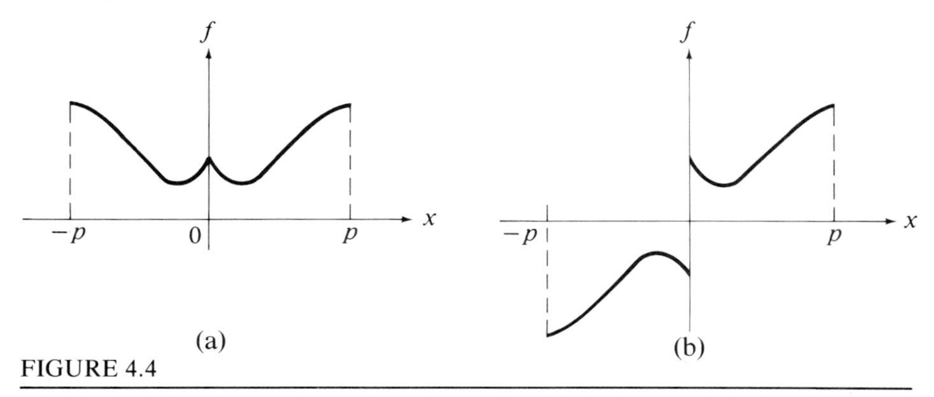

FIGURE 4.4

(a) Even function. (b) Odd function.

EXAMPLE 3

Determine whether the following functions are even, odd, or neither: (a) $\cos nx$, (b) $\sin nx$, (c) 1, (d) x^4, and (e) $1 - x$.

SOLUTION Applying the definition, we find that replacing x by $-x$ yields

$$\begin{aligned}
&\text{(a)} \quad \cos(-nx) = \cos nx \\
&\text{(b)} \quad \sin(-nx) = -\sin nx \\
&\text{(c)} \quad 1 = 1 \\
&\text{(d)} \quad (-x)^4 = x^4 \\
&\text{(e)} \quad 1 - (-x) = 1 + x,
\end{aligned}$$

and thus conclude that (a), (c), and (d) are *even* functions, (b) is an *odd* function, and (e) is *neither*.

Provided the integrals exist, it follows from the geometric interpretation of integrals that if f is an even function, then

$$\int_{-p}^{p} f(x)\,dx = 2\int_{0}^{p} f(x)\,dx, \tag{10}$$

and if f is an odd function,

$$\int_{-p}^{p} f(x)\,dx = 0, \tag{11}$$

for any p (see Problems 33 and 34 in Exercises 4.2).

If f is an even function, the product $f(x)\cos nx$ is an even function and the product $f(x)\sin nx$ is an odd function. (See Problem 35 in Exercises 4.2.) Hence, from (10) and (11) we see that the Fourier coefficients of f satisfy

$$a_n = \frac{1}{\pi}\int_{-\pi}^{\pi} f(x)\cos nx\,dx = \frac{2}{\pi}\int_{0}^{\pi} f(x)\cos nx\,dx, \qquad n = 0, 1, 2, \ldots \tag{12}$$

and

$$b_n = \frac{1}{\pi}\int_{-\pi}^{\pi} f(x)\sin nx\,dx = 0, \qquad n = 1, 2, 3, \ldots. \tag{13}$$

The Fourier series of an even function therefore reduces to one containing only cosine terms,

$$f(x) = \frac{1}{2}a_0 + \sum_{n=1}^{\infty} a_n \cos nx, \tag{14}$$

called a *cosine series*. The periodic function in Example 1 is an example of an even function, and its Fourier series is indeed a cosine series.

On the other hand, if f is an odd function, its Fourier series reduces to the *sine series*

$$f(x) = \sum_{n=1}^{\infty} b_n \sin nx, \tag{15}$$

where $a_n = 0$, $n = 0, 1, 2, \ldots$, and

$$b_n = \frac{2}{\pi} \int_0^{\pi} f(x) \sin nx \, dx, \qquad n = 1, 2, 3, \ldots. \tag{16}$$

It is probably fair to say that most functions are neither even nor odd. However, in generating Fourier series, as well as when working in other areas of analysis, it can be helpful to recognize whether a given function is either even or odd since certain computations are simplified. For example, in computing Fourier coefficients for even and odd functions we find that either the a's or b's are zero for all n.

EXERCISES 4.2

1. Verify that the constant function $f(x) = 1$ is periodic. What is its period?
2. Verify the integral relations (2) and (3).
Hint: It may be helpful to recall the trigonometric identities

$$\sin A \sin B = \frac{1}{2}[\cos(A - B) - \cos(A + B)],$$

$$\cos A \cos B = \frac{1}{2}[\cos(A - B) + \cos(A + B)],$$

$$\sin A \cos B = \frac{1}{2}[\sin(A - B) + \sin(A + B)].$$

3. Sketch two periods or more of the following periodic functions:
(a) $f(x) = x, \quad -\pi \leqslant x < \pi, \quad f(x + 2\pi) = f(x)$
(b) $f(x) = x^2, \quad -\pi \leqslant x < \pi, \quad f(x + 2\pi) = f(x)$
(c) $f(x) = e^x, \quad -\pi \leqslant x < \pi, \quad f(x + 2\pi) = f(x)$
4. If $f(x)$ and $g(x)$ are periodic functions with common period T, show that their sum $[f(x) + g(x)]$ and product $f(x)g(x)$ are also periodic functions with the same period.
5. If $f(x) = x^2$ for $-\pi \leqslant x < \pi$, and $f(x + 2\pi) = f(x)$, find a formula for $f(x)$ in the interval $\pi \leqslant x < 3\pi$.

In Problems 6–15, find the Fourier series of the periodic function f defined over one period by the following formulas. Graph one period of the function.

6. $f(x) = \begin{cases} 0, & -\pi \leqslant x < 0 \\ -1, & 0 \leqslant x < \pi \end{cases}$
7. $f(x) = \begin{cases} -1, & -\pi \leqslant x < 0 \\ 2, & 0 \leqslant x < \pi \end{cases}$
8. $f(x) = \sin x - 4 \sin 3x + 7$
9. $f(x) = 2 \sin^2 x$
10. $f(x) = \begin{cases} 0, & -\pi \leqslant x < 0 \\ \cos x, & 0 \leqslant x < \pi \end{cases}$
11. $f(x) = e^x, \quad -\pi \leqslant x < \pi$

12. $f(x) = \begin{cases} x, & -\pi \leqslant x < 0 \\ 0, & 0 \leqslant x < \pi \end{cases}$

13. $f(x) = \begin{cases} 0, & -\pi \leqslant x < 0 \\ x^2, & 0 \leqslant x < \pi \end{cases}$

14. $f(x) = x(1 - x), \quad -\pi \leqslant x < \pi$

15. $f(x) = \begin{cases} x, & -\pi \leqslant x < 0 \\ \pi - x, & 0 \leqslant x < \pi \end{cases}$

16. A sinusoidal voltage $E \sin t$ is passed through a *half-wave rectifier,* which clips the negative portion of the wave. Find the Fourier series of the resulting waveform

$$f(t) = \begin{cases} 0, & -\pi \leqslant t < 0 \\ E \sin t, & 0 \leqslant t < \pi. \end{cases}$$

In Problems 17–26, identify the given function as *even, odd,* or *neither.*

17. $f(x) = x$

18. $f(x) = \sin^5 x$

19. $f(x) = \tan x$

20. $f(x) = x^3 \sin 2x$

21. $f(x) = \cos(1 - x)$

22. $f(x) = e^{-x^2}$

23. $f(x) = (5x - x^3)^2$

24. $f(x) = x^5 - 7x + 3$

25. $f(x) = \log |\cos x|$

26. $f(x) = \log |\sin x|$

In Problems 27–32, find a cosine or sine series of the periodic function f defined over one period by the following formulas. Graph one period of the function.

27. $f(x) = x, \quad -\pi \leqslant x < \pi$

28. $f(x) = x^2, \quad -\pi \leqslant x < \pi$

29. $f(x) = \begin{cases} x + \pi, & -\pi \leqslant x < 0 \\ x - \pi, & 0 \leqslant x < \pi \end{cases}$

30. $f(x) = \begin{cases} \pi/2, & -\pi \leqslant x < -\pi/2 \\ |x|, & -\pi/2 \leqslant x < \pi/2 \\ \pi/2, & \pi/2 \leqslant x < \pi \end{cases}$

31. $f(x) = e^{-|x|}, \quad -\pi \leqslant x < \pi$

32. $f(x) = \begin{cases} -e^{x}, & -\pi \leqslant x < 0 \\ e^{-x}, & 0 \leqslant x < \pi \end{cases}$

33. If f is an even function, prove that (for any $p > 0$)

$$\int_{-p}^{p} f(x)\, dx = 2\int_{0}^{p} f(x)\, dx.$$

34. If f is an odd function, prove that (for any $p > 0$)

$$\int_{-p}^{p} f(x)\, dx = 0.$$

35. Prove that
   (a) the product of two odd functions is an even function.
   (b) the product of two even functions is an even function.
   (c) the product of an even and an odd function is an odd function.

36. Any function f can be expressed as the sum of an even function and an odd function. Prove this statement by verifying that the following identity is the sum of an even and an odd function:

$$f(x) = \frac{1}{2}\underbrace{[f(x) + f(-x)]}_{\text{even}} + \frac{1}{2}\underbrace{[f(x) - f(-x)]}_{\text{odd}}.$$

37. Using the results of Problem 36, write the following functions as the sum of an even and an odd function.
    (a) $f(x) = x^5 - 7x + 3$
    (b) $f(x) = e^x$
    (c) $f(x) = \dfrac{x}{x - 1}, \qquad x \neq 1$

★38. Prove that the derivative of an even function is odd, and that the derivative of an odd function is even.

★39. Let

$$F(x) = \int_0^x f(t)\, dt.$$

Show that F is even when f is odd, and that F is odd when f is even.

40. If a_n and b_n are the Fourier coefficients associated with the function $f(x)$, show that ka_n and kb_n are the Fourier coefficients associated with the function $kf(x)$, where k is constant.

41. If a_n and b_n are the Fourier coefficients associated with the function $f(x)$, and A_n and B_n are the Fourier coefficients associated with $g(x)$, show that $[f(x) + g(x)]$ has the Fourier coefficients $(a_n + A_n)$ and $(b_n + B_n)$.

★42. Observe that the sine series (15) always sums to zero when $x = 0$ or when x is a multiple of π. Does this mean that the function f must be zero at these points? If the answer is no, provide an example.

4.3 CONVERGENCE OF THE SERIES

Now that we have identified the constants a_n and b_n occurring in the Fourier series

$$f(x) = \frac{1}{2} a_0 + \sum_{n=1}^{\infty} (a_n \cos nx + b_n \sin nx), \tag{17}$$

it seems natural to inquire whether the series actually represents the function f. What we mean is—if a value of x is selected and each term of the series is evaluated for this value of x, will the sum of the series be equal to the function value $f(x)$? If so, we say the series (17) *converges pointwise* to $f(x)$ for that particular value of x. In order to establish pointwise convergence of the series, we need to obtain an expression for the partial sum

$$S_N(x) = \frac{1}{2} a_0 + \sum_{n=1}^{N} (a_n \cos nx + b_n \sin nx) \tag{18}$$

and then, for a fixed value of x, show that

$$\lim_{N \to \infty} S_N(x) = f(x). \tag{19}$$

The Fourier coefficients a_n and b_n appearing in (17) always exist provided that the functions $f(x) \cos nx$ and $f(x) \sin nx$ are integrable on $[-\pi, \pi]$. The mere existence of the Fourier coefficients, and hence the existence of the series (17), in no way by itself implies convergence of the series. Indeed, examples have been constructed to show that the Fourier series associated with a given function f may not converge to $f(x)$ or converge at all! To be sure that the series converges to the function that generates the series, it is essential to place certain restrictions (other than integrability) on the function f. From a practical point of view, such conditions should be broad enough to cover most situations of concern and still simple enough to be easily checked for the given function. Over the years, several sets of conditions have been put forth, each of which is a slight modification of the others.

DEFINITION 4.1 A function f is said to be *piecewise continuous* on the interval $[a, b]$ provided
(a) f is defined and continuous at all but a finite number of points on $[a, b]$, and
(b) the left-hand and right-hand limits exist at each point on the interval $[a, b]$.*

REMARK: *The left-hand and right-hand limits are defined, respectively, by*

$$\lim_{\epsilon \to 0^+} f(x-\epsilon) = f(x^-) \quad \text{and} \quad \lim_{\epsilon \to 0^+} f(x+\epsilon) = f(x^+).$$

Furthermore, when x is a point of continuity, $f(x^-) = f(x^+) = f(x)$.

A piecewise continuous function f need not be defined at every point in the interval of interest. In particular, it is often not defined at a point of discontinuity, and even when it is, the functional value assigned at this point really doesn't matter. Also, the interval of interest may be open or closed, or open at one end and closed at the other (see Figure 4.5).

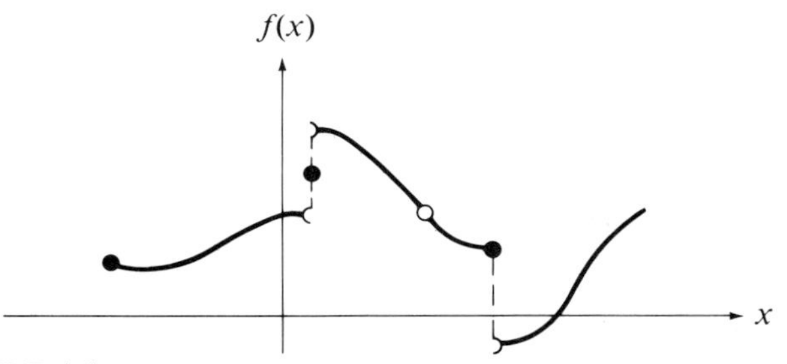

FIGURE 4.5

A piecewise continuous function.

The function

$$f(x) = \begin{cases} -1, & -\pi < x < 0 \\ 1, & 0 < x < \pi \end{cases}$$

is clearly not continuous on $[-\pi, \pi]$ because it is not defined at $x = 0$ and at $x = \pm\pi$. Since the jump discontinuity at $x = 0$ is finite, and

$$\lim_{x \to -\pi^+} f(x) = -1 \quad \text{and} \quad \lim_{x \to \pi^-} f(x) = 1,$$

*Only one of these limits is required at the endpoints a and b.

we deduce from Definition 4.1 that the function is piecewise continuous. On the other hand, the function

$$f(x) = \frac{1}{x}$$

cannot be piecewise continuous on any interval containing the point $x = 0$ since it is not bounded there, i.e., the left-hand and right-hand limits do not exist at $x = 0$.

The following properties of piecewise continuous functions are of particular importance to us:

1. Every continuous function is also piecewise continuous.
2. If f is piecewise continuous on $[-\pi, \pi]$ and $f(x + 2\pi) = f(x)$, then f is piecewise continuous for all x.
3. If f and g are piecewise continuous functions on $[-\pi, \pi]$, so are their sum $f + g$ and product fg.
4. If f is piecewise continuous on $[-\pi, \pi]$, then

$$\int_{-\pi}^{\pi} f(x)\, dx$$

exists and is independent of the values f assumes at points of discontinuity. Specifically, if f and g are identical on the interval $[-\pi, \pi]$, except for points of (finite) discontinuity, then

$$\int_{-\pi}^{\pi} f(x)\, dx = \int_{-\pi}^{\pi} g(x)\, dx.$$

DEFINITION 4.2 If f and f' are both continuous on $[a, b]$, we say that f is *smooth* on this interval. We say that f is *piecewise smooth* on the interval $[a, b]$ if f and/or f' are merely piecewise continuous on $[a, b]$.

EXAMPLE 4

Classify the following functions in Figure 4.6 as smooth, piecewise smooth, or neither on $[-\pi, \pi]$:

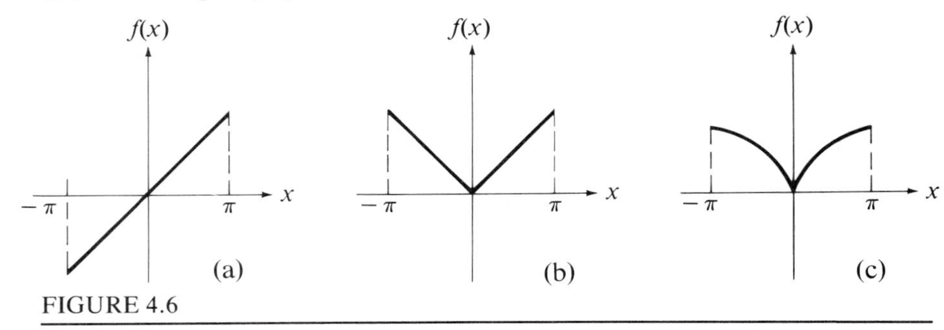

FIGURE 4.6

(a) $f(x) = x$. (b) $f(x) = |x|$. (c) $f(x) = x^{1/2}$.

SOLUTION In (a), the function $f(x) = x$ and its derivative $f'(x) = 1$ are both continuous, and thus f is *smooth*. The function in (b) is also continuous, but because the derivative is piecewise continuous, i.e.,

$$f'(x) = \begin{cases} -1, & -\pi < x < 0 \\ 1, & 0 < x < \pi, \end{cases}$$

the function $f(x) = |x|$ is not smooth but only *piecewise smooth*. In (c) the function is once again continuous, but

$$f'(x) = \begin{cases} -\dfrac{1}{2}(-x)^{-1/2}, & -\pi < x < 0 \\ \dfrac{1}{2}x^{-1/2}, & 0 < x < \pi, \end{cases}$$

so that $|f'(x)| \to \infty$ as $x \to 0$. Hence, the function is *neither* smooth nor piecewise smooth.

We are now prepared to state our main theorem, the rather lengthy proof of which is given in Section 4.4.

THEOREM 4.1

Pointwise Convergence. If $f(x + 2\pi) = f(x)$ for all x, and f is piecewise smooth on $[-\pi, \pi]$, then the Fourier series

$$f(x) = \frac{1}{2}a_0 + \sum_{n=1}^{\infty}(a_n \cos nx + b_n \sin nx),$$

$$a_n = \frac{1}{\pi}\int_{-\pi}^{\pi} f(x) \cos nx \, dx, \qquad n = 0, 1, 2, \ldots,$$

$$b_n = \frac{1}{\pi}\int_{-\pi}^{\pi} f(x) \sin nx \, dx, \qquad n = 1, 2, 3, \ldots,$$

converges pointwise for all values of x. The sum of the series equals $f(x)$ at all points of continuity of f and equals the average value $[f(x^+) + f(x^-)]/2$ at every point of discontinuity of f.

We should emphasize that the conditions listed in Theorem 4.1 are sufficient conditions for the convergence of a Fourier series, but not necessary conditions. For example, the function $f(x) = |x|^{1/2}$ is not piecewise smooth on any interval containing $x = 0$, and yet it has a convergent Fourier series. Furthermore, the conditions stated in Theorem 4.1 are not the most general set of sufficient conditions that have been established over the years, but still are adequate for most of the functions met in practice. Finally, it should come as no surprise that the series does not converge to the function value $f(x)$ at points of discontinuity. That is, we can change the functional values at these points without changing the Fourier series.

In order to illustrate the convergence of a Fourier series at a point of discontinuity, let us consider the periodic function

$$f(x) = \begin{cases} -K, & -\pi < x < 0 \\ K, & 0 < x < \pi, \end{cases} \qquad f(x + 2\pi) = f(x). \tag{20}$$

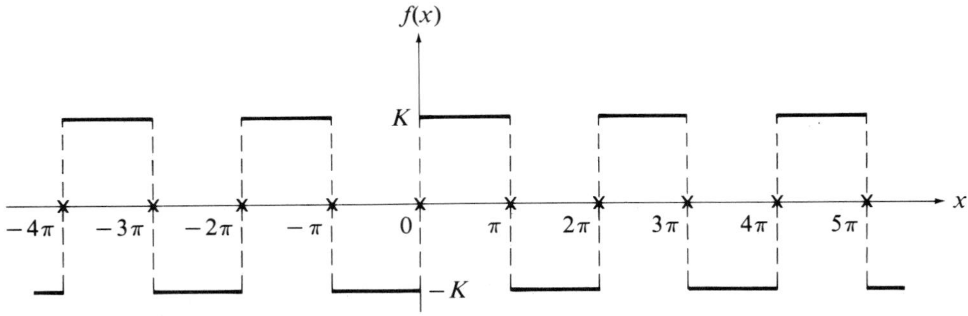

FIGURE 4.7

A square wave function.

Such a function is called a *square wave* for obvious reasons (see Figure 4.7). Because it is an *odd function,* it has a sine series representation. A simple computation reveals that

$$b_n = \frac{2}{\pi}\int_0^{\pi} K \sin nx\, dx = \begin{cases} \dfrac{4K}{n\pi}, & n = 1, 3, 5, \ldots, \\ 0, & n = 2, 4, 6, \ldots \end{cases}$$

and therefore

$$f(x) = \frac{4K}{\pi}\sum_{n=1}^{\infty}\frac{\sin(2n-1)x}{2n-1}. \tag{21}$$

Because of discontinuities in the function f at $x = 0$ and multiples of π, the series (21) converges to the average value 0 (marked by $\times$ in Figure 4.7) of the left-hand and right-hand limits of f at these points. Elsewhere the function f is smooth and thus the series (21) converges to $f(x)$. For instance, setting $x = \pi/2$ in (21) leads to

$$f(\pi/2) = K = \frac{4K}{\pi}\left(1 - \frac{1}{3} + \frac{1}{5} - \frac{1}{7} + \cdots\right),$$

from which we deduce the interesting result

$$1 - \frac{1}{3} + \frac{1}{5} - \frac{1}{7} + \cdots = \frac{\pi}{4}.$$

Leibniz (around 1673) was the first to discover this relation between π and the reciprocal of the odd integers, but he obtained it from geometrical considerations alone.

By taking only a finite number of terms of the series (21), we get the partial sums

$$S_N(x) = \frac{4K}{\pi} \sum_{n=1}^{N} \frac{\sin(2n-1)x}{2n-1},$$

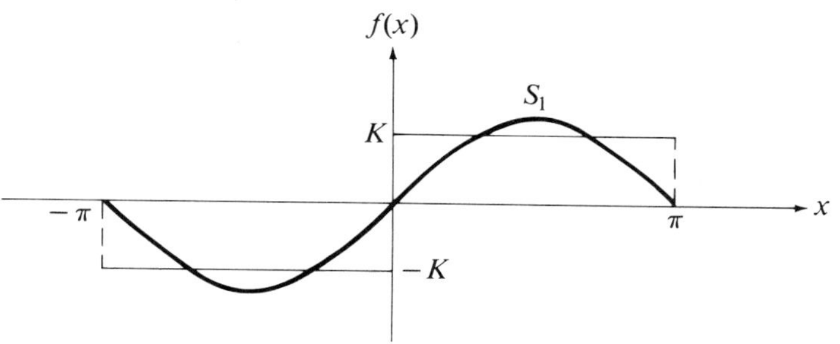

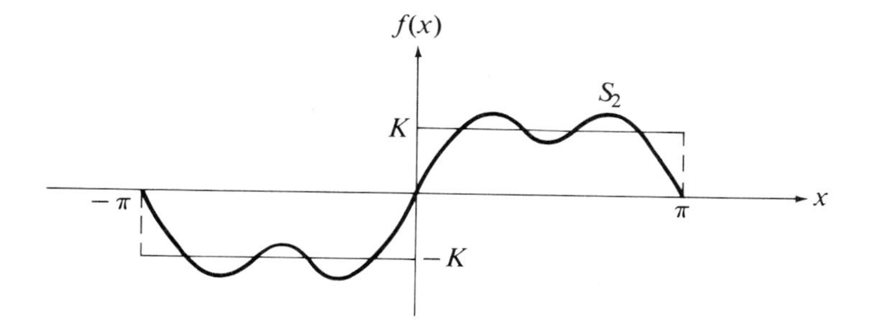

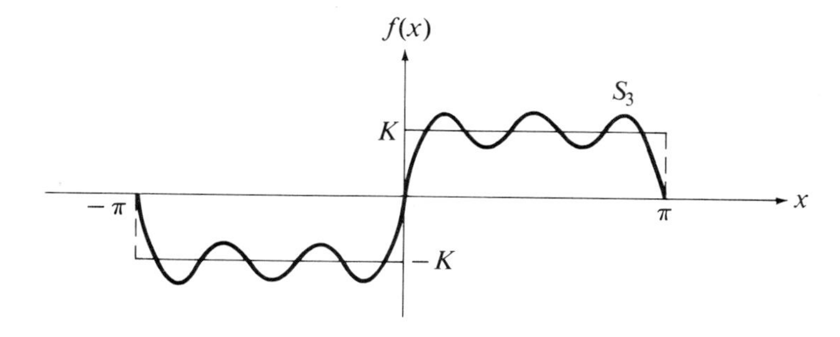

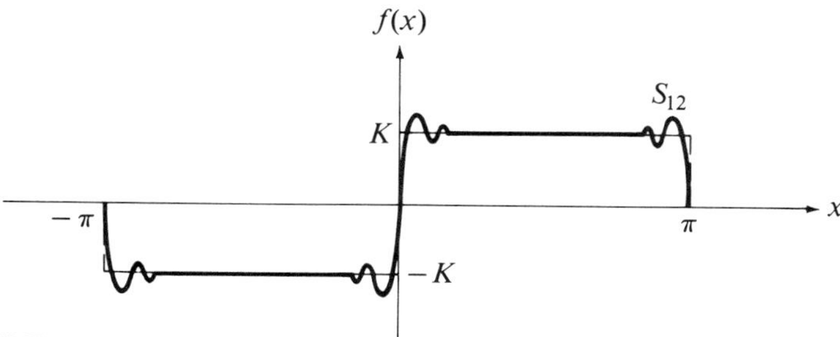

FIGURE 4.8

Gibbs' phenomenon.

which are plotted over one period in Figure 4.8 for $N = 1, 2, 3$, and 12. Here the convergence of the series to $f(x)$ for increasing N becomes evident. However, at the discontinuities of f at $x = 0$ and $x = \pm\pi$ the partial sums tend to overshoot their mark at first and then approach the value K or $-K$. This feature is typical of Fourier series in the vicinity of a discontinuity of f, and is known as the *Gibbs' phenomenon*.* The Gibbs' phenomenon was observed by Gibbs in 1899 when he pointed out in a letter he had written that the approximation curves $S_N(x)$ for a given function f (different from our example) behave in quite a distinct way at the points of discontinuity of the function. He even gave approximate values for how far the approximation curves would differ from the actual functional values in the neighborhood of the discontinuity. Curiously, this statement of Gibbs' in his letter went practically unnoticed for several years until 1906 when Bôucher rediscovered the Gibbs' phenomenon.

If we examine the Fourier series obtained thus far, we will see that the Fourier coefficients decrease at a rate proportional to $1/n$ or $1/n^2$. In general, the Fourier coefficients approach zero at least as rapidly as c/n, where c is a constant independent of n. However, when the terms of the series drop off as $1/n$ as in (21), the convergence of the series is woefully slow. Continuous functions have Fourier series that converge at a faster rate than series associated with functions having discontinuities, and the rate is hastened if the function is also differentiable everywhere.

4.3.1 UNIFORM CONVERGENCE

Pointwise covergence treats convergence of a series at individual points in an interval. Although it is an important type of convergence, it is inadequate for the kinds of problems that require convergence of a series for all x in a specified interval. For example, pointwise convergence is not adequate to permit termwise differentiation or even termwise integration of a series. The simplest notion of convergence that implies pointwise convergence throughout an interval is called *uniform convergence*.

DEFINITION 4.3 If, given some $\epsilon > 0$, there exists a number $M = M(\epsilon)$, independent of x on the interval $[-\pi, \pi]$, such that

$$|f(x) - S_N(x)| < \epsilon$$

for all $N \geq M$, we say that $S_N(x)$ *converges uniformly* to $f(x)$ on the interval $[-\pi, \pi]$ as $N \to \infty$.

The key word in uniform convergence is *continuity* of the function f. It can be shown that if a Fourier series converges uniformly on some interval, it must converge to a continuous function, and in particular, that function that generates the series. The converse of this statement is not necessarily

*Named after JOSIAH WILLARD GIBBS (1839–1903), who is more famous for his work on vector analysis and statistical mechanics.

true; that is, the Fourier series of a continuous function need not converge. The first example of such a function was presented in 1873. Fortunately, continuous functions of this nature are not commonly found in practice.

One of the ways in which we prove that a Fourier series converges uniformly is given by the following theorem, which we state without proof.*

THEOREM 4.2

If the series $\sum_{n=1}^{\infty} (|a_n| + |b_n|)$ converges, then the Fourier series

$$\frac{1}{2} a_0 + \sum_{n=1}^{\infty} (a_n \cos nx + b_n \sin nx)$$

converges uniformly on every finite interval.

To illustrate Theorem 4.2, consider the periodic function from Example 1, whose Fourier series is

$$\frac{\pi}{2} - \frac{4}{\pi} \sum_{\substack{n=1 \\ (\text{odd})}}^{\infty} \frac{\cos nx}{n^2}.$$

The Fourier coefficients $a_n = -4/\pi n^2$, $n = 1, 3, 5, \ldots, a_n = 0$, $n = 2, 4, 6, \ldots$, and $b_n = 0$, $n = 1, 2, 3, \ldots$, lead to the convergent series

$$\sum_{n=1}^{\infty} (|a_n|) + |b_n|) = \frac{4}{\pi} \sum_{\substack{n=1 \\ (\text{odd})}}^{\infty} \frac{1}{n^2},$$

and hence, by Theorem 4.2 we conclude that the above Fourier series converges uniformly on every finite interval.

By adding the requirement of continuity to Theorem 4.1, we are now led to the following important theorem.

THEOREM 4.3

Uniform Convergence. The Fourier series of a continuous, piecewise smooth function f of period 2π converges uniformly to $f(x)$ on every finite interval.

One of the consequences of continuity of a periodic function f (with period 2π) is that, in particular,

$$f(-\pi) = f(\pi). \tag{22a}$$

---

* For a proof of Theorem 4.2, see pp. 80–81 of G. P. Tolstov, *Fourier Series* (Prentice-Hall: New Jersey), 1962.

All even continuous functions certainly have this property, but furthermore, if f is odd and continuous, it must be true that

$$f(-\pi) = f(\pi) = f(0) = 0. \tag{22b}$$

□ 4.3.2 OPERATIONS ON FOURIER SERIES

In the use of Fourier series it is often necessary to perform certain operations on the series, such as *differentiation* and *integration*. For example, in deriving formulas for the Fourier coefficients it was necessary to integrate the series termwise, although conditions under which this is valid had not been stated. While the conditions for termwise integration are almost always satisfied in practice, the same is not true of termwise differentiation.

To illustrate some of the difficulties that arise in termwise differentiation of Fourier series, consider the square–wave function defined by (20) whose (convergent) Fourier series is

$$f(x) = \frac{4K}{\pi} \sum_{n=1}^{\infty} \frac{\sin(2n-1)x}{2n-1}.$$

Since $f(x)$ is either K or $-K$, its derivative at every point of continuity is clearly zero. Term-by-term differentiation of the above series then leads to the relation

$$0 = \frac{4K}{\pi} \sum_{n=1}^{\infty} \cos(2n-1)x,$$

which is an absurd result.* It follows, therefore, that convergence alone is not sufficient to justify termwise differentiation of the series.

THEOREM 4.4

If f is a continuous function of period 2π and f' is piecewise smooth on $[-\pi, \pi]$, then the Fourier series

$$f(x) = \frac{1}{2} a_0 + \sum_{n=1}^{\infty} (a_n \cos nx + b_n \sin nx),$$

$$a_n = \frac{1}{\pi} \int_{-\pi}^{\pi} f(x) \cos nx \, dx, \qquad n = 0, 1, 2, \ldots,$$

$$b_n = \frac{1}{\pi} \int_{-\pi}^{\pi} f(x) \sin nx \, dx, \qquad n = 1, 2, 3, \ldots,$$

can be differentiated termwise to produce the series

$$f'(x) = \sum_{n=1}^{\infty} n(-a_n \sin nx + b_n \cos nx),$$

which converges pointwise to $f'(x)$ wherever $f''(x)$ exists.

---

* For example, when $x = 0$, we have $0 = \frac{4K}{\pi}(1 + 1 + 1 + \ldots)$.

PROOF Because f' satisfies the conditions of Theorem 4.1, it has a convergent Fourier series. If x is a point of continuity of f', then we can represent the Fourier series of f' by

$$f'(x) = \frac{1}{2} A_0 + \sum_{n=1}^{\infty} (A_n \cos nx + B_n \sin nx).$$

By definition,

$$A_0 = \frac{1}{\pi} \int_{-\pi}^{\pi} f'(x)\, dx = \frac{1}{\pi} [f(\pi) - f(-\pi)] = 0,$$

where we are using the result of (22a) for the continuous function f. Also, using integration by parts, we find

$$\begin{aligned} A_n &= \frac{1}{\pi} \int_{-\pi}^{\pi} f'(x) \cos nx\, dx \\ &= \frac{1}{\pi} f(x) \overset{0}{\cancel{\cos nx}} \Big|_{-\pi}^{\pi} + \frac{n}{\pi} \int_{-\pi}^{\pi} f(x) \sin nx\, dx \\ &= nb_n, \qquad n = 1, 2, 3, \ldots, \end{aligned}$$

and similarly, it can be shown that

$$B_n = -na_n, \qquad n = 1, 2, 3, \ldots.$$

Hence, we deduce that

$$f'(x) = \sum_{n=1}^{\infty} n(b_n \cos nx - a_n \sin nx),$$

which is the same result as obtained by differentiating the series for f.

The conditions under which termwise integration of a Fourier series is justified are not so stringent as for differentiation, as indicated in our next theorem.

THEOREM 4.5

If f is a piecewise continuous function of period 2π, then the Fourier series of f can be integrated termwise to yield a series that converges pointwise to the integral of f. In particular, if

$$f(x) = \frac{1}{2} a_0 + \sum_{n=1}^{\infty} (a_n \cos nx + b_n \sin nx),$$

then

$$\int_{-\pi}^{x} f(t)\, dt = \frac{1}{2} a_0(x + \pi) + \sum_{n=1}^{\infty} \frac{1}{n} [a_n \sin nx - b_n \cos nx + (-1)^n b_n].$$

PROOF Let us first define the function

$$g(x) = \int_{-\pi}^{x} f(t)\,dt - \frac{1}{2}a_0 x.$$

It follows that g is continuous and

$$g'(x) = f(x) - \frac{1}{2}a_0.$$

Hence, g' is piecewise continuous and therefore the function g is at least piecewise smooth. Consequently it has the convergent Fourier series

$$g(x) = \frac{1}{2}A_0 + \sum_{n=1}^{\infty}(A_n \cos nx + B_n \sin nx).$$

For $n \geqslant 1$, we have (using integration by parts)

$$\begin{aligned} A_n &= \frac{1}{\pi}\int_{-\pi}^{\pi} g(x)\cos nx\,dx \\ &= \frac{1}{n\pi}\, g(x)\overset{0}{\cancel{\sin nx}}\Big|_{-\pi}^{\pi} - \frac{1}{n\pi}\int_{-\pi}^{\pi} g'(x)\sin nx\,dx \\ &= -\frac{1}{n\pi}\int_{-\pi}^{\pi}\left[f(x) - \frac{1}{2}a_0\right]\sin nx\,dx, \end{aligned}$$

from which we deduce

$$A_n = -\frac{b_n}{n}, \qquad n = 1, 2, 3, \ldots.$$

Likewise, it can readily be shown that

$$B_n = \frac{a_n}{n}, \qquad n = 1, 2, 3, \ldots,$$

and hence, it follows that

$$g(x) = \frac{1}{2}A_0 + \sum_{n=1}^{\infty}\frac{1}{n}(a_n \sin nx - b_n \cos nx).$$

Finally, to determine A_0 we note that

$$\begin{aligned} g(\pi) &= \int_{-\pi}^{\pi} f(t)\,dt - \frac{1}{2}\pi a_0 \\ &= \pi a_0 - \frac{1}{2}\pi a_0 \\ &= \frac{1}{2}\pi a_0, \end{aligned}$$

and so by setting $x = \pi$ in the series for g, we have

$$g(\pi) = \frac{1}{2}\pi a_0 = \frac{1}{2}A_0 - \sum_{n=1}^{\infty}\frac{(-1)^n}{n}b_n$$

[recall that $\cos n\pi = (-1)^n$] or

$$\frac{1}{2}A_0 = \frac{1}{2}\pi a_0 + \sum_{n=1}^{\infty}\frac{(-1)^n}{n}b_n,$$

and we have proved the theorem.

Unless the constant term $a_0/2$ of the series for f is zero, the integrated series in Theorem 4.5 is not truly a Fourier series. Also we find that in some cases it is best to simply take an indefinite integral of the Fourier series for f and determine the constant of integration by a technique such as illustrated in Example 5.

Finally, we observe that in Theorem 4.5 we merely assumed the existence of the Fourier series of f and nothing was said about its convergence to $f(x)$. In fact, it can happen that the trigonometric series for f diverges, and yet the term-by-term integration of that series could yield a convergent series.

EXAMPLE 5

Starting with the known Fourier series

$$f(x) = \frac{4}{\pi}\sum_{n=1}^{\infty}\frac{\sin(2n-1)x}{2n-1},$$

where

$$f(x) = \begin{cases} -1, & -\pi < x \leqslant 0 \\ 1, & 0 < x \leqslant \pi, \end{cases} \qquad f(x+2\pi) = f(x),$$

derive the Fourier series for the integral of f, which over one period is defined by*

$$g(x) = |x|, \qquad -\pi < x \leqslant \pi.$$

SOLUTION Performing an indefinite integration of the series, we get

$$g(x) = C - \frac{4}{\pi}\sum_{n=1}^{\infty}\frac{\cos(2n-1)x}{(2n-1)^2},$$

where C is a constant of integration. Because this last result should be a cosine series of $g(x)$, we recognize that

$$C = \frac{1}{2}A_0 = \frac{1}{2\pi}\int_{-\pi}^{\pi}|x|\,dx = \frac{1}{2}\pi,$$

---

*Note that $g'(x) = f(x)$, $x \neq 0$, $-\pi < x < \pi$.

and thus,

$$g(x) = \frac{1}{2}\pi - \frac{4}{\pi}\sum_{n=1}^{\infty}\frac{\cos(2n-1)x}{(2n-1)^2}.$$

EXERCISES 4.3

In Problems 1–8, discuss whether the given function is smooth, piecewise smooth, continuous, piecewise continuous, or none of these on the interval $[-\pi, \pi]$.

1. $f(x) = \sin x$
2. $f(x) = \dfrac{x^2 - 1}{x - 1}, \quad x \neq 1$
3. $f(x) = \tan x$
4. $f(x) = \begin{cases} 1, & \text{if } x \text{ is irrational} \\ 0, & \text{if } x \text{ is rational} \end{cases}$
5. $f(x) = \dfrac{\sin x}{x}, \quad x \neq 0, \quad f(0) = 1$
6. $f(x) = \dfrac{\sin x}{|x|}, \quad x \neq 0$
7. $f(x) = |x| + x$
8. $f(x) = x^{1/3}$
9. Verify that the functions in odd-numbered Problems 7–15 in Exercises 4.2 satisfy the conditions of Theorem 4.1.
10. Verify that the functions in even-numbered Problems 6–14 in Exercises 4.2 satisfy the conditions of Theorem 4.1.
11. If $f(x + 2\pi) = f(x)$ and

$$f(x) = \begin{cases} 0, & -\pi < x < 0 \\ 1, & 0 < x < \pi/2 \\ 4, & \pi/2 < x < \pi \end{cases}$$

to what numerical value will the Fourier series converge at

(a) $x = 0$?
(b) $x = \pi/2$?
(c) $x = \pi$?
(d) $x = 2\pi$?
(e) $x = \pi/3$?

12. Use the Fourier series in Example 1 (Section 4.2) to deduce that

$$1 + \frac{1}{3^2} + \frac{1}{5^2} + \cdots + \frac{1}{(2n-1)^2} + \cdots = \frac{\pi^2}{8}.$$

13. Given that the periodic function $f(x) = x^2$, $-\pi \leq x < \pi$, has the Fourier cosine series

$$f(x) = \frac{\pi^2}{3} + 4\sum_{n=1}^{\infty}\frac{(-1)^n}{n^2}\cos nx,$$

use this result to show that

(a) $1 + \frac{1}{2^2} + \frac{1}{3^2} + \cdots + \frac{1}{n^2} + \cdots = \frac{\pi^2}{6}$.

(b) $1 - \frac{1}{2^2} + \frac{1}{3^2} - \cdots + \frac{(-1)^{n+1}}{n^2} + \cdots = \frac{\pi^2}{12}$.

(c) From (a) and (b), deduce that

$$1 + \frac{1}{3^2} + \frac{1}{5^2} + \cdots + \frac{1}{(2n-1)^2} + \cdots = \frac{\pi^2}{8}.$$

14. Use Equation (21) to derive the result

$$\frac{\sqrt{2}}{2}\left(1 + \frac{1}{3} - \frac{1}{5} - \frac{1}{7} + \cdots\right) = \frac{\pi}{4}.$$

15. Based on Theorem 4.3, determine which functions in odd-numbered Problems 7–15 in Exercises 4.2 have uniformly convergent Fourier series.

16. Based on Theorem 4.3, determine which functions in even-numbered Problems 6–14 in Exercises 4.2 have uniformly convergent Fourier series.

17. The odd periodic function $f(x) = x$, $-\pi \leq x < \pi$, has the Fourier sine series

$$f(x) = 2\sum_{n=1}^{\infty} \frac{(-1)^{n-1}}{n} \sin nx.$$

Use this result and termwise integration to develop the Fourier cosine series for the periodic function $f(x) = x^2$, $-\pi \leq x < \pi$.

★18. Suppose we wish to approximate a given function $f(x)$ by the trigonometric polynomial

$$S_N(x) = \frac{1}{2}\alpha_0 + \sum_{n=1}^{N} (\alpha_n \cos nx + \beta_n \sin nx),$$

where $\alpha_0, \alpha_1, \ldots, \alpha_N, \beta_1, \ldots, \beta_N$ are not necessarily the Fourier coefficients of $f(x)$. In this case we define the *mean square error* of the approximation by the expression

$$E_N = \int_{-\pi}^{\pi} [f(x) - S_N(x)]^2 \, dx.$$

Show that E_N is minimized by choosing the constants $\alpha_0, \alpha_1, \ldots, \alpha_N, \beta_1, \ldots, \beta_N$ to be the Fourier coefficients $a_0, a_1, \ldots, a_N, b_1, \ldots, b_N$.
Hint: Set

$$\frac{\partial E_N}{\partial \alpha_k} = 0 \ (k = 0, 1, \ldots, N) \quad \text{and} \quad \frac{\partial E_N}{\partial \beta_k} = 0 \ (k = 1, 2, \ldots, N)$$

by differentiating under the integral sign.

★19. If $f(x + 2\pi) = f(x)$ and f is piecewise continuous, derive the following results for f and its partial sum $S_N(x)$:

(a) $\int_{-\pi}^{\pi} f(x)S_N(x)\, dx = \frac{\pi}{2} a_0^2 + \pi \sum_{n=1}^{N} (a_n^2 + b_n^2)$.

(b) $\displaystyle\int_{-\pi}^{\pi} S_N^2(x)\,dx = \frac{\pi}{2}a_0^2 + \pi\sum_{n=1}^{N}(a_n^2 + b_n^2).$

★20. Using the results of Problem 19, show that

(a) $\displaystyle\int_{-\pi}^{\pi} [f(x) - S_N(x)]^2\,dx = \int_{-\pi}^{\pi} f^2(x)\,dx - \frac{\pi}{2}a_0^2 - \pi\sum_{n=1}^{N}(a_n^2 + b_n^2).$

(b) From (a), deduce *Bessel's inequality* (for all N)

$$\frac{1}{2}a_0^2 + \sum_{n=1}^{N}(a_n^2 + b_n^2) \leqslant \frac{1}{\pi}\int_{-\pi}^{\pi} f^2(x)\,dx.$$

★21. A Fourier series is said to *converge in the mean* if and only if

$$\lim_{N\to\infty}\int_{-\pi}^{\pi} [f(x) - S_N(x)]^2\,dx = 0.$$

(a) Under this assumption and the results of Problem 20, derive *Parseval's equality*

$$\frac{1}{2}a_0^2 + \sum_{n=1}^{\infty}(a_n^2 + b_n^2) = \frac{1}{\pi}\int_{-\pi}^{\pi} f^2(x)\,dx.$$

(b) Use the results of part (a) to obtain the result of Problem 12.

★22. The function $f(x) = -\log|2\sin x/2|$ is unbounded at $x = 2n\pi$ but is still an integrable function with period 2π.

(a) Show that the Fourier coefficients of f are defined by

$$a_n = -\frac{2}{\pi}\int_0^{\pi}\log\left(2\sin\frac{1}{2}x\right)\cos nx\,dx, \qquad n = 0, 1, 2, \ldots,$$

$$b_n = 0, \qquad n = 1, 2, 3, \ldots.$$

(b) Show that the integrtal $I = \int_0^{\pi}\log(2\sin x/2)\,dx$ satisfies the relation $I = \pi\log x + J$, where $J = \int_0^{\pi}\log(\sin x/2)\,dx$, and that J satisfies the relation $J = \pi\log 2 + 2J$. From this result deduce the value of a_0.
Hint: Let $x = 2t$ and show that

$$J = \pi\log 2 + 2\int_0^{\pi/2}\log\left(\sin\frac{1}{2}t\right)dt + 2\int_0^{\pi/2}\log\left(\cos\frac{1}{2}t\right)dt.$$

(c) Assuming the relation*

$$\frac{\sin[(n + 1/2)x]}{2\sin x/2} = \frac{1}{2} + \sum_{k=1}^{n}\cos kx,$$

obtain the Fourier series for $f(x)$, and use this result to sum the alternating harmonic series

$$1 - \frac{1}{2} + \frac{1}{3} - \frac{1}{4} + \cdots.$$

Hint: Use integration by parts.

---

*The proof of this relation is given in Section 4.4.

★23. Given the function $f(x) = \cos kx$, where k is not an integer,

(a) find its Fourier series.

(b) Letting $k = z$ in (a), and substituting $x = 0$ and $x = \pi$, obtain the series expansions

$$\csc \pi z = \frac{1}{\pi z} + \frac{2z}{\pi} \sum_{n=1}^{\infty} \frac{(-1)^n}{z^2 - n^2}, \qquad \cot \pi z = \frac{1}{\pi z} + \frac{2z}{\pi} \sum_{n=1}^{\infty} \frac{1}{z^2 - n^2}.$$

(c) Assume $0 < z < 1$ and integrate the series in (b) for $\cot \pi z$ from $z = 0$ to $z = x$, $0 < x < 1$, thus showing that

$$\log\left(\frac{\sin \pi x}{\pi x}\right) = \sum_{n=1}^{\infty} \log\left(1 - \frac{x^2}{n^2}\right),$$

or equivalently,*

$$\sin \pi x = \pi x \prod_{n=1}^{\infty} \left(1 - \frac{x^2}{n^2}\right).$$

(d) Finally, assuming $0 < z < 1$, show that

$$\int_0^{\infty} \frac{x^{z-1}}{1 + x}\, dx = \frac{\pi}{\sin \pi z}.$$

Hint: Express the integral as a sum of two integrals, the first having $(0, 1)$ as the interval of integration and the second $(1, \infty)$. Let $x = 1/t$ in the second integral and use the series relation

$$\frac{1}{1 + x} = \sum_{n=0}^{\infty} (-1)^n x^n, \qquad |x| < 1$$

together with (b) to obtain your answer.

□ 4.4 A PROOF OF POINTWISE CONVERGENCE

In order to prove Theorem 4.1 we wish to express the Nth partial sum $S_N(x)$ of the Fourier series of f in a form such that its limit as N tends to infinity is easily established. To do so, we begin by substituting the Euler formulas for the Fourier coefficients a_n and b_n into this sum to get

$$\begin{aligned} S_N(x) &= \frac{1}{2} a_0 + \sum_{n=1}^{N} (a_n \cos nx + b_n \sin nx) \\ &= \frac{1}{2\pi} \int_{-\pi}^{\pi} f(t)\, dt + \sum_{n=1}^{N} \left[\frac{1}{\pi} \int_{-\pi}^{\pi} f(t) \cos nt \cos nx\, dt \right. \\ &\quad \left. + \frac{1}{\pi} \int_{-\pi}^{\pi} f(t) \sin nt \sin nx\, dt\right] \end{aligned}$$

* The pi notation $\Pi_{n=1}^{\infty}$ denotes an infinite product.

$$= \frac{1}{\pi} \int_{-\pi}^{\pi} f(t) \left[\frac{1}{2} + \sum_{n=1}^{N} \cos[n(t - x)] \right] dt \tag{23}$$

(using the dummy variable t), where we have used the trigonometric identity

$$\cos(A - B) = \cos A \cos B + \sin A \sin B.$$

The finite series

$$D_N(u) = \frac{1}{2} + \sum_{n=1}^{N} \cos nu \tag{24}$$

can be summed through use of the trigonometric identity

$$\sin A \cos B = \frac{1}{2} [\sin(A + B) + \sin(A - B)].$$

That is, we first multiply (24) by $2 \sin u/2$ and then use the above identity to obtain

$$2 \sin \frac{1}{2} u D_N(u) = \sin \frac{1}{2} u + \sum_{n=1}^{N} 2 \sin \frac{1}{2} u \cos nu$$

$$= \sin \frac{1}{2} u + \sum_{n=1}^{N} \left[\sin\left(n + \frac{1}{2}\right) u - \sin\left(n - \frac{1}{2}\right) u \right].$$

We recognize the right-hand side of this expression as a "telescoping sum" which sums to $\sin(N + 1/2)u$. Hence, we have shown that

$$2 \sin \frac{1}{2} u D_N(u) = \sin\left(n + \frac{1}{2}\right) u,$$

from which we deduce

$$D_N(u) = \frac{\sin(N + 1/2)u}{2 \sin u/2}. \tag{25}$$

The function $D_N(u)$, called the *Dirichlet kernel,* is introduced primarily for notational convenience. It is left to the reader to verify that it has the properties

$$D_N(u + 2\pi) = D_N(u), \tag{26}$$

$$D_N(-u) = D_N(u), \tag{27}$$

$$\int_{-\pi}^{\pi} D_N(u)\, du = \pi \tag{28}$$

(see Problems 1–3 in Exercises 4.4).

In terms of the Dirichlet kernel, Equation (23) takes the form

$$S_N(x) = \frac{1}{\pi} \int_{-\pi}^{\pi} f(t) D_N(t - x)\, dt, \tag{29}$$

and by making the change of variable $u = t - x$, this becomes

$$S_N(x) = \frac{1}{\pi} \int_{-\pi-x}^{\pi-x} f(x+u) D_N(u)\, du.$$

However, by using the identity

$$\int_{c-\pi}^{c+\pi} f(x)\, dx = \int_{-\pi}^{\pi} f(x)\, dx \tag{30}$$

(see Problem 4 in Exercises 4.4), where f has period 2π and c is any constant, we get

$$S_N(x) = \frac{1}{\pi} \int_{-\pi}^{\pi} f(x+u) D_N(u)\, du. \tag{31}$$

In order to continue the proof of convergence, we need to establish certain properties of the Fourier coefficients that are significant in their own right.

4.4.1 BESSEL'S INEQUALITY AND RIEMANN'S THEOREM

Let us suppose that f is any piecewise continuous function that has period 2π. It follows directly from the definition of the Fourier coefficients that

$$\int_{-\pi}^{\pi} f(x) S_N(x)\, dx = \frac{1}{2}\pi a_0^2 + \pi \sum_{n=1}^{N} (a_n^2 + b_n^2) \tag{32}$$

(see Problem 19 in Exercises 4.3). In a similar fashion it can be shown that

$$\int_{-\pi}^{\pi} [S_N(x)]^2\, dx = \frac{1}{2}\pi a_0^2 + \pi \sum_{n=1}^{N} (a_n^2 + b_n^2), \tag{33}$$

and so consequently,

$$\begin{aligned} \int_{-\pi}^{\pi} [f(x) - S_N(x)]^2\, dx &= \int_{-\pi}^{\pi} f^2(x)\, dx - 2\int_{-\pi}^{\pi} f(x) S_N(x)\, dx \\ &\quad + \int_{-\pi}^{\pi} [S_n(x)]^2\, dx \\ &= \int_{-\pi}^{\pi} f^2(x)\, dx - \pi\left[\frac{1}{2} a_0^2 + \sum_{n=1}^{N} (a_n^2 + b_n^2)\right]. \end{aligned} \tag{34}$$

The left-hand side of (34) is clearly non-negative. Thus we obtain the inequality

$$\frac{1}{2} a_0^2 + \sum_{n=1}^{N} (a_n^2 + b_n^2) \leq \frac{1}{\pi} \int_{-\pi}^{\pi} f^2(x)\, dx,$$

which is valid for all N. Therefore, if we formally pass to the limit as $N \to \infty$, we obtain *Bessel's inequality*

$$\frac{1}{2} a_0^2 + \sum_{n=1}^{\infty} (a_n^2 + b_n^2) \leq \frac{1}{\pi} \int_{-\pi}^{\pi} f^2(x)\, dx. \tag{35}$$

Because the integral on the right is finite, the infinite series on the left-hand side of Bessel's inequality is *bounded,* and hence *converges.* Moreover, this series converges regardless of the convergence of the associated Fourier series for f! Hence, as a consequence,

$$\lim_{n\to\infty} (a_n^2 + b_n^2) = 0,$$

or equivalently,

$$\lim_{n\to\infty} a_n = \lim_{n\to\infty} b_n = 0, \tag{36}$$

and we have proved the following theorem.

THEOREM 4.6

> *Riemann's Theorem.* If f is piecewise continuous on the interval $[-\pi, \pi]$, then
>
> $$\lim_{n\to\infty} \int_{-\pi}^{\pi} f(x) \cos nx\, dx = \lim_{n\to\infty} \int_{-\pi}^{\pi} f(x) \sin nx\, dx = 0.$$

As a corollary to Theorem 4.6, we have the result

$$\lim_{N\to\infty} \int_{-\pi}^{\pi} f(x) \sin\left(N + \frac{1}{2}\right)x\, dx = 0 \tag{37}$$

(see Problem 5 in Exercises 4.4), which follows directly from Riemann's theorem and the trigonometric identity

$$f(x) \sin\left(N + \frac{1}{2}\right)x = \left[f(x) \sin \frac{1}{2} x\right] \cos Nx + \left[f(x) \cos \frac{1}{2} x\right] \sin Nx.$$

4.4.2 CONVERGENCE AT A POINT OF CONTINUITY

Let us now rewrite Equation (31) as

$$\begin{aligned} S_N(x) &= \frac{1}{\pi} \int_{-\pi}^{\pi} f(x + u) D_N(u)\, du \\ &= \frac{1}{\pi} f(x) \int_{-\pi}^{\pi} D_N(u)\, du + \frac{1}{\pi} \int_{-\pi}^{\pi} [f(x + u) - f(x)] D_N(u)\, du^* \end{aligned}$$

---

* Here we are simply adding and subtracting the function $f(x)$, which is constant with respect to the integration variable.

$$= f(x) + \frac{1}{\pi}\int_{-\pi}^{\pi} [f(x+u) - f(x)]D_N(u)\,du, \tag{38}$$

where we are using the result of Equation (28). Our task at this point is to show that the integral in (38) vanishes as N tends to infinity.

Recalling (25), the integrand in (38) becomes

$$[f(x+u) - f(x)]D_N(u) = \frac{f(x+u) - f(x)}{2\sin\frac{1}{2}u}\sin\left(N + \frac{1}{2}\right)u.$$

If we assume that x is a point of continuity of f, then the function

$$g(u) = \frac{f(x+u) - f(x)}{2\sin\frac{1}{2}u} = \frac{f(x+u) - f(x)}{u}\cdot\frac{\frac{1}{2}u}{\sin\frac{1}{2}u}, \qquad -\pi \leqslant u \leqslant \pi$$

is piecewise continuous, at least. That is, for $u \neq 0$ the function g is certainly continuous, since it is composed of continuous functions, and for $u = 0$ we have

$$\lim_{u\to 0} g(u) = f'(x),$$

which is piecewise continuous by hypothesis. Therefore, from (37) it now follows that

$$\lim_{N\to\infty} S_N(x) = f(x) + \lim_{N\to\infty}\frac{1}{\pi}\int_{-\pi}^{\pi} g(u)\sin\left(N + \frac{1}{2}\right)u\,du$$

$$= f(x), \tag{39}$$

and Theorem 4.1 is proved for points of continuity of f.

The corresponding proof for points of discontinuity of f is very similar and is outlined in the following exercises (see Problems 6, 7, and 8).

EXERCISES 4.4

In Problems 1–3, verify the given property for $D_N(u) = \frac{1}{2} + \sum_{n=1}^{N} \cos nu$.

1. $D_N(u + 2\pi) = D_N(u)$
2. $D_N(-u) = D_N(u)$
3. $\int_{-\pi}^{\pi} D_N(u)\,du = \pi$
4. If $f(x + 2\pi) = f(x)$, then show that for any constant c it follows that

$$\int_{c-\pi}^{c+\pi} f(x)\,dx = \int_{-\pi}^{\pi} f(x)\,dx.$$

Hint: Write

$$\int_{c-\pi}^{c+\pi} f(x)\,dx = \int_{c-\pi}^{-\pi} f(x)\,dx + \int_{-\pi}^{c+\pi} f(x)\,dx$$

and let $x = t + 2\pi$ in the first integral on the right-hand side.

5. Use Theorem 4.6 to prove the result of Equation (37).

★6. By writing

$$\begin{aligned} I &= \frac{1}{\pi}\int_0^{\pi} f(x+u)D_N(u)\,du \\ &= \frac{1}{\pi} f(x^+)\int_0^{\pi} D_N(u)\,du + \frac{1}{\pi}\int_0^{\pi} [f(x+u) - f(x^+)]D_N(u)\,du, \end{aligned}$$

show that, if f satisfies the hypothesis of Theorem 4.1,

$$\lim_{N\to\infty} I = \frac{1}{2} f(x^+).$$

★7. By writing

$$\begin{aligned} J &= \frac{1}{\pi}\int_{-\pi}^{0} f(x+u)D_N(u)\,du \\ &= \frac{1}{\pi} f(x^-)\int_{-\pi}^{0} D_N(u)\,du + \frac{1}{\pi}\int_{-\pi}^{0} [f(x+u) - f(x^-)]D_N(u)\,du, \end{aligned}$$

show that, if f satisfies the hypothesis of Theorem 4.1,

$$\lim_{N\to\infty} J = \frac{1}{2} f(x^-).$$

★8. Show that $S_N(x) = I + J$, where I and J are defined in Problems 6 and 7, and hence deduce that

$$\lim_{N\to\infty} S_N(x) = \frac{1}{2}[f(x^+) + f(x^-)],$$

whether x is a point of continuity or discontinuity.

4.5 GENERAL PERIODS AND INTERVALS

Thus far we have dealt entirely with periodic functions of period 2π, but the application of Fourier trigonometric series extends to more general functions than these. For instance, the period T is often required to be something other than 2π, and in many cases the function being considered may not be periodic at all.

Suppose we first consider a function that has arbitrary period $T = 2p$. A Fourier series for such a function can be found formally by a simple change of the independent variable. To show this, let f be a function of period 2π that has the Fourier series representation

$$f(x) = \frac{1}{2} a_0 + \sum_{n=1}^{\infty} (a_n \cos nx + b_n \sin nx), \tag{40}$$

where

$$\begin{aligned} a_n &= \frac{1}{\pi} \int_{-\pi}^{\pi} f(x) \cos nx \, dx, \qquad n = 0, 1, 2, \ldots, \\ b_n &= \frac{1}{\pi} \int_{-\pi}^{\pi} f(x) \sin nx \, dx, \qquad n = 1, 2, 3, \ldots. \end{aligned} \tag{41}$$

If we make the variable change $x = \pi t/p$ and set $f(x) \equiv f(\pi t/p) \equiv g(t)$, the new function g is periodic with period $2p$, and from (40) and (41) we deduce that

$$g(t) = \frac{1}{2} a_0 + \sum_{n=1}^{\infty} \left(a_n \cos \frac{n\pi t}{p} + b_n \sin \frac{n\pi t}{p} \right), \tag{42}$$

where

$$\begin{aligned} a_n &= \frac{1}{p} \int_{-p}^{p} g(t) \cos \frac{n\pi t}{p} \, dt, \qquad n = 0, 1, 2, \ldots, \\ b_n &= \frac{1}{p} \int_{-p}^{p} g(t) \sin \frac{n\pi t}{p} \, dt, \qquad n = 1, 2, 3, \ldots. \end{aligned} \tag{43}$$

The expression (42) is a Fourier trigonometric series for the periodic function g with period $T = 2p$, whose Fourier coefficients are defined by (43). These expressions can also be derived from first principles in the same way we did in Section 4.2 for functions with period 2π. In subsequent discussions we will revert back to the function symbol f and variable x for consistency of notation.

More general yet is when the interval of interest is not symmetrical about the origin. Suppose the function f is defined throughout the interval $[c, c + T]$, where c is any real number and T is the period of f. The function is then defined for all other values of x by the relation $f(x + T) = f(x)$. By a procedure analogous to the above argument, it can be shown that such a function f has the representation

$$f(x) = \frac{1}{2} a_0 + \sum_{n=1}^{\infty} \left(a_n \cos \frac{2n\pi x}{T} + b_n \sin \frac{2n\pi x}{T} \right), \tag{44}$$

where

$$\begin{aligned} a_n &= \frac{2}{T} \int_{c}^{c+T} f(x) \cos \frac{2n\pi x}{T} \, dx, \qquad n = 0, 1, 2, \ldots, \\ b_n &= \frac{2}{T} \int_{c}^{c+T} f(x) \sin \frac{2n\pi x}{T} \, dx, \qquad n = 1, 2, 3, \ldots. \end{aligned} \tag{45}$$

The formal derivation of these expressions is based upon the result of Problem 4 in Exercises 4.4, and is left to the reader. Note, however, that (44) and (45) reduce to (42) and (43) when $c = -p$ and $T = 2p$.

EXAMPLE 6

Find the Fourier series of $f(x) = x$, $0 < x < 1$, and $f(x + 1) = f(x)$ for all x.

SOLUTION Referring to (44) and (45), here we see that $c = 0$ and $T = 1$. Therefore

$$a_n = 2\int_0^1 x \cos 2n\pi x \, dx = \begin{cases} 1, & n = 0 \\ 0, & n = 1, 2, 3, \ldots, \end{cases}$$

and

$$b_n = 2\int_0^1 x \sin 2n\pi x \, dx = -\frac{1}{n\pi}, \qquad n = 1, 2, 3, \ldots .$$

Hence, the Fourier series is

$$f(x) = \frac{1}{2} - \frac{1}{\pi}\sum_{n=1}^{\infty} \frac{\sin 2n\pi x}{n}.$$

It follows that the convergence conditions for series of the type (42) or (44) are the same as stated in the theorems in Section 4.3 for periodic functions having period 2π. Similar comments also apply to the theorems concerning termwise differentiation and integration of Fourier series, and thus we will not restate them in their more general forms.

4.5.1 NONPERIODIC FUNCTIONS OVER FINITE INTERVALS

In many of the applications involving partial differential equations that occur later in the text it is necessary to represent by a Fourier series a function f that is defined only on a finite interval, and nothing is said about this function outside this interval. Let us first suppose the finite interval is $[-p, p]$, and let f^* denote the periodic extension of f over the entire x-axis (see Figure 4.9). On the interval $[-p, p]$, f and f^* assume the same functional values. If f^* is a piecewise smooth function, it has a convergent Fourier series over the entire x-axis, and in particular, over the interval $[-p, p]$. Hence, the series representation of f^* is also the representation of f, but for the function f we restrict the interval of convergence to the finite interval $[-p, p]$. If $f(-p) = f(p)$, then the periodic extension of f leads to a function that is continuous at $x = \pm p$, and hence the Fourier series converges to $f(p)$ at these points. However, if $f(-p) \neq f(p)$, the periodic extension of f is discontinuous at these points and the Fourier series converges to the average value $[f(-p) + f(p)]/2$ of the endpoints, marked $\times$ in Figure 4.9.

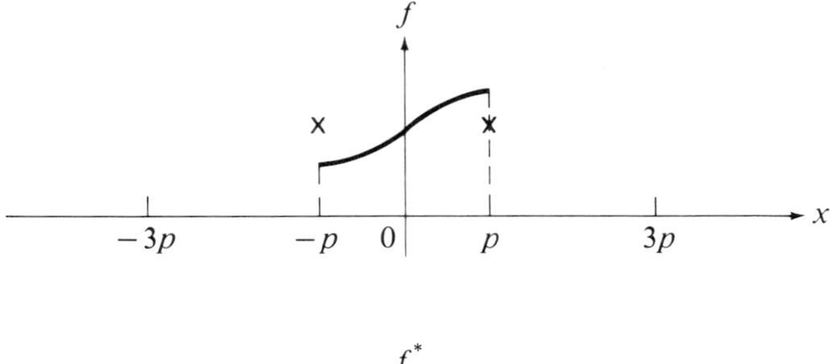

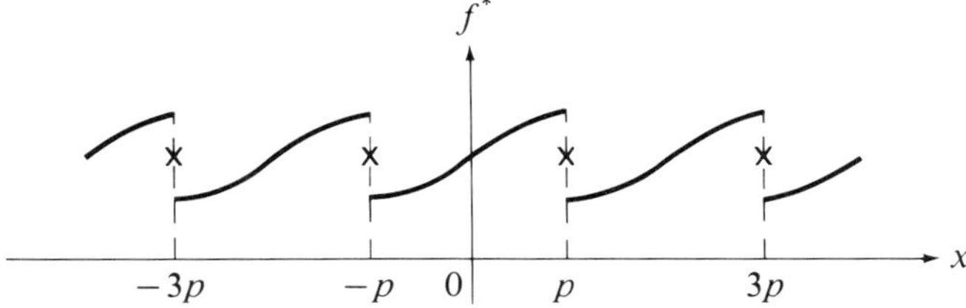

FIGURE 4.9

Periodic extension of $f(x)$.

4.5.2 HALF-RANGE EXPANSIONS

If the function f is defined only on the interval $[0, p]$, then we have several choices of representing this function in a Fourier series. First, we can use (44) and (45) with $T = p$ to generate the complete trigonometric series. Also, by considering the even and odd extensions of f to the interval $[-p, p]$, we can produce either a cosine series or sine series representation of f. It is these latter possibilities that come up most often in later chapters of this text.

If we desire a cosine series representation of a function f defined only on the interval $[0, p]$, we can define f_e as the *even extension* of f over the larger interval $[-p, p]$ as shown in Figure 4.10. Thus, f_e has a cosine series representation over the interval $[-p, p]$, and this series also represents the original function f on the smaller interval $[0, p]$. Similar arguments can be made for finding a sine series of f on the same interval, by first considering the *odd extension* of f shown in Figure 4.10. For purposes of computing the Fourier coefficients in either case, it is not necessary to actually make either the even or odd extensions of f, since all computations are performed on the half-interval $[0, p]$ anyway. Series of this nature are called *half-range expansions*.

The cosine and sine series representations for these half-range expansions are given respectively by

$$f(x) = \frac{1}{2} a_0 + \sum_{n=1}^{\infty} a_n \cos \frac{n\pi x}{p}, \qquad 0 < x < p$$

$$a_n = \frac{2}{p} \int_0^p f(x) \cos \frac{n\pi x}{p}\, dx, \qquad n = 0, 1, 2, \ldots, \tag{46}$$

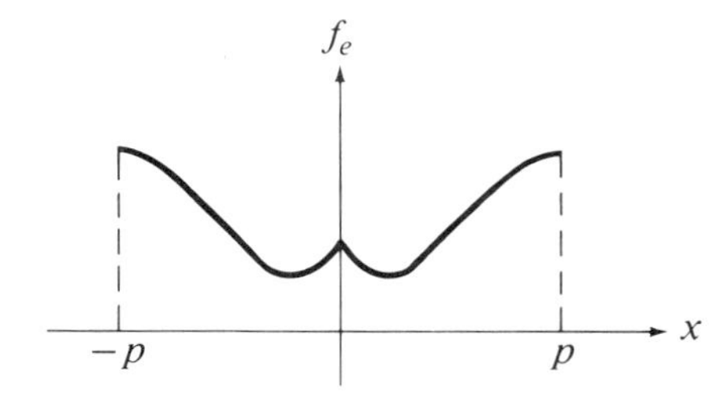

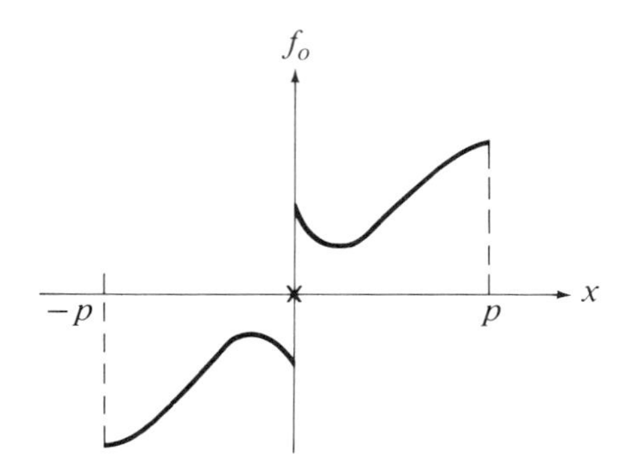

FIGURE 4.10

Even and odd extensions of $f(x)$.

and

$$
\begin{aligned}
f(x) &= \sum_{n=1}^{\infty} b_n \sin \frac{n\pi x}{p}, \qquad 0 < x < p \\
b_n &= \frac{2}{p} \int_0^p f(x) \sin \frac{n\pi x}{p}\, dx, \qquad n = 1, 2, 3, \ldots .
\end{aligned}
\tag{47}
$$

EXAMPLE 7

Find the half-range cosine and sine series expansions of the function

$$
f(x) = \begin{cases} x, & 0 < x < \dfrac{\pi}{2} \\[2ex] \pi - x, & \dfrac{\pi}{2} < x < \pi \end{cases}
$$

(see Figure 4.11).

SOLUTION The even and odd extensions of f are shown in Figure 4.12.

From (46), we calculate

$$
a_0 = \frac{2}{\pi} \left[\int_0^{\pi/2} x\, dx + \int_{\pi/2}^{\pi} (\pi - x)\, dx \right] = \frac{\pi}{2},
$$

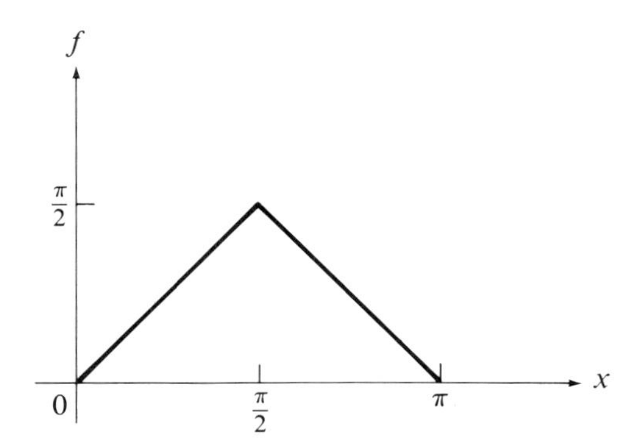

FIGURE 4.11

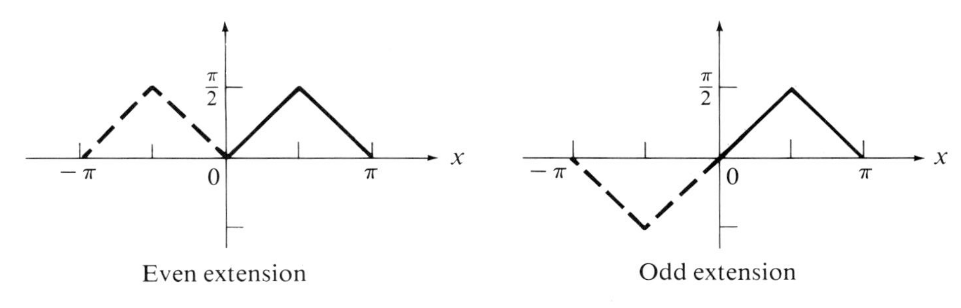

FIGURE 4.12

and, using integration by parts,

$$a_n = \frac{2}{\pi}\left[\int_0^{\pi/2} x \cos nx \, dx + \int_{\pi/2}^{\pi} (\pi - x) \cos nx \, dx\right]$$

$$= \frac{2}{\pi}\left\{\left(\frac{x}{n} \sin nx + \frac{1}{n^2} \cos nx\right)\Bigg|_0^{\pi/2}\right.$$

$$\left. + \left[\frac{1}{n} (\pi - x) \sin nx - \frac{1}{n^2} \cos nx\right]\Bigg|_{\pi/2}^{\pi}\right\}$$

$$= \frac{2}{n^2\pi}\left(2 \cos \frac{n\pi}{2} - \cos n\pi - 1\right), \qquad n = 1, 2, 3, \ldots .$$

Thus,

$$a_1 = 0, \qquad a_2 = -\frac{8}{2^2\pi}, \qquad a_3 = a_4 = a_5 = 0,$$

$$a_6 = -\frac{8}{6^2\pi}, \qquad a_7 = a_8 = a_9 = 0, \qquad a_{10} = -\frac{8}{10^2\pi}, \ldots ,$$

and the cosine half-range expansion is

$$f(x) = \frac{\pi}{4} - \frac{8}{\pi}\left(\frac{1}{2^2}\cos 2x + \frac{1}{6^2}\cos 6x + \frac{1}{10^2}\cos 10x + \cdots\right)$$

$(0 < x < \pi)$.

Similarly, from (47) we find

$$b_n = \frac{2}{\pi}\int_0^{\pi} f(x)\sin nx\,dx = \frac{4}{n^2\pi}\sin\frac{n\pi}{2}, \qquad n = 1, 2, 3, \ldots$$

and hence the sine series half-range expansion is

$$f(x) = \frac{4}{\pi}\left(\sin x - \frac{1}{3^2}\sin 3x + \frac{1}{5^2}\sin 5x - \cdots\right)$$

$(0 < x < \pi)$.

EXERCISES 4.5

In Problems 1–6, sketch each function and determine its Fourier series.

1. $f(x) = x^2, \quad -1 \leqslant x \leqslant 1$

2. $f(x) = 1 + 2x, \quad -3 \leqslant x \leqslant 3$

3. $f(x) = \begin{cases} 0, & -2 < x < 0 \\ 1, & 0 < x < 1 \\ 2, & 1 < x < 2 \end{cases}$

4. $f(x) = \sin x, \quad -\frac{1}{2} < x < \frac{1}{2}$

5. $f(x) = x, \quad 1 < x < 2$

6. $f(x) = |x|, \quad -1 < x < 2$

★7. Show that the function defined by

$$f(x) = \begin{cases} 1, & 1 < x < 4 \\ -1, & 4 < x < 7 \end{cases}$$

has the Fourier series

$$f(x) = \frac{4}{\pi}\sum^{\infty} \frac{\sin\left[\frac{1}{3}(2n-1)\pi(x-1)\right]}{2n-1}.$$

8. Verify the integral identities where n and k are integers:

$$\int_{-p}^{p}\cos\frac{n\pi x}{p}\,dx = \int_{-p}^{p}\sin\frac{n\pi x}{p}\,dx = \int_{-p}^{p}\sin\frac{n\pi x}{p}\cos\frac{k\pi x}{p}\,dx = 0,$$

$$\int_{-p}^{p}\cos\frac{n\pi x}{p}\cos\frac{k\pi x}{p}\,dx = \int_{-p}^{p}\sin\frac{n\pi x}{p}\sin\frac{k\pi x}{p}\,dx = \begin{cases} 0, & k \neq n \\ p, & k = n \end{cases}$$

9. Find the Fourier series of the periodic function resulting from passing the voltage $v(t) = E\cos 100\pi t$ through a *half-wave rectifier* (see Problem 16 in Exercises 4.2).

10. Find the Fourier series of the periodic function resulting from passing the voltage $v(t) = E\sin\omega t$ through a *half-wave rectifier* (see Problem 16 in Exercises 4.2).

11. A certain type of *full-wave rectifier* converts the input voltage $v(t)$ to its absolute value at the output, i.e., $|v(t)|$. Assuming the input voltage is given by $v(t) = E \sin \omega t$, determine the Fourier series of the periodic output voltage.

12. Find the Fourier series of the periodic function resulting from passing the voltage $v(t) = E \cos \omega t$ through a *full-wave rectifier* (see Problem 11).

13. Given the function $f(x) = \pi - x$, $0 < x < \pi$, find
    (a) the Fourier cosine series (with period 2π).
    (b) the Fourier sine series (with period 2π).
    (c) the complete Fourier series (with period π).

★14. If f is a periodic function with period $T > 0$ and g is likewise a periodic function with period $S > 0$,
    (a) under what condition on T and S is the sum $f + g$ a periodic function?
    (b) If $f(x) = \cos \pi x$ and $g(x) = \cos \sqrt{2}\pi x$, is the sum $f + g$ periodic?

In Problems 15–22, find (a) the half-range cosine series and (b) the half-range sine series for the given function.

15. $f(x) = 1, \quad 0 \leq x \leq 1$

16. $f(x) = \begin{cases} 1, & 0 < x < 1 \\ 0, & 1 < x < 2 \end{cases}$

17. $f(x) = x, \quad 0 < x < 2$

18. $f(x) = x - x^2, \quad 0 < x < 1$

19. $f(x) = \sin x, \quad 0 \leq x \leq \pi$

20. $f(x) = \cos x, \quad 0 \leq x \leq \pi$

21. $f(x) = e^{2x}, \quad 0 < x < 1$

22. $f(x) = \cosh x, \quad 0 \leq x \leq \pi$

4.6 GENERALIZED FOURIER SERIES

In addition to sines and cosines, one may well wonder if other functions can be used in developing infinite series like Fourier series. The answer is yes. Series of this nature have the general form

$$f(x) = \sum_{n=1}^{\infty} c_n \phi_n(x), \qquad x_1 < x < x_2, \tag{48}$$

where the set of functions $\{\phi_n(x)\}$ is *orthogonal* on the specified interval with respect to a given weighting function $r(x) > 0$, i.e.,

$$\int_{x_1}^{x_2} r(x)\phi_n(x)\phi_k(x)\,dx = 0, \qquad \text{for all } k \neq n. \tag{49}$$

Such series are commonly called *generalized Fourier series* since they contain Fourier trigonometric series as a special case. Because the set of orthogonal functions is usually composed of eigenfunctions of some Sturm-Liouville system, we also refer to (48) as a *Sturm-Liouville series*.*

---

*The Fourier trigonometric series are actually Sturm-Liouville series whose eigenfunctions belong to the periodic Sturm-Liouville system discussed in Example 21 in Section 1.6.

When $k = n$ in the integral in (49), the non-negative number defined by

$$\| \phi_n(x) \| = \sqrt{\int_{x_1}^{x_2} r(x)[\phi_n(x)]^2 \, dx}, \qquad n = 1, 2, 3, \ldots \tag{50}$$

is called the *norm* of $\phi_n(x)$. This is comparable to the norm or magnitude of a vector in three-dimensional vector space. If $\| \phi_n(x) \| = 1$ for all n, we say the orthogonal set $\{\phi_n(x)\}$ is *normalized*, or that the set is an *orthonormal* set. Although we choose not to do so here, it is a common practice in many textbooks to develop the theory of generalized Fourier series in terms of orthonormal functions. This poses no real restriction since any orthogonal set $\{\psi_n(x)\}$ can be transformed into an orthonormal set $\{\phi_n(x)\}$ by simply setting

$$\phi_n(x) = \frac{\psi_n(x)}{\| \psi_n(x) \|}, \qquad n = 1, 2, 3, \ldots . \tag{51}$$

EXAMPLE 8

Transform the orthogonal set of functions $\{\sin nx\}$ on the interval $(0, \pi)$ into an orthonormal set on this interval.

SOLUTION From previous results we know that for $r(x) = 1$, we have

$$\int_0^\pi \sin nx \sin kx \, dx = 0, \qquad k \neq n.$$

Thus, since

$$\| \sin nx \|^2 = \int_0^\pi \sin^2 nx \, dx = \frac{\pi}{2}, \qquad n = 1, 2, 3, \ldots ,$$

it follows that the orthonormal set is

$$\phi_n(x) = \frac{\sin nx}{\| \sin nx \|} = \sqrt{\frac{2}{\pi}} \sin nx, \qquad n = 1, 2, 3, \ldots .$$

In order to obtain an expression for the coefficients c_n in (48), we make use of the orthogonality property of the ϕ's. That is, we begin by multiplying both sides of (48) by the term $r(x)\phi_k(x)$ and integrating the result termwise (assuming that this is justified), which yields

$$\int_{x_1}^{x_2} r(x)f(x)\phi_k(x) \, dx = \sum_{n=1}^{\infty} c_n \int_{x_1}^{x_2} \underbrace{r(x)\phi_n(x)\phi_k(x) \, dx}_{0\,(n \neq k)}.$$

Based upon (49) and (50), we see that all terms on the right-hand side vanish except the term corresponding to $n = k$. Thus the expression is reduced to

$$\int_{x_1}^{x_2} r(x)f(x)\phi_k(x) \, dx = c_k \int_{x_1}^{x_2} r(x)[\phi_k(x)]^2 \, dx = c_k \| \phi_k(x) \|^2,$$

which leads to the formal result

$$c_k = \|\phi_k(x)\|^{-2} \int_{x_1}^{x_2} r(x)f(x)\phi_k(x)\, dx, \qquad k = 1, 2, 3, \ldots. \tag{52}$$

The constants defined by (52) are called the *Fourier coefficients* of f. We will refer to $\|\phi_k(x)\|^2$ in subsequent discussions as the *normalization factor*.

The theory of generalized Fourier series is very similar to that of Fourier trigonometric series. For example, we have the following convergence theorem for regular Sturm-Liouville systems, which we state without proof.*

THEOREM 4.7

Let $\{\phi_n(x)\}$ denote the set of eigenfunctions of the regular Sturm-Liouville system

$$L[y] + \lambda r(x)y = 0, \qquad x_1 < x < x_2$$

$$B_1[y] \equiv a_{11}y(x_1) + a_{12}y'(x_1) = 0,$$

$$B_2[y] \equiv a_{21}y(x_2) + a_{22}y'(x_2) = 0,$$

where $L = D[p(x)D] + q(x)$. If f is piecewise smooth on $[x_1, x_2]$, then the series

$$f(x) = \sum_{n=1}^{\infty} c_n\phi_n(x), \qquad x_1 < x < x_2$$

where

$$c_n = \|\phi_n(x)\|^{-2} \int_{x_1}^{x_2} r(x)f(x)\phi_n(x)\, dx, \qquad n = 1, 2, 3, \ldots,$$

converges pointwise to $f(x)$ at each point of continuity of f in the interval (x_1, x_2), and to the average value $[f(x^+) + f(x^-)]/2$ at each point of discontinuity of f in (x_1, x_2).

EXAMPLE 9

Find a generalized Fourier series of the function $f(x) = 1, 0 < x < 1$, in terms of the eigenfunctions of

$$y'' + y' + \lambda y = 0, \qquad 0 < x < 1, \qquad y(0) = 0, \qquad y(1) = 0.$$

SOLUTION We first solve the eigenvalue problem to identify the eigenfunctions $\{\phi_n(x)\}$. The DE has the auxiliary equation

* For a proof of Theorem 4.7, see E. L. Ince, *Ordinary Differential Equations*, Chapter XI (Dover: New York), 1956.

$$m^2 + m + \lambda = 0,$$

with roots

$$m = -\frac{1}{2} \pm ik,$$

where we have set $\lambda = k^2 + 1/4$. Thus, the general solution is

$$y = e^{-x/2}(C_1 \cos kx + C_2 \sin kx),$$

and applying the boundary conditions, we find $C_1 = 0$ and

$$C_2 e^{-1/2} \sin k = 0.$$

For $C_2 \neq 0$, it follows that $k = n\pi$ ($n = 1, 2, 3, \ldots$), and therefore the eigenfunctions are

$$\phi_n(x) = e^{-x/2} \sin n\pi x, \qquad n = 1, 2, 3, \ldots$$

(there are no further eigenfunctions).

By multiplying the DE by e^x, we obtain the self-adjoint form (see Section 1.6)

$$e^x y'' + e^x y' + \lambda e^x y = 0,$$

and hence we see that the weighting function is $r(x) = e^x$. Thus, the normalization factor of the eigenfunctions is

$$\| \phi_n(x) \|^2 = \int_0^1 e^x [e^{-x/2} \sin n\pi x]^2 \, dx = \frac{1}{2}, \qquad n = 1, 2, 3, \ldots.$$

For this set of eigenfunctions, the series for f is

$$f(x) = \sum_{n=1}^{\infty} c_n e^{-x/2} \sin n\pi x = e^{-x/2} \sum_{n=1}^{\infty} c_n \sin n\pi x, \qquad 0 < x < 1,$$

where

$$c_n = 2 \int_0^1 e^x f(x) e^{-x/2} \sin n\pi x \, dx$$

$$= 2 \int_0^1 e^{x/2} \sin n\pi x \, dx$$

$$= \frac{2n\pi}{n^2\pi^2 + \frac{1}{4}} [(-1)^{n-1} e^{1/2} + 1], \qquad n = 1, 2, 3, \ldots.$$

Therefore,

$$f(x) = e^{-x/2} \sum_{n=1}^{\infty} \frac{2n\pi}{n^2\pi^2 + \frac{1}{4}} [(-1)^{n-1} e^{1/2} + 1] \sin n\pi x.$$

4.6.1 NONHOMOGENEOUS BOUNDARY VALUE PROBLEMS

Generalized Fourier series can be used in an interesting and important manner to solve certain nonhomogeneous boundary value problems. For the purpose of discussion, let us consider the problem described by the *nonhomogeneous DE* and *homogeneous boundary conditions*

$$L[y] = f(x), \qquad x_1 < x < x_2, \qquad B_1[y] = 0, \qquad B_2[y] = 0, \tag{53}$$

where $L = D[p(x)D] + q(x)$ is a self-adjoint operator, and where B_1 and B_2 are the unmixed boundary operators defined by

$$B_1[y] \equiv a_{11}y(x_1) + a_{12}y'(x_1),$$

$$B_2[y] \equiv a_{21}y(x_2) + a_{22}y'(x_2).$$

Closely associated with this problem is the regular Sturm-Liouville system

$$L[y] + \lambda r(x)y = 0, \qquad B_1[y] = 0, \qquad B_2[y] = 0 \tag{54}$$

(see Section 1.6.2), with eigenvalues λ_n and eigenfunctions $\phi_n(x)$, $n = 1, 2, 3, \ldots$. The function $r(x)$ is chosen so that the Sturm-Liouville problem (54) is reasonably tractable. We will clarify this comment shortly.

We start by assuming that the solution of (53) can be expressed as an infinite series expansion of the form

$$y = \sum_{n=1}^{\infty} c_n\phi_n(x), \qquad x_1 < x < x_2. \tag{55}$$

In this setting we call (55) an *eigenfunction expansion,* which is also the name of the technique we are about to describe. Clearly y, given by (55), satisfies the prescribed homogeneous boundary conditions, since each eigenfunction $\phi_n(x)$ does. To determine the c's, we recognize that (55) is a generalized Fourier series, and therefore it follows that

$$c_n = \|\phi_n(x)\|^{-2} \int_{x_1}^{x_2} r(x)y(x)\phi_n(x)\,dx, \qquad n = 1, 2, 3, \ldots. \tag{56}$$

Unfortunately, since y itself is unknown, we cannot use (56) to calculate the c's. Instead, we try to determine the c's by satisfying the DE in (53) with the series (55). Direct substitution of (55) into the DE in (53) leads to

$$L[y] = \sum_{n=1}^{\infty} c_n L[\phi_n(x)] = f(x), \tag{57}$$

where we are assuming that it is permissible to interchange the operations of summation and differentiation. Because the eigenfunctions and eigenvalues must satisfy the relation

$$L[\phi_n(x)] = -\lambda_n r(x)\phi_n(x), \qquad n = 1, 2, 3, \ldots, \tag{58}$$

we find that (57) becomes

$$-\sum_{n=1}^{\infty} c_n\lambda_n r(x)\phi_n(x) = f(x),$$

which can be rearranged as

$$\frac{f(x)}{r(x)} = \sum_{n=1}^{\infty} a_n \phi_n(x), \qquad x_1 < x < x_2, \tag{59}$$

where we have set $-c_n\lambda_n = a_n$ for notational convenience. The constants a_n in (59) are simply the Fourier coefficients of the function $f(x)/r(x)$, and therefore are given by the formula

$$a_n = \| \phi_n(x) \|^{-2} \int_{x_1}^{x_2} r(x) \frac{f(x)}{r(x)} \phi_n(x)\, dx,$$

which simplifies to

$$a_n = \| \phi_n(x) \|^{-2} \int_{x_1}^{x_2} f(x)\phi_n(x)\, dx, \qquad n = 1, 2, 3, \ldots . \tag{60}$$

In summary, our formal solution of (53) is given by the eigenfunction expansion

$$y = -\sum_{n=1}^{\infty} \left(\frac{a_n}{\lambda_n}\right) \phi_n(x), \qquad x_1 < x < x_2, \tag{61}$$

where the a's are defined by (60). Let us illustrate the technique with an example.

REMARK: *It is important to recognize that (60) and (61) are based upon the assumption that the original DE (53) is in self-adjoint form. This permits proper identification of the function $f(x)$ and weighting function $r(x)$.*

EXAMPLE 10

The small displacement of an elastic string stretched tightly with tension T between fixed supports at $x = 0$ and $x = \pi$, and subject to the external load $q(x) = Tx$, is governed by the boundary value problem

$$y'' = -x, \qquad 0 < x < \pi, \qquad y(0) = 0, \qquad y(\pi) = 0$$

(see Section 1.5.1). Use the eigenfunction expansion method to determine the displacment y. Note that this DE is already in self-adjoint form.

SOLUTION An associated Sturm-Liouville problem is

$$y'' + \lambda y = 0, \qquad y(0) = 0, \qquad y(\pi) = 0,$$

with eigenvalues and eigenfunctions given by

$$\lambda_n = n^2, \qquad \phi_n(x) = \sin nx, \qquad n = 1, 2, 3, \ldots .$$

The solution we seek is then of the form

$$y = -\sum_{n=1}^{\infty} \frac{a_n}{n^2} \sin nx, \qquad 0 < x < \pi.$$

To determine the a's, we first calculate

$$\| \phi_n(x) \|^2 = \int_0^{\pi} \sin^2 nx \, dx = \frac{\pi}{2}, \qquad n = 1, 2, 3, \ldots,$$

[note that $r(x) = 1$] and then using (60), we obtain

$$a_n = -\frac{2}{\pi} \int_0^{\pi} x \sin nx \, dx = (-1)^n \frac{2}{n}, \qquad n = 1, 2, 3, \ldots.$$

Thus, we get the series solution for the displacement

$$y = 2 \sum_{n=1}^{\infty} \frac{(-1)^{n-1}}{n^3} \sin nx, \qquad 0 < x < \pi.$$

The Sturm-Liouville system (54) associated with the nonhomogeneous problem (53) is not uniquely defined. That is, more than one suitable eigenvalue problem may be found for the same operator L and boundary conditions, but with different weighting functions $r(x)$. As a general rule, when the nonhomogeneous DE is not in self-adjoint form but is given by

$$M[y] = F(x), \qquad x_1 < x < x_2, \tag{62}$$

where $M = A_2(x)D^2 + A_1(x)D + A_0(x)$, then the associated eigenequation can often be formed by simply adding the term λy to the left-hand side and equating the result to zero, i.e.,

$$M[y] + \lambda y = 0, \qquad x_1 < x < x_2. \tag{63}$$

In some cases, however, it may be either necessary or convenient to assume the more general form

$$M[y] + \lambda R(x) y = 0, \qquad x_1 < x < x_2, \tag{64}$$

where $R(x)$ is not unity. The choice of $R(x)$ is usually suggested by the form of the prescribed boundary conditions. Subtleties of this kind are more commonly associated with variable–coefficient equations such as those discussed in Chapter 9.

Equation (61) is a unique solution of (53) provided $\lambda_n \neq 0$ for all n. But if one of the eigenvalues, say λ_j, is zero, the problem has either no solution (when $a_j \neq 0$) or infinitely many solutions (when $a_j = 0$). Let us formalize these remarks in the following theorem.*

---

*Theorem 4.8 (in part) is another restatement of Theorems 1.2 and 2.3.

THEOREM 4.8

If L is a self-adjoint operator and $\lambda_j = 0$ is an eigenvalue of the regular Sturm-Liouville system

$$L[y] + \lambda r(x)y = 0, \qquad x_1 < x < x_2, \qquad B_1[y] = 0, \qquad B_2[y] = 0,$$

then the associated nonhomogeneous problem

$$L[y] = f(x), \qquad x_1 < x < x_2, \qquad B_1[y] = 0, \qquad B_2[y] = 0,$$

has a solution if and only if $a_j = 0$, where

$$a_n = \| \phi_n(x) \|^{-2} \int_{x_1}^{x_2} f(x)\phi_n(x)\, dx, \qquad n = 1, 2, 3, \ldots, j, \ldots.$$

In this case the solution is given by

$$y = C\phi_j(x) - \sum_{\substack{n=1 \\ n \neq j}}^{\infty} \left(\frac{a_n}{\lambda_n}\right) \phi_n(x),$$

where C is an arbitrary constant, and λ_n and $\phi_n(x)$ are the eigenvalues and eigenfunctions, respectively, of the given regular Sturm-Liouville system.

Theorem 4.8 may be paraphrased as saying that if $\phi_j(x)$ is a nontrivial solution of the homogeneous boundary value problem

$$L[y] = 0, \qquad B_1[y] = 0, \qquad B_2[y] = 0, \tag{65}$$

then the related nonhomogeneous problem

$$L[y] = f(x), \qquad B_1[y] = 0, \qquad B_2[y] = 0, \tag{66}$$

has a solution if and only if the function $f(x)$ is (simply) orthogonal to $\phi_j(x)$, i.e., if and only if

$$\int_{x_1}^{x_2} f(x)\phi_j(x)\, dx = 0. \tag{67}$$

EXAMPLE 11

Use the eigenfunction expansion method to solve

$$y'' + 4y = \sin x, \qquad 0 < x < \pi, \qquad y(0) = 0, \qquad y(\pi) = 0.$$

SOLUTION We first solve the associated Sturm-Liouville system

$$y'' + (\lambda + 4)y = 0, \qquad y(0) = 0, \qquad y(\pi) = 0,$$

which has eigenvalues and eigenfunctions given by

$$\lambda_n = n^2 - 4, \qquad \phi_n(x) = \sin nx, \qquad n = 1, 2, 3, \ldots .$$

Here $\lambda_2 = 0$, and so a solution exists if and only if $a_2 = 0$, and in this case it has the form

$$y = C \sin 2x - \sum_{\substack{n=1 \\ n \neq 2}}^{\infty} \frac{a_n}{n^2 - 4} \sin nx.$$

Calculating the a's, we find

$$a_n = \frac{2}{\pi} \int_0^{\pi} \sin x \sin nx \, dx = \begin{cases} 1, & n = 1 \\ 0, & n \neq 1. \end{cases}$$

Therefore, $a_2 = 0$ as required for a solution to exist, and accordingly,

$$y = C \sin 2x + \frac{1}{3} \sin x,$$

where C can be any constant.

Although we have limited our discussion in this section to regular Sturm-Liouville systems, the technique of eigenfunction expansions carries over to periodic and singular systems as well.

Finally, if the nonhomogeneous problem should be of the more general form

$$L[y] = f(x), \qquad x_1 < x < x_2, \qquad B_1[y] = \alpha, \qquad B_2[y] = \beta, \tag{68}$$

where both the DE and boundary conditions are nonhomogeneous, we simply split (68) into two problems:

$$L[y] = 0, \qquad B_1[y] = \alpha, \qquad B_2[y] = \beta, \tag{69}$$

and

$$L[y] = f(x), \qquad B_1[y] = 0, \qquad B_2[y] = 0. \tag{70}$$

The problem described by (69), which features a homogeneous DE, can be solved by techniques discussed in Chapter 1, and the eigenfunction expansion technique can be applied to (70). The solution of (68) is then the sum of solutions of (69) and (70).

EXAMPLE 12

Use the eigenfunction expansion method to solve the Cauchy-Euler problem

$$x \frac{d}{dx}(xy') - y = \log x, \qquad 1 < x < 2, \qquad y'(1) = 0, \qquad y(2) = 5.$$

SOLUTION The two problems as described by (69) and (70) are given by

$$x^2 y'' + xy' - y = 0, \qquad y'(1) = 0, \qquad y(2) = 5$$

and

$$\frac{d}{dx}(xy') - \frac{1}{x}y = \frac{\log x}{x}, \qquad y'(1) = 0, \qquad y(2) = 0,$$

solutions of which will be designated by y_H and y_P, respectively. We have expressed the DE in the second problem in self-adjoint form so that we can properly identify the forcing function $f(x) = (\log x)/x$.

The first problem above has the solution

$$y_H = 2(x + x^{-1}),$$

which is readily found by techniques discussed in Chapter 1.* To solve the second problem above, we must first solve the associated eigenvalue problem

$$x^2y'' + xy' + (\lambda - 1)y = 0, \qquad y'(1) = 0, \qquad y(2) = 0.$$

Setting $\lambda - 1 = k^2 > 0$, we obtain the general solution

$$y = C_1 \cos(k \log x) + C_2 \sin(k \log x),$$

and by imposing the homogeneous boundary conditions, we find that $C_2 = 0$ and

$$C_1 \cos(k \log 2) = 0.$$

Therefore, for $C_1 \neq 0$ it follows that $k \log 2 = (2n - 1)\pi/2$, $n = 1, 2, 3, \ldots$, and consequently,

$$\lambda_n = 1 + \frac{(2n-1)^2\pi^2}{4(\log 2)^2}, \qquad \phi_n(x) = \cos\left[\frac{(2n-1)\pi \log x}{2 \log 2}\right], \qquad n = 1, 2, 3, \ldots.$$

We leave it to the reader to verify that there are no further eigenvalues.

The solution y_P that we are seeking can now be expressed in the form

$$y_P = -\sum_{n=1}^{\infty}\left(\frac{a_n}{\lambda_n}\right)\cos\left[\frac{(2n-1)\pi \log x}{2 \log 2}\right],$$

where λ_n denotes the set of eigenvalues cited above, none of which is zero. In self-adjoint form, the preceding eigenequation becomes

$$\frac{d}{dx}(xy') + \frac{(\lambda - 1)y}{x} = 0.$$

Thus, we identify $r(x) = 1/x$, which we use to calculate

$$\|\phi_n(x)\|^2 = \int_1^2 \frac{1}{x}\cos^2\left[\frac{(2n-1)\pi \log x}{2 \log 2}\right] dx = \frac{1}{2}\log 2, \qquad n = 1, 2, 3, \ldots,$$

and hence,

---

*In particular, see Problem 34 in Exercises 1.2.

$$a_n = \frac{2}{\log 2}\int_1^2 \frac{\log x}{x}\cos\left[\frac{(2n-1)\pi\log x}{2\log 2}\right]dx$$

$$= \begin{cases} -\dfrac{16\log 2}{(2n-1)^2\pi^2}, & n = 1, 3, 5, \ldots \\ 0, & n = 2, 4, 6, \ldots. \end{cases}$$

Lastly, combining solutions $y = y_H + y_P$ leads to the result

$$y = 2(x + x^{-1}) + \frac{16\log 2}{\pi^2}\sum_{n=1}^{\infty}\frac{\cos\left[\dfrac{(2n-1)\pi\log x}{2\log 2}\right]}{\lambda_n(2n-1)^2},$$

where

$$\lambda_n = 1 + \frac{(2n-1)^2\pi^2}{4(\log 2)^2}.$$

□ 4.6.2 BILINEAR FORMULA FOR THE GREEN'S FUNCTION

The eigenfunction expansion method provides us with an alternative technique for constructing the Green's function discussed in Section 2.4. To see this, we first note that (70) above has the formal solution

$$y = -\int_{x_1}^{x_2} g(x, s)f(s)\,ds, \tag{71}$$

where $g(x, s)$ is the Green's function. On the other hand, if we substitute Equation (60) for the a's into the solution (61), using s as a dummy variable of integration, we are led to the result

$$y = -\int_{x_1}^{x_2}\left(\sum_{n=1}^{\infty}\frac{\phi_n(x)\phi_n(s)}{\|\phi_n(x)\|^2\lambda_n}\right)f(s)\,ds. \tag{72}$$

Because (71) and (72) are unique solutions of the same problem,* we deduce that the Green's function has the series representation

$$g(x, s) = \sum_{n=1}^{\infty}\frac{\phi_n(x)\phi_n(s)}{\|\phi_n(x)\|^2\lambda_n}, \tag{73}$$

called the *bilinear formula* for the Green's function. Its construction depends only upon the eigenvalues and eigenfunctions of the associated Sturm-Liouville system. It offers an interesting alternative for constructing the Green's function in those cases where it may be difficult to construct by the method of Section 2.4.

---

* We are assuming that none of the eigenvalues λ_n are zero.

EXAMPLE 13

Find a bilinear representation of the Green's function associated with

$$y'' = f(x), \quad 0 < x < 1, \quad y(0) = \alpha, \quad y(1) = \beta.$$

SOLUTION An associated eigenvalue problem is

$$y'' + \lambda y = 0, \quad y(0) = 0, \quad y(1) = 0,$$

with eigenvalues and eigenfunctions given by

$$\lambda_n = n^2\pi^2, \quad \phi_n(x) = \sin n\pi x, \quad n = 1, 2, 3, \ldots .$$

The normalization factor is found to be

$$\| \phi_n(x) \|^2 = \int_0^1 \sin^2 n\pi x \, dx = \frac{1}{2}, \quad n = 1, 2, 3, \ldots$$

$[r(x) = 1]$, so that upon substituting these expressions into (73), we obtain

$$g(x, s) = \frac{2}{\pi^2} \sum_{n=1}^{\infty} \frac{\sin n\pi x \sin n\pi s}{n^2}.$$

Recall from Equation (68) in Section 2.4.2 that the Green's function for this problem was previously shown to have the form

$$g(x, s) = \begin{cases} s(1 - x), & 0 < s \leq x \\ x(1 - s), & x < s < 1. \end{cases}$$

EXERCISES 4.6

1. Show that $L_0(x) = 1$, $L_1(x) = 1 - x$, and $L_2(x) = 1 - 2x + x^2/2$ are mutually orthogonal on the interval $(0, \infty)$ with respect to the weighting function $r(x) = e^{-x}$. These polynomials are the first three members of the set of *Laguerre polynomials*.

2. Show that $T_0(x) = 1$, $T_1(x) = x$, and $T_2(x) = 2x^2 - 1$ are mutually orthogonal on the interval $(-1, 1)$ with respect to the weighting function $r(x) = (1 - x^2)^{-1/2}$. These polynomials are the first three members of the set of *Chebyshëv polynomials*.
Hint: Let $x = \cos \theta$.

In Problems 3–6, determine the *normalized* eigenfunctions.

3. $y'' + \lambda y = 0, \quad y'(0) = 0, \quad y'(\pi) = 0$

4. $y'' + \lambda y = 0, \quad y(0) = 0, \quad y'(1) = 0$

★5. $y'' + \lambda y = 0, \quad y(0) = 0, \quad y'(1) + y(1) = 0$

6. $y'' - 2y' + \lambda y = 0, \quad y(0) = 0, \quad y(1) = 0$

In Problems 7–14, find a generalized Fourier series for the function f in terms of the eigenfunctions of the given Sturm-Liouville system.

7. $f(x) = 1, \quad 0 < x < 1: \quad y'' + \lambda y = 0, \quad y(0) = 0, \quad y'(1) = 0$

8. $f(x) = x, \quad 0 < x < \pi: \quad y'' + \lambda y = 0, \quad y'(0) = 0, \quad y'(\pi) = 0$

9. $f(x) = 1 + x, \quad -\pi < x < \pi: \quad y'' + \lambda y = 0, \quad y(-\pi) = y(\pi),$
$y'(-\pi) = y'(\pi)$

10. $f(x) = 1, \quad 0 < x < 2: \quad y'' + 4y' + (4 + 9\lambda)y = 0, \quad y(0) = 0, \quad y(2) = 0$

11. $f(x) = 1, \quad 0 < x < 1: \quad y'' - 3y' + 2\lambda y = 0, \quad y(0) = 0, \quad y(1) = 0$

★12. $f(x) = 1, \quad 0 < x < 2: \quad y'' + 2y' + \lambda y = 0, \quad y(0) = 0, \quad y'(2) = 0$

13. $f(x) = \log x, \quad 1 < x < e: \quad x^2y'' + xy' + \lambda y = 0, \quad y(1) = 0, \quad y(e) = 0$

14. $f(x) = x^2, \quad 1 < x < e: \quad x^2y'' - xy' + \lambda y = 0, \quad y(1) = 0, \quad y(e) = 0$

In Problems 15–26, solve by the method of eigenfunction expansions.

15. $y'' = -x, \quad y(0) = 0, \quad y(1) = 0$

★16. $y'' = -x, \quad y(0) = 0, \quad y(1) + 2y'(1) = 0$

17. $y'' + y = 1, \quad y'(0) = 1, \quad y'(1) = 1$

18. $y'' - y = x^2, \quad y(0) = 0, \quad y(1) = 0$

19. $y'' - y = e^x, \quad y(0) = 2, \quad y'(1) = 0$

★20. $y'' + 4y = 1 - \delta(x - 1), \quad y(0) = 0, \quad y(2) = 1$

21. $y'' + 2y' + y = e^x, \quad y(0) = 3, \quad y(1) = 0$

22. $xy'' + y' = x, \quad y(1) = 5, \quad y(e) = 2$

★23. $x^2y'' - 3xy' + 3y = 2x^2, \quad y(1) = 1, \quad y(2) = -1$

24. $\dfrac{d}{dx}(xy') - \dfrac{y}{x} = \dfrac{1}{x}, \quad y(1) = 5, \quad y(e) = 0$

25. $x\dfrac{d}{dx}(xy') + y = \log x, \quad y'(1) = 0, \quad y'(e^2) = 1$

★26. $x\dfrac{d}{dx}(xy') + 4y = \sin(2 \log x), \quad y(1) = 1, \quad y(3) = 0$

In Problems 27–32, solve by Theorem 4.8 (if possible).

27. $y'' = \cos \pi x, \quad y'(0) = 0, \quad y'(1) = 0$

28. $y'' = -x, \quad y'(0) = 0, \quad y'(1) = 0$

29. $y'' + 9y = 1, \quad y(0) = 0, \quad y(\pi) = 0$

30. $y'' + 9y = 5 \sin 7x, \quad y(0) = 0, \quad y(\pi) = 0$

31. $x^2y'' + xy' + 16y = \sin(3 \log x), \quad y(1) = 0, \quad y(e^\pi) = 0$

32. $x^2y'' + xy' + y = \dfrac{1}{x}, \quad y(1) = 0, \quad y(e^\pi) = 0$

In Problems 33–40, find bilinear formulas for the Green's function associated with the specified operators.

33. $L = D^2; \quad y(a) = 0, \quad y(b) = 0, \quad a \neq b$

34. $L = D^2 + 1; \quad y(0) = 0, \quad y'(1) = 0$

35. $L = D^2 + 1; \quad y'(0) = 0, \quad y(1) = 0$

36. $L = D^2 + 1; \quad y'(0) = 0, \quad y'(1) = 0$

37. $L = D^2 - 1; \quad y'(0) = \alpha, \quad y'(\pi) = \beta$

38. $L = D(e^{3x}D) + 2e^{3x}; \quad y(0) = \alpha, \quad y(1) = \beta$

39. $L = D(xD) + \dfrac{1}{x}; \quad y(1) = \alpha, \quad y(e) = \beta$

40. $L = D(xD) + \dfrac{1}{x}; \quad y'(1) = \alpha, \quad y(e) = \beta$

41. A simply supported beam of length c has a constant load $q(x) = q_0/c$ distributed over its length. The small deflections of the beam are governed by the boundary value problem

$$EIy^{(4)} = \frac{q_0}{c}, \quad y(0) = y''(0) = 0, \quad y(c) = y''(c) = 0,$$

where EI is constant. Show that the deflections are given by the series

$$y = \frac{4q_0c^4}{EI\pi^5} \sum_{\substack{n=1 \\ (\text{odd})}}^{\infty} \frac{(-1)^{n-1}}{n^5} \sin \frac{n\pi x}{c}.$$

★42. Verify directly that the solution given in Theorem 4.8 satisfies the specified DE.

★43. Suppose we wish to approximate a given function $f(x)$ by the series

$$S_N(x) = \sum_{n=1}^{N} \alpha_n \phi_n(x), \quad x_1 < x < x_2$$

where $\{\phi_n(x)\}$ is an orthogonal set of functions with weight function $r(x)$. Show that the *mean–square error* of the approximation

$$E_N = \int_{x_1}^{x_2} r(x)[f(x) - S_N(x)]^2\, dx$$

is minimized by choosing the constants α_n to be the Fourier constants c_n defined by (52), i.e., show that $\alpha_n = c_n$, $n = 1, 2, \ldots, N$.

Hint: Set $\dfrac{\partial E_N}{\partial \alpha_k} = 0, \quad k = 1, 2, \ldots, N.$

★44. If $E_N \to 0$ as $N \to \infty$ in Problem 43, we say that the series $S_N(x)$ *converges in the mean* to $f(x)$. Show that this condition leads to *Parseval's equality*

$$\int_{x_1}^{x_2} r(x)[f(x)]^2\, dx = \sum_{n=1}^{\infty} c_n^2.$$

4.7 FOURIER INTEGRAL REPRESENTATIONS

In previous sections we demonstrated that a periodic function satisfying minimal requirements can be represented by an infinite sum of sinusoidal functions (Fourier series). Nonperiodic functions defined on the entire axis, however, cannot be represented by Fourier series since such series necessarily define periodic functions. Nonetheless, it is instructive to think of nonperiodic functions as if they were limiting cases of periodic functions for which the period is "infinite." In this fashion we find that the formal limit of a Fourier series representation as the period tends to infinity can be used to introduce the notion of a *Fourier integral representation.*

If f is a piecewise smooth function with period $2p$, it has the Fourier series representation

$$f(x) = \frac{1}{2} a_0 + \sum_{n=1}^{\infty} \left(a_n \cos \frac{n\pi x}{p} + b_n \sin \frac{n\pi x}{p} \right), \tag{74}$$

where

$$a_n = \frac{1}{p} \int_{-p}^{p} f(t) \cos \frac{n\pi t}{p} \, dt, \qquad n = 0, 1, 2, \ldots \tag{75}$$

and

$$b_n = \frac{1}{p} \int_{-p}^{p} f(t) \sin \frac{n\pi t}{p} \, dt, \qquad n = 1, 2, 3, \ldots. \tag{76}$$

Substitution of the integral formulas for a_0, a_n, and b_n into (74) leads to the result

$$\begin{aligned} f(x) &= \frac{1}{2p} \int_{-p}^{p} f(t) \, dt + \sum_{n=1}^{\infty} \left[\frac{1}{p} \int_{-p}^{p} f(t) \cos \frac{n\pi t}{p} \cos \frac{n\pi x}{p} \, dt \right. \\ &\quad \left. + \frac{1}{p} \int_{-p}^{p} f(t) \sin \frac{n\pi t}{p} \sin \frac{n\pi x}{p} \, dt \right] \\ &= \frac{1}{2p} \int_{-p}^{p} f(t) \, dt + \frac{1}{p} \int_{-p}^{p} f(t) \sum_{n=1}^{\infty} \cos \left[\frac{n\pi(t - x)}{p} \right] dt, \end{aligned} \tag{77}$$

where we have used the trigonometric identity

$$\cos(A - B) = \cos A \cos B + \sin A \sin B$$

and interchanged the order of summation and integration.

We now wish to examine what happens when we let p tend to infinity. Here we further require f to be *absolutely integrable*. That is, we require

$$\int_{-\infty}^{\infty} |f(t)| \, dt < \infty, \tag{78}$$

so that clearly*

$$\lim_{p\to\infty} \frac{1}{2p} \int_{-p}^{p} f(t)\, dt = 0. \tag{79}$$

For the remaining infinite sum in (77), it is convenient to define $\Delta s = \pi/p$ and then consider the equivalent limit

$$f(x) = \lim_{\Delta s\to 0} \frac{1}{\pi} \int_{-p}^{p} f(t) \sum_{n=1}^{\infty} \cos[n\Delta s(t-x)]\Delta s\, dt. \tag{80}$$

Observe that $\Delta s \to 0$ as $p \to \infty$. When Δs is a small positive number, the points $n\Delta s$ are equally spaced along the s-axis. In such a case we may expect the series in (80) to approximate the integral

$$\int_0^{\infty} \cos[s(t-x)]\, ds$$

in the limit as $\Delta s \to 0$. However, the limit of this series is not the definition of an integral but merely suggests that, under appropriate conditions on f, Equation (80) tends to the integral form

$$f(x) = \frac{1}{\pi} \int_{-\infty}^{\infty} f(t) \int_0^{\infty} \cos[s(t-x)]\, ds\, dt \tag{81}$$

in the limit. Upon switching the order of integration, we get the equivalent form

$$f(x) = \frac{1}{\pi} \int_0^{\infty} \int_{-\infty}^{\infty} f(t) \cos[s(t-x)]\, dt\, ds. \tag{82}$$

The formal procedure we just went through is invalid since the passage to the limit cannot be justified, but it leads to a correct and important result known as *Fourier's integral theorem*. Although we won't do so here, this result can be proved analogous to the proof of Theorem 4.1 (see Problems 16–19 in Exercises 4.7). In summary, we state the following theorem.

THEOREM 4.9

Fourier Integral Theorem. If f is a piecewise smooth function and absolutely integrable on $(-\infty, \infty)$, then at points of continuity of f it satisfies the identity

$$f(x) = \frac{1}{\pi} \int_0^{\infty} \int_{-\infty}^{\infty} f(t) \cos[s(t-x)]\, dt\, ds.$$

At points of discontinuity of f, the integrals converge to the average value $[f(x^+) + f(x^-)]/2$.

---

*It follows that if f is absolutely integrable, then $\int_{-\infty}^{\infty} f(t)\, dt < \infty$.

Once again we caution that the conditions stated in Theorem 4.9 are only sufficient conditions, and they are not stated in the most general form known. Nonetheless, the conditions put forth are broad enough to embrace most of the funtions commonly occurring in practice.

To emphasize the analogy between Fourier series and the Fourier integral theorem, we rewrite (82) in the form

$$f(x) = \frac{1}{\pi}\int_0^\infty \int_{-\infty}^\infty f(t)(\cos st \cos sx + \sin st \sin sx)\,dt\,ds,$$

or, equivalently,

$$f(x) = \int_0^\infty [A(s) \cos sx + B(s) \sin sx]\,ds, \tag{83}$$

where

$$A(s) = \frac{1}{\pi}\int_{-\infty}^\infty f(t) \cos st\,dt \tag{84}$$

and

$$B(s) = \frac{1}{\pi}\int_{-\infty}^\infty f(t) \sin st\,dt. \tag{85}$$

In this setting we refer to (83) as the *Fourier integral representation* of the function f with coefficients defined by (84) and (85). The general theory concerning such representations closely parallels that of Fourier series.

EXAMPLE 14

Find an integral representation of the form (83) for the rectangle function $f(x) = h(1 - |x|)$, where h is the Heaviside unit function (see Figure 4.13).*

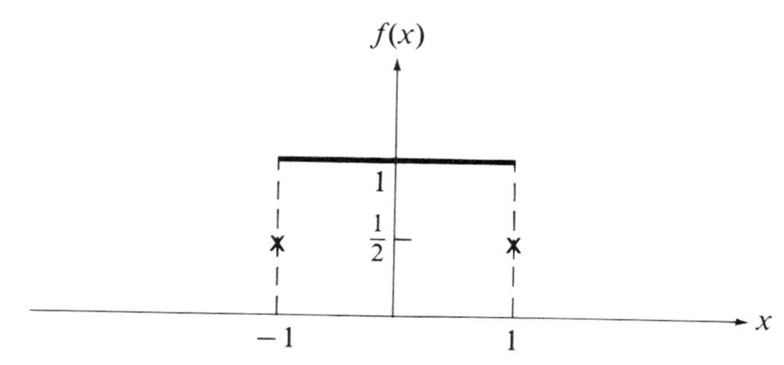

FIGURE 4.13

*The function h has the property that $h(t) = 1$, $t > 0$, and $h(t) = 0$, $t < 0$.

SOLUTION The coefficients $A(s)$ and $B(s)$ are given by

$$A(s) = \frac{1}{\pi}\int_{-\infty}^{\infty} h(1 - |x|) \cos sx\, dx = \frac{1}{\pi}\int_{-1}^{1} \cos sx\, dx = \frac{2 \sin s}{\pi s}$$

and

$$B(s) = \frac{1}{\pi}\int_{-\infty}^{\infty} h(1 - |x|) \sin sx\, dx = \frac{1}{\pi}\int_{-1}^{1} \sin sx\, dx = 0.$$

Thus the Fourier integral representation becomes

$$f(x) = \frac{2}{\pi}\int_{0}^{\infty} \left(\frac{\sin s}{s}\right) \cos sx\, ds.$$

Since $x = 0$ is a point of continuity of f in Example 14, we can use the Fourier integral theorem to deduce that

$$f(0) = 1 = \frac{2}{\pi}\int_{0}^{\infty} \frac{\sin s}{s}\, ds,$$

which leads to the interesting result

$$\int_{0}^{\infty} \frac{\sin s}{s}\, ds = \frac{\pi}{2}. \tag{86}$$

At $x = \pm 1$, however, we have a jump discontinuity in f, and the integral for f converges to the average functional value 1/2 at these points (see Figure 4.13). Therefore, we have that

$$\frac{2}{\pi}\int_{0}^{\infty} \left(\frac{\sin s}{s}\right) \cos sx\, ds = \begin{cases} \dfrac{1}{2}, & x = -1 \\ 1, & -1 < x < 1 \\ \dfrac{1}{2}, & x = 1. \end{cases} \tag{87}$$

Finally, it may be of interest to plot the "partial integral" of the function f in Example 14, defined by

$$S_{\mu}(x) = \frac{2}{\pi}\int_{0}^{\mu} \left(\frac{\sin s}{s}\right) \cos sx\, ds, \tag{88}$$

to see how it tends to $f(x)$ as $\mu \to \infty$. Recalling the identity

$$2 \sin A \cos B = \sin(A + B) + \sin(A - B),$$

we have

$$S_{\mu}(x) = \frac{1}{\pi}\int_{0}^{\mu} \frac{\sin[s(1 + x)]}{s}\, ds + \frac{1}{\pi}\int_{0}^{\mu} \frac{\sin[s(1 - x)]}{s}\, ds$$

$$= \frac{1}{\pi}\int_0^{\mu(1+x)} \frac{\sin t}{t}\,dt + \frac{1}{\pi}\int_0^{\mu(1-x)} \frac{\sin t}{t}\,dt$$

$$= \frac{1}{\pi}\{\mathrm{Si}[\mu(1+x)] + \mathrm{Si}[\mu(1-x)]\}, \tag{89}$$

where $\mathrm{Si}(z)$ is the *sine integral* defined by

$$\mathrm{Si}(z) = \int_0^z \frac{\sin t}{t}\,dt. \tag{90}$$

Equation (89) is plotted in Figure 4.14 for values $\mu = 4, 16, 128$.

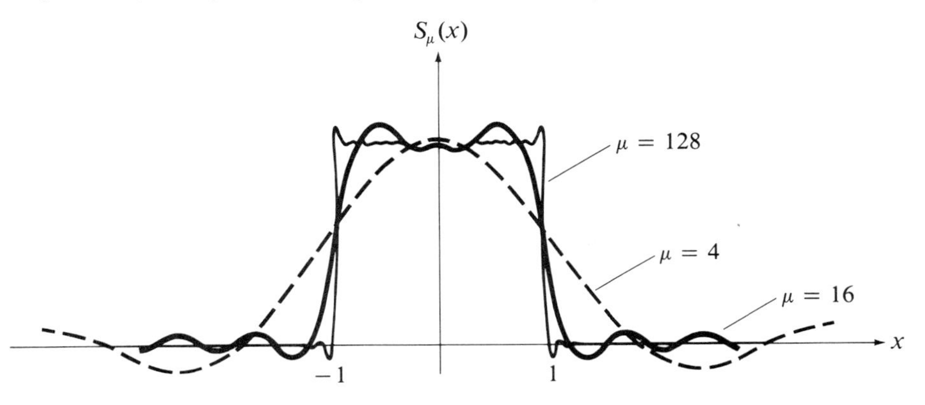

FIGURE 4.14

The partial integral of f.

4.7.1 COSINE AND SINE INTEGRAL REPRESENTATIONS

If the function f is an even function, it follows that*

$$A(s) = \frac{1}{\pi}\int_{-\infty}^{\infty} f(x)\cos sx\,dx = \frac{2}{\pi}\int_0^{\infty} f(x)\cos sx\,dx \tag{91}$$

and

$$B(s) = \frac{1}{\pi}\int_{-\infty}^{\infty} f(x)\sin sx\,dx = 0, \tag{92}$$

from which we deduce

$$f(x) = \int_0^{\infty} A(s)\cos sx\,ds, \tag{93}$$

* Since x is a dummy variable of integration, it makes little difference whether we use x or t in (91) and (92).

called a *Fourier cosine integral representation*. In a similar manner, if f is an odd function, we obtain the *Fourier sine integral representation*

$$f(x) = \int_0^\infty B(s) \sin sx \, ds, \tag{94}$$

where $A(s) = 0$ and

$$B(s) = \frac{2}{\pi} \int_0^\infty f(x) \sin sx \, dx. \tag{95}$$

Finally, if f should be a function defined only on the interval $0 < x < \infty$, we can represent it over this interval by either a Fourier cosine integral or a Fourier sine integral, analogous to the *half-range expansions* of Fourier series (see Problems 6 and 8 in Exercises 4.7).

EXERCISES 4.7

1. By using the result of Equation (86), show directly that

$$\int_0^\infty \frac{\sin s \cos s}{s} \, ds = \frac{\pi}{4}.$$

2. If

$$f(x) = \begin{cases} -1, & -1 < x < 0 \\ 1, & 0 < x < 1 \\ 0, & \text{otherwise,} \end{cases}$$

show that

$$f(x) = \frac{2}{\pi} \int_0^\infty \left(\frac{1 - \cos s}{s} \right) \sin sx \, ds.$$

3. Show that

$$e^{-|x|} = \frac{2}{\pi} \int_0^\infty \frac{\cos sx}{1 + s^2} \, ds.$$

4. If

$$f(x) = \begin{cases} 0, & x < 0 \\ e^{-x}, & x > 0, \end{cases}$$

(a) show that

$$f(x) = \frac{1}{\pi} \int_0^\infty \frac{\cos sx + s \sin sx}{1 + s^2} \, ds.$$

(b) Verify directly that the above integral representation converges to the value 1/2 at $x = 0$.

5. Find an integral representation for

$$f(x) = \begin{cases} 1 - x^2, & |x| < 1 \\ 0, & |x| > 1 \end{cases}$$

and deduce the value of the integral

$$I = \int_0^\infty \left(\frac{\sin x - x \cos x}{x^3} \right) \cos \frac{1}{2} x \, dx.$$

6. Show that e^{-kx} has the half-range representations

(a) $$e^{-kx} = \frac{2k}{\pi} \int_0^\infty \frac{\cos sx}{s^2 + k^2} \, ds, \qquad x > 0, \qquad k > 0.$$

(b) $$e^{-kx} = \frac{2}{\pi} \int_0^\infty \frac{s \sin sx}{s^2 + k^2} \, ds, \qquad x > 0, \qquad k > 0.$$

7. Using the results of Problem 6, establish the formula

$$e^{-x} - e^{-2x} = \frac{6}{\pi} \int_0^\infty \frac{s \sin sx}{(s^2 + 1)(s^2 + 4)} \, ds, \qquad x > 0.$$

8. Show that $e^{-x} \cos x$ has the half-range representations

(a) $$e^{-x} \cos x = \frac{2}{\pi} \int_0^\infty \frac{s^3 \sin sx}{s^4 + 4} \, ds, \qquad x > 0.$$

(b) $$e^{-x} \cos x = \frac{2}{\pi} \int_0^\infty \frac{s^2 + 2}{s^4 + 4} \cos sx \, ds, \quad x > 0.$$

★9. Use the result of Problem 5 to deduce that

(a) $$\int_0^\infty \left(\frac{1 - \cos x}{x} \right)^2 dx = \frac{\pi}{2}.$$

(b) $$\int_0^\infty \frac{\sin^4 x}{x^2} \, dx = \frac{\pi}{4}.$$

In Problems 10–13, express the given integral in terms of the *sine integral* Si(z).

10. $$\int_a^b \frac{\sin x}{x} \, dx$$

11. $$\int_a^b \frac{\sin x^2}{x} \, dx$$

Hint: Let $x^2 = t$.

★12. $$\int_a^b \frac{\sin x}{x^3} \, dx$$

Hint: Use integration by parts.

★13. $$\int_0^\mu \left(\frac{1 - \cos s}{s} \right) \sin sx \, ds$$

In Problems 14 and 15, express the given integral in terms of the sine integral and/or *cosine integral* defined by

$$\text{Ci}(z) = -\int_z^\infty \frac{\cos t}{t} \, dt.$$

14. $\displaystyle\int_a^b \frac{\cos x^2}{x}\,dx$

Hint: Let $x^2 = t$.

★15. $\displaystyle\int_2^3 \frac{\cos x}{1 - x^2}\,dx$

Hint: Use partial fractions.

★16. *(Riemann lemma)* Prove that if f is continuous, has a bounded derivative f', and is absolutely integrable, then

$$\lim_{\lambda\to\infty} \int_{-\infty}^{\infty} f(t)e^{i\lambda t}\,dt = 0.$$

(Actually, the result is valid for f only piecewise continuous and absolutely integrable, but the corresponding proof is much more difficult.)
Hint: Show that

$$\int_{-p}^{p} f(t)e^{i\lambda t}\,dt = \frac{f(t)e^{i\lambda t}}{i\lambda}\bigg|_{-p}^{p} - \frac{1}{i\lambda}\int_{-p}^{p} f'(t)e^{i\lambda t}\,dt,$$

and then let $\lambda \to \infty$, and finally $p \to \infty$.

★17. Using the result of Problem 16, show that

$$\lim_{\lambda\to\infty} \int_{-\infty}^{\infty} f(t)\cos\lambda t\,dt = \lim_{\lambda\to\infty} \int_{-\infty}^{\infty} f(t)\sin\lambda t\,dt = 0.$$

★18. If f is piecewise smooth and absolutely integrable, and x is a point of continuity of f, show that

$$\lim_{\lambda\to\infty} \frac{1}{\pi}\int_{-\infty}^{\infty} f(x+t)\,\frac{\sin\lambda t}{t}\,dt = f(x).$$

Hint: First show that

$$\lim_{\lambda\to\infty} \frac{1}{\pi}\int_{-p}^{p} \frac{\sin\lambda t}{t}\,dt = 1,$$

and then, assuming the validity of Problem 17, establish the result

$$\lim_{\lambda\to\infty} \frac{1}{\pi}\int_{-p}^{p} \left[\frac{f(x+t) - f(x)}{t}\right]\sin\lambda t\,dt = 0.$$

★19. Prove Theorem 4.9 for points of continuity of f.
Hint: First show that

$$\frac{1}{\pi}\int_0^{\lambda}\int_{-\infty}^{\infty} f(t)\cos[s(t-x)]\,dt\,ds = \frac{1}{\pi}\int_{-\infty}^{\infty} f(x+t)\,\frac{\sin\lambda t}{t}\,dt,$$

and then use Problem 18.

★20. Prove Theorem 4.9 for points of discontinuity of f.

CHAPTER 5

INTEGRAL TRANSFORMS OF LAPLACE AND FOURIER

Integral transforms provide an efficient tool for solving linear, constant–coefficient DEs with prescribed auxiliary conditions. The transform method offers the advantage of solving the problem directly without first producing the general solution of the DE, or even having to solve separately for the homogeneous and particular solutions. In this chapter we will illustrate the use of integral transforms in solving ordinary DEs, while in later chapters we will extend the method to certain partial DEs.

The *Laplace transform* is introduced in Section 5.2 and some transforms are computed directly from the defining integral. The basic properties of the Laplace transform are developed in Section 5.3, including the important *convolution theorem*. These *operational properties* are particularly useful in the evaluation of *inverse Laplace transforms*. In Section 5.4 the Laplace transform is used in the solution of *initial value problems* and we also illustrate the technique of constructing the *one-sided Green's function* by this transform.

The *Fourier transform* and *inverse Fourier transform* are developed in Section 5.5 as simple consequences of the Fourier integral theorem of Section 4.7. Some elementary transforms are computed using standard integration techniques. For functions defined only for positive arguments, the *Fourier cosine transform* and *Fourier sine transform* may be used. Following our treatment of Laplace transforms, we develop the basic properties of all three Fourier transforms, and in Section 5.6 we use Fourier transforms to solve certain types of *boundary value problems* on infinite domains.

5.1 INTRODUCTION

In general, an *integral transform* is a relation of the form

$$F(s) = \int_{-\infty}^{\infty} K(s, t)f(t)\,dt, \tag{1}$$

such that a given function $f(t)$ is transformed into another function $F(s)$ by means of an integral. The new function $F(s)$ is the *transform* of $f(t)$, and $K(s, t)$ is the *kernel* of this transformation.

When the kernel is

$$K(s, t) = \begin{cases} 0, & t < 0 \\ e^{-st}, & t \geq 0 \end{cases} \tag{2}$$

the resulting integral transform is the *Laplace transform*, and when

$$K(s, t) = \frac{1}{\sqrt{2\pi}} e^{ist} \tag{3}$$

we obtain the *Fourier transform*.* While many other integral transforms have been developed over the years, the transforms of Laplace and Fourier are by far the most prominent and useful in applications.

Integral transforms are very useful in solving certain DEs on infinite domains where conventional methods are either difficult to apply or fail to work. The Laplace transform is generally applied to only initial value problems where time is the independent variable. On the other hand, we use the Fourier transform for solving problems where the spatial variable is of infinite extent. The basic purpose of an integral transform in solving DEs is to "transform" the given problem into an algebraic problem that incorporates the prescribed auxiliary conditions. If the algebraic problem can be solved and an inverse transformation is possible, the solution of the original problem may be obtained.

5.2 LAPLACE TRANSFORM

The *Laplace transform*, defined by†

$$F(p) = \int_0^\infty e^{-pt} f(t)\, dt, \tag{4}$$

is formally equivalent to the operational calculus devised by Oliver Heaviside (1850–1925) for the solution of transient problems in physics and electrical engineering. The notation

$$\mathscr{L}\{f(t); p\} = F(p) \tag{5}$$

is also used to denote this transform. The function f that appears in (4) is ordinarily assumed to be zero for $t < 0$.‡ We will follow this assumption in our further discussion of the Laplace transform and its properties.

---

*Other variations of $K(s, t)$ involving either e^{ist} or e^{-ist} are also used to define the Fourier transform. For example, in engineering literature it is common to see the transform defined by

$$F(s) = \frac{1}{2\pi} \int_{-\infty}^{\infty} e^{-ist} f(t)\, dt.$$

†We will use the letter p in defining the Laplace transform and reserve the letter s for the Fourier transform.

‡Such functions are called *causal functions* in engineering literature.

Not all functions f have a Laplace transform, even if they are continuous, because the defining integral is improper. That is, to evaluate (4) we must consider the limit

$$\int_0^\infty e^{-pt}f(t)\,dt = \lim_{b\to\infty}\int_0^b e^{-pt}f(t\,dt.$$

If the limit on the right exists for certain values of p, we say the integral *converges* to the function $F(p)$; otherwise, it *diverges*. The basic requirement for the existence of a Laplace transform is that f be of *exponential order*.

DEFINITION 5.1 A function f is said to be of *exponential order* if there exists real constants c, M, and t_0, such that

$$|f(t)| < Me^{ct}, \qquad t > t_0,$$

or equivalently, such that $\lim_{t\to\infty} f(t)e^{-ct} = 0$.

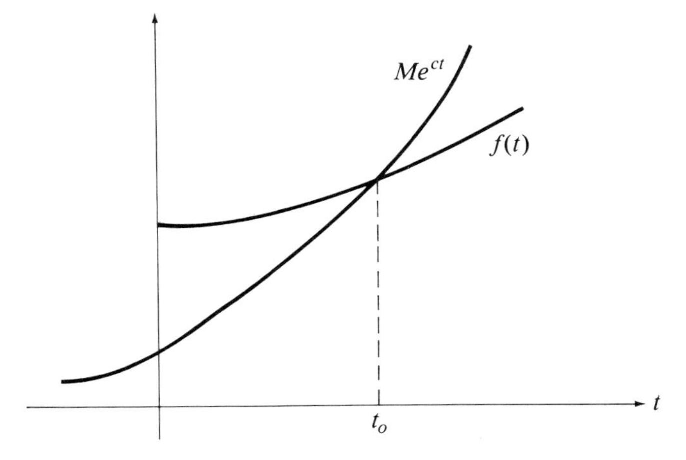

FIGURE 5.1

A function of exponential order.

Saying a function is of exponential order means that its graph on the interval $t > t_0$ does not grow faster than the graph of Me^{ct} for appropriate values of M and c (see Figure 5.1). For instance, the functions t, e^{at}, and $\sin t$ are all of exponential order, whereas the function e^{t^2} is not, since for any value of c

$$\lim_{t\to\infty} e^{t^2}e^{-ct} \neq 0$$

(see Figure 5.2).

Along with requiring f to be of exponential order, we normally specify that f be at least piecewise continuous* in order to ensure the existence of its Laplace transform.

* See Definition 4.1 for a description of piecewise continuous functions.

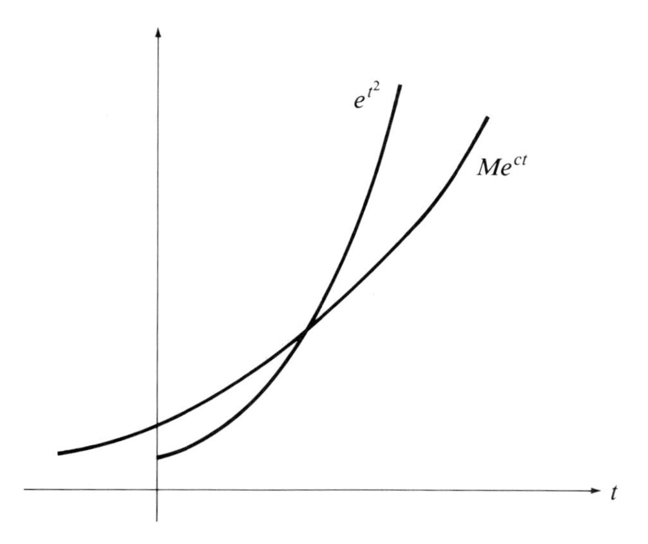

FIGURE 5.2

$f(t) = e^{t^2}$.

THEOREM 5.1

Existence Theorem. If f is piecewise continuous on $t \geqslant 0$ and of exponential order, it has a Laplace transform.

PROOF We begin by expressing the Laplace transform integral as

$$\int_0^\infty e^{-pt} f(t)\,dt = \int_0^{t_0} e^{-pt} f(t)\,dt + \int_{t_0}^\infty e^{-pt} f(t)\,dt.$$

The first integral on the right exists since f is assumed to be piecewise continuous. Because f is also assumed to be of exponential order, the second integral on the right satisfies the inequality

$$\left| \int_{t_0}^\infty e^{-pt} f(t)\,dt \right| \leqslant \int_{t_0}^\infty e^{-pt} |f(t)|\,dt < M \int_{t_0}^\infty e^{-(p-c)t}\,dt.$$

Hence, by direct integration of the last integral, we have

$$\left| \int_{t_0}^\infty e^{-pt} f(t)\,dt \right| < \frac{Me^{-(p-c)t_0}}{p-c}, \qquad p > c.$$

This last expression vanishes in the limit as t_0 tends to infinity, so we say the integral is absolutely convergent. Hence, the Laplace transform of $f(t)$ exists.

Many functions met in practice satisfy the conditions of Theorem 5.1. However, these conditions are sufficient rather than necessary to ensure that a function has a Laplace transform. For example, neither t^{-1} nor $t^{-1/2}$ is piecewise continuous on $t \geqslant 0$, but $t^{-1/2}$ has a Laplace transform (Example 3) while t^{-1} does not.

From the proof of Theorem 5.1, we have the relation

$$F(p) = \int_0^{t_0} e^{-pt}f(t)\,dt + \int_{t_0}^{\infty} e^{-pt}f(t)\,dt$$

$$< \int_0^{t_0} e^{-pt}f(t)\,dt + \frac{Me^{-(p-c)t_0}}{p-c},$$

of which the following theorem is an immediate consequence.

THEOREM 5.2

> If f is piecewise continuous on $t \geqslant 0$, of exponential order, and $\mathscr{L}\{f(t); p\} = F(p)$, then
>
> $$\lim_{p\to\infty} F(p) = 0.$$

The real significance of Theorem 5.2 is that if $F(p)$ is any function for which $\lim_{p\to\infty} F(p) \neq 0$, then it does not represent the Laplace transform of any piecewise continuous function of exponential order. This condition rules out many functions as possible Laplace transforms, such as polynomials in p, e^p, $\cos p$, and so forth. This theorem will become more significant in Section 5.3.1 where we discuss inverse Laplace transforms.

5.2.1 EVALUATING TRANSFORMS

Routine integration techniques will produce the Laplace transform of many elementary functions. In some cases, however, we have to resort to various tricks and manipulations in the evaluation of the transform.

EXAMPLE 1

Find the Laplace transform of $f(t) = e^{at}$.

SOLUTION By definition,

$$\mathscr{L}\{e^{at}; p\} = \int_0^{\infty} e^{-pt}e^{at}\,dt$$

$$= \int_0^{\infty} e^{-(p-a)t}\,dt$$

$$= \frac{e^{-(p-a)t}}{-(p-a)}\Bigg|_0^\infty, \quad *$$

from which we deduce†

$$\mathscr{L}\{e^{at}; p\} = \frac{1}{p-a}, \qquad p > a.$$

By allowing $a \to 0^+$ in the result of Example 1, we get the limiting case

$$\mathscr{L}\{1; p\} = \frac{1}{p}, \qquad p > 0. \tag{6}$$

EXAMPLE 2

Find the Laplace transform of $f(t) = t^n$, $n = 1, 2, 3, \ldots$.

SOLUTION Integration by parts yields

$$\begin{aligned}
\mathscr{L}\{t^n; p\} &= \int_0^\infty e^{-pt} t^n \, dt \\
&= -\frac{1}{p} e^{-pt} t^n \Bigg|_0^\infty + \frac{n}{p}\int_0^\infty e^{-pt} t^{n-1}\, dt \\
&= \frac{n}{p}\int_0^\infty e^{-pt} t^{n-1}\, dt,
\end{aligned}$$

(the boundary term is marked as $\nearrow 0$)

from which we deduce the relation

$$\mathscr{L}\{t^n; p\} = \frac{n}{p}\mathscr{L}\{t^{n-1}; p\}, \qquad n = 1, 2, 3, \ldots.$$

Thus, since $\mathscr{L}\{1; p\} = 1/p$ from (6), we find successively

$$\begin{aligned}
\mathscr{L}\{t; p\} &= \frac{1}{p}\mathscr{L}\{1; p\} = \frac{1}{p^2}, \\
\mathscr{L}\{t^2; p\} &= \frac{2}{p}\mathscr{L}\{t; p\} = \frac{2}{p^3}, \\
\mathscr{L}\{t^3; p\} &= \frac{3}{p}\mathscr{L}\{t^2; p\} = \frac{3 \cdot 2}{p^4} = \frac{3!}{p^4},
\end{aligned}$$

---

* For abbreviation, $\Big|_0^\infty$ will denote $\lim\limits_{b\to\infty} (\;)\Big|_0^b$

† The integral clearly diverges for $p \leqslant a$.

whereas in general, it can be shown that*

$$\mathcal{L}\{t^n; p\} = \frac{n!}{p^{n+1}}, \qquad n = 1, 2, 3, \ldots,$$

which is also valid for $n = 0$.

EXAMPLE 3

Find the Laplace transform of $f(t) = t^{-1/2}$.

SOLUTION From the defining integral, we have

$$\begin{aligned}\mathcal{L}\{t^{-1/2}; p\} &= \int_0^\infty e^{-pt} t^{-1/2}\, dt \\ &= 2\int_0^\infty e^{-px^2}\, dx,\end{aligned}$$

where we have made the change of variable $t = x^2$. This is a nonelementary integral that can be evaluated by an interesting indirect method. To do so, we first consider the square of the transform, which we write as

$$\begin{aligned}[\mathcal{L}\{t^{-1/2}; p\}]^2 &= 2\int_0^\infty e^{-px^2}\, dx \cdot 2\int_0^\infty e^{-py^2}\, dy \\ &= 4\int_0^\infty\int_0^\infty e^{-p(x^2+y^2)}\, dx\, dy.\end{aligned}$$

The term $(x^2 + y^2)$ in the integrand suggests a change to polar coordinates. Thus, by introducing

$$x = r\cos\theta, \qquad y = r\sin\theta,$$

and performing the integration with respect to θ in the resulting integrals, we obtain

$$\begin{aligned}[\mathcal{L}\{t^{-1/2}; p\}]^2 &= 4\int_0^\infty\int_0^{\pi/2} e^{-pr^2} r\, d\theta\, dr \\ &= 2\pi\int_0^\infty e^{-pr^2} r\, dr.\end{aligned}$$

Finally, the substitution $s = r^2$ reduces this last integral to a routine integral to evaluate, which yields

---

*To rigorously justify this result requires mathematical induction.

$$[\mathscr{L}\{t^{-1/2}; p\}]^2 = \pi \int_0^\infty e^{-ps}\, ds$$
$$= \frac{\pi}{p},$$

and by taking the positive square root (since the transform integral must be positive), we obtain the result

$$\mathscr{L}\{t^{-1/2}; p\} = \sqrt{\frac{\pi}{p}}, \qquad p > 0.$$

Nonelementary integrals such as that in Example 3 arise quite often in the evaluation of Laplace transforms. A useful generalization of that integral in the development of several transforms is

$$\int_0^\infty e^{-a^2x^2 - b^2x^{-2}}\, dx = \frac{\sqrt{\pi}}{2a}\, e^{-2ab}, \qquad a > 0, \qquad b \geqslant 0 \tag{7}$$

(see Problem 13 in Exercises 5.2).

EXAMPLE 4

Find the Laplace transform of $f(t) = t^{-1/2}e^{-c/t}$, $c > 0$.

SOLUTION From definition,

$$\mathscr{L}\{t^{-1/2}e^{-c/t}; p\} = \int_0^\infty e^{-pt}t^{-1/2}e^{-c/t}\, dt$$
$$= \int_0^\infty t^{-1/2}e^{-pt-c/t}\, dt$$
$$= 2\int_0^\infty e^{-px^2 - cx^{-2}}\, dx,$$

where we have set $t = x^2$ in the last step. Now using (7), we deduce that

$$\mathscr{L}\{t^{-1/2}e^{-c/t}; p\} = \sqrt{\frac{\pi}{p}}\, e^{-2\sqrt{cp}}, \qquad c > 0, \qquad p > 0.$$

Observe that by letting $c \to 0^+$ in the result of Example 4, we obtain the result of Example 3, i.e.,

$$\mathscr{L}\{t^{-1/2}; p\} = \sqrt{\frac{\pi}{p}}, \qquad p > 0. \tag{8}$$

Our final example of Laplace transforms concerns the *error function*

$$\text{erf}(t) = \frac{2}{\sqrt{\pi}} \int_0^t e^{-x^2}\,dx \tag{9}$$

and *complementary error function*

$$\text{erfc}(t) = \frac{2}{\sqrt{\pi}} \int_t^\infty e^{-x^2}\,dx. \tag{10}$$

Both of these functions are prominent in the solution of certain heat conduction problems on infinite domains (e.g., see Section 6.5). By writing

$$\text{erfc}(t) = \frac{2}{\sqrt{\pi}} \int_0^\infty e^{-x^2}\,dx - \frac{2}{\sqrt{\pi}} \int_0^t e^{-x^2}\,dx,$$

and evaluating the first integral by specializing (7), we obtain the important relation

$$\text{erfc}(t) = 1 - \text{erf}(t). \tag{11}$$

EXAMPLE 5

Find the Laplace transform of $f(t) = \text{erf}(t)$.

SOLUTION From the defining integral,

$$\begin{aligned}\mathscr{L}\{\text{erf}(t); p\} &= \int_0^\infty e^{-pt}\,\text{erf}(t)\,dt \\ &= \int_0^\infty e^{-pt} \frac{2}{\sqrt{\pi}} \int_0^t e^{-x^2}\,dx\,dt.\end{aligned}$$

Recharacterizing the region of integration $0 \leqslant x \leqslant t$, $0 \leqslant t < \infty$, by $x \leqslant t < \infty$, $0 \leqslant x < \infty$,we can interchange the order of integration above to get

$$\begin{aligned}\mathscr{L}\{\text{erf}(t); p\} &= \frac{2}{\sqrt{\pi}} \int_0^\infty e^{-x^2} \int_x^\infty e^{-pt}\,dt\,dx \\ &= \frac{2}{p\sqrt{\pi}} \int_0^\infty e^{-(x^2+px)}\,dx \\ &= \frac{2}{p\sqrt{\pi}} e^{\frac{1}{4}p^2} \int_0^\infty e^{-(x+\frac{1}{2}p)^2}\,dx\end{aligned}$$

(see Figure 5.3), where we have written

$$x^2 + px = \left(x + \frac{1}{2}p\right)^2 - \frac{1}{4}p^2.$$

Finally, making the change of variable $u = x + \frac{1}{2}p$ leads to

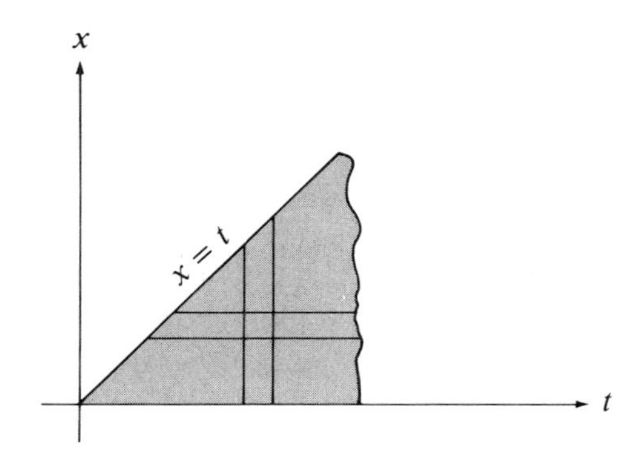

FIGURE 5.3

$$\mathcal{L}\{\text{erf}(t); p\} = \frac{2}{p\sqrt{\pi}} e^{\frac{1}{4}p^2} \int_{p/2}^{\infty} e^{-u^2}\, du,$$

and therefore we deduce that

$$\mathcal{L}\{\text{erf}(t); p\} = \frac{1}{p} e^{\frac{1}{4}p^2} \operatorname{erfc}(p/2), \qquad p > 0.$$

EXERCISES 5.2

In Problems 1–10, evaluate the Laplace transform of each function directly from the defining integral.

1. $f(t) = t^2$
2. $f(t) = \sin kt$
3. $f(t) = \cos kt$
4. $f(t) = \cosh kt$
★5. $f(t) = t^{1/2}$
6. $f(t) = e^{-at} - e^{-bt}$
7. $f(t) = te^{2t}$
★8. $f(t) = t \sin kt$
9. $f(t) = \cos^2 kt$
10. $f(t) = e^{at} \cosh kt$
11. Find the Laplace transform of the *Heaviside unit function* $f(t) = h(t - a)$, $a \geqslant 0$, where $h(t - a) = 1$, $t > a$ and $h(t - a) = 0$, $t < a$.
12. Using Definition 5.1, determine which of the following functions are of exponential order:

    (a) $f(t) = t^{100}$
    (b) $f(t) = e^{-t^2}$
    (c) $f(t) = te^t$
    (d) $f(t) = 3e^{t^2-t}$
    (e) $f(t) = 5\sin(e^{t^2})$
    (f) $f(t) = \dfrac{\sin t}{t}$

★13. Considering the integral

$$I(b) = \int_0^{\infty} e^{-a^2x^2 - b^2x^{-2}}\, dx, \qquad a > 0, \qquad b \geqslant 0$$

as a function of the parameter b,

(a) show that I satisfies the first-order linear DE

$$\frac{dI}{db} + 2aI = 0.$$

(b) Evaluate $I(0)$ directly from the integral.

(c) Solve the DE in (a) subject to the initial condition in (b) to deduce the result

$$I(b) = \frac{\sqrt{\pi}}{2a} e^{-2ab}.$$

★14. Considering the integral

$$I(b) = \int_0^\infty e^{-pu^2} \cos bu \, du, \qquad b \geqslant 0, \qquad p > 0$$

as a function of the parameter b,

(a) show that I satisfies the first-order linear DE

$$\frac{dI}{db} - \frac{b}{2p} I = 0.$$

(b) Evaluate $I(0)$ directly from the integral.
Hint: Use Problem 13.

(c) Solve the DE in (a) subject to the initial condition in (b) to deduce the result

$$I(b) = \frac{1}{2}\sqrt{\frac{\pi}{p}} e^{-b^2/4p}.$$

In Problems 15–24, verify the Laplace transform relations.

15. $\mathscr{L}\{f(t)h(t-a); p\} = e^{-pt}\mathscr{L}\{f(t+a); p\}$
Hint: See Problem 11.

16. $\mathscr{L}\{e^{-t^2/4}; p\} = \sqrt{\pi}\, e^{p^2} \operatorname{erfc}(p), \qquad p > 0$

17. $\mathscr{L}\{\operatorname{erfc}(t^{-1/2}); p\} = \dfrac{1}{p} e^{-2\sqrt{p}}, \qquad p > 0$

18. $\mathscr{L}\{\operatorname{erf}(t^{-1/2}); p\} = \dfrac{1}{p}(1 - e^{-2\sqrt{p}}), \qquad p > 0$

19. $\mathscr{L}\{(t+a)^{-1/2}; p\} = \sqrt{\dfrac{\pi}{p}}\, e^{ap} \operatorname{erfc}(\sqrt{ap}), \qquad a > 0, \qquad p > 0$

★20. $\mathscr{L}\{(t+a)^{-3/2}; p\} = \dfrac{2}{\sqrt{a}} - 2\sqrt{\pi p}\, e^{ap} \operatorname{erfc}(\sqrt{ap}), \qquad a > 0, \qquad p > 0$
Hint: Differentiate both sides of the result in Problem 19 with respect to a.

21. $\mathscr{L}\{t^{-1/2} \cos(at^{1/2}); p\} = \sqrt{\dfrac{\pi}{p}}\, e^{-a^2/4p}, \qquad p > 0$
Hint: Use the result of Problem 14.

★22. $\mathscr{L}\{\sin(at^{1/2}); p\} = \dfrac{a}{2p}\sqrt{\dfrac{\pi}{p}}\, e^{-a^2/4p}, \qquad p > 0$
Hint: Differentiate both sides of the result in Problem 21 with respect to a.

★23. $\mathscr{L}\{t^{-1}\sin(t^{1/2}); p\} = \pi \operatorname{erf}(1/2\sqrt{p}), \qquad p > 0$
Hint: Integrate both sides of the result of Problem 21 with respect to a from 0 to 1.

★24. $\mathscr{L}\{J_0(t^{1/2}); p\} = \dfrac{1}{p} e^{-1/p}, \qquad p > 0$

Hint: Use the fact that the *Bessel function* $J_0(x)$ has the series representation (see Section 9.4)

$$J_0(x) = \sum_{n=0}^{\infty} \frac{(-1)^n (x/2)^{2n}}{(n!)^2}.$$

5.3 OPERATIONAL PROPERTIES AND INVERSE TRANSFORMS

The evaluation of Laplace transforms from definition often involves cumbersome calculations. To avoid direct evaluations of the defining integral, we can sometimes use certain known Laplace transforms together with various *operational properties* of the transform to produce the desired results. Basically, operational properties are consequences of certain integral properties. For example, if C_1 and C_2 are any constants, and f and g both have Laplace transforms, then

$$\mathscr{L}\{C_1 f(t) + C_2 g(t); p\} = C_1 \mathscr{L}\{f(t); p\} + C_2 \mathscr{L}\{g(t); p\}, \tag{12}$$

which is the *linearity property*. To verify this property, we simply observe that

$$\begin{aligned} \mathscr{L}\{C_1 f(t) + C_2 g(t); p\} &= \int_0^{\infty} e^{-pt}[C_1 f(t) + C_2 g(t)]\, dt \\ &= C_1 \int_0^{\infty} e^{-pt} f(t)\, dt + C_2 \int_0^{\infty} e^{-pt} g(t)\, dt, \end{aligned}$$

from which (12) now follows. An interesting application of this property is given in Example 6.

EXAMPLE 6

Find $\mathscr{L}\{\sin kt; p\}$ and $\mathscr{L}\{\cos kt; p\}$.

SOLUTION While both of these transforms can be found through routine integration methods, another approach using the results of Example 1 and Equation (12) is available to us. We start by setting $a = ik$ in the transform relation (Example 1)

$$\mathscr{L}\{e^{at}; p\} = \frac{1}{p - a}$$

to find

$$\mathscr{L}\{e^{ikt}; p\} = \frac{1}{p - ik} = \frac{p + ik}{p^2 + k^2},$$

where $i^2 = -1$. Then, by Euler's formula* this becomes

$$\mathscr{L}\{\cos kt + i \sin kt; p\} = \frac{p + ik}{p^2 + k^2},$$

or, using the linearity property, we obtain

$$\mathscr{L}\{\cos kt; p\} + i\mathscr{L}\{\sin kt; p\} = \frac{p}{p^2 + k^2} + i\frac{k}{p^2 + k^2}.$$

Finally, by matching up the real and imaginary parts of this last expression, we deduce that

$$\mathscr{L}\{\cos kt; p\} = \frac{p}{p^2 + k^2}, \qquad \mathscr{L}\{\sin kt; p\} = \frac{k}{p^2 + k^2}.$$

Functions multiplied by exponentials are easily handled because of the exponential function occurring in the defining integral of the Laplace transform. For example, if $\mathscr{L}\{f(t); p\} = F(p)$, then

$$\mathscr{L}\{e^{at}f(t); p\} = \int_0^\infty e^{-pt}[e^{at}f(t)]\,dt$$

$$= \int_0^\infty e^{-(p-a)t}f(t)\,dt,$$

from which we obtain the *shift property*

$$\mathscr{L}\{e^{at}f(t); p\} = F(p - a). \tag{13}$$

EXAMPLE 7

Evaluate $\mathscr{L}\{e^{-2t} \cos 3t; p\}$.

SOLUTION From Example 6 we have the result $\mathscr{L}\{\cos 3t; p\} = p/(p^2 + 9)$, and hence, using (13) it now follows that

$$\mathscr{L}\{e^{-2t} \cos 3t; p\} = \frac{p + 2}{(p + 2)^2 + 9}.$$

Perhaps the real merit of the Laplace transform is revealed by its effect on derivatives. Suppose that f is continuous and of exponential order with a

---

* $e^{ix} = \cos x + i \sin x$.

piecewise continuous derivative f' for all $t \geq 0$. Then, using integration by parts, we are led to

$$\mathscr{L}\{f'(t); p\} = \int_0^\infty e^{-pt} f'(t)\, dt$$

$$= e^{-pt} f(t) \Big|_0^\infty + p \int_0^\infty e^{-pt} f(t)\, dt,$$

but, since $\lim_{t\to\infty} e^{-pt} f(t) = 0$ for functions of exponential order, we see that

$$\mathscr{L}\{f'(t); p\} = pF(p) - f(0). \tag{14}$$

Similarly, if f and f' are continuous and f'' is piecewise continuous, and all three have transforms, it follows from (14) that

$$\mathscr{L}\{f''(t); p\} = p\mathscr{L}\{f'(t); p\} - f'(0),$$

which we can express as

$$\mathscr{L}\{f''(t); p\} = p^2 F(p) - pf(0) - f'(0). \tag{15}$$

Generalizations to higher-order derivatives are taken up in the exercises (see Problem 7 in Exercises 5.3). We will soon see that relations such as (14) and (15) are very useful in solving initial value problems.

5.3.1 INVERSE LAPLACE TRANSFORMS

Generally the use of Laplace transforms is effective only if we can also solve the inverse problem. In other words, given $F(p)$, what is $f(t)$? In symbols, we write

$$f(t) = \mathscr{L}^{-1}\{F(p); t\}. \tag{16}$$

We might well wonder if an explicit representation for the *inverse Laplace transform* (16) exists analogous to the integral representation of the transform itself. Such a representation does exist (see Section 5.5.4), but it requires integrations performed in the complex plane. Hence, we will rely on other methods for constructing the inverse transform.

In the process of finding inverse Laplace transforms we might first question whether a given function $F(p)$ has an inverse or not. That is, is there a function $f(t)$ whose Laplace transform is $F(p)$? If $F(p)$ does not satisfy the conditions of Theorem 5.2, it is not the transform of a piecewise continuous function of exponential order. Also, because a discontinuous function can have a Laplace transform, it is possible for two functions $f(t)$ and $g(t)$ that are identical everywhere except for a finite number of points to have the same transform, say $F(p)$. Hence, either $f(t)$ or $g(t)$, or perhaps some other function, can be considered the inverse transform of $F(p)$. For instance, both $f(t) = 1$ and

$$g(t) = \begin{cases} 1, & 0 < t < 5 \\ 2, & t = 5 \\ 1, & t > 5 \end{cases}$$

have the same Laplace transform $F(p) = 1/p$. Therefore, either $f(t)$ or $g(t)$ is the inverse transform. Because of this, we say the inverse Laplace transform of a given function is uniquely determined only up to an additive *null function*. This result is known as *Lerch's theorem*. Null functions are normally of little consequence in applications and so the difficulty of finding unique inverse Laplace transforms is of no practical concern. If we can find a continuous function $f(t)$ that is the inverse of $F(p)$, that is the one we use.

REMARK: *A null function $n(t)$ is one for which $\int_0^t n(u)\,du = 0$ for all t.*

When constructing inverse Laplace transforms, in many cases the inverse Laplace transform can be obtained directly from existing tables of transforms (see Appendix A). Also, many of the operational properties used in finding the transform itself can likewise be used in constructing the inverse transform. For instance, the *linearity* and *shift properties* become, respectively,

$$\mathscr{L}^{-1}\{C_1F(p) + C_2G(p); t\} = C_1\mathscr{L}^{-1}\{F(p); t\} + C_2\mathscr{L}^{-1}\{G(p); t\} \tag{17}$$

and

$$\mathscr{L}^{-1}\{F(p-a); t\} = e^{at}\mathscr{L}^{-1}\{F(p); t\}. \tag{18}$$

EXAMPLE 8

Find the inverse Laplace transform of

$$F(p) = \frac{p-5}{p^2+6p+13}.$$

SOLUTION Completing the square in the denominator, we get

$$\frac{p-5}{p^2+6p+13} = \frac{p-5}{(p+3)^2+4} = \frac{(p+3)-8}{(p+3)^2+4}.$$

Thus, using (17) and (18), we obtain

$$\begin{aligned}\mathscr{L}^{-1}\{F(p); t\} &= \mathscr{L}^{-1}\left\{\frac{(p+3)-8}{(p+3)^2+4}; t\right\} \\ &= e^{-3t}\mathscr{L}^{-1}\left\{\frac{p-8}{p^2+4}; t\right\} \\ &= e^{-3t}\left[\mathscr{L}^{-1}\left\{\frac{p}{p^2+4}; t\right\} - 4\mathscr{L}^{-1}\left\{\frac{2}{p^2+4}; t\right\}\right],\end{aligned}$$

or, recalling Example 6,

$$\mathscr{L}^{-1}\{F(p); t\} = e^{-3t}(\cos 2t - 4 \sin 2t).$$

In many cases of practical importance we wish to find the inverse transform of a rational function, i.e., a function having the form

$$F(p) = \frac{R(p)}{Q(p)},$$

where $R(p)$ and $Q(p)$ are polynomials in p. The inverse transform in such cases can most easily be obtained by representing $F(p)$ in terms of its *partial fractions*. The partial fraction representation is the same as that found in the calculus, for example, as a means of integrating certain rational functions. It is assumed that $R(p)$ and $Q(p)$ have no common factors and that the degree of $R(p)$ is lower than that of $Q(p)$.* Let us illustrate the technique with an example.

EXAMPLE 9

Find the inverse Laplace transform of

$$F(p) = \frac{2}{(p+1)(p^2+1)}.$$

SOLUTION Using partial fraction expansions, we write

$$\frac{2}{(p+1)(p^2+1)} = \frac{A}{p+1} + \frac{Bp + C}{p^2+1},$$

and clearing fractions yields

$$2 = A(p^2+1) + (Bp + C)(p+1).$$

Setting $p = -1$, we find $A = 1$, and equating like coefficients of p^2 and p^0 gives the equations

$$0 = A + B,$$

$$2 = A + C,$$

from which we deduce $B = -1$ and $C = 1$. Thus, we find

$$\begin{aligned}\mathscr{L}^{-1}\{F(p); t\} &= \mathscr{L}^{-1}\left\{\frac{1}{p+1}; t\right\} - \mathscr{L}^{-1}\left\{\frac{p}{p^2+1}; t\right\} + \mathscr{L}^{-1}\left\{\frac{1}{p^2+1}; t\right\} \\ &= e^{-t} - \cos t + \sin t.\end{aligned}$$

*The partial fraction method is generally not used unless $Q(p)$ is at least a third degree polynomial.

5.3.2 CONVOLUTION THEOREM

In applications we often must find the inverse Laplace transform of a function that is the simple product of two other transforms. Unfortunately it is true that

$$\mathscr{L}^{-1}\{F(p)G(p); t\} \neq \mathscr{L}^{-1}\{F(p); t\}\mathscr{L}^{-1}\{G(p); t\}.$$

Hence, we must find the inverse by another method.

To start, let us introduce the *Laplace convolution* of two functions, f and g, which is defined by

$$(f * g)(t) = \int_0^t f(t-u)g(u)\,du. \tag{19}$$

By taking the Laplace transform of both sides of (19), we obtain

$$\mathscr{L}\{(f * g)(t); p\} = \int_0^\infty e^{-pt} \int_0^t f(t-u)g(u)\,du\,dt,$$

which we can write as a double integral

$$\mathscr{L}\{(f * g)(t); p\} = \int_0^\infty \int_0^t e^{-pt} f(t-u)g(u)\,du\,dt. \tag{20}$$

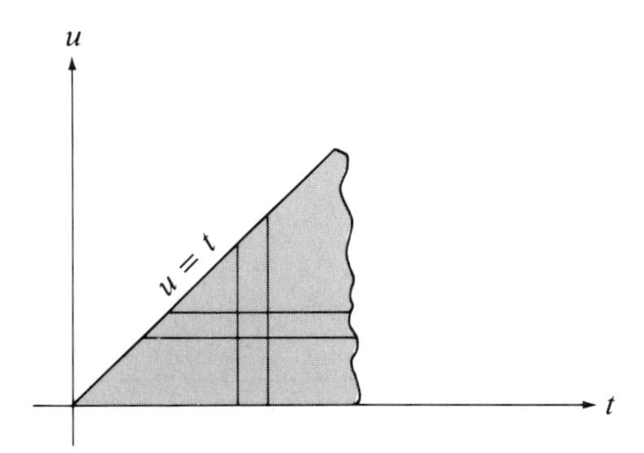

FIGURE 5.4

We can interpret the integrals on the right in (20) as an iterated integral over the region $0 \leqslant u \leqslant t$, $0 \leqslant t < \infty$, as shown in Figure 5.4. If we interchange the order of integration, we find that the region is characterized by $u \leqslant t < \infty$, $0 \leqslant u < \infty$, and thus

$$\mathscr{L}\{(f * g)(t); p\} = \int_0^\infty \int_u^\infty e^{-pt} f(t-u)g(u)\,dt\,du.$$

The change of variables $x = t - u$ leads to the expression

$$\mathscr{L}\{(f * g)(t); p\} = \int_0^\infty \int_0^\infty e^{-(x+u)p} f(x)g(u)\,dx\,du$$

$$= \int_0^\infty e^{-px} f(x)\, dx \cdot \int_0^\infty e^{-pu} g(u)\, du,$$

from which we deduce

$$\mathscr{L}\{(f * g)(t); p\} = F(p)G(p). \tag{21}$$

Finally, by taking the inverse transform of (21), we get

$$\mathscr{L}^{-1}\{F(p)G(p); t\} = \int_0^t f(t-u)g(u)\, du = (f * g)(t), \tag{22}$$

known as the *Laplace convolution theorem.*

Before illustrating the use of (22) in obtaining inverse transforms, let us take note of some properties of the convolution integral (19) that are often useful. Foremost among these is the *commutative law*

$$(f * g)(t) = (g * f)(t), \tag{23}$$

which can be verified by making the change of variable $v = t - u$ in (19); i.e.,

$$(f * g)(t) = -\int_t^0 f(v)g(t-v)\, dv = \int_0^t g(t-v)f(v)\, dv,$$

from which (23) now follows. Other properties of the convolution integral that readily follow from definition are

$$f * (Cg) = (Cf) * g = C(f * g), \qquad C \text{ constant}, \tag{24}$$

$$f * (g + k) = f * g + f * k, \qquad \textit{distributive law}, \tag{25}$$

$$f * (g * k) = (f * g) * k, \qquad \textit{associative law}, \tag{26}$$

the proofs of which are left to the exercises.

EXAMPLE 10

Find $\mathscr{L}^{-1}\left\{\dfrac{1}{p^2(p^2+k^2)}; t\right\}$.

SOLUTION Let us select $F(p) = 1/p^2$ and $G(p) = 1/(p^2 + k^2)$, whose inverse transforms are

$$\mathscr{L}^{-1}\left\{\frac{1}{p^2}; t\right\} = f(t) = t$$

and

$$\mathscr{L}^{-1}\left\{\frac{1}{p^2+k^2}; t\right\} = g(t) = \frac{1}{k}\sin kt.$$

Thus, using the convolution theorem (22), we write

$$\mathscr{L}^{-1}\left\{\frac{1}{p^2(p^2+k^2)}; t\right\} = (f * g)(t) = \int_0^t (t-u)\frac{1}{k}\sin ku\, du,$$

which leads to the result

$$\mathcal{L}^{-1}\left\{\frac{1}{p^2(p^2+k^2)};t\right\} = \frac{1}{k^2}\left(t - \frac{1}{k}\sin kt\right).$$

EXERCISES 5.3

In Problems 1–5, evaluate the Laplace transform of the given function using known results and operational properties.

1. $f(t) = 3te^{2t}$
2. $f(t) = t^3e^{-2t}$
3. $f(t) = 2e^t \sin 3t$
4. $f(t) = e^{-t}(t^2 - 2t + 7)$
5. $f(t) = \cosh kt \cos kt$
6. Given

$$f(t) = \begin{cases} t+1, & 0 \leqslant t \leqslant 2 \\ 3, & t > 2, \end{cases}$$

(a) find $\mathcal{L}\{f(t); p\}$.
(b) find $\mathcal{L}\{f'(t); p\}$ in two ways.

7. If all transforms exist, show that

$$\mathcal{L}\{f^{(n)}(t); p\} = p^nF(p) - p^{n-1}f(0) - p^{n-2}f'(0) - \ldots - f^{(n-1)}(0)$$

$(n = 1, 2, 3, \ldots)$, where $F(p) = \mathcal{L}\{f(t); p\}$.

★8. Use the result of Problem 7 to deduce that

$$\mathcal{L}\{t^n; p\} = \frac{n!}{p^{n+1}}, \qquad n = 1, 2, 3, \ldots .$$

Hint: Observe that $f(0) = f'(0) = \ldots = f^{(n-1)}(0) = 0$, $f^{(n)}(t) = n!$, where $f(t) = t^n$.

★9. The *Laguerre polynomials* are defined by $L_n(t) = \dfrac{e^t}{n!}\dfrac{d^n}{dt^n}(t^ne^{-t})$, $n = 0, 1, 2, \ldots$

Show that

$$\mathcal{L}\{L_n(t); p\} = \frac{1}{p}\left(\frac{p-1}{p}\right)^n.$$

Hint: First find $\mathcal{L}\{e^{-t}L_n(t); p\}$.

10. If $\mathcal{L}\{f(t); p\} = F(p)$, show that

$$\mathcal{L}\{f(at); p\} = \frac{1}{a}F\left(\frac{p}{a}\right), \qquad a > 0.$$

11. Using Problem 10 and $\mathcal{L}\{\cos t; p\} = p/(p^2 + 1)$, find $\mathcal{L}\{\cos 4t; p\}$.
12. Formally differentiate both sides of

$$F(p) = \int_0^\infty e^{-pt}f(t)\,dt$$

with respect to p and deduce that
(a) $\mathcal{L}\{tf(t); p\} = -F'(p)$.
(b) $\mathcal{L}\{t^n f(t); p\} = (-1)^n F^{(n)}(p), \qquad n = 1, 2, 3, \ldots .$

In Problems 13–17, use the result of Problem 12 and any operational properties to evaluate the Laplace transform of the given function.

13. $f(t) = t \sin t$

14. $f(t) = t^2 \sin t$

15. $f(t) = te^{-2t} \cos t$

16. $f(t) = 5te^{3t} \sin^2 t$

17. $f(t) = 3e^{-4t} (\cos 4t - t \sin 4t)$

★18. If $\mathcal{L}\{f(t); p\} = F(p)$, show that

$$\mathcal{L}\left\{\frac{f(t)}{t}; p\right\} = \int_p^\infty F(u)\, du.$$

Hint: Start with $\int_0^\infty e^{-pt} f(t)\, dt = F(p)$ and integrate both sides from p to ∞.

★19. If $\mathcal{L}\{f(t); p\} = F(p)$, show that

(a) $$\mathcal{L}\left\{\int_0^t \frac{f(u)}{u}\, du; p\right\} = \frac{1}{p} \int_p^\infty F(x)\, dx.$$

(b) $$\mathcal{L}\left\{\int_t^\infty \frac{f(u)}{u}; p\right\} = \frac{1}{p} \int_0^p F(x)\, dx.$$

(c) $$\mathcal{L}\left\{\int_0^\infty \frac{f(u)}{u}; p\right\} = \frac{1}{p} \int_0^\infty F(x)\, dx.$$

In Problems 20–23, use the results of Problems 18 and 19 to evaluate the Laplace transform of the given function.

20. $f(t) = \dfrac{\sin t}{t}$

21. $f(t) = \dfrac{e^t - e^{-t}}{t}$

22. $f(t) = \displaystyle\int_0^t (u^2 - u + e^{-u})\, du$

23. $f(t) = \text{Si}(t) = \displaystyle\int_0^t \frac{\sin u}{u}\, du$

In Problems 24–35, determine the inverse Laplace transform using the tables in Appendix A and various opertional properties.

24. $F(p) = \dfrac{7}{p^3}$

25. $F(p) = \dfrac{3p + 7}{p^2 + 5}$

26. $F(p) = \dfrac{2}{(p - 3)^5}$

27. $F(p) = \dfrac{1}{p^2 - 6p + 10}$

28. $F(p) = \dfrac{2p^2 + 5p - 1}{p^3 - p}$

29. $F(p) = \dfrac{3p - 2}{p^3(p^2 + 4)}$

30. $F(p) = \dfrac{p}{p^2 - 6p + 13}$

31. $F(p) = \dfrac{p^2 + 1}{(p^2 - 1)(p^2 - 4)}$

32. $F(p) = \dfrac{3p + 1}{(p + 1)^5}$

★33. $F(p) = \dfrac{5p - 2}{3p^2 + 4p + 8}$

34. $F(p) = \dfrac{4p^2 - 16}{p^3(p+2)^2}$

35. $F(p) = \dfrac{p+1}{p^3 + p^2 - 6p}$

36. Given that $F(p) = \mathcal{L}^{-1}\{f(t); p\}$, show for constants a, b, and k that

(a) $\mathcal{L}^{-1}\{F(kp); t\} = \dfrac{1}{k} f\left(\dfrac{t}{k}\right), \quad k > 0.$

Hint: See Problem 10.

(b) $\mathcal{L}^{-1}\{F(ap+b); t\} = \dfrac{1}{a} e^{-bt/a} f\left(\dfrac{t}{a}\right), \quad a > 0.$

37. If it is known that

$$\mathcal{L}^{-1}\{p^{-1/2}e^{-1/p}; t\} = (\pi t)^{-1/2}\cos(2t^{1/2}),$$

find $\mathcal{L}^{-1}\{p^{-1/2}e^{-a/p}; t\}$, $a > 0$.
Hint: Use Problem 36.

38. Show that the Laplace convolution (19) satisfies
(a) $f * (Cg) = (Cf) * g = C(f * g)$, C constant.
(b) $f * (g + k) = (f * g) + (f * k)$.

★39. Show that

$$f * (g * k) = (f * g) * k.$$

In Problems 40 and 41, find the Laplace transform of each convolution integral.

40. $\displaystyle\int_0^t (t-u)^2 e^{-2u}\, du$

41. $\displaystyle e^{-t}\int_0^t e^u \cos u\, du$

In Problems 42–45, find the inverse Laplace transform of each function using the convolution theorem.

42. $Y(p) = \dfrac{1}{p^2(p+1)}$

43. $Y(p) = \dfrac{1}{(p^2+1)^2}$

44. $Y(p) = \dfrac{1}{(p+1)^2(p^2+4)}$

45. $Y(p) = \dfrac{1}{(p^2+a^2)(p^2+b^2)}, \quad a \neq b$

46. Given that $\mathcal{L}\{f(t); p\} = (p^2 - a^2)^{-1/2}$, evaluate $\displaystyle\int_0^t f(t-u)f(u)\, du$.

★47. Show that

$$\int_0^t J_0(t-u)J_0(u)\, du = \sin t,$$

where J_0 is the *Bessel function* of the first kind of order zero.
Hint: $\mathcal{L}\{J_0(t); p\} = (p^2+1)^{-1/2}$.

★48. *(Translation property)* If $\mathcal{L}\{f(t); p\} = F(p)$, show that

$$\mathcal{L}\{f(t-a)h(t-a); p\} = e^{-ap}F(p),$$

where h is the Heaviside unit function defined by $h(t) = 1$, $t > 0$ and $h(t) = 0$, $t < 0$.*

★49. Use Problem 48 to find the Laplace transform of

$$f(t) = t^2 + (3 - t^2)h(t - 2).$$

★50. Use Problem 48 to find the inverse Laplace transform of

$$F(p) = \frac{1 - 3e^{-5p}}{p^2}.$$

★51. Use the convolution theorem to derive the inverse of the translation property in Problem 48, i.e., show that

$$\mathcal{L}^{-1}\{e^{-ap}F(p); t\} = f(t - a)h(t - a).$$

Hint: See Problem 11 in Exercises 5.2.

5.4 INITIAL VALUE PROBLEMS

The Laplace transform is a powerful tool for solving linear DEs with constant coefficients—in particular, *initial value problems*. The usefulness of the transform rests primarily on the fact that the transform of the DE together with prescribed initial conditions reduces the differential system to an algebraic equation in the transformed function $Y(p) = \mathcal{L}\{y(t); p\}$. Such an algebraic equation is readily solved, and the inverse transform of its solution then yields the solution of the initial value problem, i.e., $y(t) = \mathcal{L}^{-1}\{Y(p); t\}$. Furthermore, the solution of the DE satisfying certain initial conditions is found directly without first producing a general solution and then solving for the arbitrary constants. Let us illustrate with some examples.

EXAMPLE 11

Solve $y'' + 2y' + 5y = 0$, $y(0) = 2$, $y'(0) = -4$.

SOLUTION The transform of the DE term-by-term yields

$$\mathcal{L}\{y''(t); p\} + 2\mathcal{L}\{y'(t); p\} + 5\mathcal{L}\{y(t); p\} = \mathcal{L}\{0; p\}.$$

However, from (14) and (15) it follows that

$$\mathcal{L}\{y'(t); p\} = pY(p) - y(0) = pY(p) - 2,$$

$$\mathcal{L}\{y''(t); p\} = p^2Y(p) - py(0) - y'(0) = p^2Y(p) - 2p + 4.$$

Hence the DE leads to the algebraic equation

$$[p^2Y(p) - 2p + 4] + 2[pY(p) - 2] + 5Y(p) = 0,$$

* See also Problem 15 in Exercises 5.2 for an alternate (and possibly more convenient) version of the translation property.

which has incorporated the initial conditions. By simplifying this last expression, we get

$$(p^2 + 2p + 5)Y(p) = 2p,$$

and solving for $Y(p)$ gives us

$$Y(p) = \frac{2p}{p^2 + 2p + 5}.$$

To make use of the shift property (18) in finding the inverse transform, we first observe

$$\frac{2p}{p^2 + 2p + 5} = \frac{2p}{(p + 1)^2 + 4} = \frac{2(p + 1) - 2}{(p + 1)^2 + 4}.$$

Thus

$$\begin{aligned} y(t) &= \mathscr{L}^{-1}\left\{\frac{2p}{p^2 + 2p + 5}; t\right\} \\ &= \mathscr{L}^{-1}\left\{\frac{2(p + 1) - 2}{(p + 1)^2 + 4}; t\right\} \\ &= e^{-t}\mathscr{L}^{-1}\left\{\frac{2p - 2}{p^2 + 4}; t\right\} \\ &= e^{-t}\left(2\mathscr{L}^{-1}\left\{\frac{p}{p^2 + 4}; t\right\} - \mathscr{L}^{-1}\left\{\frac{2}{p^2 + 4}; t\right\}\right), \end{aligned}$$

from which we finally deduce

$$y(t) = e^{-t}(2 \cos 2t - \sin 2t).$$

EXAMPLE 12

Solve $y'' - 6y' + 9y = t^2e^{3t}$, $y(0) = 2$, $y'(0) = 6$.

SOLUTION Taking the transform of the DE gives us

$$[p^2Y(p) - 2p - 6] - 6[pY(p) - 2] + 9Y(p) = \frac{2}{(p - 3)^3},$$

which simplifies to

$$(p^2 - 6p + 9)Y(p) = 2(p - 3) + \frac{2}{(p - 3)^3}.$$

Hence,

$$Y(p) = \frac{2}{p - 3} + \frac{2}{(p - 3)^5},$$

and by taking inverse transforms, we find

$$y(t) = 2\mathscr{L}^{-1}\left\{\frac{1}{p-3}; t\right\} + \frac{2}{4!}\mathscr{L}^{-1}\left\{\frac{4!}{(p-3)^5}; t\right\}$$

$$= 2e^{3t} + \frac{1}{12}t^4 e^{3t}.$$

It may be of interest to see exactly where each input parameter ends up in the transform domain by considering the general initial value problem

$$y'' + ay' + by = f(t), \qquad y(0) = k_0, \qquad y'(0) = k_1, \tag{27}$$

where a and b are known constants. If we apply the Laplace transform to each term in (27), we get

$$[p^2 Y(p) - pk_0 - k_1] + a[pY(p) - k_0] + bY(p) = F(p),$$

or

$$(p^2 + ap + b)Y(p) = (p + a)k_0 + k_1 + F(p),$$

where $Y(p) = \mathscr{L}\{y(t); p\}$ and $F(p) = \mathscr{L}\{f(t); p\}$. Solving this algebraic equation, we have

$$Y(p) = \frac{(p + a)k_0 + k_1}{p^2 + ap + b} + \frac{F(p)}{p^2 + ap + b}, \tag{28}$$

and by taking the inverse Laplace transform, we obtain

$$y(t) = \underbrace{\mathscr{L}^{-1}\left\{\frac{(p + a)k_0 + k_1}{p^2 + ap + b}; t\right\}}_{y_H(t)} + \underbrace{\mathscr{L}^{-1}\left\{\frac{F(p)}{p^2 + ap + b}; t\right\}}_{y_P(t)}. \tag{29}$$

Here it is interesting to observe that the solution (29) has naturally split into two parts—the function y_H, which is a solution of the initial value problem

$$y'' + ay' + by = 0, \qquad y(0) = k_0, \qquad y'(0) = k_1, \tag{30}$$

and y_P, which satisfies

$$y'' + ay' + by = f(t), \qquad y(0) = 0, \qquad y'(0) = 0. \tag{31}$$

We see, therefore, that the transform method divides the solution into two parts, much like we did in Section 2.2 because of physical significance. This property makes the Laplace transform an effective tool for analyzing the basic characteristics of a system in response to each of the input parameters: k_0 and k_1, or $f(t)$.

□ 5.4.1 ONE-SIDED GREEN'S FUNCTION

The Laplace transform can also be used to construct the one-sided Green's function (see Section 2.2) for constant–coefficient DEs. To see this, consider the initial value problem

$$y'' + ay' + by = f(t), \qquad y(0) = 0, \qquad y'(0) = 0. \tag{32}$$

From Equation (29), we know the solution of this problem is

$$y(t) = \mathcal{L}^{-1}\left\{\frac{F(p)}{p^2 + ap + b}; t\right\}, \tag{33}$$

where $F(p) = \mathcal{L}\{f(t); p\}$. Thus, if we define the function

$$k(t) = \mathcal{L}^{-1}\left\{\frac{1}{p^2 + ap + b}; t\right\}, \tag{34}$$

the convolution formula (22) yields the solution

$$y(t) = \int_0^t k(t - \tau)f(\tau)\, d\tau. \tag{35}$$

On the other hand, the solution of (32) obtained through use of the one-sided Green's function is

$$y(t) = \int_0^t g_1(t, \tau)f(\tau)\, d\tau. \tag{36}$$

Hence, by comparing (35) and (36), we deduce that

$$k(t - \tau) \equiv g_1(t, \tau). \tag{37}$$

EXAMPLE 13

Find the one-sided Green's function associated with the differential operator $D^2 - 2D + 5$.

SOLUTION We first determine

$$k(t) = \mathcal{L}^{-1}\left\{\frac{1}{p^2 - 2p + 5}; t\right\} = \mathcal{L}^{-1}\left\{\frac{1}{(p - 1)^2 + 4}; t\right\}$$

which leads to

$$k(t) = \frac{1}{2} e^t \sin 2t.$$

Hence, from (37) we deduce that the one-sided Green's function is

$$k(t - \tau) \equiv g_1(t, \tau) = \frac{1}{2} e^{t-\tau} \sin 2(t - \tau).$$

Comparison of the convolution theorem of the Laplace transform and the Green's function method show that they are equivalent for constant–coefficient equations. In fact, it now follows that the one-sided Green's function for a constant coefficient DE can always be expressed as a function of the variable $t - \tau$. The method of Green's function is more general than that of Laplace transforms, however, since it can also be applied to variable–coefficient equations (at least in theory) and is more readily adapted to problems when the initial data is prescribed at a point other than $t = 0$.

EXERCISES 5.4

In Problems 1–10, use the Laplace transform to solve the initial value problem.

1. $y' + y = e^t, \quad y(0) = 0$
2. $y'' + y = 0, \quad y(0) = 1, \quad y'(0) = 0$
3. $y'' + y = 1, \quad y(0) = 0, \quad y'(0) = 0$
4. $y'' + y' - 2y = -4, \quad y(0) = 2, \quad y'(0) = 3$
5. $y'' + 2y' + 2y = \sin 2t - 2\cos 2t, \quad y(0) = 0, \quad y'(0) = 0$
6. $y'' + 2y' + y = 3te^{-t}, \quad y(0) = 4, \quad y'(0) = 2$
7. $y'' + 4y' + 6y = 1 + e^{-t}, \quad y(0) = 1, \quad y'(0) = -4$
8. $y'' + 16y = \cos 4t, \quad y(0) = 0, \quad y'(0) = 1$
9. $y'' - 4y' + 4y = t, \quad y(0) = 1, \quad y'(0) = 0$
10. $y'' - y' = e^t \cos t, \quad y(0) = 0, \quad y'(0) = 0$
11. Solve the initial value problem
$$y'' + 2ay' + b^2y = 0, \qquad y(0) = y_0, \qquad y'(0) = v_0,$$
for each of the cases
(a) $b^2 - a^2 = -\alpha^2 < 0$.
(b) $b = a$.
(c) $b^2 - a^2 = \mu^2 > 0$.
12. Solve the initial value problem
$$y'' + \omega_0^2 y = A\cos \omega t, \qquad y(0) = 0, \qquad y'(0) = 0,$$
for the case when
(a) $\omega \neq \omega_0$.
(b) $\omega = \omega_0$.

In Problems 13–20, use the Laplace transform to construct the one-sided Green's function for the given operator.

13. D^2
14. $D^2 - 5$
15. $D^2 + 5$
16. $D^2 + 4D + 4$
17. $4D^2 - 8D + 5$
18. $D^2 - D - 2$
19. $D^2 + 2D + 10$
20. $(D - a)(D - b), \quad a \neq b$

*21. Show that the one-sided Green's function for the nth-order linear DE
$$y^{(n)} + a_{n-1}y^{(n-1)} + \ldots + a_1y' + a_0y = f(t)$$
is given by $k(t - \tau)$, where $k(t) = \mathscr{L}^{-1}\{1/w(p); t\}$ and
$$w(p) = p^n + a_{n-1}p^{n-1} + \ldots + a_1p + a_0.$$

22. Use the result of Problem 21 to determine the one-sided Green's function associated with the following differential operators:

(a) D^n, $n = 2, 3, 4, \ldots$
(b) $D^2(D^2 - 1)$
(c) $D^4 - 1$
(d) $D^3 - 6D^2 + 11D - 6$

★23. Show that the Laplace transform of the differential system

$$ty'' + 2(t-1)y' - 2y = 0, \qquad y(0) = 0,$$

leads to the first-order DE in the transform domain

$$(p^2 + 2p)\frac{dY}{dp} + (4p + 4)Y = 0.$$

Solve this first-order DE to obtain $Y(p)$, and invert it to find the solution $y(t)$. This problem is one of the few variable–coefficient DEs for which the Laplace transform method proves fruitful.

★24. Apply the method of Problem 23 to *Bessel's equation* of order zero

$$ty'' + y' + ty = 0,$$

(a) and show that $Y(p)$ satisfies the first-order DE

$$(1 + p^2)\frac{dY}{dp} + pY = 0.$$

(b) Show that the general solution of the DE in (a) is

$$Y(p) = C(1 + p^2)^{-1/2},$$

where C is an arbitrary constant.

(c) Express the term $(1 + p^2)^{-1/2}$ in a binomial series valid for $p > 1$. Assuming it is permissible to take the inverse Laplace transform termwise, deduce that

$$y = CJ_0(t) = C\sum_{n=0}^{\infty}\frac{(-1)^n(t/2)^{2n}}{(n!)^2},$$

where $J_0(t)$ is the *Bessel function* of the first kind.*

★25. Abel (1802–1829) studied a particular Volterra integral equation that has some important applications. In particular, suppose a particle of mass m is constrained to move without friction along a certain path in a vertical plane under the influence of gravity alone. Given the time T required for the particle to descend the curve, we wish to determine the equation of the curve. This problem reduces to finding the solution of the *Volterra integral equation of the first kind*

$$T = k(y) = \int_0^y \frac{f(u)\,du}{\sqrt{2g(y-u)}},$$

where g is the gravitational constant and $f(y)$ is the length of the path.

(a) Take the Laplace transform of the integral equation and deduce that $K(p) = (\pi/2gp)^{1/2}F(p)$, where $K(p)$ and $F(p)$ denote the Laplace transforms, respectively, of $k(t)$ and $f(t)$.

---

* See Section 9.4 for a discussion of Bessel functions.

(b) Solve (a) for $F(p)$, and by taking inverse Laplace transforms, show that

$$f(y) = \frac{(2g)^{1/2}}{\pi} \int_0^y (y-u)^{-1/2} k'(u)\, du.$$

Hint: Write $F(p) = (2g)^{1/2}(\pi p)^{-1/2} pK(p)$.

5.5 FOURIER TRANSFORM PAIRS

In Section 4.7 we derived the Fourier integral relation

$$f(x) = \frac{1}{\pi} \int_0^\infty \int_{-\infty}^\infty f(t) \cos[s(t-x)]\, dt\, ds. \tag{38}$$

Through the use of Euler's formula for the cosine function,

$$\cos x = \frac{1}{2}(e^{ix} + e^{-ix}),$$

we can write Fourier's integral (38) in terms of complex exponential functions. That is,

$$\begin{aligned} f(x) &= \frac{1}{\pi} \int_0^\infty \int_{-\infty}^\infty f(t) \cos[s(t-x)]\, dt\, ds \\ &= \frac{1}{2\pi} \int_0^\infty \int_{-\infty}^\infty f(t)[e^{is(t-x)} + e^{-is(t-x)}]\, dt\, ds \\ &= \frac{1}{2\pi} \int_{-\infty}^\infty \int_{-\infty}^\infty f(t) e^{is(t-x)}\, dt\, ds,^* \end{aligned} \tag{39}$$

or

$$f(x) = \frac{1}{2\pi} \int_{-\infty}^\infty e^{-isx} \int_{-\infty}^\infty e^{ist} f(t)\, dt\, ds, \tag{40}$$

which is the exponential form of Fourier's integral theorem.

What we have established by the integral formula (40) is the pair of transform formulas

$$F(s) = \frac{1}{\sqrt{2\pi}} \int_{-\infty}^\infty e^{ist} f(t)\, dt \tag{41}$$

---

*Note that

$$\begin{aligned} \int_{-\infty}^\infty \int_{-\infty}^\infty f(t) e^{isz}\, dt\, ds &= \int_{-\infty}^0 \int_{-\infty}^\infty f(t) e^{isz}\, dt\, ds + \int_0^\infty \int_{-\infty}^\infty f(t) e^{isz}\, dt\, ds \\ &= \int_0^\infty \int_{-\infty}^\infty f(t) e^{-isz}\, dt\, ds + \int_0^\infty \int_{-\infty}^\infty f(t) e^{isz}\, dt\, ds, \end{aligned}$$

where $z = t - x$.

and

$$f(t) = \frac{1}{\sqrt{2\pi}} \int_{-\infty}^{\infty} e^{-ist} F(s)\, ds. \tag{42}$$

We define $F(s)$ as the *Fourier transform* of $f(t)$, also written

$$F(s) = \mathscr{F}\{f(t); s\}, \tag{43}$$

and $f(t)$ as the *inverse Fourier transform* of $F(s)$, or

$$f(t) = \mathscr{F}^{-1}\{F(s); t\}. \tag{44}$$

The location of the constant $1/2\pi$ in the definition of the transform pairs is arbitrarily selected as long as (40) is satisfied. For reasons of symmetry we have split the constant (i.e., $1/\sqrt{2\pi}$ and $1/\sqrt{2\pi}$) between the transform pairs in (41) and (42). It is also common to see the constant $1/2\pi$ in front of one integral or the other in much of the literature.* There is also some variation as to which integral represents the transform and which one represents the inverse transform. In practice, such differences in definition are of little consequence except when consulting tables of transforms or referring to their properties.

5.5.1 FOURIER COSINE AND SINE TRANSFORMS

In Section 4.7.1 we found that when the function f is even, the Fourier integral representation reduces to

$$\begin{aligned} f(x) &= \int_0^{\infty} A(s) \cos sx\, ds \\ &= \frac{2}{\pi} \int_0^{\infty} \cos sx \int_0^{\infty} f(t) \cos st\, dt\, ds. \end{aligned} \tag{45}$$

Based on this integral relation, we can define the *Fourier cosine transform*

$$\mathscr{F}_C\{f(t); s\} = \sqrt{\frac{2}{\pi}} \int_0^{\infty} f(t) \cos st\, dt = F_C(s), \qquad s > 0 \tag{46}$$

and *inverse cosine transform*

$$\mathscr{F}_C^{-1}\{F_c(s); t\} = \sqrt{\frac{2}{\pi}} \int_0^{\infty} F_C(s) \cos st\, ds = f(t), \qquad t > 0. \tag{47}$$

These results are interesting in that they imply the equivalence of the operators $\mathscr{F}_C$ and $\mathscr{F}_C^{-1}$. In other words, the cosine transform and its inverse are exactly the same in functional form.

Similarly, when f is an odd function the Fourier integral representation becomes

---

* See the first footnote on p. 194.

$$f(x) = \frac{2}{\pi} \int_0^\infty \sin sx \int_0^\infty f(t) \sin st \, dt \, ds, \tag{48}$$

which leads to the *Fourier sine transform*

$$\mathcal{F}_S\{f(t); s\} = \sqrt{\frac{2}{\pi}} \int_0^\infty f(t) \sin st \, dt = F_S(s), \qquad s > 0 \tag{49}$$

and *inverse sine transform*

$$\mathcal{F}_S^{-1}\{F_S(s); t\} = \sqrt{\frac{2}{\pi}} \int_0^\infty F_S(s) \sin st \, ds = f(t), \qquad t > 0. \tag{50}$$

Hence, the sine transform and its inverse are also exactly the same in functional form.

If the function f is neither even nor odd, but defined only for $t \geqslant 0$, then it may have both a cosine transform and a sine transform. Moreover, the even and odd extensions of f will have exponential Fourier transforms. To see the relations between these various transforms, let us construct the even extension of f by setting

$$f_e(t) = f(|t|), \qquad -\infty < t < \infty. \tag{51}$$

The Fourier transform of $f_e(t)$ leads to

$$\begin{aligned}
\mathcal{F}\{f_e(t); s\} &= \frac{1}{\sqrt{2\pi}} \int_{-\infty}^\infty f_e(t) e^{ist} \, dt \\
&= \frac{1}{\sqrt{2\pi}} \int_{-\infty}^\infty f_e(t) \cos st \, dt + i \frac{1}{\sqrt{2\pi}} \int_{-\infty}^\infty \underbrace{f_e(t) \sin st}_{\to 0} \, dt \\
&= \sqrt{\frac{2}{\pi}} \int_0^\infty f(t) \cos st \, dt,
\end{aligned}$$

from which we deduce, for all s,

$$\mathcal{F}\{f_e(t); s\} = \mathcal{F}_C\{f(t); s\}. \tag{52}$$

Based on (52), it is clear that the Fourier transform and cosine transform of an even function give identical results. In particular, their transforms are even functions of s. The odd extension of f is constructed by setting

$$f_0(t) = f(|t|) \operatorname{sgn}(t), \qquad -\infty < t < \infty, \tag{53}$$

where the *signum function* is defined by

$$\operatorname{sgn}(t) = \begin{cases} -1, & t < 0 \\ 1, & t > 0. \end{cases} \tag{54}$$

In this case, we find

$$\mathscr{F}\{f_0(t); s\} = \frac{1}{\sqrt{2\pi}} \int_{-\infty}^{\infty} f_0(t) e^{ist}\, dt$$

$$= \frac{1}{\sqrt{2\pi}} \int_{-\infty}^{\infty} f_0(t) \cos st\, dt + i \frac{1}{\sqrt{2\pi}} \int_{-\infty}^{\infty} f_0(t) \sin st\, dt$$

(the first integral is marked as equal to 0)

$$= i \sqrt{\frac{2}{\pi}} \int_0^{\infty} f(t) \sin st\, dt.$$

Because the Fourier transform of an odd function is also an odd function, we make the conclusion that the Fourier transform and sine transform are related for all s by

$$\mathscr{F}\{f_0(t); s\} = i\mathscr{F}_S\{f(t); |s|\} \operatorname{sgn}(s). \tag{55}$$

5.5.2 EVALUATING TRANSFORMS

Many of the elementary functions like sines, cosines, polynomials, and in general any periodic function, do not have Fourier transforms. This is so because such functions are not absolutely integrable. A special class of elementary functions that do have Fourier transforms, and can be calculated by basic methods, are those involving exponential functions. Several such transforms are related to the integrals

$$I = \int_0^{\infty} e^{-at} \cos st\, dt = a(s^2 + a^2)^{-1}, \qquad a > 0 \tag{56}$$

and

$$J = \int_0^{\infty} e^{-at} \sin st\, dt = s(s^2 + a^2)^{-1}, \qquad a > 0, \tag{57}$$

the derivations of which are left to the exercises (see Problem 3 in Exercises 5.5).

EXAMPLE 14

Find the Fourier transform of $f(t) = e^{-a|t|}, \quad a > 0.$

SOLUTION Since f is an even fuction, we can use (52) to write

$$\mathscr{F}\{e^{-a|t|}; s\} = \mathscr{F}_C\{e^{-at}; s\}$$

$$= \sqrt{\frac{2}{\pi}} \int_0^{\infty} e^{-at} \cos st\, dt,$$

or, using the result (56),

$$\mathscr{F}\{e^{-a|t|}; s\} = \sqrt{\frac{2}{\pi}}\, a(s^2 + a^2)^{-1}, \qquad a > 0.$$

EXAMPLE 15

Find the Fourier sine and cosine transforms of

$$f(t) = te^{-at}, \qquad a > 0.$$

SOLUTION With differentiation of both sides of (57), first with respect to a and then with respect to s, we have

$$\frac{\partial J}{\partial a} = -\int_0^\infty te^{-at} \sin st\, dt = -2as(s^2 + a^2)^{-2}$$

and

$$\frac{\partial J}{\partial s} = \int_0^\infty te^{-at} \cos st\, dt = (a^2 - s^2)(s^2 + a^2)^{-2}.$$

Thus we deduce that

$$\mathscr{F}_S\{te^{-at}; s\} = \sqrt{\frac{2}{\pi}}\, 2as(s^2 + a^2)^{-2}, \qquad a, s > 0,$$

$$\mathscr{F}_C\{te^{-at}; s\} = \sqrt{\frac{2}{\pi}}\, (a^2 - s^2)(s^2 + a^2)^{-2}, \qquad a, s > 0.$$

Although there are several other transforms that can be evaluated by similar means, we find that, in general, more sophisticated methods of evaluating the integrals must be found. Such methods do exist, but they often involve tricky manipulations combined with techniques from complex variables. Fortunately, many of the transforms we are interested in have been calculated over the years and extensive transform tables are now available, including one by A. Erdelyi, et al., *Tables of Integral Transforms,* McGraw-Hill, New York (1953). A short table of transforms is provided in Appendix B.

As our final example here, let us illustrate how the Fourier transform of a function can be constructed from knowledge of its Fourier sine transform.

EXAMPLE 16

Given that $\mathscr{F}_S\{f(t); s\} = \sqrt{\dfrac{2}{\pi}}(2e^{-bs} - 1)$, $s > 0$, where

$$f(t) = (t^2 - b^2)t^{-1}(t^2 + b^2)^{-1}, \qquad b > 0,$$

determine the Fourier transform of $f(t)$.

SOLUTION We note that f is an odd function. Therefore, by the use of (55), we deduce that

$$\mathscr{F}\{f(t); s\} = i\sqrt{\frac{2}{\pi}}\,(2e^{-b|s|} - 1)\,\text{sgn}(s), \qquad -\infty < s < \infty.$$

5.5.3 OPERATIONAL PROPERTIES

The calculation of integral transforms directly from definition is often tedious and quite complex. However, once we have developed the transforms of several functions, other transforms can be obtained from these through use of certain *operational properties*. We wish to derive some of the basic properties of the Fourier transform, while other properties will be taken up in the exercises as well as similar properties involving the sine and cosine transforms.

Foremost among the properties of the Fourier transform is the *linearity property*

$$\mathscr{F}\{C_1 f(t) + C_2 g(t); s\} = C_1\mathscr{F}\{f(t); s\} + C_2\mathscr{F}\{g(t); s\}, \tag{58}$$

where C_1 and C_2 are any constants. To derive (58), we merely observe that

$$\begin{aligned}\mathscr{F}\{C_1 f(t) + C_2 g(t); s\} &= \frac{1}{\sqrt{2\pi}}\int_{-\infty}^{\infty} e^{its}[C_1 f(t) + C_2 g(t)]\,dt \\ &= C_1\frac{1}{\sqrt{2\pi}}\int_{-\infty}^{\infty} e^{ist} f(t)\,dt + C_2\frac{1}{\sqrt{2\pi}}\int_{-\infty}^{\infty} e^{ist} g(t)\,dt,\end{aligned}$$

from which (58) follows.

Other than the linearity property, perhaps the most useful property of the Fourier transform is its effect on derivatives. If f is continuous and f' is piecewise smooth, and both f and f' are absolutely integrable, then by definition

$$\begin{aligned}\mathscr{F}\{f'(t); s\} &= \frac{1}{\sqrt{2\pi}}\int_{-\infty}^{\infty} e^{ist} f'(t)\,dt \\ &= \frac{1}{\sqrt{2\pi}} f(t)e^{ist}\Big|_{-\infty}^{\infty} - is\frac{1}{\sqrt{2\pi}}\int_{-\infty}^{\infty} e^{ist} f(t)\,dt,\end{aligned}$$

where we have employed an integration by parts. Now if f satisfies the limit condition

$$\lim_{|t|\to\infty} f(t) = 0,$$

we then obtain

$$\mathscr{F}\{f'(t); s\} = -isF(s), \tag{59}$$

where $F(s)$ is the Fourier transform of f. Likewise, if $f, f', \ldots, f^{(n-1)}$ are continuous, $f^{(n)}$ is piecewise smooth, and all are absolutely integrable, it can be shown that

$$\mathscr{F}\{f^{(n)}(t); s\} = (-is)^n F(s), \qquad n = 1, 2, 3, \ldots. \tag{60}$$

In the case of the cosine and sine transforms, the above results are somewhat different. For example, in the case of the cosine transform we use integration by parts to obtain*

$$\begin{aligned}\mathscr{F}_C\{f'(t); s\} &= \sqrt{\frac{2}{\pi}} \int_0^\infty f'(t) \cos st\, dt \\ &= -\sqrt{\frac{2}{\pi}}\, f(0) + s\sqrt{\frac{2}{\pi}} \int_0^\infty f(t) \sin st\, dt,\end{aligned}$$

from which we deduce

$$\mathscr{F}_C\{f'(t); s\} = sF_S(s) - \sqrt{\frac{2}{\pi}}\, f(0). \tag{61}$$

Similarly, it can be shown that

$$\mathscr{F}_S\{f'(t); s\} = -sF_C(s) \tag{62}$$

(see Problem 20 in Exercises 5.5). For second derivatives, we are led to the relations

$$\mathscr{F}_C\{f''(t); s\} = -s^2 F_C(s) - \sqrt{\frac{2}{\pi}}\, f'(0) \tag{63}$$

and

$$\mathscr{F}_S\{f''(t); s\} = -s^2 F_S(s) + \sqrt{\frac{2}{\pi}}\, sf(0) \tag{64}$$

(see Problem 21 in Exercises 5.5). These last two formulas give us some indication of which transform—sine or cosine—to use in a particular application. That is, in any problem in which $f(0)$ is known but $f'(0)$ is not known, we should use the Fourier sine transform of $f''(t)$. In the same way, if $f'(0)$ is known rather than $f(0)$, the Fourier cosine transform should be used.

The final property that we wish to develop involves the calculation of the inverse transform of a function that is the simple product of two other functions. To develop the required property, we introduce the Fourier *convolution integral*

$$(f \circ g)(t) = \frac{1}{\sqrt{2\pi}} \int_{-\infty}^\infty f(u) g(t-u)\, du. \tag{65}$$

We first observe that the change of variable $v = t - u$ leads to

$$(f \circ g)(t) = -\frac{1}{\sqrt{2\pi}} \int_\infty^{-\infty} f(v) g(t-v)\, dv$$

---

*Recall that $\lim_{t\to\infty} f(t) = 0$.

$$= \frac{1}{\sqrt{2\pi}} \int_{-\infty}^{\infty} g(t-v)f(v)\, dv,$$

from which we deduce the *commutative relation*

$$(f \circ g)(t) = (g \circ f)(t). \tag{66}$$

The important result we need concerns the Fourier transform of the convolution. Let us consider

$$\mathcal{F}\{(f \circ g)(t); s\} = \frac{1}{2\pi} \int_{-\infty}^{\infty} e^{ist} \int_{-\infty}^{\infty} f(u)g(t-u)\, du\, dt$$

$$= \frac{1}{2\pi} \int_{-\infty}^{\infty} \int_{-\infty}^{\infty} e^{ist} f(u)g(t-u)\, du\, dt. \tag{67}$$

We can interpret (67) as an iterated integral for which the order of integration can be interchanged. Hence, by changing the order of integration and making the change of variable $t = x + u$, we find

$$\mathcal{F}\{(f \circ g)(t); s\} = \frac{1}{2\pi} \int_{-\infty}^{\infty} \int_{-\infty}^{\infty} e^{is(x+u)} f(u)g(x)\, dx\, du$$

$$= \frac{1}{\sqrt{2\pi}} \int_{-\infty}^{\infty} e^{isu} f(u)\, du \cdot \frac{1}{\sqrt{2\pi}} \int_{-\infty}^{\infty} e^{isx} g(x)\, dx,$$

and thus conclude that

$$\mathcal{F}\{(f \circ g)(t); s\} = F(s)G(s). \tag{68}$$

Finally, we recognize that (68) leads to the result

$$\mathcal{F}^{-1}\{F(s)G(s); t\} = \frac{1}{\sqrt{2\pi}} \int_{-\infty}^{\infty} f(u)g(t-u)\, du, \tag{69}$$

which is called the *convolution theorem of Fourier*.

EXAMPLE 17

Use the convolution theorem (69) to evaluate the integral

$$I = \int_{-\infty}^{\infty} \frac{dx}{(x^2+a^2)(x^2+b^2)}, \qquad a, b > 0.$$

SOLUTION The integral has the form

$$I = \int_{-\infty}^{\infty} F(x)G(x)\, dx,$$

where $F(x) = (x^2 + a^2)^{-1}$ and $G(x) = (x^2 + b^2)^{-1}$ are recognized from the result of Example 14 as Fourier transforms, respectively, of

$$f(t) = \sqrt{\frac{\pi}{2}}\,\frac{1}{a}\,e^{-a|t|}$$

and

$$g(t) = \sqrt{\frac{\pi}{2}}\,\frac{1}{b}\,e^{-b|t|}.$$

Hence, using (69) leads to

$$\frac{1}{\sqrt{2\pi}}\int_{-\infty}^{\infty} e^{-itx}F(x)G(x)\,dx = \frac{1}{\sqrt{2\pi}}\int_{-\infty}^{\infty} f(u)g(t-u)\,du.$$

By cancelling the common factor $1/\sqrt{2\pi}$ and setting $t = 0$, the left-hand side becomes I and we find

$$I = \int_{-\infty}^{\infty} f(u)g(-u)\,du.$$

Lastly, by using the fact that f and g are even functions, we get

$$\begin{aligned} I &= 2\int_0^{\infty} f(u)g(u)\,du \\ &= \frac{\pi}{ab}\int_0^{\infty} e^{-(a+b)u}\,du, \end{aligned}$$

from which we now deduce

$$\int_{-\infty}^{\infty} \frac{dx}{(x^2+a^2)(x^2+b^2)} = \frac{\pi}{ab(a+b)}.$$

□ 5.5.4 RELATION BETWEEN FOURIER AND LAPLACE TRANSFORMS

If g is piecewise smooth and absolutely integrable, it satisfies the conditions of the Fourier integral theorem and we can write

$$g(t) = \frac{1}{2\pi}\int_{-\infty}^{\infty} e^{-ist}\int_{-\infty}^{\infty} e^{isx}g(x)\,dx\,ds \tag{70}$$

[see Equation (40)]. Now, if g is related to another function f according to

$$g(t) = e^{-ct}f(t)h(t), \tag{71}$$

where c is a positive constant and $h(t)$ is the Heaviside unit function,* then as a consequence of the absolute integrability of g it follows that f satisfies

---

*Recall that $h(t) = 1$ for $t > 0$, and $h(t) = 0$ for $t < 0$.

$$\int_0^\infty e^{-ct}|f(t)|\,dt < \infty.$$

This last relation is satisfied by functions of exponential order. The substitution of (71) into (70) leads to

$$e^{-ct}f(t)h(t) = \frac{1}{2\pi}\int_{-\infty}^\infty e^{-ist}\int_{-\infty}^\infty e^{-(c-is)x}f(x)h(x)\,dx\,ds,$$

or equivalently,

$$f(t)h(t) = \frac{1}{2\pi}\int_{-\infty}^\infty e^{(c-is)t}\int_0^\infty e^{-(c-is)x}f(x)\,dx\,ds. \tag{72}$$

By introducing the change of variable $p = c - is$, we find that (72) becomes

$$f(t)h(t) = \frac{1}{2\pi i}\int_{c-i\infty}^{c+i\infty} e^{pt}\int_0^\infty e^{-px}f(x)\,dx\,dp. \tag{73}$$

Thus, in an heuristic manner, we have derived the following pair of transform formulas

$$F(p) = \int_0^\infty e^{-px}f(x)\,dx \tag{74}$$

and

$$f(t)h(t) = \frac{1}{2\pi i}\int_{c-i\infty}^{c+i\infty} e^{pt}F(p)\,dp. \tag{75}$$

Equation (74) is recognized as the Laplace transform of f, whereas (75) is the *inverse formula* for the Laplace transform. Although (74) can usually be evaluated by elementary integration techniques, the same is not true of (75). That is, the evaluation of (75) requires techniques of complex variables and therefore we will not use it.

REMARK: *The Heaviside unit function $h(t)$ is usually left out of the expression (75) when it is understood that $f(t) = 0$ for $t < 0$.*

EXERCISES 5.5

1. If $f(t)$ is an odd function, show that the Fourier integral representation becomes

$$f(x) = \frac{2}{\pi}\int_0^\infty \sin sx \int_0^\infty f(t)\sin st\,dt\,ds.$$

2. Given the following functions, develop the even and odd extensions, $f_e(t)$ and $f_0(t)$, respectively:
   (a) $f(t) = e^{-at}$
   (b) $f(t) = e^{-t^2}\sin t$
   (c) $f(t) = (1 + t)e^{-at}$

3. Using integration by parts, verify that

(a) $$\int_0^\infty e^{-at} \cos st \, dt = a(s^2 + a^2)^{-1}, \qquad a > 0.$$

(b) $$\int_0^\infty e^{-at} \sin st \, dt = s(s^2 + a^2)^{-1}, \qquad a > 0.$$

4. Show that

(a) $$\mathscr{F}_C\{e^{-at}; s\} = \sqrt{\frac{2}{\pi}}\, a(s^2 + a^2)^{-1}, \qquad a, s > 0.$$

(b) $$\mathscr{F}_S\{e^{-at}; s\} = \sqrt{\frac{2}{\pi}}\, s(s^2 + a^2)^{-1}, \qquad a, s > 0.$$

In Problems 5–8, determine the Fourier transform of the given function.

5. $f(t) = \begin{cases} 0, & t < 0 \\ e^{-at}, & t > 0, \quad a > 0 \end{cases}$

6. $f(t) = \begin{cases} 1, & 0 < t < 1 \\ 0, & \text{otherwise} \end{cases}$

7. $f(t) = \begin{cases} 0, & t < 0, \quad t > \pi \\ \sin t, & 0 < t < \pi \end{cases}$

8. $f(t) = \delta(t - a)$, where δ is the delta function.

9. Use the result of Example 15 to determine
   (a) $\mathscr{F}\{te^{-a|t|}; s\}$, $\quad a > 0$.
   (b) $\mathscr{F}\{|t|e^{-a|t|}; s\}$, $\quad a > 0$.

10. Use the integral formula (7) to show that

$$\mathscr{F}\{e^{-t^2}; s\} = 2^{-1/2} e^{-s^2/4}.$$

Hint: $t^2 - ist = (t - \frac{1}{2}is)^2 + \frac{1}{4}s^2$.

11. If $\mathscr{F}\{f(t); s\} = F(s)$, show that

$$\mathscr{F}\{f(at); s\} = a^{-1}F(s/a), \qquad a > 0.$$

12. Use the results of Problems 10 and 11 to deduce that
    (a) $\mathscr{F}\{e^{-a^2t^2}; s\} = 2^{-1/2}a^{-1}e^{-s^2/4a^2}$, $\quad a>0$
    (b) $\mathscr{F}\{e^{-t^2/2}; s\} = e^{-s^2/2}$.

13. Given the triangle function $f(t) = (1 - |t|)h(1 - |t|)$, where h is the Heaviside unit function, show that

(a) $$\mathscr{F}\{f(t); s\} = 2\sqrt{\frac{2}{\pi}}\, s^{-2} \sin^2(s/2).$$

(b) From (a), deduce that

$$\int_0^\infty \left(\frac{\sin x}{x}\right)^2 dx = \frac{\pi}{2}.$$

★14. Putting $f(t) = t^{-1/2}$ in the sine and cosine forms of Fourier's integral theorem, show that

(a) $$\int_0^\infty t^{-1/2} \sin t \, dt = \int_0^\infty t^{-1/2} \cos t \, dt = \sqrt{\frac{\pi}{2}}.$$

(b) From (a), deduce that

$$\mathscr{F}_S\{t^{-1/2}; s\} = \mathscr{F}_C\{t^{-1/2}; s\} = s^{-1/2}.$$

15. Use the result of Problem 12(b) to derive the relations

(a) $\mathscr{F}\{\cos(\frac{1}{2}t^2); s\} = 2^{-1/2}[\cos(\frac{1}{2}s^2) + \sin(\frac{1}{2}s^2)]$.

(b) $\mathscr{F}\{\sin(\frac{1}{2}t^2); s\} = 2^{-1/2}[\cos(\frac{1}{2}s^2) - \sin(\frac{1}{2}s^2)]$.

Hint: Set $a = \frac{1}{2}(1 - i)$ in Problem 12(b) and compare real and imaginary parts.

★16. Starting with the integral formula in Problem 3(b), integrate both sides with respect to a from b to ∞ and derive the formula

(a) $$\mathscr{F}_S\{t^{-1}e^{-bt}; s\} = \sqrt{\frac{2}{\pi}} \arctan(s/b), \qquad b > 0.$$

(b) Let $b \to 0^+$ in (a) and deduce that

$$\mathscr{F}\{t^{-1}; s\} = i\sqrt{\frac{\pi}{2}} \operatorname{sgn}(s).$$

(c) From (b), determine $\mathscr{F}\{\operatorname{sgn}(t); s\}$.

17. Given that $F(s) = \mathscr{F}\{f(t); s\}$, show that

(a) $\mathscr{F}\{e^{iat}f(t); s\} = F(s + a)$.

(b) $\mathscr{F}\{f(t - a); s\} = e^{ias}F(s)$.

18. Formally differentiate both sides of

$$F(s) = \frac{1}{\sqrt{2\pi}} \int_{-\infty}^\infty e^{ist} f(t) \, dt$$

with respect to s and show that

(a) $\mathscr{F}\{tf(t); s\} = -iF'(s)$.

(b) $\mathscr{F}\{t^m f(t); s\} = (-i)^m F^{(m)}(s), \qquad m = 1, 2, 3, \ldots.$

19. Use the results of Problems 12, 17, and 18 to evaluate

(a) $\mathscr{F}\{te^{-t^2/2}; s\}$.

(b) $\mathscr{F}\{t^2e^{-t^2/2}; s\}$.

(c) $\mathscr{F}\{e^{-t^2+bt}; s\}$.

20. Verify that the sine transform satisfies the relation

$$\mathscr{F}_S\{f'(t); s\} = -sF_C(s).$$

21. Derive the derivative relations

(a) $$\mathscr{F}_S\{f''(t); s\} = -s^2F_S(s) + \sqrt{\frac{2}{\pi}}\, sf(0).$$

(b) $$\mathscr{F}_C\{f''(t); s\} = -s^2F_C(s) - \sqrt{\frac{2}{\pi}}\, f'(0).$$

22. Show that
(a) $\mathcal{F}_S\{f(at); s\} = a^{-1}F_S(s/a), \qquad a > 0.$
(b) $\mathcal{F}_C\{f(at); s\} = a^{-1}F_C(s/a), \qquad a > 0.$

23. Show that
(a) $\mathcal{F}_S\{\cos(at)f(t); s\} = (1/2)[F_S(s + a) + F_S(s - a)].$
(b) $\mathcal{F}_C\{\cos(at)f(t); s\} = (1/2)[F_C(s + a) + F_C(s - a)].$

24. Show that
(a) $\mathcal{F}_S\{\sin(at)f(t); s\} = (1/2)[F_C(s - a) - F_C(s + a)].$
(b) $\mathcal{F}_C\{\sin(at)f(t); s\} = (1/2)[F_S(s + a) - F_S(s - a)].$

25. Use the results of Problems 4, 23, and 24 to show that

(a) $\mathcal{F}_C\{e^{-at}\cos at; s\} = \sqrt{\dfrac{2}{\pi}}\,(as^2 + 2a^3)(s^4 + 4a^4)^{-1}, \qquad a > 0.$

(b) $\mathcal{F}_C\{e^{-at}\sin at; s\} = \sqrt{\dfrac{2}{\pi}}\,(2a^3 - as^2)(s^4 + 4a^4)^{-1}, \qquad a > 0.$

★26. From the results of Problem 25, deduce that

$$\mathcal{F}^{-1}\{(s^4 + k^4)^{-1}; t\} = \frac{\sqrt{\pi}}{2k^3}\,e^{-k|t|/\sqrt{2}}\left[\cos\left(\frac{kt}{\sqrt{2}}\right) + \sin\left(\frac{k|t|}{\sqrt{2}}\right)\right], \qquad k > 0.$$

27. Use the results of Problems 4, 23, and 24 to evaluate
(a) $\mathcal{F}_S\{e^{-at}\cos at; s\}, \qquad a > 0.$
(b) $\mathcal{F}_S\{e^{-at}\sin at; s\}, \qquad a > 0.$

28. From the results of Problem 27, evaluate
(a) $\mathcal{F}_S\{t^3(t^4 + k^4)^{-1}; s\}.$
(b) $\mathcal{F}_S\{t(t^4 + k^4)^{-1}; s\}.$
(c) $\mathcal{F}_S\{t^3(t^4 + k^4)^{-2}; s\}.$
Hint: Differentiate the result in (a) with respect to k.

29. Show that the Fourier convolution (65) satisfies
(a) $f \circ (Cg) = (Cf) \circ g = C(f \circ g), \qquad C$ constant.
(b) $f \circ (g + k) = (f \circ g) + (f \circ k).$
(c) $(f \circ \delta)(t) = f(t), \qquad$ where δ is the delta function.

★30. Show that the Fourier convolution (65) satisfies

$$f \circ (g \circ k) = (f \circ g) \circ k.$$

31. Verify the convolution theorem (69) for
(a) $f(t) = g(t) = h(1 - |t|).$
(b) $f(t) = g(t) = e^{-t^2/2}.$

★32. Show that the convolution theorem can also be expressed in the form

$$\int_{-\infty}^{\infty} e^{ist}f(t)g(t)\,dt = \int_{-\infty}^{\infty} F(u)G(s - u)\,du,$$

and use this result to evaluate $\mathcal{F}\left\{e^{-|t|}\dfrac{\sin t}{t}; s\right\}.$

Hint: Recall Example 14 in this chapter and Example 14 in Section 4.7.

33. For a, $b > 0$, show that

$$\int_{-\infty}^{\infty} \frac{x^2 dx}{(x^2 + a^2)(x^2 + b^2)} = \frac{\pi}{a + b}.$$

★34. Given that

$$\mathcal{F}\{(a^2 - t^2)^{-1/2} h(a - |t|); s\} = \sqrt{\frac{\pi}{2}}\, J_0(as), \qquad a > 0,$$

where J_0 is the *Bessel function* of order zero, show that

$$\int_0^{\infty} J_0(ax) J_0(bx)\, dx = \frac{2}{\pi b} K\left(\frac{a}{b}\right), \qquad 0 < a < b,$$

where $K(m)$ denotes the *complete elliptic integral*

$$K(m) = \int_0^{\pi/2} (1 - m^2 \sin^2 \theta)^{-1/2}\, d\theta.$$

★35. If $F_C(s)$ and $G_C(s)$ denote the cosine transforms of $f(t)$ and $g(t)$, respectively, show that the convolution theorem leads to

(a) $\displaystyle\int_{-\infty}^{\infty} e^{-ist} F_C(s) G_C(s)\, ds = \int_{-\infty}^{\infty} f(|u|) g(|t - u|)\, du.$

(b) From (a), deduce that

$$\int_0^{\infty} \cos st F_C(s) G_C(s)\, ds = \frac{1}{2} \int_0^{\infty} f(u)[g(|t - u|) + g(t + u)]\, du.$$

(c) For $g(t) \equiv f(t)$, show that (b) leads to the *Parseval relation*

$$\int_0^{\infty} |F_C(s)|^2\, ds = \int_0^{\infty} |f(u)|^2\, du.$$

★36. If $F_S(s)$ and $G_S(s)$ denote the sine transforms of $f(t)$ and $g(t)$, respectively, show that the convolution theorem leads to

(a) $\displaystyle\int_0^{\infty} \cos st F_S(s) G_S(s)\, ds = \frac{1}{2} \int_0^{\infty} f(u)[g(u + t) + g(u - t)]\, du.$

(b) For $g(t) \equiv f(t)$, show that (a) leads to the *Parseval relation*

$$\int_0^{\infty} |F_S(s)|^2\, ds = \int_0^{\infty} |f(u)|^2\, du.$$

5.6 BOUNDARY VALUE PROBLEMS

The Fourier transforms discussed in the last section are particularly useful in solving boundary value problems on infinite domains. According to Definition 1.4, problems formulated on infinite domains are classified as *singular*.

For such problems, the boundary conditions that are normally prescribed are of the general form

$$y(x), \qquad y'(x) \quad \text{finite as} \quad |x| \to \infty. \tag{76}$$

The use of Fourier transforms to solve this class of problems, however, often forces us to (at least initially) impose the more stringent requirements

$$y(x) \to 0, \qquad y'(x) \to 0 \quad \text{as} \quad |x| \to \infty. \tag{77}$$

These last conditions are necessary to ensure that the Fourier transforms of $y'(x)$ and $y''(x)$ exist (see Section 5.5.3). Nonetheless, in some situations the final solution obtained from the transform method may not satisfy (77). We do require that $y(x)$ at least satisfy (76), and therefore in practice we often take the point of view that the transform method produces a "tentative solution" which must be scrutinized to see if it indeed satisfies the basic physical conditions of the problem.

EXAMPLE 18

Use the Fourier transform to solve

$$y'' - y = -h(1 - |x|), \qquad -\infty < x < \infty,$$

$$y(x) \to 0, \qquad y'(x) \to 0 \quad \text{as} \quad |x| \to \infty.$$

SOLUTION By introducing the Fourier transforms

$$\mathscr{F}\{y(x); s\} = Y(s),$$

$$\mathscr{F}\{y''(x); s\} = -s^2 Y(s),$$

and

$$\mathscr{F}\{h(1 - |x|); s\} = \frac{1}{\sqrt{2\pi}} \int_{-1}^{1} e^{isx}\, dx = \sqrt{\frac{2}{\pi}} \frac{\sin s}{s},$$

the DE is converted into the algebraic equation

$$-s^2 Y(s) - Y(s) = -\sqrt{\frac{2}{\pi}} \frac{\sin s}{s},$$

with solution

$$Y(s) = \sqrt{\frac{2}{\pi}} \frac{\sin s}{s(s^2 + 1)}.$$

Finally, the solution $y(x)$ that we seek is simply the inverse Fourier transform of $Y(s)$, which we can obtain through use of the convolution theorem. That is, we have

$$\mathscr{F}^{-1}\left\{\sqrt{\frac{2}{\pi}} \frac{\sin s}{s}; x\right\} = h(1 - |x|)$$

and

$$\mathscr{F}^{-1}\{(s^2+1)^{-1}; x\} = \sqrt{\frac{\pi}{2}}\, e^{-|x|},$$

where this last result follows from Example 14. Thus, from Equation (69), we deduce that

$$\begin{aligned} y(x) &= \mathscr{F}^{-1}\{Y(s); x\} \\ &= \frac{1}{2}\int_{-\infty}^{\infty} e^{-|u|} h(1 - |x-u|)\, du \\ &= \frac{1}{2}\int_{x-1}^{x+1} e^{-|u|}\, du. \end{aligned}$$

The evaluation of this last integral leads to

$$y(x) = \begin{cases} \sinh(1)e^{x}, & -\infty < x < -1 \\ 1 - e^{-1}\cosh x, & -1 \leq x \leq 1 \\ \sinh(1)e^{-x}, & 1 < x < \infty \end{cases}$$

(see Problem 1 in Exercises 5.6).

This one example illustrates the basic technique used in solving boundary value problems by the transform method. That is, the transform applied term-by-term to the DE leads to an algebraic equation in the transform function $Y(s)$. This algebraic equation can be readily solved for $Y(s)$, the inversion of which yields the desired solution $y(x)$. It is perhaps the final step of inverting $Y(s)$ that is generally the most difficult part of the process. In many cases, we simply represent our solution as an integral whose evaluation is left to numerical procedures.

If the domain of interest in Example 18 were modified to the semi-infinite region $0 < x < \infty$, either the sine or cosine transform would be used to solve the problem. The decision as to which of these two transforms to use will depend upon the type of boundary condition specified at the finite boundary $x = 0$. That is, from Equations (63) and (64), we recall

$$\mathscr{F}_S\{y''(x); s\} = -s^2 Y_S(s) + \sqrt{\frac{2}{\pi}}\, s y(0), \tag{78}$$

$$\mathscr{F}_C\{y''(x); s\} = -s^2 Y_C(s) - \sqrt{\frac{2}{\pi}}\, y'(0), \tag{79}$$

and so if $y(0)$ is specified we use the sine transform, whereas the cosine transform is called for when $y'(0)$ is specified. Problems involving these transforms are taken up in the exercises.

Finally, we remark that the technique illustrated in Example 18 can readily be generalized to boundary value problems of the form

$$\begin{aligned} &y'' + ay' + by = f(x), \qquad -\infty < x < \infty \\ &y(x) \to 0, \qquad y'(x) \to 0 \quad \text{as} \quad |x| \to \infty, \end{aligned} \tag{80}$$

where a and b are constants. Proceeding as before, the Fourier transform applied to (80) leads to the algebraic problem

$$-(s^2 + ias - b)Y(s) = F(s), \tag{81}$$

with solution

$$Y(s) = -\frac{F(s)}{s^2 + ias - b}. \tag{82}$$

The inversion of (82) by use of the convolution theorem gives us the solution formula

$$y(x) = -\int_{-\infty}^{\infty} f(u)g(x-u)\,du, \tag{83}$$

where we define

$$\begin{aligned} g(x) &= \frac{1}{\sqrt{2\pi}}\mathscr{F}^{-1}\left\{\frac{1}{s^2 + ias - b}; x\right\} \\ &= \frac{1}{\sqrt{2\pi}}\mathscr{F}^{-1}\left\{\frac{1}{\left(s + \frac{ia}{2}\right)^2 + \left(\frac{a^2}{4} - b\right)}; x\right\} \\ &= \frac{1}{\sqrt{2\pi}}e^{-ax/2}\mathscr{F}^{-1}\left\{\frac{1}{s^2 + \left(\frac{a^2}{4} - b\right)}; x\right\}, \end{aligned} \tag{84}$$

the last step of which follows from Problem 17(a) in Exercises 5.5. Finally, we recall Example 14 to obtain the inversion

$$g(x) = \frac{1}{\sqrt{a^2 - 4b}}\exp\left[-\frac{1}{2}\left(ax + \sqrt{a^2 - 4b}\,|x|\right)\right]. \tag{85}$$

□ 5.6.1 GREEN'S FUNCTION

The form of the solution (83) suggests that the function $g(x-u)$ is a kind of Green's function for the boundary value problem (80). This is indeed correct, but $g(x-u)$ is a variation of the Green's function discussed in Section 2.4, unless $a = 0$. In particular, if $a \neq 0$ the function $g(x-u)$ will not possess the symmetry property where the roles of x and u can be interchanged.

EXAMPLE 19

Find a solution formula for the boundary value problem ($k > 0$)

$$y'' - k^2 y = f(x), \qquad -\infty < x < \infty,$$

$$y(x) \to 0, \qquad y'(x) \to 0 \quad \text{as} \quad |x| \to \infty,$$

and identify the Green's function.

SOLUTION Application of the Fourier transform yields

$$-(s^2 + k^2)Y(s) = F(s),$$

which has the solution

$$Y(s) = -\frac{F(s)}{(s^2 + k^2)}.$$

Recalling the inverse transform relation (Example 14)

$$\mathscr{F}^{-1}\{(s^2 + k^2)^{-1}; x\} = \sqrt{\frac{\pi}{2}}\,\frac{1}{k}\,e^{-k|x|},$$

the inversion of $Y(s)$ leads to the formal solution

$$y(x) = -\frac{1}{2k}\int_{-\infty}^{\infty} f(u)e^{-k|x-u|}\,du.$$

Clearly, the Green's function is, in the notation of Section 2.4,

$$g(x, u) = \frac{1}{2k}\,e^{-k|x-u|}, \qquad k > 0.$$

EXERCISES 5.6

1. Given the integral

$$I = \int_{x-1}^{x+1} e^{-|u|}\,du,$$

show that

(a) $I = \displaystyle\int_{x-1}^{x+1} e^{u}\,du, \qquad -\infty < x < -1.$

(b) $I = \displaystyle\int_{x-1}^{0} e^{u}\,du + \int_{0}^{x+1} e^{-u}\,du, \qquad -1 \leqslant x \leqslant 1.$

(c) $I = \displaystyle\int_{x-1}^{x+1} e^{-u}\,du, \qquad 1 < x < \infty.$

(d) From (a), (b), and (c), deduce the value of I.

In Problems 2–6, use an appropriate integral transform to solve the given problem. If necessary, use the tables in Appendix B.

2. $y'' - y = e^{-|x|}, \qquad -\infty < x < \infty,$
$y(x) \to 0, \qquad y'(x) \to 0 \quad \text{as} \quad |x| \to \infty$

3. $y'' - y = e^{-x}, \qquad 0 < x < \infty$
$y(0) = 0, \qquad y(x) \to 0, \qquad y'(x) \to 0 \quad \text{as} \quad x \to \infty$

4. $y'' - y = e^{-x}, \quad 0 < x < \infty$
$y'(0) = 0, \quad y(x) \to 0, \quad y'(x) \to 0 \text{ as } x \to \infty$

5. $y'' - k^2 y = -h(1 - x), \quad k > 0, \quad 0 < x < \infty$
$y'(0) = 0, \quad y(x) \to 0, \quad y'(x) \to 0 \text{ as } x \to \infty$

★6. $y'' - y = xe^{-x}, \quad 0 < x < \infty$
$y(0) = 0, \quad y(x) \to 0, \quad y'(x) \to 0 \text{ as } x \to \infty$

Hint: $\mathcal{F}_S\{t^2 e^{-t}; s\} = \sqrt{\frac{2}{\pi}}\, 2s(3 - s^2)/(s^2 + 1)^3$.

7. Use the Fourier sine transform to formally solve

$$y'' - k^2 y = f(x), \quad 0 < x < \infty$$
$$y(0) = 1, \quad y(x) \to 0, \quad y'(x) \to 0 \text{ as } x \to \infty$$

$(k > 0)$.

★8. Show that the Green's function for Problem 7 is

$$g(x, u) = \begin{cases} \frac{1}{k} e^{-kx} \sinh ku, & 0 < u \leq x \\ \frac{1}{k} e^{-ku} \sinh kx, & x < u < \infty. \end{cases}$$

9. Use the Fourier cosine transform to formally solve

$$y'' - k^2 y = f(x), \quad 0 < x < \infty$$
$$y'(0) = 1, \quad y(x) \to 0, \quad y'(x) \to 0 \text{ as } x \to \infty$$

$(k > 0)$.

10. The small deflections of an "infinitely long" beam resting on an elastic foundation, such as a railroad track on a road bed, supporting a distributed load are governed by the DE $(k > 0)$

$$y^{(4)} + k^4 y = f(x), \quad -\infty < x < \infty,$$

where $f(x)$ is proportional to the distributed load. Show that the solution can be expressed in the form

$$y = \frac{1}{\sqrt{2\pi}} \int_{-\infty}^{\infty} f(u) g(x - u)\, du,$$

where

$$g(x) = \frac{\sqrt{\pi}}{2k^3} e^{-k|x|/\sqrt{2}}[\cos(kx/\sqrt{2}) + \sin(k|x|/\sqrt{2})].$$

Hint: Recall Problem 26 in Exercises 5.5.

CHAPTER 6

THE HEAT EQUATION

In the study of the flow of heat in thermally conducting regions, the governing PDE is the *heat equation*. This same equation arises, however, in other diffusion processes and is, therefore, also widely known as the *diffusion equation*. The flow of electricity in a long cable or transmission line can lead to the telegraph equations, which also have the same functional form as the heat equation. For one-dimensional problems, the mathematical treatment of all of these applications is basically the same, and for that reason we will discuss the one-dimensional heat equation primarily in terms of heat flow in a thin rod or wire without any real loss of generality.

One-dimensional heat flow in a rod is characterized by constant temperature throughout any cross section of the rod while allowing for variations in temperature from cross section to cross section. In Section 6.2 we discuss the flow of heat in a *finite rod* devoid of any heat sources. Our method of solution is *separation of variables*. The technique requires that we solve a related *eigenvalue problem* and perform a *Fourier series* expansion of the known initial temperature distribution in the rod. The solution so obtained is generally in the form of an infinite series. Several special cases of basic heat flow are considered, each of which involves a different set of boundary conditions. General *nonhomogeneous problems* including *heat sources* as well as *time-varying end conditions* are treated in Section 6.3.

By modeling the rod as *infinite in extent,* we can often obtain solutions relatively simple in form as compared with the infinite series solutions for a finite rod. Of course, the solutions in such cases are valid only in restricted intervals. For example, we may model the rod as infinite if it is very long and we are interested in the temperature distribution only in the center portion of the rod. The method of *Fourier integrals* is developed in Section 6.4 for this class of problems, and in Section 6.5 we solve problems of this type by using *Laplace* and *Fourier transforms*. A short section on the *finite difference numerical method* ends the chapter.

6.1 INTRODUCTION

The *heat equation*

$$\nabla^2 u = a^{-2} u_t, \tag{1}$$

where a^2 is a physical constant, arises in problems concerning the temperature distribution in homogeneous solids, electromagnetic theory, diffusion processes, and the propagation of current in transmission lines. A properly-posed problem consists of (1) coupled with a *single* boundary condition and a *single* initial condition, an example of which is given by*

$$\nabla^2 u = a^{-2} u_t, \qquad u \text{ in } R$$

$$u = g \quad \text{on} \quad \partial R$$

$$u = f \quad \text{for} \quad t = 0.$$

REMARK: *In some problems the boundary ∂R of the region of interest may not be a simple smooth curve or surface. For example, the boundary of a finite rod consists of two isolated points (endpoints) and a boundary condition is specified at each end. Even in such cases, the specified conditions collectively count as* one condition at each boundary point.

The fundamental problem in the mathematical theory of heat conduction is to solve Equation (1) for the temperature u in a homogeneous solid when the distribution of temperature throughout the solid is known at time $t = 0$ and a certain boundary condition is prescribed at each exposed point of the solid. There are three distinct kinds of boundary conditions that might ordinarily be prescribed for such problems. The first is to specify the temperature u along each finite surface of the solid *(Dirichlet condition);* the second condition is to prescribe the flux of heat across the surface, which is accomplished by specifying the *normal derivative* of u at the surface *(Neumann condition);* and the third boundary condition is to prescribe the rate at which heat is lost from the solid due to surface radiation into the surrounding medium *(Newton's law of cooling)*. This last boundary condition is sometimes called *Robin's condition*.

6.1.1 ONE-DIMENSIONAL MODEL

In this chapter we will concern ourselves with only the one-dimensional analogue of Equation (1). Among other areas of application, the one-dimensional heat equation governs the temperature distribution in a long rod or wire whose lateral surface is impervious to heat (i.e., insulated). For modeling purposes we assume the rod coincides with a portion of the x-axis, is made of uniform material, and has uniform cross section. Further, we will assume that the temperature $u(x, t)$ is the same at any point of a particular cross section of the rod, but may change from cross section to cross section.

To derive the governing equation of heat flow in such a rod from first principles, we apply the *law of conservation of thermal energy* to a slice of the rod between x and a nearby point $(x + \Delta x)$ as shown in Figure 6.1. The law of conservation of thermal energy states that *the rate of heat entering a region plus that which is generated inside the region equals the rate of heat leaving the region plus that which is stored.*

---

*R denotes the domain of the function u, and ∂R is the boundary of R.

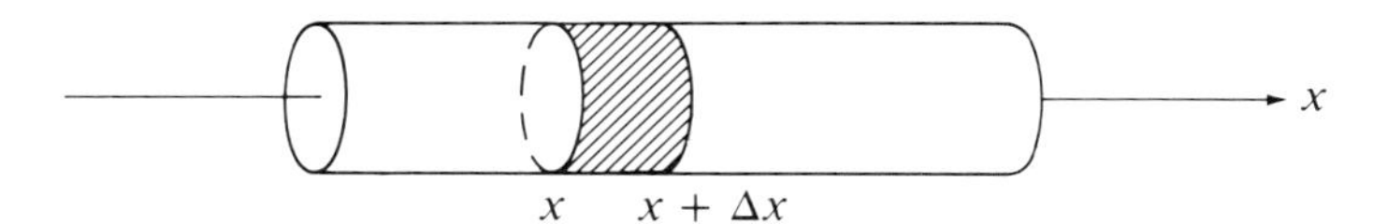

FIGURE 6.1

Rod of finite length.

Suppose we let $Q(x, t)$ be the rate of heat flow at the point x at time t, also known as the *heat flux*. If A denotes the cross-sectional area of the uniform rod, then $AQ(x, t)$ represents the rate at which heat enters the slice of the rod through the flat surface at x. Accordingly, the rate at which heat vacates the slice at $(x + \Delta x)$ is $AQ(x + \Delta x, t)$. If a heat source is present within the region, the rate at which heat is generated in the slice is $A\Delta x F(x, t)$, where F is the rate of heat generation per unit volume. In some instances F may also depend upon the temperature u inside the slice. Lastly, the rate of heat energy storage in the slice is proportional to the time rate change of temperature u_t. To find the constant of proportionality necessitates the introduction of the *specific heat* constant c,* defined as the heat energy that must be supplied to a unit mass of the rod to raise it one unit of temperature. Thus, if ρ is the constant *mass density* of the rod (i.e., mass per unit volume), the rate of change of heat energy is approximately $A\Delta x \rho c u_t$.

If we now invoke the law of conservation of energy, it follows that

$$AQ(x, t) + A\Delta x F(x, t) = AQ(x + \Delta x, t) + A\Delta x \rho c u_t,$$

which can be rearranged as

$$-\left(\frac{Q(x + \Delta x, t) - Q(x, t)}{\Delta x}\right) = \rho c u_t - F(x, t). \tag{2}$$

By allowing $\Delta x \to 0$, we find that (2) becomes

$$-Q_x = \rho c u_t - F(x, t). \tag{3}$$

We can relate Q and the temperature u by use of *Fourier's law of heat conduction,* which for the present problem reads

$$Q = -k u_x, \tag{4}$$

where k is a constant.† In words, Fourier's law says that the heat flux Q at any point x is proportional to the temperature gradient u_x at that point. The proportionality constant k is characteristic of the material of the rod, called the *heat conductivity,* but for a nonuniform rod is actually *not* a constant. Fourier's law (4) combined with (3) yields the governing equation

* The specific heat c is actually dependent upon temperature, but for the problems we consider in this text it can reasonably be approximated by a constant.

† The negative sign in (4) reflects the fact that heat flows from hotter to cooler regions.

$$u_{xx} = a^{-2}u_t - \frac{F(x, t)}{k}, \tag{5}$$

where $a^2 = k/\rho c$ is a positive constant known as the *diffusivity* of the material forming the rod. When the heat source F is not present, this PDE reduces to the homogeneous *one-dimensional heat equation*

$$u_{xx} = a^{-2}u_t. \tag{6}$$

Using the classification scheme of Section 3.2, it can be shown that the heat equation (6) is of the *parabolic* type.

The diffusivity constant $a^2 = k/\rho c$ has units of (length)^2/time. Its value depends only upon the material from which the rod is made. Some typical values of a^2 are shown in Table 6.1.

TABLE 6.1 Some Values of Thermal Diffusivity

MATERIAL	a^2 (cm^2/sec)
Silver	1.71
Copper	1.14
Aluminum	0.86
Cast iron	0.12

6.2 HEAT FLOW IN A ROD WITH NO SOURCES

Let us consider a thin rod or wire made of homogeneous material, the diameter of which coincides with the x-axis from $x = 0$ to $x = p$. It is assumed that the initial temperature f of the rod is specified as a function of the distance x from one end of the rod ($x = 0$). The temperature distribution $u(x, t)$ at some later time in the absence of any heat source is then a solution of the one-dimensional heat equation

$$u_{xx} = a^{-2}u_t, \qquad 0 < x < p, \qquad t > 0 \tag{7}$$

and the prescribed initial condition (I.C.)

$$\text{I.C.:} \qquad u(x, 0) = f(x), \qquad 0 < x < p. \tag{8}$$

The temperature inside the rod will also be affected by how the ends of the rod exchange heat energy with the surrounding medium. We describe such heat exchanges by boundary conditions (B.C.) of the general form

$$\text{B.C.:} \qquad \begin{cases} B_1[u] \equiv a_{11}u(0, t) + a_{12}u_x(0, t) = \alpha(t), \\ B_2[u] \equiv a_{21}u(p, t) + a_{22}u_x(p, t) = \beta(t). \end{cases} \qquad t > 0 \tag{9}$$

Rather than trying to solve the problem at this time with such general boundary conditions, it is helpful instead to consider several special cases.

6.2.1 ROD WITH ENDS HELD AT ZERO TEMPERATURE

Suppose the two ends of the rod described above are at time $t = 0$ suddenly placed in contact with ice packs at 0°C, and that this temperature at the ends is maintained at all later times. Under the assumption that the initial temperature distribution in the rod is $f(x)$, the problem we wish to solve is characterized by

$$u_{xx} = a^{-2}u_t, \quad 0 < x < p, \quad t > 0$$

$$\text{B.C.:} \quad u(0, t) = 0, \quad u(p, t) = 0, \quad t > 0 \tag{10}$$

$$\text{I.C.:} \quad u(x, 0) = f(x), \quad 0 < x < p$$

(see Figure 6.2).

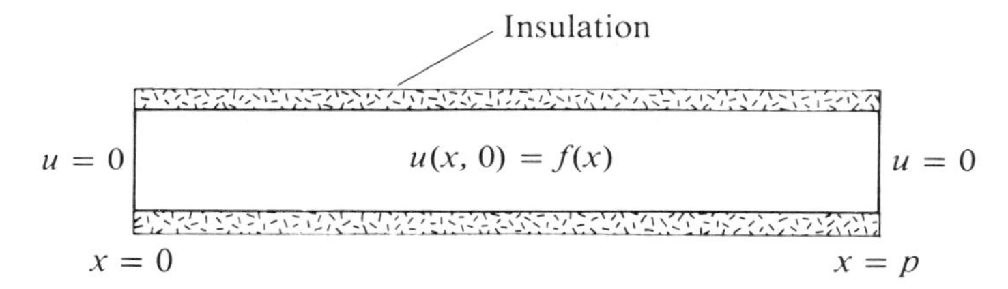

FIGURE 6.2

REMARK: *Because the ends of the rod are maintained at temperature zero, it seems plausible that after a long period of time ($t \to \infty$) the temperature $u(x, t)$ throughout the rod will eventually approach zero, regardless of the initial temperature distribution $f(x)$. Experience indicates that this is so, and in general all variations of temperature with time will die out if the prescribed boundary conditions are not time dependent.*

To solve (10), we use the technique of separating the variables (Section 3.3.1). That is, we seek solutions of the heat equation that exhibit the product form

$$u(x, t) = X(x)W(t). \tag{11}$$

The substitution of (11) into the heat equation in (10), followed by some algebraic manipulation, leads to the separated form

$$\frac{X''(x)}{X(x)} = \frac{W'(t)}{a^2W(t)} = -\lambda,$$

where λ is the separation constant. Hence we deduce that X and W satisfy, respectively, the ODEs

$$X'' + \lambda X = 0, \quad 0 < x < p \tag{12}$$

and

$$W' + \lambda a^2 W = 0, \quad t > 0. \tag{13}$$

Under the assumption (11), the boundary conditions in (10) take the form

$$u(0, t) = X(0)W(t) = 0,$$

$$u(p, t) = X(p)W(t) = 0,$$

which are satisfied by setting

$$\text{B.C.:} \qquad X(0) = 0, \qquad X(p) = 0. \tag{14}$$

Instead of (14), the other alternative is to set $W(t) \equiv 0$, but this condition produces only the *trivial solution* $u(x, t) \equiv 0$.

Equation (12) has various solution forms depending upon the value of λ, i.e.,

$$X(x) = \begin{cases} C_1 \cosh \sqrt{-\lambda}x + C_2 \sinh \sqrt{-\lambda}x, & \lambda < 0 \\ C_1 + C_2 x, & \lambda = 0 \\ C_1 \cos \sqrt{\lambda}x + C_2 \sin \sqrt{\lambda}x, & \lambda > 0. \end{cases} \tag{15}$$

We need to examine each of these solution forms for nontrivial solutions [i.e., solutions other than $X(x) \equiv 0$] that also satisfy the boundary conditions (14). In this analysis it is sometimes convenient to introduce an auxiliary parameter that eliminates the square root in the general solution forms found in (15). For instance, when $\lambda < 0$ we can write $\lambda = -k^2$, and then the general solution for this case becomes*

$$X(x) = C_1 \cosh kx + C_2 \sinh kx.$$

The boundary conditions (14) lead to the set of equations

$$X(0) = C_1 \cdot 1 + C_2 \cdot 0 = 0,$$

$$X(p) = C_1 \cosh kp + C_2 \sinh kp = 0,$$

which requires that $C_1 = C_2 = 0$. Hence, in this case we obtain only the trivial solution $X(x) \equiv 0$. For $\lambda = 0$ we reach the same conclusion, the details of which are left to the reader.

To test the remaining solution form in (15) we set $\lambda = k^2 > 0$, so that

$$X(x) = C_1 \cos kx + C_2 \sin kx.$$

This time the boundary conditions lead to

$$X(0) = C_1 = 0,$$

$$X(p) = C_2 \sin kp = 0.$$

Clearly $C_1 = 0$, but if we also choose $C_2 = 0$, then $X(x)$ is again the trivial solution. Instead we will require that $C_2 \neq 0$, and then the second equation above is satisfied provided we choose $k = n\pi/p$, where $n = 1, 2, 3, \ldots$. Selecting $n = -1, -2, -3, \ldots$ does not give new solutions since we are

---

* Although we could also express the general solution as $y = C_1 e^{kx} + C_2 e^{-kx}$, the use of hyperbolic functions is often more convenient when satisfying boundary conditions.

only interested in $\lambda = k^2$. Thus we see that $k^2 = n^2\pi^2/p^2$, which restricts our choice of separation constant λ to the set of values

$$\lambda = \lambda_n = \frac{n^2\pi^2}{p^2}, \qquad n = 1, 2, 3, \ldots. \tag{16}$$

These "allowed" values of λ are called *eigenvalues*. The corresponding nontrivial solutions are called *eigenfunctions*, which are found to be

$$X(x) \equiv \phi_n(x) = \sin\frac{n\pi x}{p}, \qquad n = 1, 2, 3, \ldots \tag{17}$$

(here we set $C_2 = 1$ for mathematical convenience).* With λ restricted to the values (16), Equation (13) yields the collection of solutions

$$W_n(t) = c_n e^{-a^2n^2\pi^2t/p^2}, \qquad n = 1, 2, 3, \ldots, \tag{18}$$

where the c's are arbitrary constants.

Combining (17) and (18) leads to the family of solutions

$$\begin{aligned} u_n(x, t) &= \phi_n(x)W_n(t) \\ &= c_n \sin\left(\frac{n\pi x}{p}\right)e^{-a^2n^2\pi^2t/p^2}, \qquad n = 1, 2, 3, \ldots, \end{aligned} \tag{19}$$

where each $u_n(x, t)$, $n = 1, 2, 3, \ldots$, is itself a solution of the heat equation and boundary conditions in (10). Unfortunately, individual solutions like (19) will not usually satisfy the specified initial condition in (10). For instance, setting $t = 0$ demands that

$$u_n(x, 0) = f(x) = c_n \sin\frac{n\pi x}{p},$$

which is impossible unless $f(x)$ is a multiple of an eigenfunction. However, new solution forms can be derived from (19) by the *principle of superposition*.†

THEOREM 6.1

> *Superposition principle*. If u_1 and u_2 are two solutions of a homogeneous, linear PDE in some domain R, then
>
> $$u = C_1u_1 + C_2u_2$$
>
> is also a solution in R for any constants C_1 and C_2. More generally, if $u_1, u_2, \ldots, u_n, \ldots$ are all solutions in R, then
>
> $$u = \sum_{n=1}^{\infty} C_nu_n$$
>
> is also a solution in R for any set of constants $C_1, C_2, \ldots, C_n \ldots$ for which the series converges.

* See Chapter 1 for a general discussion of eigenvalue problems.

† The superposition principle is actually an extension of the linearity of ODEs.

By invoking the superposition principle on the set of solutions (19), we obtain

$$u(x, t) = \sum_{n=1}^{\infty} c_n \sin\left(\frac{n\pi x}{p}\right) e^{-a^2n^2\pi^2 t/p^2}. \tag{20}$$

Since each term of the series (20) satisfies the heat equation and boundary conditions in (10), we will assume the same is true of the complete series. Moreover, by letting $t = 0$, we now have that

$$u(x, 0) = f(x) = \sum_{n=1}^{\infty} c_n \sin\frac{n\pi x}{p}, \qquad 0 < x < p. \tag{21}$$

We recognize (21) as a *Fourier sine series* of the function f, for which

$$c_n = \frac{2}{p}\int_0^p f(x) \sin\frac{n\pi x}{p}\, dx, \qquad n = 1, 2, 3, \ldots \tag{22}$$

(see Chapter 4). Finally, we take note of the fact that whatever values are assumed by the constants c_n, Equation (20) leads to the result

$$\lim_{t\to\infty} u(x, t) = 0, \tag{23}$$

in agreement with physical considerations. Hence we claim that the formal* solution of the problem described by (10) is that given by (20) and (22).

EXAMPLE 1

Solve

$$u_{xx} = a^{-2}u_t, \qquad 0 < x < \pi, \qquad t > 0$$

B.C.: $u(0, t) = 0, \quad u(\pi, t) = 0, \quad t > 0$

I.C.: $u(x, 0) = T_0$ (constant), $\quad 0 < x < \pi$.

SOLUTION By separating variables, we obtain the two ODEs

$$X'' + \lambda X = 0, \qquad X(0) = 0, \qquad X(\pi) = 0$$

and

$$W' + \lambda a^2 W = 0.$$

The problem for X is a standard eigenvalue problem for which

$$\lambda_n = n^2, \qquad \phi_n(x) = \sin nx, \qquad n = 1, 2, 3, \ldots,$$

and for these values of λ, it follows that the second equation above has solutions

$$W_n(t) = c_n e^{-a^2n^2t}, \qquad n = 1, 2, 3, \ldots,$$

---

* By formal, we mean a solution that has not been rigorously verified.

where the c's are arbitrary constants. By calling upon the superposition principle, we get the solution form

$$u(x, t) = \sum_{n=1}^{\infty} c_n \sin(nx) e^{-a^2 n^2 t}.$$

Imposing the prescribed initial condition leads to

$$u(x, 0) = T_0 = \sum_{n=1}^{\infty} c_n \sin nx,$$

from which we deduce

$$c_n = \frac{2}{\pi} \int_0^{\pi} T_0 \sin nx \, dx = \frac{2T_0}{n\pi} [1 - (-1)^n], \qquad n = 1, 2, 3, \ldots .$$

Clearly, $c_n = 0$ for $n = 2, 4, 6, \ldots$, and from the remaining c_n values we obtain the solution

$$u(x, t) = \frac{4T_0}{\pi} \sum_{\substack{n=1 \\ (\text{odd})}}^{\infty} \left(\frac{\sin nx}{n} \right) e^{-a^2 n^2 t}.$$

It often happens in problems like Example 1 that the boundary conditions at $x = 0$ and $x = p$ do not agree with the values at these points prescribed by the initial condition. Of course, in practice such discontinuities cannot occur. Discontinuities of this nature are part of the math model simply because the resulting problem is easier to solve than the actual problem. For instance, in Example 1 we might imagine that the rod is initially heated to temperature T_0 everywhere, and then at time $t = 0$ the ends of the rod are covered with ice packs at 0°C. Since the ends of the rod cannot instantaneously assume zero temperature, there will be a short interval of time for this to take place that is normally ignored in the model. Also, as the end of the rod is being cooled to 0°C, the temperature near the end of the rod will change to some extent, and this too is usually neglected in the model. For most situations, except possibly for very short rods, it turns out that these effects are so minimal that it is reasonable to neglect them in lieu of solving a more difficult problem.

6.2.2 ROD WITH ENDS HELD AT CONSTANT TEMPERATURE

Let us now consider the case where the temperature at the end $x = 0$ of the rod is maintained at T_1 while the temperature at $x = p$ is maintained at T_2, where T_1 and T_2 are constants. Thus, we wish to solve

$$u_{xx} = a^{-2} u_t, \qquad 0 < x < p, \qquad t > 0$$

$$\text{B.C.:} \quad u(0, t) = T_1, \qquad u(p, t) = T_2, \qquad t > 0 \tag{24}$$

$$\text{I.C.:} \quad u(x, 0) = f(x), \qquad 0 < x < p$$

(see Figure 6.3).

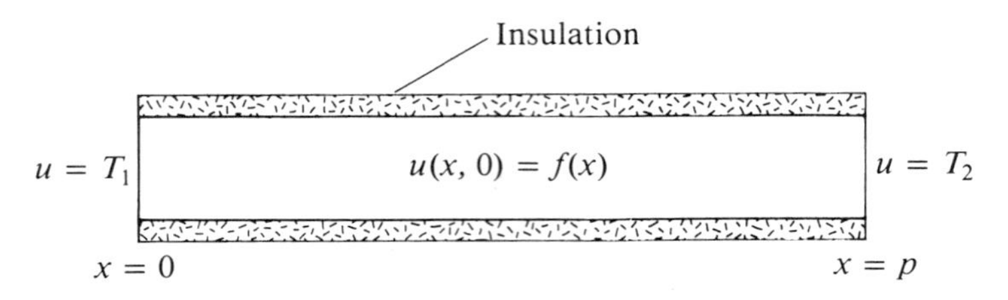

FIGURE 6.3

When $T_1 = T_2 = 0$, we say the boundary conditions are *homogeneous,* and *nonhomogeneous* otherwise. For nonhomogeneous boundary conditions the separation of variables technique will not usually work in a direct manner.* To alleviate this apparent difficulty, we rely on physical intuition, which suggests that after a long time ($t \to \infty$) the temperature in the rod should depend only upon the end temperatures T_1 and T_2, and upon the position x along the rod. Accordingly, it must be true that there is some nonzero limit temperature distribution $S(x)$ such that

$$\lim_{t \to \infty} u(x, t) = S(x), \tag{25}$$

where $S(x)$ is independent of time. Based upon this conjecture, it seems plausible that the temperature distribution $u(x, t)$ can then be expressed as a sum of functions of the form

$$u(x, t) = S(x) + v(x, t), \tag{26}$$

where $v(x, t)$ dies out with increasing time. It turns out this is indeed correct, and we refer to $v(x, t)$ as the *transient solution* of (24) while $S(x)$ is called the *steady-state* or *equilibrium solution.*

The substitution of (26) into (24) leads to

$$S'' + v_{xx} = a^{-2}v_t, \qquad 0 < x < p, \qquad t > 0$$

$$\text{B.C.:} \qquad S(0) + v(0, t) = T_1, \qquad S(p) + v(p, t) = T_2, \tag{27}$$

$$\text{I.C.:} \qquad S(x) + v(x, 0) = f(x).$$

By allowing $t \to \infty$ in the DE and boundary conditions in (27), it follows that $v(x, t)$ and its derivatives must vanish, and in this case we obtain in the limit the steady-state problem

$$S'' = 0, \qquad S(0) = T_1, \qquad S(p) = T_2. \tag{28}$$

* For instance, the boundary conditions in (24) would lead to

$$X(0)W(t) = T_1, \qquad X(p)W(t) = T_2,$$

which can be satisfied only if $W(t)$ is a *constant,* i.e., independent of time.

Two successive integrations of $S'' = 0$ lead to

$$S(x) = C_1 + C_2 x,$$

and by imposing the boundary conditions, we find $S(0) = C_1 = T_1$ and $S(p) = C_1 + C_2 p = T_2$, from which we deduce the steady-state solution

$$S(x) = T_1 + (T_2 - T_1)\frac{x}{p}. \tag{29}$$

By choosing $S(x)$ in this fashion, the problem described by (27) reduces to

$$v_{xx} = a^{-2}v_t, \qquad 0 < x < p, \qquad t > 0$$

$$\text{B.C.:} \qquad v(0, t) = 0, \qquad v(p, t) = 0, \qquad t > 0$$

$$\text{I.C.:} \qquad v(x, 0) = f(x) - S(x) \tag{30}$$

$$= f(x) - T_1 - (T_2 - T_1)\frac{x}{p}, \qquad 0 < x < p.$$

Here we see that $v(x, t)$ is the solution of a heat conduction problem of the type discussed in Section 6.2.1, except that now the initial temperature distribution is the difference between the actual initial temperature $f(x)$ and the steady-state solution $S(x)$.

Because the boundary conditions in (30) are homogeneous, we can solve it directly by the method of separation of variables. This leads to the formal solution

$$v(x, t) = \sum_{n=1}^{\infty} c_n \sin\left(\frac{n\pi x}{p}\right) e^{-a^2 n^2 \pi^2 t/p^2}, \tag{31}$$

where the constants c_n are given by

$$c_n = \frac{2}{p}\int_0^p [f(x) - S(x)] \sin\frac{n\pi x}{p}\, dx$$

$$= \frac{2}{p}\int_0^p f(x)\sin\frac{n\pi x}{p}\, dx - \frac{2}{p}\int_0^p \left[T_1 + (T_2 - T_1)\frac{x}{p}\right]\sin\frac{n\pi x}{p}\, dx,$$

or

$$c_n = \frac{2}{p}\int_0^p f(x)\sin\frac{n\pi x}{p}\, dx + \frac{2}{n\pi}[(-1)^n T_2 - T_1], \qquad n = 1, 2, 3, \ldots. \tag{32}$$

Combining results, our formal solution of (24) becomes

$$u(x, t) = S(x) + v(x, t)$$

$$= T_1 + (T_2 - T_1)\frac{x}{p} + \sum_{n=1}^{\infty} c_n \sin\left(\frac{n\pi x}{p}\right) e^{-a^2 n^2 \pi^2 t/p^2}, \tag{33}$$

where the c's are given by (32).

EXAMPLE 2

Solve

$$u_{xx} = a^{-2}u_t, \qquad 0 < x < 10, \qquad t > 0$$

B.C.: $u(0, t) = 10, \quad u(10, t) = 30, \quad t > 0$

I.C.: $u(x, 0) = 0, \quad 0 < x < 10.$

SOLUTION This problem is a special case of (24) for which the initial temperature distribution throughout the rod is zero. Because the boundary conditions are nonhomogeneous, we start with the assumption

$$u(x, t) = S(x) + v(x, t),$$

where S and v are solutions, respectively, of

$$S'' = 0, \qquad S(0) = 10, \qquad S(10) = 30$$

and

$$v_{xx} = a^{-2}v_t, \qquad 0 < x < 10, \qquad t > 0$$

B.C.: $v(0, t) = 0, \quad v(10, t) = 0$

I.C.: $v(x, 0) = -S(x).$

The solution of the steady-state problem is readily found to be

$$S(x) = 10 + 2x,$$

while the transient solution leads to

$$v(x, t) = \sum_{n=1}^{\infty} c_n \sin\left(\frac{n\pi x}{10}\right) e^{-a^2n^2\pi^2 t/100}.$$

Lastly, by imposing the initial condition in the transient problem, we find that

$$c_n = -\frac{1}{5}\int_0^{10} (10 + 2x) \sin \frac{n\pi x}{10}\, dx$$

$$= \frac{20}{n\pi}[(-1)^n 3 - 1], \qquad n = 1, 2, 3, \ldots,$$

and by combining results, we have

$$u(x, t) = 10 + 2x + \frac{20}{\pi}\sum_{n=1}^{\infty} \frac{(-1)^n 3 - 1}{n} \sin\left(\frac{n\pi x}{10}\right) e^{-a^2n^2\pi^2 t/100}.$$

In Figure 6.4 we show how this solution converges from a 0°C initial temperature distribution to the steady-state solution $S(x) = 10 + 2x$ for increasing time. Because of the exponential function in the solution, $u(x, t)$ will normally converge quite rapidly to the steady-state solution, depending somewhat of course on the constant a^2.

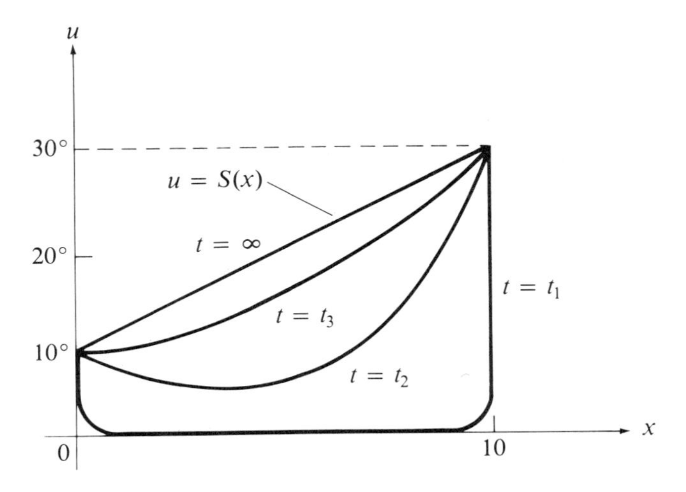

FIGURE 6.4

Temperature approaching steady-state.

6.2.3 ROD WITH ENDS IMPERVIOUS TO HEAT

If the ends of the rod are insulated, no heat flow takes place; i.e., $u_x = 0$ at each end of the bar. Under this condition, the problem is characterized by

$$u_{xx} = a^{-2}u_t, \qquad 0 < x < p, \qquad t > 0$$

$$\text{B.C.:} \qquad u_x(0, t) = 0, \qquad u_x(p, t) = 0, \qquad t > 0 \tag{34}$$

$$\text{I.C.:} \qquad u(x, 0) = f(x), \qquad 0 < x < p$$

(see Figure 6.5).

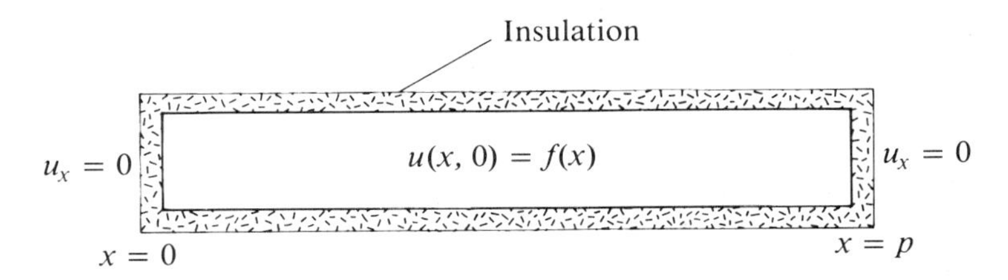

FIGURE 6.5

Here again the boundary conditions are homogeneous, and thus we can proceed immediately with separating the variables. By writing $u(x, t) = X(x)W(t)$, we are led to

$$X'' + \lambda X = 0, \qquad X'(0) = 0, \qquad X'(p) = 0, \tag{35}$$

$$W' + \lambda a^2 W = 0. \tag{36}$$

For $\lambda = 0$, the DE in (35) has the general solution

$$X(x) = C_1 + C_2x.$$

The prescribed boundary conditions demand that $C_2 = 0$ while C_1 is left arbitrary. Hence, $\lambda = 0$ is an eigenvalue with corresponding eigenfunction $\phi_0(x) = 1$. For positive λ we set $\lambda = k^2 > 0$, and thus

$$X(x) = C_1 \cos kx + C_2 \sin kx.$$

This time the boundary conditions require that $C_2 = 0$ and also

$$C_1k \sin kp = 0.$$

To avoid the trivial solution, we assume $C_1 \neq 0$, and thus we must choose $kp = n\pi$, $n = 1, 2, 3, \ldots$. It follows that λ is then restricted to the set of values $\lambda_n = n^2\pi^2/p^2$, $n = 1, 2, 3, \ldots$, with corresponding eigenfunctions $\phi_n(x) = \cos(n\pi x/p)$, $n = 1, 2, 3, \ldots$. It can be shown that there are no further eigenvalues of this problem. We conclude, therefore, that the eigenvalue problem (35) has the solutions

$$\begin{aligned} &\lambda_0 = 0, \qquad \phi_0(x) = 1, \\ &\lambda_n = \frac{n^2\pi^2}{p^2}, \qquad \phi_n(x) = \cos\frac{n\pi x}{p}, \qquad n = 1, 2, 3, \ldots, \end{aligned} \tag{37}$$

and for these values of λ, (36) has the set of solutions

$$W_n(t) = \begin{cases} \dfrac{1}{2}c_0, & n = 0 \\ c_n e^{-a^2n^2\pi^2t/p^2}, & n = 1, 2, 3, \ldots, \end{cases} \tag{38}$$

where we write the constant term as $\frac{1}{2}c_0$ for mathematical convenience. Combining (37) and (38) by the superposition principle yields

$$u(x, t) = \frac{1}{2}c_0 + \sum_{n=1}^{\infty} c_n \cos\left(\frac{n\pi x}{p}\right)e^{-a^2n^2\pi^2t/p^2}. \tag{39}$$

Equation (39) represents a solution of the heat equation satisfying the prescribed boundary conditions in (34). The remaining condition to be satisfied is the initial condition, which leads to

$$u(x, 0) = f(x) = \frac{1}{2}c_0 + \sum_{n=1}^{\infty} c_n \cos\frac{n\pi x}{p}, \qquad 0 < x < p. \tag{40}$$

This expression is simply a *Fourier cosine series* for f, and hence the Fourier coefficients are given by

$$c_n = \frac{2}{p}\int_0^p f(x) \cos\frac{n\pi x}{p}\,dx, \qquad n = 0, 1, 2, \ldots. \tag{41}$$

As a final observation here we note that as $t \to \infty$, the solution (39) approaches the constant value $c_0/2$. Hence, this value denotes the *steady-state solution* in this case, which also happens to be the average initial temperature of the rod. To understand why this is so, let us consider the steady-state problem directly, which is

$$S'' = 0, \qquad S'(0) = 0, \qquad S'(p) = 0.$$

The solution is $S(x) = C$ (any constant). Of course, physical considerations do not permit C to assume any arbitrary value. Because the lateral surface and ends of the rod are insulated in this problem, there can be no heat loss. This means that whatever heat is present at time $t = 0$ is still present at all later times, but due to the diffusion process it tends to distribute itself evenly throughout the rod. Hence, it finally becomes constant, which must equal the average initial temperature of the rod, $c_0/2$.

6.2.4 CONVECTIVE HEAT TRANSFER AT ONE ENDPOINT

Thus far we have only considered heat conduction problems in a rod where the boundary conditions are of either the first or second kind; i.e., where either u or u_x is specified at the endpoints. A third boundary condition, which involves a linear combination of u and u_x, arises when heat is lost from the end of the rod due to radiation into the surrounding medium. In this case we say the end (boundary) of the rod is exposed to *convective heat transfer*. For illustrative purposes, let us consider the specific problem described by

$$u_{xx} = a^{-2}u_t, \qquad 0 < x < p, \qquad t > 0$$

$$\text{B.C.:} \quad u(0, t) = 0, \qquad hu(p, t) + u_x(p, t) = 0, \qquad t > 0, \tag{42}$$

$$\text{I.C.:} \quad u(x, 0) = f(x), \qquad 0 < x < p$$

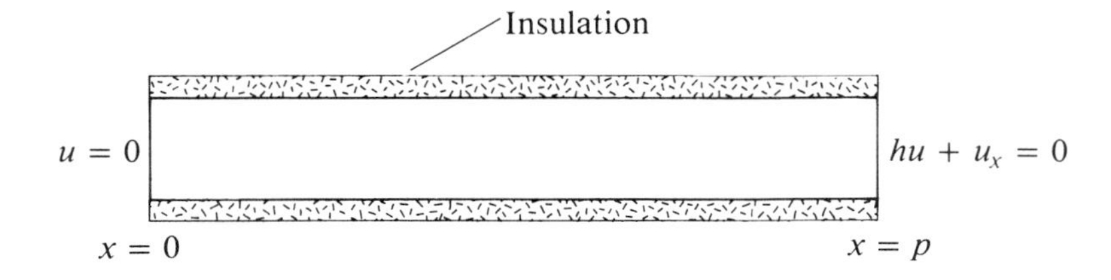

FIGURE 6.6

where $h > 0$ (see Figure 6.6). Negative values of h would physically correspond to thermal energy constantly put into the rod through the right end.

By separating variables, we obtain the system of DEs

$$X'' + \lambda X = 0, \qquad X(0) = 0, \qquad hX(p) + X'(p) = 0, \tag{43}$$

$$W' + \lambda a^2 W = 0. \tag{44}$$

For the special case $h = 1$ and $p = 1$, the eigenvalue problem (43) is the same as solved in Example 12 in Section 1.4. Following a similar analysis here, we set $\lambda = k^2 > 0$, which yields the general solution

$$X(x) = C_1 \cos kx + C_2 \sin kx. \tag{45}$$

The boundary condition $X(0) = 0$ requires that $C_1 = 0$, whereas the second boundary condition leads to

$$C_2(h \sin kp + k \cos kp) = 0. \tag{46}$$

Hence, denoting the nth solution of $h \sin kp + k \cos kp = 0$ by k_n, the eigenvalues and eigenfunctions of (43) are represented by*

$$\lambda_n = k_n^2, \qquad \phi_n(x) = \sin k_n x, \qquad n = 1, 2, 3, \ldots. \tag{47}$$

Returning now to (44), we obtain

$$W_n(t) = c_n e^{-a^2 k_n^2 t}, \qquad n = 1, 2, 3, \ldots, \tag{48}$$

which, combined with (47), gives the formal solution

$$u(x, t) = \sum_{n=1}^{\infty} c_n \sin(k_n x) e^{-a^2 k_n^2 t}. \tag{49}$$

Setting $t = 0$ in (49) yields the relation

$$u(x, 0) = f(x) = \sum_{n=1}^{\infty} c_n \sin k_n x, \qquad 0 < x < p, \tag{50}$$

which is a generalized Fourier series (see Section 4.6). In this case the Fourier coefficients are calculated from

$$c_n = \| \phi_n(x) \|^{-2} \int_0^p f(x) \sin k_n x \, dx, \qquad n = 1, 2, 3, \ldots, \tag{51}$$

where

$$\begin{aligned} \| \phi_n(x) \|^2 &= \int_0^p \sin^2 k_n x \, dx \\ &= \frac{1}{2}\left(p - \frac{\sin 2k_n p}{2k_n}\right) \\ &= \frac{1}{2}\left(p - \frac{\sin k_n p \cos k_n p}{k_n}\right), \end{aligned}$$

but since $\sin k_n p = -(k_n/h) \cos k_n p$, we have

$$\| \phi_n(x) \|^2 = \frac{1}{2}\left(p + \frac{1}{h} \cos^2 k_n p\right), \qquad n = 1, 2, 3, \ldots. \tag{52}$$

---

*We leave it to the reader to show that there are no further eigenvalues (see Problem 16 in Exercises 6.2).

EXAMPLE 3

Solve

$$u_{xx} = a^{-2}u_t, \qquad 0 < x < 1, \qquad t > 0$$

B.C.: $\quad u(0, t) = T_1, \qquad u(1, t) + u_x(1, t) = T_2, \qquad t > 0$

I.C.: $\quad u(x, 0) = T_1, \qquad 0 < x < 1.$

SOLUTION Because the boundary conditions are nonhomogeneous, we start by writing

$$u(x, t) = S(x) + v(x, t),$$

where the steady-state temperature distribution $S(x)$ satisfies

$$S'' = 0, \qquad 0 < x < 1, \qquad S(0) = T_1, \qquad S(1) + S'(1) = T_2,$$

and $v(x, t)$ is a solution of

$$v_{xx} = a^{-2}v_t, \qquad 0 < x < 1, \qquad t > 0$$

B.C.: $\quad v(0, t) = 0, \qquad v(1, t) + v_x(1, t) = 0, \qquad t > 0$

I.C.: $\quad v(x, 0) = T_1 - S(x), \qquad 0 < x < 1.$

The steady-state solution is readily found to be

$$S(x) = T_1 + \frac{1}{2}(T_2 - T_1)x,$$

and the transient solution has the form

$$v(x, t) = \sum_{n=1}^{\infty} c_n \sin(k_n x)e^{-a^2k_n^2t},$$

where k_n denotes the solutions of

$$\sin k_n + k_n \cos k_n = 0, \qquad n = 1, 2, 3, \ldots$$

(see also Example 12 in Section 1.4). The initial condition for $v(x, t)$ requires that

$$v(x, 0) = -\frac{1}{2}(T_2 - T_1)x = \sum_{n=1}^{\infty} c_n \sin k_n x$$

and thus, based on (51) and (52),

$$c_n = -\frac{T_2 - T_1}{1 + \cos^2 k_n}\int_0^1 x \sin k_n x\, dx$$

$$= \frac{2(T_2 - T_1)\cos k_n}{k_n(1 + \cos^2 k_n)}, \qquad n = 1, 2, 3, \ldots,$$

where we have made use of the relation $\sin k_n + k_n \cos k_n = 0$. Combining steady-state and transient solutions, we have

$$u(x, t) = T_1 + \frac{1}{2}(T_2 - T_1)x + 2(T_2 - T_1)\sum_{n=1}^{\infty} \frac{\cos k_n \sin k_n x}{k_n(1 + \cos^2 k_n)} e^{-a^2 k_n^2 t}.$$

6.2.5 SUMMARY AND DISCUSSION

All of the problems discussed up to now are special cases of the general heat conduction problem

$$u_{xx} = a^{-2}u_t, \qquad 0 < x < p, \qquad t > 0$$

$$\text{B.C.:} \qquad B_1[u] = T_1, \qquad B_2[u] = T_2, \qquad t > 0 \tag{53}$$

$$\text{I.C.:} \qquad u(x, 0) = f(x), \qquad 0 < x < p,$$

where there are no heat sources within the rod and the boundary conditions are independent of time. When T_1 and T_2 are not both zero, we look for the possibility of a steady-state solution by writing

$$u(x, t) = S(x) + v(x, t), \tag{54}$$

where $S(x)$ is the steady-state solution and $v(x, t)$ the transient solution. The substitution of (54) into (53) leads to the steady-state problem*

$$S'' = 0, \qquad 0 < x < p, \qquad B_1[S] = T_1, \qquad B_2[S] = T_2, \tag{55}$$

and the time-dependent problem

$$v_{xx} = a^{-2}v_t, \qquad 0 < x < p, \qquad t > 0$$

$$\text{B.C.:} \qquad B_1[v] = 0, \qquad B_2[v] = 0, \qquad t > 0 \tag{56}$$

$$\text{I.C.:} \qquad v(x, 0) = f(x) - S(x), \qquad 0 < x < p.$$

When a solution exists, the steady-state problem (55) can be solved by elementary methods, while (56) is solved by separation of variables, which yields†

$$v(x, t) = \sum_{n=1}^{\infty} c_n \phi_n(x) e^{-a^2 \lambda_n t}, \tag{57}$$

---

* In some cases it happens that there is no steady-state solution of (53). For example, if the boundary conditions in (53) are

$$u_x(0, t) = T_1, \qquad u_x(p, t) = T_2,$$

then (55) becomes

$$S'' = 0, \qquad S'(0) = T_1, \qquad S'(p) = T_2,$$

which has no solution unless $T_1 = T_2$.

† In some cases the summation in (57) may start with $n = 0$ (see Section 6.2.3).

where λ_n and $\phi_n(x)$ are eigenvalues and eigenfunctions, respectively, of

$$X'' + \lambda X = 0, \qquad B_1[X] = 0, \qquad B_2[X] = 0. \tag{58}$$

Subjecting (57) to the initial condition in (56), we find that the Fourier coefficients are defined by

$$c_n = \| \phi_n(x) \|^{-2} \int_0^p [f(x) - S(x)]\phi_n(x)\,dx, \qquad n = 1, 2, 3, \ldots, \tag{59}$$

where $\| \phi_n(x) \|$ is the norm of the nth eigenfunction (see Section 4.6). The solution of (53) is then simply $u(x, t) = S(x) + v(x, t)$.

Our treatment here of heat conduction problems has been purely formal. That is, we have provided only *formal solutions* of the problems and have not discussed their validity as *strict solutions*. To validate our formal results we must demonstrate the convergence of each of the series for u, u_x, u_{xx}, and u_t. From Chapter 4 we know that the Fourier series representation of the function f appearing in the initial condition converges pointwise to $f(x)$ provided f is at least piecewise smooth on the interval $[0, p]$. It turns out that this restriction on the function f is all that is required to also ensure convergence of all the above named series involving u and its derivatives. What is surprising, however, is that the series for u will converge to a strict solution even when the function f is only *piecewise continuous*. Moreover, the solutions obtained when f is at least piecewise continuous are analytic functions* of x, in spite of the fact that f may contain discontinuities. The physical implication of this result is that heat conduction, or diffusion, is a *smoothing process*. In fact, discontinuities in the initial data are instantaneously smoothed out for $t > 0$. Hence, we say that *thermal energy propagates with infinite speed*. This is a fundamental property of the heat equation and points out a basic difference from the wave equation for which the *propagation speed is finite* (see Section 3.2.2). Finally, we remark that a solution of the heat equation, obtained under the condition that f is piecewise continuous, is *unique*.

EXERCISES 6.2

In Problems 1–3, solve the heat conduction problem with zero temperatures on the boundary.

1. $u_{xx} = a^{-2}u_t, \quad 0 < x < 1, \quad t > 0$
   B.C.: $u(0, t) = 0, \quad u(1, t) = 0$
   I.C.: $u(x, 0) = 3 \sin \pi x - 5 \sin 4\pi x$

---

* An *analytic function* is one that is infinitely differentiable and has a convergent power series expansion.

2. $u_{xx} = a^{-2}u_t, \quad 0 < x < \pi, \quad t > 0$
B.C.: $u(0, t) = 0, \quad u(\pi, t) = 0$
I.C.: $u(x, 0) = \begin{cases} x, & 0 < x < \pi/2 \\ \pi - x, & \pi/2 < x < \pi \end{cases}$

3. $u_{xx} = a^{-2}u_t, \quad 0 < x < p, \quad t > 0$
B.C.: $u(0, t) = 0, \quad u(p, t) = 0$
I.C.: $u(x, 0) = x(p - x)$

In Problems 4–6, find the steady-state solution.

4. $u_{xx} = a^{-2}u_t, \quad 0 < x < p, \quad t > 0$
B.C.: $u(0, t) - u_x(0, t) = T_1, \quad u(p, t) = 0$

5. $u_{xx} = a^{-2}u_t, \quad 0 < x < p, \quad t > 0$
B.C.: $u(0, t) = T_1, \quad u(p, t) + u_x(p, t) = T_2$

★6. $\frac{\partial}{\partial x}\left[(1 + x)\frac{\partial u}{\partial x}\right] = \frac{\partial u}{\partial t}, \quad 0 < x < p, \quad t > 0$
B.C.: $u(0, t) = T_1, \quad u(p, t) = T_2$

In Problems 7–15, solve the given heat conduction problem.

7. $u_{xx} = a^{-2}u_t, \quad 0 < x < 1, \quad t > 0$
B.C.: $u(0, t) = 1, \quad u(1, t) = 0$
I.C.: $u(x, 0) = 1 + x$

8. $u_{xx} = a^{-2}u_t, \quad 0 < x < 2, \quad t > 0$
B.C.: $u(0, t) = T_1, \quad u(2, t) = T_2$
I.C.: $u(x, 0) = T_0$

9. $u_{xx} = a^{-2}u_t, \quad 0 < x < p, \quad t > 0$
B.C.: $u_x(0, t) = 0, \quad u_x(p, t) = 0$
I.C.: $u(x, 0) = T_0 \sin^2(\pi x/p)$

10. $u_{xx} = a^{-2}u_t, \quad 0 < x < p, \quad t > 0$
B.C.: $u_x(0, t) = 0, \quad u(p, t) = T_0$
I.C.: $u(x, 0) = T_0$

11. $u_{xx} = a^{-2}u_t, \quad 0 < x < 1, \quad t > 0$
B.C.: $u(0, t) = T_0, \quad u(1, t) = T_0$
I.C.: $u(x, 0) = T_0 + x(1 - x)$

★12. $u_{xx} = a^{-2}u_t, \quad 0 < x < 1, \quad t > 0$
B.C.: $u(0, t) = T_1, \quad u_x(1, t) = 0$
I.C.: $u(x, 0) = x$

13. $u_{xx} = a^{-2}u_t, \quad -\pi < x < \pi, \quad t > 0$
B.C.: $u(-\pi, t) = u(\pi, t), \quad u_x(-\pi, t) = u_x(\pi, t)$
I.C.: $u(x, 0) = |x|$

14. $u_{xx} = a^{-2}u_t, \quad 0 < x < 1, \quad t > 0$
B.C.: $u(0, 1) = 0, \quad 2u(1, t) + u_x(1, t) = 0$
I.C.: $u(x, 0) = T_0$

★15. $u_{xx} = a^{-2}u_t, \quad 0 < x < \pi, \quad t > 0$
B.C.: $u(0, t) - u_x(0, t) = 100, \quad u_x(\pi, t) = 0$
I.C.: $u(x, 0) = 200$

★16. Given the eigenvalue problem

$$X'' + \lambda X = 0, \quad 0 < x < p, \quad X(0) = 0, \quad hX(p) + X'(p) = 0,$$

(a) show that for $h > 0$ there are only positive eigenvalues.
(b) If $h < 0$, what additional restriction on h is necessary to ensure that the smallest positive eigenvalue is less than $\pi/2$?
(c) What value of h will lead to the eigenvalue $\lambda = 0$?
(d) If $h < 0$, what additional restriction on h is necessary for the existence of a negative eigenvalue?

Hint: Use graphical techniques for (a), (b), and (d) like Example 12 in Section 1.4.

17. The ends $x = 0$ and $x = 100$ of a rod 100 cm in length, with insulated lateral surfaces, are held at temperatures 0° and 100°C, respectively, until steady-state conditions prevail. Then, at time $t = 0$, the temperatures of the two ends are interchanged. Find the subsequent temperature distribution throughout the rod.

18. Both ends of a rod 10 cm long are held at 0°C and the initial temperature is $300 \sin(\pi x/10)$. Assuming the lateral surface is insulated, calculate the times required for the midpoint of the rod to reach 200°, 100°, and 50°C, respectively, if the rod is made of
(a) aluminum.
(b) cast iron.
Hint: See Table 6.1 on p. 242.

19. A rod 2 m long is given the initial temperature distribution

$$u(x, 0) = \begin{cases} 50x, & 0 < x < 1 \\ 100 - 50x, & 1 < x < 2. \end{cases}$$

If both ends of the rod are maintained at 0°C, approximately how long will it take for the center of the rod to reach a temperature of 30°C? Assume the lateral surface is insulated.
Hint: Use only the first nonzero term of the series.

20. Repeat Problem 19 if the ends of the rod are insulated. What is the temperature in the rod after a long time?

21. Suppose an aluminum rod is initially heated to a uniform temperature distribution of 25°C. At time $t = 0$, the end $x = 0$ is cooled to 0°C while the end $x = 20$ cm is heated to 60°C, and both are maintained at these temperatures thereafter.
(a) Find the temperature distribution in the rod for any time t.
Hint: See Table 6.1 on p. 242.
(b) Approximate the temperature of the rod at $x = 5$ cm and $t = 30$ s by using only the first term of the series in (a).
(c) Approximate your answer to (b) using the first two terms of the series in (a).
(d) Use only the first term of the series in (a) to estimate the time elapse necessary for the temperature at $x = 5$ cm to come within 1% of the steady-state value.

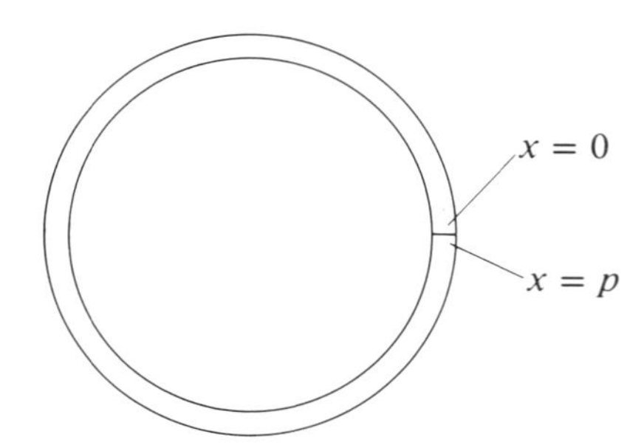

PROBLEM 22

22. If an insulated wire of length p is bent into a loop so that the ends are joined (see figure), the boundary conditions are given by

$$u(0, t) = u(p, t), \qquad u_x(0, t) = u_x(p, t),$$

which states that the temperature and heat flow at the points where the ends are joined must be the same. Find a formal solution for the temperature distribution in the wire at all later times if the initial temperature distribution in the wire is given by

$$u(x, 0) = f(x), \qquad 0 < x < p.$$

23. Solve Problem 22 for the special cases when
(a) $f(x) = T_0$ (constant).
(b) $f(x) = x(p - x)$.

24. If the lateral surface of the rod is not insulated, there is a heat exchange by convection into the surrounding medium. If the surrounding medium has constant temperature T_0, the rate at which heat is lost from the rod is proportional to the difference $u - T_0$. The governing PDE in this situation is

$$a^2 u_{xx} = u_t + b(u - T_0), \qquad b > 0.$$

Show that the change of variables

$$u(x, t) = T_0 + z(x, t)e^{-bt}$$

leads to the heat equation in z.

In Problems 25–28, use the result of Problem 24 to solve the given heat conduction problem.

25. $u_{xx} = u_t + u, \qquad 0 < x < 1, \qquad t > 0$
B.C.: $u(0, t) = 0, \qquad u(1, t) = 0$
I.C.: $u(x, 0) = T_0$

26. $u_{xx} = u_t + 4u - 20, \qquad 0 < x < \pi, \qquad t > 0$
B.C.: $u(0, t) = 5, \qquad u(\pi, t) = 5$
I.C.: $u(x, 0) = 5 + 2x$

27. $3u_{xx} = u_t + u, \qquad 0 < x < \pi, \qquad t > 0$
B.C.: $u_x(0, t) = 0, \qquad u_x(\pi, t) = 0$
I.C.: $u(x, 0) = x^2$

28. $u_{xx} = u_t + 4u, \quad 0 < x < \pi, \quad t > 0$
B.C.: $u(0, t) = 0, \quad u(\pi, t) = 1$
I.C.: $u(x, 0) = 0$

In Problems 29 and 30, solve the more general heat equation problem.

★29. $u_{xx} + 2u_x + u = u_t, \quad 0 < x < \pi, \quad t > 0$
B.C.: $u(0, t) = 0, \quad u(\pi, t) = 0$
I.C.: $u(x, 0) = e^{-x}$

★30. $u_{xx} - 4u_x = u_t, \quad 0 < x < 1, \quad t > 0$
B.C.: $u(0, t) = 0, \quad u(1, t) = 0$
I.C.: $u(x, 0) = e^{2x}$

6.3 NONHOMOGENEOUS PROBLEMS

A problem is classified as *nonhomogeneous* if either the PDE or boundary conditions are nonhomogeneous. Some special cases of nonhomogeneous boundary conditions independent of time were discussed in Section 6.2, while here we consider a more general problem involving nonhomogeneous PDE and boundary conditions, i.e.,

$$\begin{aligned} & u_{xx} = a^{-2}u_t - q(x, t), \quad 0 < x < p, \quad t > 0 \\ \text{B.C.:} \quad & B_1[u] = \alpha(t), \quad B_2[u] = \beta(t), \quad t > 0 \\ \text{I.C.:} \quad & u(x, 0) = f(x), \quad 0 < x < p, \end{aligned} \tag{60}$$

where $q(x, t)$ is proportional to the heat source [see Equation (5)]. To solve problems of this general nature we will once again consider some special cases.

6.3.1 NONHOMOGENEOUS TERMS INDEPENDENT OF TIME

When the heat source does not change in time it is called steady, and, in this case, we write

$$q(x, t) = P(x). \tag{61}$$

If the boundary conditions are also independent of time, i.e., if

$$B_1[u] = T_1, \quad B_2[u] = T_2, \tag{62}$$

where T_1 and T_2 are constants, then our problem is formalized by

$$\begin{aligned} & u_{xx} = a^{-2}u_t - P(x), \quad 0 < x < p, \quad t > 0 \\ \text{B.C.:} \quad & B_1[u] = T_1, \quad B_2[u] = T_2, \quad t > 0 \\ \text{I.C.:} \quad & u(x, 0) = f(x), \quad 0 < x < p. \end{aligned} \tag{63}$$

In solving (63), we find it convenient to use the same device that helped in Section 6.2.2. That is, we assume that u can be expressed as the sum

$$u(x, t) = S(x) + v(x, t), \tag{64}$$

where $S(x)$ denotes the steady-state solution and $v(x, t)$ is the transient solution. The substitution of (64) into (63) leads to

$$S'' + v_{xx} = a^{-2}v_t - P(x), \qquad 0 < x < p, \qquad t > 0$$

$$\text{B.C.:} \qquad B_1[S] + B_1[v] = T_1, \qquad B_2[S] + B_2[v] = T_2, \qquad t > 0 \tag{65}$$

$$\text{I.C.:} \qquad S(x) + v(x, 0) = f(x), \qquad 0 < x < p,$$

from which we deduce that if $S(x)$ is a solution of

$$S'' = -P(x), \qquad B_1[S] = T_1, \qquad B_2[S] = T_2 \tag{66}$$

and $v(x, t)$ satisfies

$$v_{xx} = a^{-2}v_t, \qquad 0 < x < p, \qquad t > 0$$

$$\text{B.C.:} \qquad B_1[v] = 0, \qquad B_2[v] = 0, \qquad t > 0 \tag{67}$$

$$\text{I.C.:} \qquad v(x, 0) = f(x) - S(x), \qquad 0 < x < p,$$

then (64) is indeed a solution of the original problem (63). Once again we have reduced our problem to a set of problems for which solution techniques have already been developed.

EXAMPLE 4

Solve the nonhomogeneous problem (A and b are constants)

$$u_{xx} = u_t - Ae^{-bx}, \qquad 0 < x < 1, \qquad t > 0$$

$$\text{B.C.:} \qquad u(0, t) = 0, \qquad u(1, t) = 0, \qquad t > 0$$

$$\text{I.C.:} \qquad u(x, 0) = \frac{A}{b^2}(1 - e^{-bx}), \qquad 0 < x < 1.$$

SOLUTION By writing $u(x, t) = S(x) + v(x, t)$, we find that the steady-state problem is

$$S'' = -Ae^{-bx}, \qquad S(0) = 0, \qquad S(1) = 0,$$

the solution of which is

$$S(x) = \frac{A}{b^2}(1 - e^{-bx}) - \frac{A}{b^2}(1 - e^{-b})x.$$

With $S(x)$ determined, the transient problem reads

$$v_{xx} = v_t, \qquad 0 < x < 1, \qquad t > 0$$

$$\text{B.C.:} \qquad v(0, t) = 0, \qquad v(1, t) = 0$$

$$\text{I.C.:} \qquad v(x, 0) = \frac{A}{b^2}(1 - e^{-b})x.$$

The solution of this PDE subject to the given boundary conditions is given by

$$v(x, t) = \sum_{n=1}^{\infty} c_n \sin(n\pi x)e^{-n^2\pi^2 t}.$$

Finally, application of the initial condition yields

$$v(x, 0) = \frac{A}{b^2}(1 - e^{-b})x = \sum_{n=1}^{\infty} c_n \sin n\pi x,$$

from which we deduce

$$c_n = \frac{2A}{\pi b^2}(1 - e^{-b})\frac{(-1)^{n+1}}{n}, \qquad n = 1, 2, 3, \ldots .$$

Combining $S(x)$ and $v(x, t)$, we have

$$u(x, t) = \frac{A}{b^2}[1 - e^{-bx} - (1 - e^{-b})x] + \frac{2A}{\pi b^2}(1 - e^{-b})\sum_{n=1}^{\infty} \frac{(-1)^{n-1}}{n} \sin(n\pi x)e^{-n^2\pi^2 t}.$$

6.3.2 EIGENFUNCTION EXPANSION METHOD

When the heat source $q(x, t)$ depends on time t, the technique of Section 6.3.1 does not apply. In order to illustrate a more general method, we will first consider the particular problem

$$u_{xx} = a^{-2}u_t - q(x, t), \qquad 0 < x < p, \qquad t > 0$$

$$\text{B.C.:} \quad B_1[u] = 0, \qquad B_2[u] = 0, \qquad t > 0 \tag{68}$$

$$\text{I.C.:} \quad u(x, 0) = f(x), \qquad 0 < x < p,$$

which features homogeneous boundary conditions.

We begin by assuming the solution of (68) can be written in the form*

$$u(x, t) = \sum_{n=1}^{\infty} E_n(t)\phi_n(x), \tag{69}$$

where the ϕ's are eigenfunctions belonging to the associated eigenvalue problem

$$X'' + \lambda X = 0, \qquad B_1[X] = 0, \qquad B_2[X] = 0, \tag{70}$$

and the E's are time-dependent coefficients to be determined. Assuming termwise differentiation is permitted, we find

---

*Observe that we are still preserving the product solution form in (69), replacing $W_n(t)$ by $E_n(t)$. Our technique here is a generalization of the *eigenfunction expansion method* of Section 4.6.1.

$$u_t(x, t) = \sum_{n=1}^{\infty} E_n'(t)\phi_n(x) \tag{71}$$

and

$$u_{xx}(x, t) = \sum_{n=1}^{\infty} E_n(t)\phi_n''(x)$$

$$= -\sum_{n=1}^{\infty} \lambda_n E_n(t)\phi_n(x), \tag{72}$$

where we are recognizing the relation [from (70)] $\phi_n''(x) = -\lambda_n\phi_n(x)$. Next, rewriting the PDE in (68) in the form

$$a^2q(x, t) = u_t - a^2u_{xx} \tag{73}$$

and substituting the expressions (71) and (72) into the right-hand side, we obtain

$$a^2q(x, t) = \sum_{n=1}^{\infty} [E_n'(t) + a^2\lambda_n E_n(t)]\phi_n(x). \tag{74}$$

For a fixed value of t, we can interpret (74) as a *generalized Fourier series* of the function $a^2q(x, t)$, with Fourier coefficients defined by

$$E_n'(t) + a^2\lambda_n E_n(t) = a^2\|\phi_n(x)\|^{-2}\int_0^p q(x, t)\phi_n(x)\,dx, \qquad n = 1, 2, 3, \ldots, \tag{75}$$

where

$$\|\phi_n(x)\|^2 = \int_0^p [\phi_n(x)]^2\,dx, \qquad n = 1, 2, 3, \ldots. \tag{76}$$

For each n, Equation (75) is a first-order linear ODE with general solution

$$E_n(t) = \left[c_n + a^2\int_0^t e^{a^2\lambda_n\tau}Q_n(\tau)\,d\tau\right]e^{-a^2\lambda_n t}, \qquad n = 1, 2, 3, \ldots \tag{77}$$

(assuming $\lambda_n \neq 0$, $n = 1, 2, 3, \ldots$) where, for notational convenience, we define

$$Q_n(t) = \|\phi_n(x)\|^{-2}\int_0^p q(x, t)\phi_n(x)\,dx, \qquad n = 1, 2, 3, \ldots, \tag{78}$$

and where the c's are arbitrary constants. Finally, the substitution of (77) into (69) provides us with the solution formula

$$u(x, t) = \sum_{n=1}^{\infty}\left[c_n + a^2\int_0^t e^{a^2\lambda_n\tau}Q_n(\tau)\,d\tau\right]\phi_n(x)e^{-a^2\lambda_n t}. \tag{79}$$

The constants c_n, $n = 1, 2, 3, \ldots$, are determined by forcing (79) to satisfy the prescribed initial condition in (68). Thus, setting $t = 0$ in (79), we get

$$u(x, 0) = f(x) = \sum_{n=1}^{\infty} c_n \phi_n(x),$$

and therefore it follows that

$$c_n = \| \phi_n(x) \|^{-2} \int_0^p f(x)\phi_n(x)\, dx, \qquad n = 1, 2, 3, \ldots . \tag{80}$$

REMARK: *The solution (79) can be interpreted as being composed of two parts, i.e.,*

$$u(x, t) = u_H(x, t) + u_P(x, t),$$

where

$$u_H(x, t) = \sum_{n=1}^{\infty} c_n \phi_n(x) e^{-a^2\lambda_n t}$$

and

$$u_P(x, t) = a^2 \sum_{n=1}^{\infty} \left[\int_0^t e^{a^2\lambda_n \tau} Q_n(\tau)\, d\tau \right] \phi_n(x) e^{-a^2\lambda_n t}.$$

Thus, when $q(x, t)$ is identically zero so is $u_P(x, t)$, and the solution is $u = u_H(x, t)$, which agrees with previous results. We can interpret the solution $u_P(x, t)$ as the "response" of the system that has zero temperature initially prescribed throughout the rod, and then is subjected to a distributed heat source at time $t = 0$.

Depending upon the nature of the boundary conditions, it sometimes happens that $\lambda_0 = 0$ is an eigenvalue of (70) with corresponding eigenfunction $\phi_0(x)$. When this is the case, Equation (69) should be modified to read

$$u(x, t) = E_0(t)\phi_0(x) + \sum_{n=1}^{\infty} E_n(t)\phi_n(x), \tag{81}$$

where $E_0(t)$ must now also be determined.

EXAMPLE 5

Solve

$$u_{xx} = u_t + (1 - x)\cos t, \qquad 0 < x < 1, \qquad t > 0$$

B.C.: $u(0, t) = 0, \quad u(1, t) = 0, \quad t > 0$

I.C.: $u(x, 0) = 0, \quad 0 < x < 1.$

SOLUTION We first identify and solve the related eigenvalue problem, which in this case is

$$X'' + \lambda X = 0, \qquad X(0) = 0, \qquad X(1) = 0;$$

hence,

$$\lambda_n = n^2\pi^2, \qquad \phi_n(x) = \sin n\pi x, \qquad n = 1, 2, 3, \ldots .$$

We now seek a solution of the form

$$u(x, t) = \sum_{n=1}^{\infty} E_n(t) \sin n\pi x,$$

which, when substituted into the PDE

$$-(1 - x) \cos t = u_t - u_{xx},$$

yields

$$-(1 - x) \cos t = \sum_{n=1}^{\infty} [E_n'(t) + n^2\pi^2 E_n(t)] \sin n\pi x.$$

Therefore, we deduce that [note that $\| \sin n\pi x \|^2 = 1/2$, $n = 1, 2, 3, \ldots$]

$$E_n'(t) + n^2\pi^2 E_n(t) = -2 \int_0^1 (1 - x) \cos t \sin n\pi x \, dx$$

$$= -\frac{2}{n\pi} \cos t, \qquad n = 1, 2, 3, \ldots,$$

and solving this DE for $E_n(t)$ leads to

$$E_n(t) = \left[c_n - \frac{2}{n\pi} \int_0^t e^{n^2\pi^2\tau} \cos \tau \, d\tau \right] e^{-n^2\pi^2 t}$$

$$= c_n e^{-n^2\pi^2 t} + \frac{2}{n\pi(1 + n^4\pi^4)} [n^2\pi^2(e^{-n^2\pi^2 t} - \cos t) - \sin t].$$

Substituting this expression for $E_n(t)$ into the above solution form for $u(x, t)$ and imposing the initial condition $u(x, 0) = 0$, we see that $c_n = 0$ for all n. Therefore, our solution is

$$u(x, t) = \frac{2}{\pi} \sum_{n=1}^{\infty} \frac{\sin n\pi x}{n(1 + n^4\pi^4)} [n^2\pi^2(e^{-n^2\pi^2 t} - \cos t) - \sin t].$$

□ 6.3.3 TIME-VARYING END CONDITIONS

General time-varying end conditions are described by

$$\text{B.C.:} \qquad B_1[u] = \alpha(t), \qquad B_2[u] = \beta(t), \tag{82}$$

where the boundary operators B_1 and B_2 are defined by (9). For illustrative purposes, we will consider only the simplified conditions

$$\text{B.C.:} \qquad u(0, t) = \alpha(t), \qquad u(p, t) = \beta(t), \tag{83}$$

which describe the temperature changes at the ends of the rod. Assuming that a heat source is present, the problem we wish to address is characterized by

$$u_{xx} = a^{-2}u_t - q(x, t), \qquad 0 < x < p, \qquad t > 0$$

$$\text{B.C.:} \quad u(0, t) = \alpha(t), \quad u(p, t) = \beta(t), \quad t > 0 \tag{84}$$

$$\text{I.C.:} \quad u(x, 0) = f(x), \quad 0 < x < p.$$

Our general procedure once again will be to decompose (84) into two simpler problems, each of which can be solved by methods already discussed. Thus, we make the assertion*

$$u(x, t) = K(x, t) + V(x, t) \tag{85}$$

which, upon substitution into (84), leads to

$$K_{xx} + V_{xx} = a^{-2}K_t + a^{-2}V_t - q(x, t), \quad 0 < x < p, \quad t > 0$$

$$\text{B.C.:} \quad \begin{cases} K(0, t) + V(0, t) = \alpha(t), \\ K(p, t) + V(p, t) = \beta(t), \end{cases} \tag{86}$$

$$\text{I.C.:} \quad K(x, 0) + V(x, 0) = f(x).$$

Now the situation is more complicated than before since there is no simple split of (86) that is easy to resolve. What is desired is one problem where the boundary conditions are homogeneous. Hence, if we judiciously select $K(x, t)$ to be any differentiable function satisfying the conditions

$$K(0, t) = \alpha(t), \quad K(p, t) = \beta(t), \tag{87}$$

then the remaining problem for $V(x, t)$ will feature homogeneous boundary conditions. Among other possibilities, an appropriate choice is†

$$K(x, t) = \alpha(t) + [\beta(t) - \alpha(t)]\,\frac{x}{p}\cdot \tag{88}$$

In this case $K_{xx}(x, t) = 0$, and the problem left to solve for $V(x, t)$ reduces to

$$V_{xx} = a^{-2}V_t + [a^{-2}K_t(x, t) - q(x, t)], \quad 0 < x < p, \quad t > 0$$

$$\text{B.C.:} \quad V(0, t) = 0, \quad V(p, t) = 0, \quad t > 0 \tag{89}$$

$$\text{I.C.:} \quad V(x, 0) = f(x) - K(x, 0), \quad 0 < x < p.$$

We can now solve for $V(x, t)$ by the method of Section 6.3.2.

EXAMPLE 6

Solve the boundary value problem

$$u_{xx} = u_t, \quad 0 < x < 1, \quad t > 0$$

$$\text{B.C.:} \quad u(0, t) = \sin t, \quad u(1, t) = 0, \quad t > 0$$

$$\text{I.C.:} \quad u(x, 0) = 0, \quad 0 < x < 1.$$

---

* Because the boundary conditions are time–dependent, there will be no steady-state solution this time.

† Other choices for $K(x, t)$ will still yield equivalent solutions for $u(x, t)$.

SOLUTION The function $K(x, t)$ as determined by (88) is

$$K(x, t) = (1 - x) \sin t.$$

The substitution of this function into (89) yields

$$V_{xx} = V_t + (1 - x) \cos t, \qquad 0 < x < 1, \qquad t > 0$$

B.C.: $V(0, t) = 0, \quad V(1, t) = 0, \quad t > 0$

I.C.: $V(x, 0) = 0, \quad 0 < x < 1,$

but this is exactly the problem we solved in Example 5, and thus the solution we seek is

$$u(x, t) = K(x, t) + V(x, t)$$

$$= (1 - x) \sin t + \frac{2}{\pi} \sum_{n=1}^{\infty} \frac{\sin n\pi x}{n(1 + n^4\pi^4)} [n^2\pi^2(e^{-n^2\pi^2 t} - \cos t) - \sin t].$$

EXERCISES 6.3

In Problems 1–4, solve by the method of Section 6.3.1.

1. $u_{xx} = a^{-2}u_t - 1, \quad 0 < x < 1, \quad t > 0$
   B.C.: $u(0, t) = 0, \quad u(1, t) = 0$
   I.C.: $u(x, 0) = x/2$

2. $u_{xx} = u_t - A, \quad 0 < x < p, \quad t > 0$ (A constant)
   B.C.: $u(0, t) = T_1, \quad u(p, t) = T_1$ (T_1 constant)
   I.C.: $u(x, 0) = T_0$ (constant)

3. $u_{xx} = a^{-2}u_t + 6x, \quad 0 < x < 1, \quad t > 0$
   B.C.: $u_x(0, t) = 0, \quad u(1, t) = 1$
   I.C.: $u(x, 0) = x^3$

4. $u_{xx} = a^{-2}u_t - e^x, \quad 0 < x < 1, \quad t > 0$
   B.C.: $u(0, t) = 0, \quad u_x(1, t) = 0$
   I.C.: $u(x, 0) = ex + 1$

In Problems 5–8, solve by the method of Section 6.3.2.

5. $u_{xx} = a^{-2}u_t - A \cos \omega t, \quad 0 < x < \pi, \quad t > 0$ (A, ω constants)
   B.C.: $u_x(0, t) = 0, \quad u_x(\pi, t) = 0$
   I.C.: $u(x, 0) = 0$

6. $u_{xx} = u_t - e^{-t}, \quad 0 < x < 1, \quad t > 0$
   B.C.: $u_x(0, t) = 0, \quad u(1, t) = 0$
   I.C.: $u(x, 0) = 0$

7. $u_{xx} = u_t + 5 \sin t - 2x, \quad 0 < x < 1, \quad t > 0$
   B.C.: $u(0, t) = 0, \quad u(1, t) = 0$
   I.C.: $u(x, 0) = 0$

8. $u_{xx} = u_t - Ae^{-bx}, \quad 0 < x < 1, \quad t > 0 \quad (A, b \text{ constants})$

B.C.: $u(0, t) = 0, \quad u(1, t) = 0$

I.C.: $u(x, 0) = (1 - e^{-bx}) \dfrac{A}{b^2}$

In Problems 9–12, solve by the method of Section 6.3.3.

9. $u_{xx} = u_t, \quad 0 < x < 1, \quad t > 0$

B.C.: $u(0, t) = \sin t, \quad u(1, t) = 0$

I.C.: $u(x, 0) = \sin 2\pi x$

10. $u_{xx} = u_t - 2(1 - x)t, \quad 0 < x < 1, \quad t > 0$

B.C.: $u(0, t) = t^2, \quad u(1, t) = 2t$

I.C.: $u(x, 0) = \sin \pi x - x(1 - x)(1 + x)/3$

11. $u_{xx} = u_t - 2xt + x, \quad 0 < x < 1, \quad t > 0$

B.C.: $u(0, t) = t, \quad u(1, t) = t^2$

I.C.: $u(x, 0) = x^2/2$

12. $u_{xx} = u_t - t, \quad 0 < x < \pi, \quad t > 0$

B.C.: $u(0, t) = \cos t, \quad u(\pi, t) = \cos t$

I.C.: $u(x, 0) = 1$

★13. Show that the solution of (s and τ fixed)

$$u_{xx} - u_t = -\delta(x - s)\delta(t - \tau), \quad 0 < x < \pi, \quad t > 0$$

B.C.: $u(0, t) = 0, \quad u(\pi, t) = 0$

I.C.: $u(x, 0) = 0,$

where δ is the delta function, leads to the solution form

$$u(x, t) = \begin{cases} 0, & t < \tau \\ \dfrac{2}{\pi} \displaystyle\sum_{n=1}^{\infty} \frac{\sin nx \sin ns}{n} e^{-n^2(t-\tau)}, & t > \tau. \end{cases}$$

[This solution can be interpreted as a type of *Green's function* $g(x, s; t, \tau)$ for the heat equation.]

★14. Using the result of Problem 13 with $u(x, t) \equiv g(x, s; t, \tau)$, show that the solution of

$$u_{xx} - u_t = -q(x, t), \quad 0 < x < \pi, \quad t > 0$$

B.C.: $u(0, t) = 0, \quad u(\pi, t) = 0$

I.C.: $u(x, 0) = 0,$

can be expressed in the form

$$u(x, t) = \int_0^{\pi} \int_0^t g(x, s; t, \tau) q(s, \tau) \, d\tau \, ds.$$

6.4 INFINITE RODS AND FOURIER INTEGRALS

For certain situations it may be convenient to think of a long rod as if it were mathematically unbounded. For example, if we are interested in the temperature distribution in the middle portion of a long rod, prior to the time when such temperatures are greatly influenced by the boundary conditions of the rod, we often model the problem as if the rod extended over the domain $-\infty < x < \infty$. In other instances, we might model the problem so that the rod extends over $0 \leq x < \infty$ when we are interested in the temperature distribution only near the finite boundary $x = 0$.

Problems on infinite domains are usually best handled by either the method of Fourier integrals (discussed in this section) or the method of integral transforms (see Section 6.5). Basically we will find that the *method of Fourier integrals* applied to rods of infinite extent is very similar to the procedure used thus far in solving for the temperature distribution in rods of finite length. The major distinction lies in the fact that for infinite rods the solution assumes the form of an *infinite integral*, rather than an infinite series as is the case for finite rods.

6.4.1 A SEMI-INFINITE ROD

Let us start by considering the specific problem of finding the temperature distribution in a long rod where one end of the rod is always kept at temperature zero and the initial temperature distribution throughout the rod is described by the function $f(x)$. Our mathematical formulation of this problem becomes

$$
\begin{aligned}
& u_{xx} = a^{-2}u_t, && 0 < x < \infty, \quad t > 0 \\
\text{B.C.:}\quad & u(0, t) = 0, && u \text{ and } u_x \text{ finite as } x \to \infty, \quad t > 0 \\
\text{I.C.:}\quad & u(x, 0) = f(x), && 0 < x < \infty.
\end{aligned}
\tag{90}
$$

For the solution technique we are about to use it is necessary that the initial temperature distribution f be *piecewise smooth* and *absolutely integrable*. Also, physical considerations of heat flow are responsible for requiring that u and u_x remain bounded for large x's (see Section 1.6.4 in regard to singular endpoint conditions).

Calling upon separation of variables with $u(x, t) = X(x)W(t)$, we are led to

$$
\begin{aligned}
& X'' + \lambda X = 0, \\
\text{B.C.:}\quad & X(0) = 0, \quad X \text{ and } X' \text{ finite as } x \to \infty
\end{aligned}
\tag{91}
$$

and

$$W' + \lambda a^2 W = 0. \tag{92}$$

The only bounded solutions of the first DE that vanish at $x = 0$ are constant multiples of

$$X(x) = \sin sx \tag{93}$$

$(\lambda = s^2)$,* where s can assume any positive value (assuming s also negative produces no new solutions). Thus, rather than obtaining a discrete set of eigenvalues as before, we now have a *continuum* of eigenvalues generated by $\lambda = s^2$. For these values of λ, the solution of (92) is given by

$$W(t) = B(s)e^{-a^2s^2t}, \tag{94}$$

where $B(s)$ is an arbitrary function of s. Therefore, we deduce that all solutions of the form

$$u(x, t; s) = B(s) \sin(sx)e^{-a^2s^2t}, \qquad s > 0 \tag{95}$$

satisfy the heat equation and prescribed boundary conditions in (90).

Because (95) represents a continuum of solutions depending on s, the *superposition principle* that we normally invoke at this point of the solution process must now take the form of an integral, i.e.,

$$u(x, t) = \int_0^\infty u(x, t; s)\, ds. \tag{96}$$

We leave it to the reader to formally verify that (96) indeed satisfies the heat equation. Finally, the substitution of (95) into (96) gives us the solution

$$u(x, t) = \int_0^\infty B(s) \sin(sx)e^{-a^2s^2t}\, ds. \tag{97}$$

Our remaining task in solving (90) is to select $B(s)$ so that (97) satisfies the initial condition. By setting $t = 0$ in (97), we see that

$$u(x, 0) = f(x) = \int_0^\infty B(s) \sin sx\, dx, \tag{98}$$

which is recognized as a *Fourier sine integral representation* of f. The function $B(s)$ is therefore given by the integral

$$B(s) = \frac{2}{\pi} \int_0^\infty f(x) \sin sx\, dx \tag{99}$$

(see Section 4.7).

EXAMPLE 7

Solve the problem described by (90) for the special case when the initial temperature distribution in the rod is

$$f(x) = \begin{cases} T_0, & 0 < x < 1 \\ 0, & x > 1. \end{cases}$$

---

* See Example 22 in Section 1.6.4.

SOLUTION Based on (97) and (99), our solution has the form

$$u(x, t) = \int_0^\infty B(s)\,\sin(sx)e^{-a^2s^2t}\,ds,$$

where the unknown function $B(s)$ is

$$B(s) = \frac{2T_0}{\pi}\int_0^1 \sin sx\,dx = \frac{2T_0}{\pi s}(1 - \cos s).$$

Therefore, the solution has the integral representation

$$u(x, t) = \frac{2T_0}{\pi}\int_0^\infty \left(\frac{1 - \cos s}{s}\right)\sin(sx)e^{-a^2s^2t}\,ds.$$

There is another solution form that can be derived from (97), which proves to be more satisfactory in certain problems. By replacing the dummy variable x in (99) with ξ and substituting the resulting integral for $B(s)$ directly into (97), we obtain

$$\begin{aligned} u(x, t) &= \frac{2}{\pi}\int_0^\infty\int_0^\infty f(\xi)\,\sin(s\xi)\,\sin(sx)e^{-a^2s^2t}\,d\xi\,ds \\ &= \frac{2}{\pi}\int_0^\infty f(\xi)\int_0^\infty \sin(s\xi)\,\sin(sx)e^{-a^2s^2t}\,ds\,d\xi, \end{aligned}$$

where the last step results from interchanging the order of integration. Next, making use of the trigonometric identity

$$\sin A\,\sin B = \frac{1}{2}[\cos(A - B) - \cos(A + B)]$$

leads to

$$u(x, t) = \frac{1}{\pi}\int_0^\infty f(\xi)\int_0^\infty [\cos s(x - \xi) - \cos s(x + \xi)]e^{-a^2s^2t}\,ds\,d\xi, \qquad (100)$$

and finally, with the aid of the integral formula

$$\int_0^\infty e^{-bx^2}\cos cx\,dx = \frac{1}{2}\sqrt{\frac{\pi}{b}}\,e^{-c^2/4b}, \qquad b > 0 \qquad (101)$$

(see Problem 5 in Exercises 6.4), we obtain the alternate solution form

$$u(x, t) = \frac{1}{2a\sqrt{\pi t}}\int_0^\infty f(\xi)\left\{\exp\left[-\frac{(x - \xi)^2}{4a^2t}\right] - \exp\left[-\frac{(x + \xi)^2}{4a^2t}\right]\right\}d\xi. \qquad (102)$$

REMARK: *Greater use of the solution formula (102) will be made in Section 6.5 in connection with integral transform methods. Here we are content with merely deriving it. However, it is interesting to make the observation that (102) does not require the function f to be absolutely integrable—a condition that must be satisfied for the convergence of (97).*

The solutions (97) or (102) may also be used to approximate the temperature distribution near the end $x = 0$ of a rod of finite length p if we restrict x such that $0 < x \ll p$. That is, for short time durations the temperature distribution near $x = 0$ will depend only upon the boundary condition at $x = 0$, but not upon the boundary condition at $x = p$ until later.

6.4.2 AN INFINITE ROD

For an infinite rod, the problem is characterized by

$$u_{xx} = a^{-2}u_t, \qquad -\infty < x < \infty, \qquad t > 0$$

$$\text{B.C.:} \qquad u \text{ and } u_x \text{ finite as } |x| \to \infty, \qquad t > 0 \tag{103}$$

$$\text{I.C.:} \qquad u(x, 0) = f(x), \qquad -\infty < x < \infty.$$

Following the procedure used in the last section, the solution becomes

$$u(x, t) = \int_0^\infty [A(s) \cos sx + B(s) \sin sx]e^{-a^2s^2t}\, ds \tag{104}$$

(see Problem 8 in Exercises 6.4), where

$$A(s) = \frac{1}{\pi}\int_{-\infty}^{\infty} f(x) \cos sx\, dx \tag{105}$$

and

$$B(s) = \frac{1}{\pi}\int_{-\infty}^{\infty} f(x) \sin sx\, dx. \tag{106}$$

EXAMPLE 8

Solve the problem described by (103) when the initial temperature distribution in the rod is

$$f(x) = \begin{cases} T_0, & |x| < b \\ 0, & |x| > b. \end{cases}$$

SOLUTION The solution is given by (104), where

$$A(s) = \frac{1}{\pi}\int_{-\infty}^{\infty} f(x) \cos sx\, dx = \frac{T_0}{\pi}\int_{-b}^{b} \cos sx\, dx = \frac{2T_0}{\pi}\frac{\sin sb}{s}$$

and

$$B(s) = \frac{T_0}{\pi}\int_{-b}^{b} \sin sx\, dx = 0.$$

Hence, it follows that

$$u(x, t) = \frac{2T_0}{\pi}\int_0^\infty \left(\frac{\sin sb}{s}\right) \cos(sx)e^{-a^2s^2t}\, ds.$$

EXERCISES 6.4

In Problems 1–4, solve the heat conduction problem for the semi-infinite rod (leave answer in integral form).

1. $u_{xx} = a^{-2}u_t, \quad 0 < x < \infty, \quad t > 0$
   B.C.: $u(0, t) = 0, \quad u$ and u_x finite as $x \to \infty$
   I.C.: $u(x, 0) = \begin{cases} 1, & 0 < x < c \\ 0, & x > c \end{cases}$

2. $u_{xx} = a^{-2}u_t, \quad 0 < x < \infty, \quad t > 0$
   B.C.: $u_x(0, t) = 0, \quad u$ and u_x finite as $x \to \infty$
   I.C.: $u(x, 0) = \begin{cases} 1, & 0 < x < 1 \\ 0, & x > 1 \end{cases}$

3. $u_{xx} = a^{-2}u_t, \quad 0 < x < \infty, \quad t > 0$
   B.C.: $u(0, t) = 0, \quad u$ and u_x finite as $x \to \infty$
   I.C.: $u(x, 0) = e^{-x}, \quad 0 < x < \infty$

4. $u_{xx} = a^{-2}u_t, \quad 0 < x < \infty, \quad t > 0$
   B.C.: $u(0, t) = 0, \quad u$ and u_x finite as $x \to \infty$
   I.C.: $u(x, 0) = e^{-x} \cos x, \quad 0 < x < \infty$

★5. Show that

   (a) $\displaystyle\int_0^\infty e^{-bx^2} \cos cx \, dx = \frac{1}{2}\sqrt{\frac{\pi}{b}}\, e^{-c^2/4b}, \qquad b > 0.$

   Hint: See Problem 14 in Exercises 5.2.

   (b) By differentiating the result of (a) with respect to c, show that

$$\int_0^\infty x e^{-b^2x^2/2} \sin cx \, dx = \sqrt{\frac{\pi}{2}}\, b^{-3} c e^{-c^2/2b^2}, \qquad b > 0.$$

6. Given

$$u_{xx} = a^{-2}u_t, \qquad 0 < x < \infty, \qquad t > 0$$

   B.C.: $u(0, t) = 0, \quad u$ and u_x finite as $x \to \infty$
   I.C.: $u(x, 0) = T_0 x e^{-x^2/4}, \quad 0 < x < \infty,$

   show that

$$u(x, t) = T_0 x (1 + a^2 t)^{-3/2} \exp\left[-\frac{x^2}{4(1 + a^2 t)}\right].$$

   Hint: See Problem 5(b).

7. Given

$$u_{xx} = a^{-2}u_t, \qquad 0 < x < \infty, \qquad t > 0$$

   B.C.: $u(0, t) = T_1, \quad u$ and u_x finite as $x \to \infty$
   I.C.: $u(x, 0) = 0, \quad 0 < x < \infty,$

   (a) find the steady-state solution $S(x)$.
   (b) What mathematical problem does the transient solution $v(x, t)$ satisfy if $u(x, t) = S(x) + v(x, t)$?

(c) Can (97) be used to solve part (b)? Explain.

8. Derive the solution formula (104)–(106) for the heat conduction problem on the infinite line characterized by (103).

In Problems 9–12, solve the heat conduction problem for the infinite rod (leave answer in integral form).

9. $u_{xx} = a^{-2}u_t, \qquad -\infty < x < \infty, \qquad t > 0$
   B.C.: u and u_x finite as $|x| \to \infty$
   I.C.: $u(x, 0) = e^{-|x|}, \qquad -\infty < x < \infty$

10. $u_{xx} = a^{-2}u_t, \qquad -\infty < x < \infty, \qquad t > 0$
    B.C.: u and u_x finite as $|x| \to \infty$
    I.C.: $u(x, 0) = \begin{cases} 0, & x < 0 \\ e^{-x}, & x > 0 \end{cases}$

11. $u_{xx} = a^{-2}u_t, \qquad -\infty < x < \infty, \qquad t > 0$
    B.C.: u and u_x finite as $|x| \to \infty$
    I.C.: $u(x, 0) = \begin{cases} T_0, & 0 < x < 1 \\ 0, & \text{otherwise} \end{cases}$

12. $u_{xx} = a^{-2}u_t, \qquad -\infty < x < \infty, \qquad t > 0$
    B.C.: u and u_x finite as $|x| \to \infty$
    I.C.: $u(x, 0) = \begin{cases} 1 - x^2, & |x| < 1 \\ 0, & |x| > 1 \end{cases}$

★13. Show that the solution given by (104) can also be expressed in the form

$$u(x, t) = \frac{1}{2a\sqrt{\pi t}} \int_{-\infty}^{\infty} f(\xi) e^{-(x-\xi)^2/4a^2t} \, d\xi.$$

14. Verify directly that the function

$$g(x, t) = \frac{1}{2a\sqrt{\pi t}} e^{-x^2/4a^2t}$$

is a solution of the heat equation.

★15. Show that $g(x, t)$ given in Problem 14 is the solution to Problem 13 when $f(x) = \delta(x)$, where $\delta(x)$ is the Dirac delta function. [The function $g(x - \xi, t)$ can be interpreted as a type of Green's function for the heat equation on an infinite line.]

6.5 METHOD OF INTEGRAL TRANSFORMS

The problems discussed in Section 6.4 concerning infinite rods can sometimes be treated more effectively by the use of *integral transforms*. The basic aim of the transform method is to transform the given problem into one that is easier to solve. In the case of ordinary DEs with constant coefficients, the transformed problem is algebraic. While this is not the case in solving PDEs, here we will find that the transformed problem is an ODE, and hence, can be solved by elementary methods.* Upon inverting the solution of this transformed problem, we obtain the solution of the original problem.

While there are many similarities between the transform method and the Fourier integral method of Section 6.4, perhaps the most notable distinction is that the transform method eliminates the need to deal with eigenvalue problems or Fourier representations. However, to be effective with integral transform methods, the practitioner should have access to extensive tables of integral transforms, such as *Tables of Integral Transforms,* Volumes 1 and 2, edited by A. Erdelyi (New York: McGraw-Hill, 1954).

6.5.1 LAPLACE TRANSFORM

Let us suppose that $u(x, t)$ is an arbitrary function that has a Laplace transform in the variable t, i.e.,

$$\mathscr{L}\{u(x, t); t \to p\} = \int_0^\infty e^{-pt} u(x, t)\, dt = U(x, p). \tag{107}$$

If $u(x, t)$ is also differentiable, it follows that

$$\mathscr{L}\{u_t(x, t); t \to p\} = pU(x, p) - u(x, 0). \tag{108}$$

On the other hand, we find

$$\begin{aligned} \mathscr{L}\{u_x(x, t); t \to p\} &= \int_0^\infty e^{-pt} u_x(x, t)\, dt \\ &= \frac{\partial}{\partial x} \int_0^\infty e^{-pt} u(x, t)\, dt, \end{aligned}$$

or†

$$\mathscr{L}\{u_x(x, t); t \to p\} = U_x(x, p), \tag{109}$$

and likewise

$$\mathscr{L}\{u_{xx}(x, t); t \to p\} = U_{xx}(x, p). \tag{110}$$

REMARK: *When the function being transformed depends upon more than one variable, we need to use a special kind of notation to designate the variable being transformed. By using the notation adopted above, we can readily distinguish between*

$$\mathscr{L}\{u(x, t); t \to p\} = \int_0^\infty e^{-pt} u(x, t)\, dt$$

---

*In general, the transform method reduces a PDE in n variables to a new PDE in $(n - 1)$ variables.

†We are assuming that conditions are met that permit the interchange of integration and differentiation, leading to (109) and (110).

and

$$\mathcal{L}\{u(x, t); x \to p\} = \int_0^\infty e^{-px} u(x, t)\, dx.$$

Using the Laplace transform, let us consider a very long rod, one end of which is exposed to a time-varying heat reservoir. We will assume the initial temperature distribution is 0°C along the rod, and thus the subsequent temperatures are solutions of

$$u_{xx} = a^{-2} u_t, \qquad 0 < x < \infty, \qquad t > 0$$

$$\text{B.C.:} \quad u(0, t) = f(t), \qquad u(x, t) \to 0 \quad \text{as} \quad x \to \infty, \qquad t > 0 \tag{111}$$

$$\text{I.C.:} \quad u(x, 0) = 0, \qquad 0 < x < \infty.$$

By applying the Laplace transform termwise to the PDE and boundary conditions in (111), using (107)–(110), we obtain the transformed problem

$$U_{xx} - \left(\frac{p}{a^2}\right) U = 0, \qquad 0 < x < \infty$$

$$\text{B.C.:} \quad U(0, p) = F(p), \qquad U(x, p) \to 0 \quad \text{as} \quad x \to \infty, \tag{112}$$

where $F(p)$ denotes the Laplace transform of $f(t)$. The general solution of (112) is*

$$U(x, p) = A(p) e^{-x\sqrt{p}/a} + B(p) e^{x\sqrt{p}/a}, \tag{113}$$

where $A(p)$ and $B(p)$ are arbitrary functions of p. Since $U(x, p)$ must vanish for large x's, we choose $B(p) = 0$. The remaining boundary condition leads to $A(p) = F(p)$, and thus the solution of (112) is

$$U(x, p) = F(p) e^{-x\sqrt{p}/a}. \tag{114}$$

The inversion of (114) will provide us with the solution of (111). Because $U(x, p)$ has the form of a product, inversion through use of the *convolution theorem* is suggested.† Thus,

$$\begin{aligned} u(x, t) &= \mathcal{L}^{-1}\{F(p) e^{-x\sqrt{p}/a}; p \to t\} \\ &= \int_0^t f(\tau) g(x, t - \tau)\, d\tau, \end{aligned} \tag{115}$$

where‡

$$\begin{aligned} g(x, t) &= \mathcal{L}^{-1}\{e^{-x\sqrt{p}/a}; p \to t\} \\ &= \frac{x}{2a\sqrt{\pi} t^{3/2}} e^{-x^2/4a^2 t}. \end{aligned} \tag{116}$$

---

* Although $U(x, p)$ is a function of two variables, we treat p as fixed and solve (112) as if it were an ODE.

† In some cases, (114) may be directly inverted without the use of the convolution theorem.

‡ See Equation (69) in Chapter 5.

The substitution of (116) into (115) leads to the formal solution

$$u(x, t) = \frac{x}{2a\sqrt{\pi}} \int_0^t \frac{f(\tau)}{(t-\tau)^{3/2}} \exp\left[-\frac{x^2}{4a^2(t-\tau)}\right] d\tau. \tag{117}$$

EXAMPLE 9

Solve (111) for the special case when $f(t) = T_1$ (constant).

SOLUTION By making the change of variable

$$z = \frac{x}{2a\sqrt{t-\tau}},$$

we find that (117) becomes

$$u(x, t) = \frac{2}{\sqrt{\pi}} \int_{x/2a\sqrt{t}}^{\infty} f(t - x^2/4a^2z^2)e^{-z^2}\, dz,$$

which, for $f(t) = T_1$, reduces to

$$u(x, t) = T_1 \frac{2}{\sqrt{\pi}} \int_{x/2a\sqrt{t}}^{\infty} e^{-z^2}\, dz.$$

Recalling the definition of the *complementary error function* (Section 5.2), i.e.,

$$\operatorname{erfc}(x) = \frac{2}{\sqrt{\pi}} \int_x^{\infty} e^{-z^2}\, dz,$$

we can express our solution as

$$u(x, t) = T_1 \operatorname{erfc}\left(\frac{x}{2a\sqrt{t}}\right).$$

The physical interpretation of the solution suggests that for any fixed value of x, the temperature in the rod at that point will eventually approach T_1 if we wait long enough ($t \to \infty$).* However, at any particular instant of time t, the temperture $u(x, t)$ satisfies the limiting condition $u(x, t) \to 0$ as $x \to \infty$,† which is in agreement with our prescribed condition. We also recognize that along any member of the family of parabolas in the xt-plane defined by

$$\frac{x}{2a\sqrt{t}} = \text{constant},$$

the temperature $u(x, t)$ remains constant.

---

* $\operatorname{erfc}(0) = 1$. See Problem 2.

† $\operatorname{erfc}(\infty) = 0$. See Problem 2.

□ 6.5.2 TEMPERATURES IN A ROD OF FINITE LENGTH

As a second example illustrating the use of the Laplace transform, let us consider a homogeneous rod of *unit length,* the diameter of which coincides with the x-axis from $x = 0$ to $x = 1$. It is assumed that the initial temperature of the rod is zero and that the end $x = 0$ is maintained at zero temperature while the end at $x = 1$ is kept at constant temperature T_2. The problem is mathematically described by

$$\begin{aligned} & u_{xx} = a^{-2}u_t, \quad 0 < x < 1, \quad t > 0 \\ \text{B.C.:} \quad & u(0, t) = 0, \quad u(1, t) = T_2, \quad t > 0 \\ \text{I.C.:} \quad & u(x, 0) = 0, \quad 0 < x < 1. \end{aligned} \tag{118}$$

Although (118) is a special case of the problem solved in Section 6.2.2, we now wish to illustrate how the problem can be solved by use of the Laplace transform. If we apply the Laplace transform to the variable t, (118) is transformed into the problem

$$\begin{aligned} & U_{xx} - \left(\frac{p}{a^2}\right)U = 0, \quad 0 < x < 1 \\ \text{B.C.:} \quad & U(0, p) = 0, \quad U(1, p) = T_2/p, \end{aligned} \tag{119}$$

the general solution of which is*

$$U(x, p) = A(p)\cosh(x\sqrt{p}/a) + B(p)\sinh(x\sqrt{p}/a). \tag{120}$$

Application of the first boundary condition in (119) requires that $A(p) = 0$, and the second condition leads to

$$U(1, p) = B(p)\sinh(\sqrt{p}/a) = \frac{T_2}{p},$$

from which we deduce the result

$$U(x, p) = \frac{T_2 \sinh(x\sqrt{p}/a)}{p \sinh(\sqrt{p}/a)}. \tag{121}$$

By taking the inverse Laplace transform of this last expression, we find

$$u(x, t) = T_2\left[x + \frac{2}{\pi}\sum_{n=1}^{\infty}\frac{(-1)^n}{n}\sin(n\pi x)e^{-a^2n^2\pi^2t}\right], \tag{122}$$

where we have made use of the inversion formula

$$\mathscr{L}^{-1}\left\{\frac{\sinh(x\sqrt{p})}{p\sinh(b\sqrt{p})}; p \to t\right\} = \frac{x}{b} + \frac{2}{\pi}\sum_{n=1}^{\infty}\frac{(-1)^n}{n}\sin\left(\frac{n\pi x}{b}\right)e^{-n^2\pi^2t/b^2}. \tag{123}$$

---

*As a general rule, we use hyperbolic functions in the general solution when the domain is finite and exponential functions [see (113)] when the domain is infinite.

Notice that (122) is a special case of the solution (33) corresponding to $T_1 = 0$ and $p = 1$.

REMARK: *The inversion formulas (116) and (123) show the importance of having access to extensive tables of Laplace transforms. In the exercises, such formulas will be supplied when needed. Also see Appendix A.*

6.5.3 FOURIER TRANSFORMS

In certain problems one of the transforms of Fourier may be easier to use than the Laplace transform. The Fourier transforms are normally applied to the spatial coordinate x. If the domain is $-\infty < x < \infty$, then the *Fourier exponential transform* is used, whereas either the *Fourier sine transform* or *cosine transform* is used when the domain is $0 \leq x < \infty$.

We begin by considering the problem of heat conduction in an infinite rod when the initial temperature distribution is known. The problem is mathematically characterized by

$$u_{xx} = a^{-2}u_t, \qquad -\infty < x < \infty, \qquad t > 0$$

$$\text{B.C.:} \quad u(x, t) \to 0, \qquad u_x(x, t) \to 0 \quad \text{as} \quad |x| \to \infty, \qquad t > 0 \tag{124}$$

$$\text{I.C.:} \quad u(x, 0) = f(x), \qquad -\infty < x < \infty.$$

Since $-\infty < x < \infty$, we consider using the Fourier exponential transform. Thus, by introducing

$$\mathscr{F}\{u(x, t); x \to s\} = \frac{1}{\sqrt{2\pi}} \int_{-\infty}^{\infty} e^{isx} u(x, t)\, dx = U(s, t), \tag{125}$$

it follows that

$$\mathscr{F}\{u_{xx}(x, t); x \to s\} = -s^2 U(s, t) \tag{126}$$

and

$$\mathscr{F}\{u_t(x, t); x \to s\} = U_t(s, t). \tag{127}$$

REMARK: *The stringent boundary conditions appearing in (124) are required so that (126) exists (see Problem 10 in Exercises 6.5). Once we have produced a solution of the problem, we may be able, in some cases, to relax these conditions to only requiring u and u_x* bounded *for large x.*

Using the above results together with $F(s) = \mathscr{F}\{f(x); s\}$, we find that the Fourier transform applied to the problem described by (124) leads to

$$U_t + a^2 s^2 U = 0, \qquad t > 0$$
$$\text{I.C.:} \quad U(s, 0) = F(s). \tag{128}$$

We recognize (128) as a first-order initial value problem whose solution is readily found to be

$$U(s, t) = F(s)e^{-a^2s^2t}, \qquad -\infty < s < \infty. \tag{129}$$

The solution of the original problem is now found by taking the inverse Fourier transform of (129), which by use of the *convolution theorem* yields*

$$u(x, t) = \frac{1}{\sqrt{2\pi}} \int_{-\infty}^{\infty} f(\xi)g(x - \xi, t)\, d\xi, \tag{130}$$

where

$$g(x, t) = \mathcal{F}^{-1}\{e^{-a^2s^2t}; s \to x\}$$

$$= \frac{1}{a\sqrt{2t}} e^{-x^2/4a^2t} \tag{131}$$

(recalling Problem 12 in Exercises 5.5). Hence, we have obtained the formal solution

$$u(x, t) = \frac{1}{2a\sqrt{\pi t}} \int_{-\infty}^{\infty} f(\xi)e^{-(x-\xi)^2/4a^2t}\, d\xi. \tag{132}$$

By making the change of variable $z = (x - \xi)/2a\sqrt{t}$, we can express (132) in the equivalent form

$$u(x, t) = \frac{1}{\sqrt{\pi}} \int_{-\infty}^{\infty} f(x - 2az\sqrt{t})e^{-z^2}\, dz. \tag{133}$$

This form of the solution is particularly useful if the initial temperature distribution is a constant, say $f(x) = T_0$. Based upon physical considerations alone, we would expect the temperature distribution to remain at T_0. A quick check on (133) with $f(x) = T_0$ gives the correct result, i.e.,

$$u(x, t) = \frac{T_0}{\sqrt{\pi}} \int_{-\infty}^{\infty} e^{-z^2}\, dz = T_0, \tag{134}$$

where the integral can be evaluated using properties of the *error function* (see Problem 2 in Exercises 6.5).

REMARK: *Note that the solution $u(x, t) = T_0$, arising when $f(x) = T_0$, does not satisfy the limiting boundary condition $u(x, t) \to 0$ as $|x| \to \infty$. Moreover, the function $f(x) = T_0$ does not have a Fourier transform, and hence neither does the solution $u(x, t) = T_0$. In spite of this, the solution formulas (132) and (133) are valid, even when $f(x) = T_0$. Once we have obtained solution formulas of this variety, it makes little difference whether they were derived through the use of integral transforms or by some other method. Therefore, in handling problems by the integral transform method we usually take the approach of finding a "tentative solution" through formal reasoning, and then test the solution to see if it satisfies all "necessary" physical requirements of the original problem.*

---

*In some cases, (129) can be inverted directly without the use of the convolution theorem. For instance, see Example 10.

EXAMPLE 10

Solve the problem described by (124) when the initial temperature distribution in the rod is given by

$$f(x) = e^{-x^2/4a^2}, \qquad -\infty < x < \infty.$$

SOLUTION Using the result of Problem 12 in Exercises 5.5, we observe that

$$\mathscr{F}\{e^{-x^2/4a^2}; x \to s\} = \sqrt{2}ae^{-a^2s^2},$$

and, therefore, the transformed problem [see (128)] has the form

$$U_t + a^2s^2U = 0, \qquad t > 0$$

I.C.: $$U(s, 0) = \sqrt{2}ae^{-a^2s^2}.$$

The solution of this transformed problem is

$$U(s, t) = \sqrt{2}ae^{-a^2s^2(1+t)},$$

and thus the solution we seek is obtained from the inverse transform

$$u(x, t) = \mathscr{F}^{-1}\{\sqrt{2}ae^{-a^2s^2(1+t)}; s \to x\},$$

which yields (again referring to Problem 12 in Exercises 5.5)

$$u(x, t) = \frac{1}{\sqrt{1+t}}\, e^{-x^2/4a^2(1+t)}.$$

Suppose that we are interested in the temperature distribution in a long rod, one end of which is held at temperature zero for all time. The problem is described by

$$u_{xx} = a^{-2}u_t, \qquad 0 < x < \infty, \qquad t > 0$$

B.C.: $u(0, t) = 0, \quad u(x, t) \to 0, \quad u_x(x, t) \to 0$ as $x \to \infty, \quad t > 0$ (135)

I.C.: $u(x, 0) = f(x), \qquad 0 < x < \infty.$

The fact that the interval is semi-infinite, together with the prescribed boundary condition at $x = 0$, suggests that the Fourier sine transform be used in this case. Hence, if we define

$$\mathscr{F}_S\{u(x, t); x \to s\} = \sqrt{\frac{2}{\pi}} \int_0^\infty u(x, t) \sin sx\, dx = U(s, t), \tag{136}$$

then

$$\begin{aligned} \mathscr{F}_S\{u_{xx}(x, t); x \to s\} &= -s^2U(s, t) + \sqrt{\frac{2}{\pi}}\, su(0, t) \\ &= -s^2U(s, t). \end{aligned} \tag{137}$$

Also, by setting $F(s) = \mathscr{F}_S\{f(x); s\}$, the transformed problem becomes

$$U_t + a^2s^2U = 0, \qquad t > 0$$

$$\text{I.C.:} \qquad U(s, 0) = F(s), \tag{138}$$

with solution

$$U(s, t) = F(s)e^{-a^2s^2t}, \qquad 0 < s < \infty. \tag{139}$$

The inverse sine transform of (139) yields the solution formula

$$u(x, t) = \sqrt{\frac{2}{\pi}} \int_0^\infty e^{-a^2s^2t} F(s) \sin sx \, ds. \tag{140}$$

By replacing $F(s)$ in (140) with its transform integral, another integral form for the solution can be derived (see Problem 14 in Exercises 6.5).

REMARK: *We have chosen not to subscript the transform functions by an S to denote a sine transform as we did in Chapter 5. Because the sine transform is the only transform under consideration in this problem, such additional notation is redundant, and possibly confusing.*

□ 6.5.4 NONHOMOGENEOUS EQUATIONS

Suppose now we consider the problem of heat flow in an infinite rod when a heat source is present, i.e.,

$$u_{xx} = a^{-2}u_t - q(x, t), \qquad -\infty < x < \infty, \qquad t > 0$$

$$\text{B.C.:} \qquad u(x, t) \to 0, \qquad u_x(x, t) \to 0 \quad \text{as } |x| \to \infty, \qquad t > 0 \tag{141}$$

$$\text{I.C.:} \qquad u(x, 0) = f(x), \qquad -\infty < x < \infty,$$

where $q(x, t)$ is proportional to the heat source. By using the Fourier transforms

$$\mathscr{F}\{u(x, t); x \to s\} = U(s, t),$$

$$\mathscr{F}\{q(x, t); x \to s\} = Q(s, t),$$

$$\mathscr{F}\{f(x); x \to s\} = F(s),$$

we obtain the nonhomogeneous first-order initial value problem

$$U_t + a^2s^2U = a^2Q(s, t), \qquad t > 0$$

$$\text{I.C.:} \qquad U(s, 0) = F(s), \qquad -\infty < s < \infty, \tag{142}$$

the solution of which is given by

$$U(s, t) = F(s)e^{-a^2s^2t} + a^2 \int_0^t e^{-a^2s^2(t-\tau)} Q(s, \tau) \, d\tau. \tag{143}$$

Taking the inverse transform of (143) by use of the convolution theorem, we arrive at

$$u(x, t) = \frac{1}{\sqrt{2\pi}} \int_{-\infty}^{\infty} f(\xi) g(x - \xi, t)\, d\xi$$
$$+ \frac{a^2}{\sqrt{2\pi}} \int_{-\infty}^{\infty} \int_0^t g(x - \xi, t - \tau) q(\xi, \tau)\, d\tau\, d\xi, \tag{144}$$

where (recalling Problem 12 in Exercises 5.5)

$$g(x, t) = \mathcal{F}^{-1}\{e^{-a^2 s^2 t}; s \to x\}$$
$$= \frac{1}{a\sqrt{2\pi}} e^{-x^2/4a^2 t}. \tag{145}$$

REMARK: *The function* $(2\pi)^{-1/2} g(x - \xi, t - \tau)$ *in (144) is a* Green's function *for the heat equation on the infinite domain* $-\infty < x < \infty$, $t > 0$. *It physically represents the temperature distribution arising in the presence of a unit heat source located at* $x = \xi$ *and activated at time* $t = \tau$.

EXERCISES 6.5

1. The *error function* and *complementary error function*, defined respectively by (see figure)

$$\text{erf}(x) = \frac{2}{\sqrt{\pi}} \int_0^x e^{-z^2}\, dz$$

and

$$\text{erfc}(x) = \frac{2}{\sqrt{\pi}} \int_x^{\infty} e^{-z^2}\, dz,$$

play important roles in many heat conduction problems. Show that

(a) $\text{erf}(0) = 0$.

(b) $\text{erfc}(\infty) = 0$.

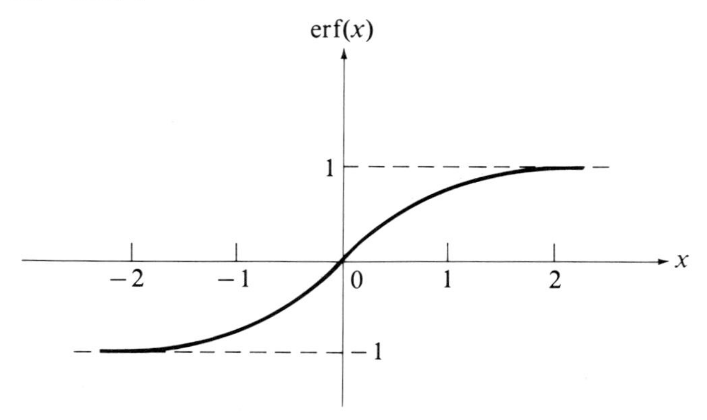

PROBLEM 1

The error function.

(c) $\displaystyle\int_{-a}^{a} e^{-z^2}\,dz = \sqrt{\pi}\,\text{erf}(a).$

(d) $\displaystyle\int_{a}^{b} e^{-z^2}\,dz = \frac{1}{2}\sqrt{\pi}\,[\text{erf}(b) - \text{erf}(a)] = \frac{1}{2}\sqrt{\pi}\,[\text{erfc}(a) - \text{erfc}(b)].$

★2. Given the integral

$$I = \int_0^{\infty} e^{-x^2}\,dx,$$

(a) show that $I = \sqrt{\pi}/2$.

Hint: Write $I^2 = \displaystyle\int_0^{\infty}\int_0^{\infty} e^{-(x^2+y^2)}\,dx\,dy$ and evaluate the double integral by changing to polar coordinates.

(b) From (a), deduce that $\text{erf}(\infty) = 1$.
(c) Show that $\text{erfc}(x) = 1 - \text{erf}(x)$.
(d) From (b) and (c), deduce that $\text{erfc}(0) = 1$ and $\text{erfc}(\infty) = 0$.
Hint: See Problem 1.

3. If the boundary condition in (111) is

$$u(0, t) = f(t) = \begin{cases} T_1, & 0 < t < b \\ 0, & t \geqslant b, \end{cases}$$

(a) show that the subsequent temperature distribution is given by

$$u(x, t) = \begin{cases} T_1\,\text{erfc}(x/2a\sqrt{t}), & 0 < t < b \\ T_1[\text{erf}(x/2a\sqrt{t-b}) - \text{erf}(x/2a\sqrt{t})], & t \geqslant b. \end{cases}$$

(b) Show that $u(x, t)$ given in (a) is continuous at $t = b$.

4. If the boundary condition in (111) is $u(0, t) = T_1/\sqrt{t}$, show that

$$u(x, t) = \frac{T_1}{\sqrt{t}}\,e^{-x^2/4a^2t}.$$

Hint: $\mathcal{L}^{-1}\{p^{-1/2}e^{-x\sqrt{p}/a};\, p \to t\} = \dfrac{1}{\sqrt{\pi t}}\,e^{-x^2/4a^2t}.$

5. Given

$$u_{xx} = a^{-2}u_t, \qquad 0 < x < \infty, \qquad t > 0$$

B.C.: $\quad u_x(0, t) = -f(t), \qquad u(x, t) \to 0 \quad \text{as} \quad x \to \infty$

I.C.: $\quad u(x, 0) = 0, \qquad 0 < x < \infty,$

show that

$$u(x, t) = \frac{a}{\sqrt{\pi}} \int_0^{t} \frac{f(\tau)}{\sqrt{t-\tau}} \exp\left[-\frac{x^2}{4a^2(t-\tau)}\right] d\tau.$$

6. By making the change of variable $z = x/2a\sqrt{t-\tau}$, show that the solution in Problem 5 takes the form

$$u(x, t) = \frac{x}{\sqrt{\pi}} \int_{x/2a\sqrt{t}}^{\infty} f(t - x^2/4a^2z^2)z^{-2}e^{-z^2}\,dz.$$

7. For the special case $f(t) = K$ (constant),
   (a) show that the solution of Problem 5 is

$$u(x, t) = K\left[2a\sqrt{\frac{t}{\pi}}\, e^{-x^2/4a^2t} - x\,\operatorname{erfc}\left(\frac{x}{2a\sqrt{t}}\right)\right].$$

   Hint: Use Problem 6 and integration by parts.
   (b) What is the temperature at the end $x = 0$ as a function of time?

8. Use the Laplace transform to solve

$$u_{xx} = a^{-2}u_t, \qquad 0 < x < 1, \qquad t > 0$$

B.C.: $\qquad u(0, t) = T_1, \qquad u_x(1, t) = 0$

I.C.: $\qquad u(x, 0) = 0, \qquad 0 < x < 1.$

Hint: $\mathscr{L}^{-1}\left\{\dfrac{\cosh x\sqrt{p}}{p\cosh b\sqrt{p}}; p \to t\right\}$

$$= 1 + \frac{4}{\pi}\sum_{n=1}^{\infty}\frac{(-1)^n}{2n-1}\cos\left[\left(n - \frac{1}{2}\right)\frac{\pi x}{b}\right]e^{-(2n-1)^2\pi^2 t/4b^2}.$$

9. If $f(t) = 100°\text{C}$ in the problem described by (111), use the linear approximation $\operatorname{erf}(x) \simeq 2x/\sqrt{\pi}$ to determine how long it will take the cross section of a copper rod at $x = 1$ cm to reach temperature
   (a) 50°C (b) 25°C (c) 1°C
   Hint: See Table 6.1.

10. Use integration by parts to show that

$$\mathscr{F}\{u_{xx}(x, t); x \to s\} = -s^2U(s, t) + \sqrt{2\pi}\,e^{ist}[u_x(x, t) - isu(x, t)]\Bigg|_{x=-\infty}^{x=\infty},$$

and thus deduce that Equation (126) is valid if we prescribe the conditions

$$u(x, t) \to 0, \qquad u_x(x, t) \to 0 \quad \text{as} \quad |x| \to \infty.$$

11. When the initial temperature in (124) is prescribed by

$$f(x) = \begin{cases} T_0, & |x| < 1 \\ 0, & |x| > 1, \end{cases}$$

   (a) show that the subsequent temperature distribution becomes

$$u(x, t) = \frac{T_0}{2}\left[\operatorname{erf}\left(\frac{x+1}{2a\sqrt{t}}\right) - \operatorname{erf}\left(\frac{x-1}{2a\sqrt{t}}\right)\right].$$

   (b) What temperature is approached as $t \to \infty$?

12. Solve the problem described by (124) when

$$f(x) = \begin{cases} 0, & x < 0 \\ T_0, & x > 0. \end{cases}$$

★13. By writing $z^2 + bz = (z + b/2)^2 - b^2/4$, show that

(a) $$\int_u^{\infty} e^{-z^2 - bz}\,dz = \frac{\sqrt{\pi}}{2}\,e^{b^2/4}\left[1 - \operatorname{erf}\left(\frac{1}{2}b + u\right)\right].$$

(b) Use the result in (a) to solve the problem described by (124) when $f(x) = e^{-|x|}$.

14. By substituting

$$F(s) = \sqrt{\frac{2}{\pi}} \int_0^\infty f(\xi) \sin s\xi \, d\xi$$

into Equation (140), show that the resulting integral is equivalent to

$$u(x, t) = \frac{1}{2a\sqrt{\pi t}} \int_0^\infty f(\xi) \left\{ \exp\left[-\frac{(x-\xi)^2}{4a^2 t} \right] - \exp\left[-\frac{(x+\xi)^2}{4a^2 t} \right] \right\} d\xi.$$

Hint: Interchange the order of integration and use the result of Problem 5 in Exercises 6.4.

15. Solve the problem described by (135) when

$$f(x) = \begin{cases} 0, & 0 < x < c \\ T_0, & x > c. \end{cases}$$

Hint: Use Problem 14.

16. Use the Fourier cosine transform to solve

$$u_{xx} = a^{-2} u_t, \qquad 0 < x < \infty, \qquad t > 0$$

B.C.: $u_x(0, t) = 0, \quad u(x, t) \to 0, \quad u_x(x, t) \to 0 \;\text{ as }\; x \to \infty$

I.C.: $u(x, 0) = f(x), \quad 0 < x < \infty.$

17. Solve Problem 16 when

$$f(x) = \begin{cases} 0, & 0 < x < c \\ T_0, & x > c. \end{cases}$$

Hint: First find a general solution formula similar to that in Problem 14.

★18. Given the boundary value problem

$$u_{xx} = u_t, \qquad 0 < x < 1, \qquad t > 0$$

B.C.: $u(0, t) = 0, \quad u(1, t) = T_0$

I.C.: $u(x, 0) = T_0, \quad 0 < x < 1,$

(a) show that the solution of the transformed problem can be expressed in the form

$$U(x, p) = \frac{T_0}{p} \left\{ 1 - e^{-\sqrt{p}} \left[\frac{1 - e^{-2(1-x)\sqrt{p}}}{1 - e^{-2\sqrt{p}}} \right] \right\}.$$

(b) By expanding $(1 - e^{-2\sqrt{p}})^{-1}$ in a series of ascending powers of $e^{-2\sqrt{p}}$, show that

$$U(x, p) = \frac{T_0}{p} [1 - e^{-x\sqrt{p}} + e^{-(2-x)\sqrt{p}} - e^{-(2+x)\sqrt{p}} + \cdots].$$

(c) Inverting the series in (b) termwise, deduce that

$$u(x, t) = T_0 \left[\operatorname{erf}\left(\frac{x}{2\sqrt{t}}\right) + \operatorname{erfc}\left(\frac{2-x}{2\sqrt{t}}\right) - \operatorname{erfc}\left(\frac{2+x}{2\sqrt{t}}\right) + \cdots \right].$$

Hint: $\mathscr{L}\left\{\operatorname{erf}\left(\frac{1}{\sqrt{t}}\right); p\right\} = \frac{1}{p}(1 - e^{-2\sqrt{p}})$.

★19. Solve the problem

$$u_{xx} = u_t, \qquad 0 < x < \infty, \qquad t > 0$$

B.C.: $u_x(0, t) - ku(0, t) = 0, \qquad k > 0$

I.C.: $u(x, 0) = f(x), \qquad 0 < x < \infty$

by first finding the function

$$v(x, t) = u_x(x, t) - ku(x, t),$$

where $v(x, t)$ satisfies a similar problem with $v(0, t) = 0$, and then showing that

$$u(x, t) = -\int_x^\infty e^{-k(\xi - x)} v(\xi, t)\, d\xi.$$

★20. A heat source of strength $q(t)h(t)$, where h is the Heaviside unit function, appears at the origin of a long rod at time $t = 0$ and moves along the positive x-axis with constant speed v. The problem is characterized by

$$u_{xx} = u_t - \delta(x - vt)q(t)h(t), \qquad -\infty < x < \infty, \qquad t > 0$$

B.C.: $u(x, t) \to 0$ as $|x| \to \infty$

I.C.: $u(x, 0) = 0, \qquad -\infty < x < \infty.$

Using the Laplace transform, show that

$$u(x, t) = \frac{1}{2\sqrt{\pi}} \int_0^t q(\tau)(t - \tau)^{-1/2} e^{-(x - v\tau)^2/4(t - \tau)}\, d\tau.$$

★21. When $f(x) = T_0$ and $q(x, t) = \delta(x)\delta(t)$ in Equations (141), show that the solution (144) can be expressed in the form

$$u(x, t) = T_0 + \frac{(\pi t)^{-1/2}}{2a} e^{-x^2/4a^2 t}.$$

★22. Given the boundary value problem

$$u_{xx} = a^{-2}u_t - q(x, t), \qquad 0 < x < \infty, \qquad t > 0$$

B.C.: $u(0, t) = 0, \qquad u(x, t) \to 0$, as $x \to \infty$

I.C.: $u(x, 0) = f(x), \qquad 0 < x < \infty,$

find a solution in the form of Equation (144) and identify the Green's function for this problem.

Hint: $\int_0^\infty \sin(st)F_S(s)G_S(s)\, ds = \frac{1}{2}\int_0^\infty f(u)[g(|u - t|) - g(u + t)]\, du.$

★23. Given the boundary value problem

$$u_{xx} = a^{-2}u_t - q(x, t), \qquad 0 < x < \infty, \qquad t > 0$$

B.C.: $u_x(0, t) = 0, \qquad u(x, t) \to 0$ as $x \to \infty$

I.C.: $u(x, 0) = f(x), \qquad 0 < x < \infty,$

find a solution in the form of Equation (144) and identify the Green's function for this problem.
Hint: See Problem 35 in Exercises 5.5.

★24. Given the nonhomogeneous heat conduction problem

$$u_{xx} = u_t - q(x, t), \qquad 0 < x < 1, \qquad t > 0$$

B.C.: $\quad u(0, t) = a(t), \qquad u(1, t) = b(t)$

I.C.: $\quad u(x, 0) = f(x), \qquad 0 < x < 1,$

(a) use the Laplace transform to derive the transformed problem

$$U_{xx} - pU = -f(x) - Q(x, p), \qquad 0 < x < 1$$

B.C.: $\quad U(0, p) = A(p), \qquad U(1, p) = B(p).$

(b) Find a bilinear formula for the Green's function $G(x, s; p)$ for the DE in (a) and thus deduce that

$$U(x, p) = A(p)\frac{\sinh(1 - x)\sqrt{p}}{\sinh \sqrt{p}} + B(p)\frac{\sinh x\sqrt{p}}{\sinh \sqrt{p}}$$
$$+ \int_0^1 G(x, s; p)[f(s) + Q(s, p)]\, ds.$$

(c) By inverting the solution in (b), find $u(x, t)$ in the form

$$u(x, t) = \int_0^t [a(\tau)w(1 - x, t - \tau) + b(\tau)w(x, t - \tau)]\, d\tau + V(x, t)$$

and identify the function $V(x, t)$.

Hint: $w(x, t) = \mathscr{L}^{-1}\left\{\dfrac{\sinh a\sqrt{p}}{\sinh \sqrt{p}}; p \to t\right\} = -\dfrac{\partial}{\partial t}\mathscr{L}^{-1}\left\{\dfrac{\sinh a\sqrt{p}}{p \sinh \sqrt{p}}; p \to t\right\}.$

★25. *(Duhamel's principle)* If $u(x, t; \tau)$ denotes the solution of

$$u_{xx} = a^{-2}u_t, \qquad u(x, 0; \tau) = f(x, \tau), \qquad -\infty < x < \infty,$$

show that

$$v(x, t) = \int_0^t u(x, t - \tau; \tau)\, d\tau$$

is a solution of

$$a^2 v_{xx} = v_t - f(x, t), \qquad v(x, 0) = 0, \qquad -\infty < x < \infty.$$

(Duhamel's principle basically points out the equivalence between solving nonhomogeneous PDEs with homogeneous initial condition and solving homogeneous PDEs with nonhomogeneous initial condition.)

6.6 NUMERICAL METHOD: FINITE DIFFERENCES

Up to now we have relied totally upon analytical techniques for solving various heat conduction problems. However, the expressions obtained by analytical methods are not always suitable for computing numerical values of the solution due to slow convergence of the resulting infinite series. In such cases we may wish to use the analytical solution only for describing some of the fundamental characteristics of the system being studied, but solve the problem again by a numerical procedure for the purpose of generating accurate solution values at various points on the domain. Also, it happens in many cases that the analytical solution technique does not apply (e.g., the DE or boundary conditions will not permit a separated solution) so that we are then forced to turn to an approximate or numerical method for solving the problem. Owing to the widespread accessibility of today's high speed computers, numerical techniques are becoming increasingly more important in applications.

6.6.1 FINITE-DIFFERENCE OPERATORS

The most commonly used numerical methods for solving DEs depend upon the notion of *finite-difference operators,* which are analogous to differential operators of DEs. Of these, the most often used difference operator is the *forward difference operator* Δ, which is defined by

$$\Delta f_i = f_{i+1} - f_i, \tag{146}$$

where the index i assumes integer values. By f_i, we mean the value of $f(x)$ at x_i, and f_{i+1} is the value of $f(x)$ at x_{i+1} (see Figure 6.7). The increment $h = x_{i+1} - x_i$ will normally be chosen of equal size for all i, but this is not necessary. Other difference operators that are also used are the *backward difference operator* ∇ and *central difference operator* δ, defined, respectively, by

$$\nabla f_i = f_i - f_{i-1} \tag{147}$$

and

$$\delta f_i = f_{i+1/2} - f_{i-1/2}. \tag{148}$$

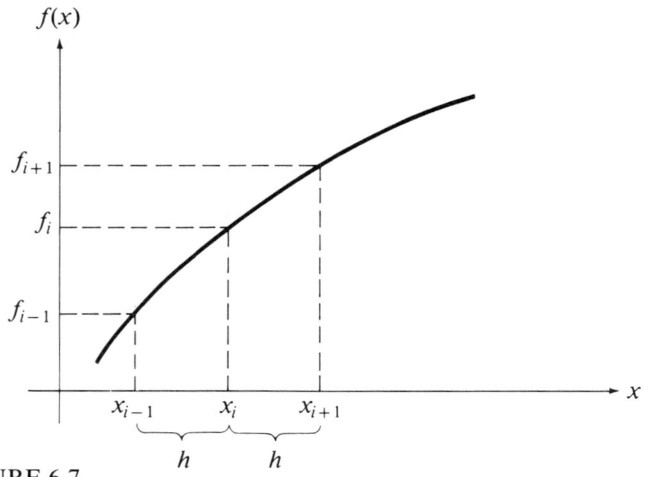

FIGURE 6.7

To relate the difference operator Δ to the differential operator $D = d/dx$, we first recall from calculus the expression

$$f(x + h) = f(x) + hf'(x) + \frac{h^2}{2!} f''(x) + \cdots, \tag{149}$$

which is the Taylor expansion of $f(x + h)$. Evaluating (149) at $x = x_i$ yields

$$f_{i+1} = f_i + hf_i' + \frac{h^2}{2!} f_i'' + \cdots, \tag{150}$$

where $x_{i+1} = x_i + h$. If h is sufficiently small, terms of the order $h^2, h^3, \ldots,$ can be reasonably neglected in (150), and thus we have the approximation

$$f_{i+1} \simeq f_i + hf_i',$$

or solving for f_i', we deduce that

$$f_i' \simeq \frac{1}{h} \Delta f_i = \frac{1}{h} (f_{i+1} - f_i). \tag{151}$$

Equation (151) gives an approximation to the first derivative of f in terms of the forward difference operator Δ. Similar approximations involving the backward and central difference operators are given respectively by

$$f_i' \simeq \frac{1}{h} \nabla f_i = \frac{1}{h} (f_i - f_{i-1}) \tag{152}$$

and

$$f_i' \simeq \frac{1}{h} \delta f_i = \frac{1}{h} (f_{i+1/2} - f_{i-1/2}). \tag{153}$$

The error of approximation for the forward and backward difference operators is about $\frac{1}{2}hf''(x)$, while for the central difference operator the error is roughly $\frac{1}{24}h^2 f'''(x)$. Hence, the approximation (153) is superior to either (151) or (152) when h is small.

REMARK: *To avoid the half-step required for the first-order central difference approximation (153), the first derivative is sometimes approximated by the central difference expression*

$$f_i' \simeq \frac{1}{2h} (f_{i+1} - f_{i-1}),$$

which in general is not as accurate as (153), but more convenient.

We define a *second-order forward difference* by the expression

$$\begin{aligned} \Delta^2 f_i &= \Delta(\Delta f_i) \\ &= \Delta(f_{i+1} - f_i) \\ &= (f_{i+2} - f_{i+1}) - (f_{i+1} - f_i), \end{aligned}$$

or upon simplifying,

$$\Delta^2 f_i = f_{i+2} - 2f_{i+1} + f_i. \tag{154}$$

Second-order backward and central differences are similarly defined by

$$\nabla^2 f_i = f_i - 2f_{i-1} + f_{i-2} \tag{155}$$

and

$$\delta^2 f_i = f_{i+1} - 2f_i + f_{i-1}. \tag{156}$$

In the case of second derivatives, we make the approximations:

$$f_i'' \simeq \frac{1}{h^2}\Delta^2 f_i = \frac{1}{h^2}(f_{i+2} - 2f_{i+1} + f_i), \tag{157}$$

$$f_i'' \simeq \frac{1}{h^2}\nabla^2 f_i = \frac{1}{h^2}(f_i - 2f_{i-1} + f_{i-2}), \tag{158}$$

$$f_i'' \simeq \frac{1}{h^2}\delta^2 f_i = \frac{1}{h^2}(f_{i+1} - 2f_i + f_{i-1}). \tag{159}$$

The error that arises, based on Taylor series expansions, is approximately $hf'''(x)$ for both the forward and backward differences, and $\frac{1}{12}h^2 f^{(4)}(x)$ for the central difference.

In the numerical procedure we are about to illustrate, we will replace the first and second derivatives that appear in the given DE by the above approximations involving either forward, backward, or central differences. We then solve the resulting *system of equations* by some elementary numerical technique, such as Gauss elimination.*

6.6.2 STEADY-STATE PROBLEMS

To illustrate the numerical method, let us solve a simple steady-state heat conduction problem characterized by

$$S'' = -x, \qquad 0 < x < 1, \qquad S(0) = 1, \qquad S(1) = 3. \tag{160}$$

We first divide the interval $0 \leqslant x \leqslant 1$ into a set (mesh) of evenly spaced points described by

$$0 = x_0, x_1, \ldots, x_i, \ldots, x_n = 1,$$

where the interval length is

$$h = x_{i+1} - x_i = \frac{1}{n}.$$

Next, we replace (160) with the difference equation (using forward differences)

---

*For an introductory discussion of Gauss elimination, see L. W. Johnson and R. D. Riess, *Numerical Analysis* (Addison-Wesley: Mass.) 1977.

$$\frac{1}{h^2}(S_{i+2} - 2S_{i+1} + S_i) = -x_i, \qquad i = 0, 1, \ldots, n-2$$
$$S_0 = 1, \qquad S_n = 3. \tag{161}$$

Specifically, for $n = 4$ we obtain the system of equations:

$$16(S_2 - 2S_1 + S_0) = 0$$
$$16(S_3 - 2S_2 + S_1) = -\frac{1}{4}$$
$$16(S_4 - 2S_3 + S_2) = -\frac{1}{2}$$

or, since $S_0 = 1$ and $S_4 = 3$, we find upon some rearranging

$$2S_1 - S_2 = 1$$
$$S_1 - 2S_2 + S_3 = -\frac{1}{64}$$
$$S_2 - 2S_3 = -\frac{97}{32}$$

which now incorporates the boundary conditions prescribed in (160). The simultaneous solution of this system of equations gives us the values

$$S_1 = \frac{97}{64}, \qquad S_2 = \frac{65}{32}, \qquad S_3 = \frac{647}{256}. \tag{162}$$

Because only the second derivative occurs in the DE in (160), it is also easy to use the central difference approximation (159), which is more accurate than the forward difference approximation. In this case, the problem (160) is replaced by

$$\frac{1}{h^2}(S_{i+1} - 2S_i + S_{i-1}) = -x_i, \qquad i = 1, 2, \ldots, n-1$$
$$S_0 = 1, \qquad S_n = 3. \tag{163}$$

Once again taking the case $n = 4$, we obtain the system of equations

$$2S_1 - S_2 = \frac{65}{64}$$
$$S_1 - 2S_2 + S_3 = -\frac{1}{32}$$
$$S_2 - 2S_3 = -\frac{195}{64},$$

with simultaneous solution

$$S_1 = \frac{197}{128}, \qquad S_2 = \frac{33}{16}, \qquad S_3 = \frac{327}{128}. \tag{164}$$

For comparison purposes, the values given by both (162) and (164) are listed in Table 6.2 along with those obtained from the exact solution,

$$S(x) = 1 + \frac{13}{6}x - \frac{x^3}{6}.$$

TABLE 6.2 Approximate Solutions of $S'' = -x$, $S(0) = 1$, $S(1) = 3$

x	FORWARD DIFFERENCE NUMERICAL SOLUTION	CENTRAL DIFFERENCE NUMERICAL SOLUTION*	EXACT SOLUTION
0	1.0000	1.0000	1.0000
0.25	1.5156	1.5391	1.5391
0.50	2.0313	2.0625	2.0625
0.75	2.5273	2.5547	2.5547
1.00	3.0000	3.0000	3.0000

*In this example there is no error in the central difference values.

In most problems the interval of solution will not be confined to $0 \leqslant x \leqslant 1$ as in the above example. When the interval is $0 \leqslant x \leqslant p$, the variable x should be replaced with the normalized variable $\xi = x/p$ so that the increment h remains small (see Problem 15 in Exercises 6.6, which concerns normalized variables, or dimensionless variables.)

When a boundary condition involves a derivative such as

$$S(0) - KS'(0) = \alpha, \qquad (K > 0) \tag{165}$$

the value of S_0 is unknown and its determination requires an extra equation. Using a forward difference, we can approximate (165) by

$$S_0 - \frac{K}{h}(S_1 - S_0) = \alpha, \tag{166}$$

which provides an extra equation in S_0 and S_1. In such cases the numerical solution will generally be less accurate than one for which S_0 is explicitly given, i.e., one for which the prescribed boundary conditions are of the first kind.

REMARK: *If we wish to represent the derivative in (165) more accurately by a central difference, it is necessary to introduce the extraneous or 'fictitious' temperature S_{-1} by imagining the left end of the rod to be slightly extended. The boundary condition in this case is*

$$S_0 - \frac{K}{2h}(S_1 - S_{-1}) = \alpha,$$

which involves another unknown S_{-1}. However, we can obtain another equation in this case by assuming that the DE is satisfied for $i = 0, 1, \ldots, n-1$, instead of $i = 1, 2, \ldots, n-1$.

EXAMPLE 11

Use finite differences to solve

$$S'' - S = -1, \qquad 0 < x < 1, \qquad S'(0) = 0, \qquad S(1) = 2.$$

SOLUTION Using central differences with mesh size $h = 1/4$, the DE is approximated by

$$16(S_{i+1} - 2S_i + S_{i-1}) - S_i = -1, \qquad i = 1, 2, 3.$$

Because the first boundary condition involves a derivative, we will approximate it by the forward difference relation

$$\frac{S_1 - S_0}{\frac{1}{4}} = 0 \quad \text{or} \quad S_1 = S_0,$$

while the second boundary condition is simply $S_4 = 2$. Upon setting $S_0 = S_1$ and $S_4 = 2$, the above system of equations reduces to

$$\begin{aligned} -17S_1 + 16S_2 &= -1 \\ 16S_1 - 33S_2 + 16S_3 &= -1 \\ 16S_2 - 33S_3 &= -33, \end{aligned}$$

with solution $S_1 = 9809/5713$, $S_2 = 10065/5713$, and $S_3 = 10593/5713$. These results are compared in Table 6.3 with the exact solution,

$$S(x) = \frac{\cosh x}{\cosh 1} + 1.$$

TABLE 6.3 Approximate Solution of $S'' - S = -1$, $S'(0) = 0$, $S(1) = 2$

x	CENTRAL DIFFERENCE NUMERICAL SOLUTION	EXACT SOLUTION	ERROR*
0	1.7170	1.6481	0.0689
0.25	1.7170	1.6684	0.0486
0.50	1.7618	1.7308	0.0310
0.75	1.8542	1.8390	0.0152
1.00	2.0000	2.0000	0.0000

*Note that the largest errors occur near $x = 0$ where the boundary condition involves a derivative.

6.6.3 HEAT EQUATION

To standardize the governing equation and interval of interest, we find it convenient in most problems to transform it to a *nondimensional form*. Essentially this eliminates certain physical constants in the DE and transforms the spatial interval to a unit interval (see Problem 15 in Exercises 6.6). Assuming this has been done, we now consider the heat conduction problem

$$u_{xx} = u_t, \qquad 0 < x < 1, \qquad t > 0$$

$$\text{B.C.:} \qquad u(0, t) = 0, \qquad u(1, t) = 0, \qquad t > 0 \tag{167}$$

$$\text{I.C.:} \qquad u(x, 0) = f(x), \qquad 0 < x < 1.$$

In this problem we must replace both the spatial derivative u_{xx} and time derivative u_t by difference approximations. Because the space derivative is second order, we will approximate it by the central difference

$$u_{xx} \simeq \frac{1}{h^2}(u_{i+1,j} - 2u_{i,j} + u_{i-1,j})$$

where $u_{i,j} = u(x_i, t_j)$ denotes the temperature at point x_i and time t_j. We use a forward difference for the time derivative, which is first order. Thus, choosing time increment $\tau = t_{j+1} - t_j$, we have

$$u_t \simeq \frac{1}{\tau}(u_{i,j+1} - u_{i,j}),$$

and (167) is approximated by*

$$\begin{aligned}
&r(u_{i+1,j} - 2u_{i,j} + u_{i-1,j}) = u_{i,j+1} - u_{i,j};\\
&\qquad\qquad i = 1, 2, \ldots, n-1; \qquad j = 0, 1, 2, \ldots\\
&u_{0,j} = 0, \qquad u_{n,j} = 0; \qquad j = 1, 2, 3, \ldots\\
&u_{i,0} = f_i; \qquad i = 0, 1, \ldots, n
\end{aligned} \tag{168}$$

where $r = \tau/h^2$. Rearranging terms in the first equation leads to

$$u_{i,j+1} = ru_{i-1,j} + (1 - 2r)u_{i,j} + ru_{i+1,j}, \tag{169}$$

which allows us to compute $u_{i,j+1}$ from values of u corresponding to earlier times, where $u_{i,0} = f_i$ was known.

Equation (169) is the difference equation that must be solved for any heat conduction problem involving the PDE $u_{xx} = u_t$, regardless of the prescribed boundary conditions. At this stage one simply has to choose a value for the ratio of increments denoted by r. Unless this value is carefully chosen, however, the numerical solution may fluctuate wildly, which we then refer to as being *unstable*. To avoid this situation, it has been determined that r must be chosen so that the coefficients of u on the right-hand side of (169) be non-negative.† Hence, we must restrict r so that $0 < r \leqslant 1/2$.

---

* Sometimes the initial condition and boundary conditions do not agree for $u_{0,0}$ and $u_{n,0}$. Fortunately, this is of little consequence in solving the heat equation and either set of values can be used.

† See Chapter 3 in G. D. Smith, *Numerical Solution of Partial Differential Equations*, 2nd ed. (Clarendon Press: Oxford) 1978.

If we choose $n = 4$ so that $h = 1/4$, then based on the above restriction on r we must select $\tau \leqslant 1/32$. Suppose we set $\tau = 1/32$ so that $r = 1/2$. In this case, Equation (169) becomes

$$\begin{aligned} u_{1,j+1} &= \frac{1}{2}(u_{0,j} + u_{2,j}) \\ u_{2,j+1} &= \frac{1}{2}(u_{1,j} + u_{3,j}) \\ u_{3,j+1} &= \frac{1}{2}(u_{2,j} + u_{4,j}) \end{aligned} \tag{170}$$

(for $j = 0, 1, 2, \ldots$). Using the fact that $u_{0,j} = u_{4,j} = 0$ for all j, we find successively for $j = 0$,

$$\begin{aligned} u_{1,1} &= \frac{1}{2} f_2 \\ u_{2,1} &= \frac{1}{2}(f_1 + f_3) \\ u_{3,1} &= \frac{1}{2} f_2, \end{aligned}$$

for $j = 1$,

$$\begin{aligned} u_{1,2} &= \frac{1}{2} u_{2,1} \\ u_{2,2} &= \frac{1}{2}(u_{1,1} + u_{3,1}) \\ u_{3,2} &= \frac{1}{2} u_{2,1}, \end{aligned}$$

and so on. In Table 6.3 we listed some numerical values for $u_{i,j}$ for the case when $f(x) = \sin \pi x$.

Examination of Equations (170) reveals that each entry in Table 6.4 (except for $x = 0$ and $x = 1$) is obtained by averaging the entries to the left and right of it in the preceeding line. Normally this is not the case, but it happened here because we chose $r = 1/2$.

The exact solution of (167) when $f(x) = \sin \pi x$ is given by

$$u(x, t) = e^{-\pi^2 t} \sin \pi x.$$

In Tables 6.5 and 6.6 we show a comparison of the exact solution and our numerical solution at $x = 1/4$ and $x = 1/2$. The error that results in our numerical solution is due primarily to the coarse division of the interval $0 \leqslant x \leqslant 1$ into four parts. By comparison, if we choose $h = r = 0.1$ (i.e., $n = 10$), we obtain the more accurate values shown in Tables 6.7 and 6.8.

TABLE 6.4 Numerical Solution of (167) when $f(x) = \sin \pi x$.

t \ x	0	$\frac{1}{4}$	$\frac{1}{2}$	$\frac{3}{4}$	1
0	0	0.7071	1.0000	0.7071	0
$\frac{1}{32}$	0	0.5000	0.7071	0.5000	0
$\frac{1}{16}$	0	0.3536	0.5000	0.3536	0
$\frac{3}{32}$	0	0.2500	0.3536	0.2500	0
$\frac{1}{8}$	0	0.1768	0.2500	0.1768	0
$\frac{5}{32}$	0	0.0884	0.1768	0.0884	0

TABLE 6.5 Solutions of (167) with $r = 0.5$ and $f(x) = \sin \pi x$.

t	NUMERICAL SOLUTION ($x = 0.25$)	EXACT SOLUTION ($x = 0.25$)	ERROR
1/32	0.5000	0.5194	0.0194
1/16	0.3536	0.3816	0.0280
3/32	0.2500	0.2803	0.0303

TABLE 6.6 Solutions of (167) with $r = 0.5$ and $f(x) = \sin \pi x$.

t	NUMERICAL SOLUTION ($x = 0.5$)	EXACT SOLUTION ($x = 0.5$)	ERROR
1/32	0.7071	0.7346	0.0275
1/16	0.5000	0.5396	0.0396
3/32	0.3536	0.3964	0.0428

TABLE 6.7 Solutions of (167) with $r = 0.1$ and $f(x) = \sin \pi x$.

t	NUMERICAL SOLUTION ($x = 0.2$)	EXACT SOLUTION ($x = 0.2$)	ERROR
0.01	0.5327	0.5325	0.0002
0.02	0.4828	0.4825	0.0003
0.10	0.2198	0.2191	0.0007

TABLE 6.8 Solutions of (167) with $r = 0.1$ and $f(x) = \sin \pi x$.

t	NUMERICAL SOLUTION ($x = 0.5$)	EXACT SOLUTION ($x = 0.5$)	ERROR
0.01	0.9063	0.9060	0.0003
0.02	0.8214	0.8209	0.0005
0.10	0.3739	0.3727	0.0012

If the prescribed boundary conditions are nonhomogeneous, e.g.,

$$u(0, t) = \alpha, \qquad u(1, t) = \beta, \tag{171}$$

the basic numerical procedure described above does not change except that we set $u_{0,j} = \alpha$ and $u_{n,j} = \beta$. For any α and β, however, if it happens that the initial condition is prescribed such that $f(0) \neq \alpha$ and/or $f(1) \neq \beta$, the numerical solution will contain more error near the endpoints than a solution for which $f(0) = \alpha$ and $f(1) = \beta$. Also, if the function $f(x)$ is not twice differentiable at some points on $(0, 1)$, the numerical solution will contain larger errors near these points. Nonetheless, because of the "smoothing effect" of the solutions of the heat equation, all such errors will tend to decrease as t increases.

When the boundary conditions involve derivatives, i.e., boundary conditions of the second or third kind, they must be approximated by a finite difference as discussed in Section 6.6.2. Consider the following example.

EXAMPLE 12

Use finite differences to solve

$$u_{xx} = u_t, \qquad 0 < x < 1, \qquad t > 0$$

B.C.: $$u(0, t) = 1, \qquad u(1, t) + u_x(1, t) = 0, \qquad t > 0$$

I.C.: $$u(x, 0) = 1, \qquad 0 < x < 1.$$

SOLUTION Our finite difference representation of the problem, using forward differences for the boundary conditions, is

$$r(u_{i+1,j} - 2u_{i,j} + u_{i-1,j}) = u_{i,j+1} - u_{i,j};$$
$$i = 1, 2, \ldots, n-1; \qquad j = 0, 1, 2, \ldots$$

$$u_{0,j} = 1, \qquad hu_{n,j} + (u_{n,j} - u_{n-1,j}) = 0; \qquad j = 1, 2, 3, \ldots$$

$$u_{i,0} = 1, \qquad i = 0, 1, \ldots, n.$$

Choosing $n = 4$ and $r = 1/2$, we have $h = 1/4$. Upon substituting these values into the above system of equations and rearranging terms, we obtain

$$u_{1,j+1} = \frac{1}{2}(u_{0,j} + u_{2,j})$$

$$u_{2,j+1} = \frac{1}{2}(u_{1,j} + u_{3,j})$$

$$u_{3,j+1} = \frac{1}{2}(u_{2,j} + u_{4,j}),$$

($j = 0, 1, 2, \ldots$) where $u_{0,j} = 1$ and $u_{4,j} = 4u_{3,j}/5$. Successively substituting $j = 0, 1, 2, \ldots$ into these equations, and using the initial condition $u_{i,0} = 1$, we obtain the values listed in Table 6.9.

The method of finite differences that we have illustrated here is computationally simple to implement. The major drawback in the technique is that the restriction $0 < r \leq \frac{1}{2}$ forces us to choose very small time increments. That is, we choose h small for reasons of accuracy, and since $\tau \leq \frac{1}{2}h^2$, it follows that τ is necessarily small. There are other finite difference techniques available, such as the Crank–Nicolson implicit method, which allow r to assume any value and yet the solution remains stable. Essentially, the idea in these other techniques is to approximate the derivative u_{xx} by another finite difference which removes the restriction on r.*

REMARK: *In the application of any numerical technique it is equally important to do an analysis on the errors that might arise due to the technique itself and round off. The theory of errors is sometimes fairly complex, however, and goes beyond the intended scope of this introductory section on numerical methods. The interested reader should consult a text on numerical analysis where the theory of errors can be discussed in detail.*

TABLE 6.9 Numerical Solution of Example 12.

t \ x	0	$\frac{1}{4}$	$\frac{1}{2}$	$\frac{3}{4}$	1
0	1.0000	1.0000	1.0000	1.0000	1.0000
$\frac{1}{32}$	1.0000	1.0000	1.0000	0.9000	0.7200
$\frac{1}{16}$	1.0000	1.0000	0.9500	0.8600	0.6880
$\frac{3}{32}$	1.0000	0.9750	0.9300	0.8190	0.6552
$\frac{1}{8}$	1.0000	0.9650	0.8970	0.7926	0.6341

---

* For a discussion of the Crank–Nicolson method, see Chapter 1 in G. D. Smith, *Numerical Solution of Partial Differential Equations*, 2nd ed. (Clarendon Press: Oxford) 1978.

EXERCISES 6.6

In Problems 1–4, solve the steady-state heat conduction problem using central differences with $n = 4$.

1. $S'' = -1, \quad S(0) = 0, \quad S(1) = 1$
2. $S'' - S = -2x, \quad S(0) = 0, \quad S(1) = 1$
3. $S'' = -1, \quad S(0) = 0, \quad S'(1) = 1$

   Hint: Write the boundary condition at $x = 1$ as

$$\frac{S_n - S_{n-1}}{h} = 1.$$

4. $S'' = -x(1 - x), \quad S(0) = 1, \quad S(1) + S'(1) = 0$
5. Use Taylor expansions to obtain the following difference approximations and find an expression for the error:

   (a) $f_i' \simeq \frac{1}{h} \nabla f_i$

   (b) $f_i' \simeq \frac{1}{h} \delta f_i$

6. Use the Taylor expansions

$$f(x + h) = f(x) + hf'(x) + \frac{h^2}{2!} f''(x) + \cdots$$

$$f(x - h) = f(x) - hf'(x) + \frac{h^2}{2!} f''(x) - \cdots$$

   to obtain the central difference approximation to $f''(x)$.

7. Use Taylor expansions to obtain expressions for the errors that occur in the approximations of $f''(x)$ by
   (a) forward differences.
   (b) backward differences.
   (c) central differences.
8. Compare the first three rows of Table 6.4 with values obtained by choosing $n = 4$ and $r = 1/4$.

In Problems 9–14, solve the heat conduction problem using finite differences with $n = 4$ and $\tau = 1/32$. Construct a table like Table 6.4 for your answer (let $j = 0, 1, 2, 3$, and 4).

9. $u_{xx} = u_t, \quad 0 < x < 1, \quad t > 0$
   B.C. $u(0, t) = 0, \quad u(1, t) = 0$
   I.C.: $u(x, 0) = x$
10. $u_{xx} = u_t, \quad 0 < x < 1, \quad t > 0$
    B.C.: $u(0, t) = 0, \quad u(1, t) = 0$
    I.C.: $u(x, 0) = x(1 - x)$

11. $u_{xx} = u_t - 1, \quad 0 < x < 1, \quad t > 0$
B.C.: $u(0, t) = 0, \quad u(1, t) = 0$
I.C.: $u(x, 0) = 0$

12. $u_{xx} = u_t, \quad 0 < x < 1, \quad t > 0$
B.C.: $u_x(0, t) = -1, \quad u(1, 0) = 0$
I.C.: $u(x, 0) = \cos(\pi x/2)$

13. $u_{xx} = u_t, \quad 0 < x < 1, \quad t > 0$
B.C.: $u_x(0, t) = 0, \quad u(1, t) = 1$
I.C.: $u(x, 0) = x$

★14. $u_{xx} - u = u_t, \quad 0 < x < 1, \quad t > 0$
B.C.: $u(0, t) = 1, \quad u(1, t) = 1$
I.C.: $u(x, 0) = x(1 - x)$

15. In solving PDEs through numerical techniques, it is convenient to normalize the equation by the use of *dimensionless variables*. Show that by introducing the new variables

$$\xi = \frac{x}{p}, \qquad \tau = \left(\frac{a}{p}\right)^2 t,$$

the homogeneous heat equation

$$u_{xx} = a^{-2}u_t, \qquad 0 < x < p, \qquad t > 0$$

assumes the normalized form

$$u_{\xi\xi} = u_\tau, \qquad 0 < \xi < 1, \qquad \tau > 0.$$

16. For $n = 4$,
(a) show that the system of equations in Example 12 can be rearranged as

$$u_{1,j+1} = r + (1 - 2r)u_{1,j} + ru_{2,j}$$

$$u_{2,j+1} = ru_{1,j} + (1 - 2r)u_{2,j} + ru_{3,j}$$

$$u_{3,j+1} = ru_{2,j} + \left(1 - \frac{6r}{5}\right)u_{3,j}.$$

(b) Find explicit expressions for $u_{1,1}$; $u_{2,1}$; and $u_{3,1}$ in terms of known quantities and r.

17. Use central finite differences to solve the steady-state problem associated with Example 12. Set $n = 4$ and compare your values with the exact solution at these points.

18. Solve Example 12 using $n = 4$ and $r = 1$. What can you say about the stability of your solution?

CHAPTER 7

THE WAVE EQUATION

Wave propagation and general vibrational phenomena lead to problems involving the *wave equation*. Historically, the wave equation evolved out of a study of the vibrating string problem. During the time following the invention of the calculus by Sir Isaac Newton (1642–1727) and Gottfried W. von Leibniz (1646–1716), many mathematicians concerned themselves with the vibrating string problem and the closely related problem of the mathematical theory of musical sounds. By the middle of the 1700s, the French mathematician J. d'Alembert and two Swiss mathematicians, D. Bernoulli and L. Euler, had advanced the mathematical theory of the vibrating string to the point where its governing PDE was known, the general solution of this equation had been found, and the fundamental vibrational modes determined. Such knowledge led these men to the notion of the superposition principle and eventually to the concept of representing an arbitrary function by a trigonometric series. A lengthy controversy erupted among the mathematicians of the time over the question of validity of the series of sine functions. About this same time, J. Fourier made a similar announcement concerning the representation of an arbitrary function in a series of sinusoidal functions in connection with his work on the theory of heat conduction. The controversy continued until 1829 when the German mathematician Peter G. L. Dirichlet (1805–1859) finally established general conditions of a function sufficient to ensure the convergence of its trigonometric series.

The techniques used in solving the wave equation with prescribed initial and boundary conditions are essentially the same as those for the heat equation. Thus, our treatment of the wave equation will mostly parallel that of the heat equation. In Section 7.2 we will examine the *free motions* of a vibrating string, and in Section 7.3 we extend our discussion to *forced motions* (nonhomogeneous PDEs). Sections 7.4 and 7.5 are concerned with vibration problems on *infinite domains* where the methods of *Fourier integrals* and *integral transforms* are once again featured. In the final section we briefly discuss numerical solutions of the wave equation by the *method of finite differences*.

7.1 INTRODUCTION

One of the most common and fundamental phenomena of nature is that of wave motion. For example, sound waves emanating from a struck bell, surface waves of displacement propagating radially outward when a pebble is dropped into a pool, and deflections of a "plucked" string are all described in terms of wave motion. Whatever the character of the wave phenomenon, the entity under consideration is usually governed by the *wave equation*

$$\nabla^2 u = c^{-2} u_{tt}, \tag{1}$$

where c is a physical constant having the dimension of velocity. To be a properly-posed problem, the solution of (1) must be subject to a *single* boundary condition at each boundary point and *two* initial conditions.

Although the wave equation arises in many physical examples, we will initially address only the one-dimensional case and discuss its solutions in terms of vibrating strings. The results are easily carried over to other phenomena.

7.1.1 EQUATION OF MOTION FOR A VIBRATING STRING

The study of free oscillations, those with no external forces other than tension, of a tightly stretched string is one of the most rudimentary problems in the theory of wave motion. Let us consider a perfectly elastic string that is tightly stretched with tension T between fixed supports which are a finite distance apart (see Figure 7.1). Imagine a violin string, for example, that is distorted at time t and then released at time $t = 0$ in such a way that it vibrates freely in the vertical plane. We want to derive the equation governing such motions.

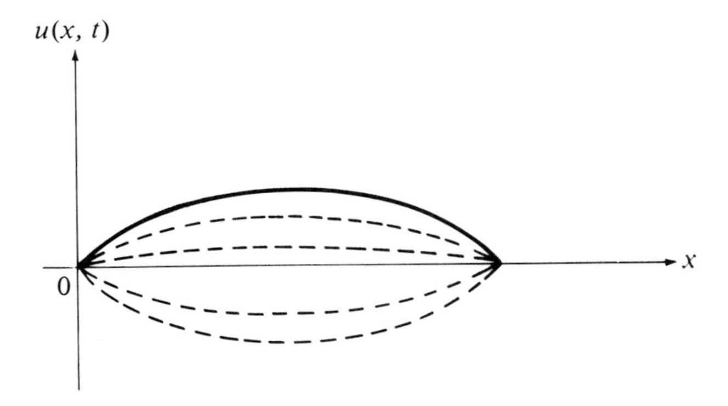

FIGURE 7.1

Freely vibrating string

To make the problem tractable, we first make some simplifying assumptions:

1. The tension T in the string is so large that gravity can be ignored.
2. The string is uniformly covered with mass of constant density ρ (i.e., mass per unit length).
3. The deflections $u(x, t)$ in the vertical plane are "small" compared with the length of the string.

Let us examine a small element of the string between x and $(x + \Delta x)$ and apply Newton's second law of motion (see Figure 7.2). Based on the

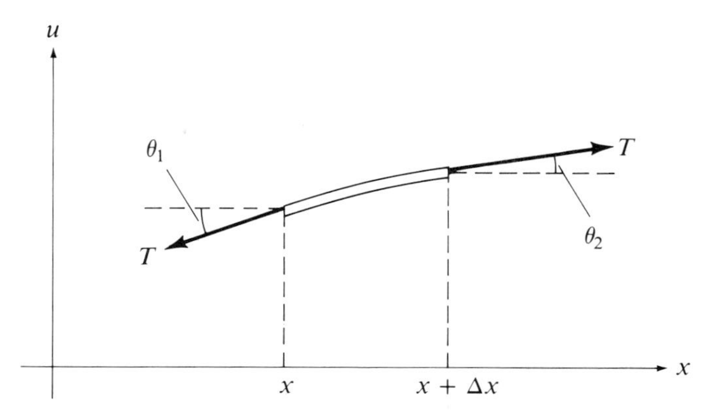

FIGURE 7.2

Small element of the string

third assumption above, we only need to consider external forces acting in the vertical direction. These forces are the vertical components of the tension at x and $(x + \Delta x)$, which are, respectively,

$$\begin{aligned} -T \sin \theta_1 &\simeq -T \tan \theta_1 = -Tu_x(x, t), \\ T \sin \theta_2 &\simeq T \tan \theta_2 = Tu_x(x + \Delta x, t). \end{aligned} \tag{2}$$

Here we are using the small-angle approximation $\sin \theta \simeq \tan \theta$, with $\tan \theta_1$ denoting the slope of the string at x and $\tan \theta_2$ is the slope of the string at $(x + \Delta x)$. Summing the forces (2) and equating the result to the inertial force $\rho \Delta x u_{tt}(x, t)$, where $\rho \Delta x$ is the mass of the small element and $u_{tt}(x, t)$ is the acceleration, we obtain

$$T[u_x(x + \Delta x, t) - u_x(x, t)] = \rho \Delta x u_{tt}(x, t). \tag{3}$$

When dividing (3) by $T\Delta x$ and taking the limit as $\Delta x \to 0$, we get the equation of motion

$$u_{xx} = c^{-2} u_{tt}, \tag{4}$$

where $c^2 = T/\rho$. We call (4) the *one-dimensional wave equation* which belongs to the class of *hyperbolic* equations (see Section 3.2).

If a distributed forcing function $F(x, t)$ is also acting on the string, then $F(x, t)$ must be summed with the other forces acting in the vertical direction. This leads to the *nonhomogeneous* equation

$$u_{xx} = c^{-2} u_{tt} - \frac{1}{T} F(x, t). \tag{5}$$

7.2 FREE MOTIONS OF A VIBRATING STRING

Let us consider the free oscillations of a taut string of length p. The transverse deflections $u(x, t)$ of such a string are solutions of the one-dimensional wave equation

$$u_{xx} = c^{-2}u_{tt}, \qquad 0 < x < p, \qquad t > 0. \tag{6}$$

Let us assume the string is stretched between *fixed supports* with zero displacements so that the boundary conditions are simply

$$\text{B.C.:} \qquad u(0, t) = 0, \qquad u(p, t) = 0, \qquad t > 0. \tag{7}$$

Of course, the motion of the string will also depend upon its deflection and velocity (speed) at the time of release. Denoting the initial deflection by the function $f(x)$ and the initial velocity by $g(x)$, the initial conditions become

$$\text{I.C.:} \qquad u(x, 0) = f(x), \qquad u_t(x, 0) = g(x), \qquad 0 < x < p. \tag{8}$$

The solution technique that we employ is the *separation of variables method,* wherein we assume that*

$$u(x, t) = X(x)W(t). \tag{9}$$

The substitution of (9) into (6) and subsequent separation of variables leads to

$$X'' + \lambda X = 0, \qquad 0 < x < p, \tag{10}$$

$$W'' + \lambda c^2 W = 0, \qquad t > 0, \tag{11}$$

where λ denotes the separation constant. Under the assumption (9), the boundary conditions (7) take the form

$$X(0)W(t) = 0, \qquad X(p)W(t) = 0,$$

from which we deduce†

$$\text{B.C.:} \qquad X(0) = 0, \qquad X(p) = 0. \tag{12}$$

The eigenvalue problem composed of (10) and (12) is one we have solved many times, and leads to the results

$$\lambda_n = \frac{n^2\pi^2}{p^2}, \qquad \phi_n(x) = \sin\frac{n\pi x}{p}, \qquad n = 1, 2, 3, \ldots. \tag{13}$$

For these values of λ, the solutions of (11) are

$$W_n(t) = a_n \cos\frac{nc\pi t}{p} + b_n \sin\frac{nc\pi t}{p}, \qquad n = 1, 2, 3, \ldots, \tag{14}$$

where the a's and b's are arbitrary constants.

By combining (13) and (14), we obtain the set of solutions

$$u_n(x, t) = \left(a_n \cos\frac{nc\pi t}{p} + b_n \sin\frac{nc\pi t}{p}\right)\sin\frac{n\pi x}{p}, \qquad n = 1, 2, 3, \ldots, \tag{15}$$

---

* See the discussions in Section 3.3.1 and Section 6.2 on separation of variables.

† We rule out the possibility that $W(t) \equiv 0$ for the same reason stated in Section 6.2 in connection with heat conduction problems.

each of which satisfies (6) and (7). These solutions are called *standing waves,* since each can be viewed as having fixed shape $\sin(n\pi x/p)$ but with varying amplitude $W_n(t)$. The points where $\sin(n\pi x/p) = 0$ are called *nodes* and physically correspond to zero displacement of the string. The number of nodes depends on the value of n. For example, when $n = 1$ there is no node on the interval $(0, p)$. When $n = 2$ there is one node, when $n = 3$ there are two nodes, and so forth (see Figure 7.3).†

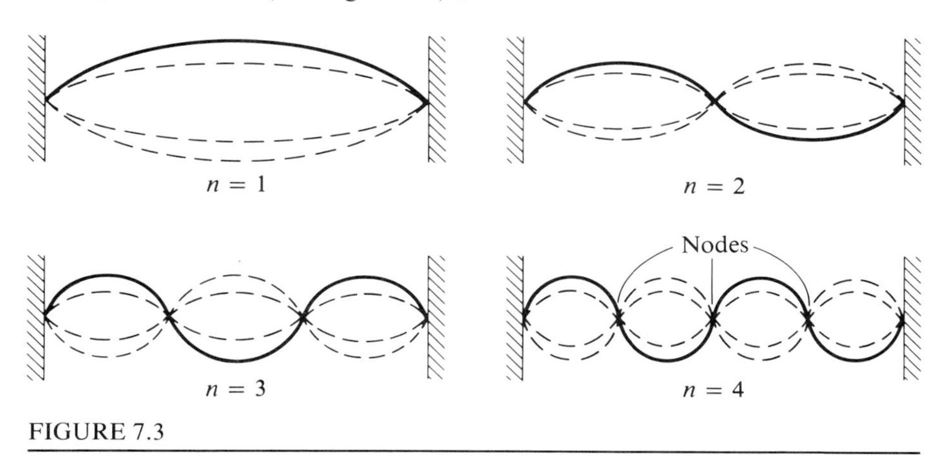

FIGURE 7.3

Harmonic motion and nodal points

We can find another interpretation of (15) if we think of x as fixed. In this case $u_n(x, t)$ represents the motion of a point on the string with abscissa x. Moreover, by writing $W_n(t)$ in the form

$$W_n(t) = \sqrt{a_n^2 + b_n^2}\cos\left(\frac{nc\pi t}{p} - \alpha_n\right), \tag{16}$$

where $\alpha_n = \tan^{-1}(b_n/a_n)$, we see that for a fixed value of x, $u_n(x, t)$ represents *simple harmonic motion* of (angular) frequency $\omega_n = nc\pi/p$ and amplitude $\sqrt{a_n^2 + b_n^2}\sin(n\pi x/p)$. The frequency ω_n is called the nth natural frequency (harmonic) of the system, whereas ω_1 denotes the fundamental frequency.

The pitch of the sound coming from a stringed musical instrument is directly related to the fundamental frequency $\omega_1 = c\pi/p$, i.e., the larger the fundamental frequency, the higher the pitch of the sound produced. Because $c = \sqrt{T/\rho}$, we see that the pitch of an instrument can be raised by increasing the tension T in the string (called tuning). A musician playing a stringed instrument also varies the pitch by varying the effective length p of the string by fretting the instrument. The "musical sounds" that are produced by a stringed instrument like a violin are due to the fact that all higher harmonics

† In general, an eigenfunction $\phi_n(x)$ has $n - 1$ zeros on the fundamental interval (excluding endpoints).

ω_n of the fundamental frequency are *integral* multiples of ω_1, i.e., $\omega_n = n\omega_1$. This is not necessarily the case for other instruments, such as a drum.

To complete the solution process, we form the linear combination of solutions *(superposition principle)*

$$u(x, t) = \sum_{n=1}^{\infty} \left(a_n \cos \frac{nc\pi t}{p} + b_n \sin \frac{nc\pi t}{p} \right) \sin \frac{n\pi x}{p}. \tag{17}$$

Finally, imposing the initial conditions (8) upon this solution, we see that

$$u(x, 0) = f(x) = \sum_{n=1}^{\infty} a_n \sin \frac{n\pi x}{p}, \qquad 0 < x < p \tag{18}$$

and

$$u_t(x, 0) = g(x) = \sum_{n=1}^{\infty} \frac{nc\pi}{p} b_n \sin \frac{n\pi x}{p}, \qquad 0 < x < p, \tag{19}$$

where we have assumed that termwise differentiation of the series (17) is justified. We recognize (18) and (19) as *Fourier sine series* of $f(x)$ and $g(x)$, respectively, whose Fourier coefficients are

$$a_n = \frac{2}{p} \int_0^p f(x) \sin \frac{n\pi x}{p} \, dx, \qquad n = 1, 2, 3, \ldots \tag{20}$$

and

$$\frac{nc\pi}{p} b_n = \frac{2}{p} \int_0^p g(x) \sin \frac{n\pi x}{p} \, dx, \qquad n = 1, 2, 3, \ldots. \tag{21}$$

REMARK: *Boundary conditions other than zero displacement at the endpoints are also possible for the vibrating string problem. For example, physical conditions leading to other types of boundary conditions were discussed in Section 1.5.1 for the related problem of an elastic string supporting a load.*

EXAMPLE 1

Solve the one-dimensional wave problem described by

$$u_{xx} = c^{-2} u_{tt}, \qquad 0 < x < 1, \qquad t > 0,$$

B.C.: $\quad u(0, t) = 0, \qquad u(1, t) = 0, \qquad t > 0,$

I.C.: $\quad u(x, 0) = f(x), \qquad u_t(x, 0) = 0, \qquad 0 < x < 1,$

where the initial deflection is the triangular shape (Figure 7.4) obtained by "plucking" the string, i.e.,

$$f(x) = \begin{cases} Ax, & 0 \le x \le \dfrac{1}{2} \\ A(1 - x), & \dfrac{1}{2} < x \le 1, \end{cases}$$

where A is constant.

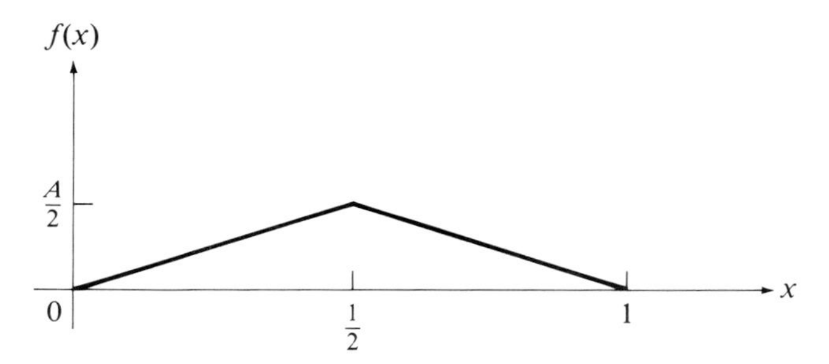

FIGURE 7.4

SOLUTION By separating variables, we get the eigenvalue problem

$$X'' + \lambda X = 0, \qquad X(0) = 0, \qquad X(1) = 0,$$

with eigenvalues and eigenfunctions given by

$$\lambda_n = n^2\pi^2, \qquad \phi_n(x) = \sin n\pi x, \qquad n = 1, 2, 3, \ldots .$$

For these values of λ, the time-dependent equation becomes

$$W'' + n^2\pi^2c^2W = 0,$$

with solutions

$$W_n(t) = a_n \cos n\pi ct + b_n \sin n\pi ct, \qquad n = 1, 2, 3, \ldots .$$

Combining all solutions by the superposition principle gives us

$$u(x, t) = \sum_{n=1}^{\infty} (a_n \cos n\pi ct + b_n \sin n\pi ct) \sin n\pi x,$$

where all that remains unfinished is the determination of a_n and b_n, $n = 1, 2, 3, \ldots$.

Since the initial velocity $g(x)$ is identically zero, it follows that $b_n = 0$, $n = 1, 2, 3, \ldots$. Thus, substituting $t = 0$ in the remaining solution yields

$$u(x, 0) = f(x) = \sum_{n=1}^{\infty} a_n \sin n\pi x,$$

where

$$\begin{aligned} a_n &= 2\int_0^1 f(x) \sin n\pi x \, dx \\ &= 2A\int_0^{1/2} x \sin n\pi x \, dx + 2A\int_{1/2}^1 (1 - x) \sin n\pi x \, dx \\ &= \frac{4A}{n^2\pi^2} \sin(n\pi/2), \qquad n = 1, 2, 3, \ldots . \end{aligned}$$

The solution then takes the form

$$u(x, t) = \frac{4A}{\pi^2} \sum_{n=1}^{\infty} \frac{\sin(n\pi/2)}{n^2} \cos n\pi ct \sin n\pi x$$

(note that all even terms are zero).

The solution form obtained in Example 1 is difficult to use for numerical computations or to determine the actual shape of the string at various times t. However, because at $t = 0$ we have

$$f(x) = \frac{4A}{\pi^2} \sum_{n=1}^{\infty} \frac{\sin(n\pi/2)}{n^2} \sin n\pi x, \qquad 0 < x < 1,$$

it follows that this series on the right converges for all x to the *odd periodic extension* of f. By using the trigonometric identity

$$\sin A \cos B = \frac{1}{2} [\sin(A + B) + \sin(A - B)],$$

we can express the solution in Example 1 in a similar form as

$$u(x, t) = \frac{2A}{\pi^2} \sum_{n=1}^{\infty} \frac{\sin(n\pi/2)}{n^2} \{\sin[n\pi(x + ct)] + \sin[n\pi(x - ct)]\}. \tag{22}$$

Thus, by comparison with the series for f, we recognize that (22) is equivalent to

$$u(x, t) = \frac{1}{2}[f_0(x + ct) + f_0(x - ct)], \tag{23}$$

where f_0 denotes the odd periodic extension of f. This result is a special case of d'Alembert's solution (see Section 3.2.2) of the wave equation for a finite string.

Expressed in the form (23), the solution $u(x, t)$ can easily be sketched for various values of t. For example, at $t = 0$ we obtain the initial deflection shape shown in Figure 7.5(a). Other values of t lead to the various shapes also shown in Figure 7.5. On the left in each case are the odd periodic extensions $f_0(x + ct)$ (solid curve) and $f_0(x - ct)$ (dashed line) plotted over the unit interval. The curves on the right are obtained by graphically averaging the curves on the left.

7.2.1 VALIDITY OF THE FORMAL SOLUTION

The solutions derived thus far are only *formal solutions*. We have not yet stated conditions under which they also become *strict solutions*. In the case of the heat equation a strict solution is obtained when the initial data are merely piecewise continuous, but for the wave equation we must impose much stronger conditions.

To begin, we first take note that the series

$$v(x, t) = \sum_{n=1}^{\infty} a_n \cos \frac{nc\pi t}{p} \sin \frac{n\pi x}{p}, \tag{24}$$

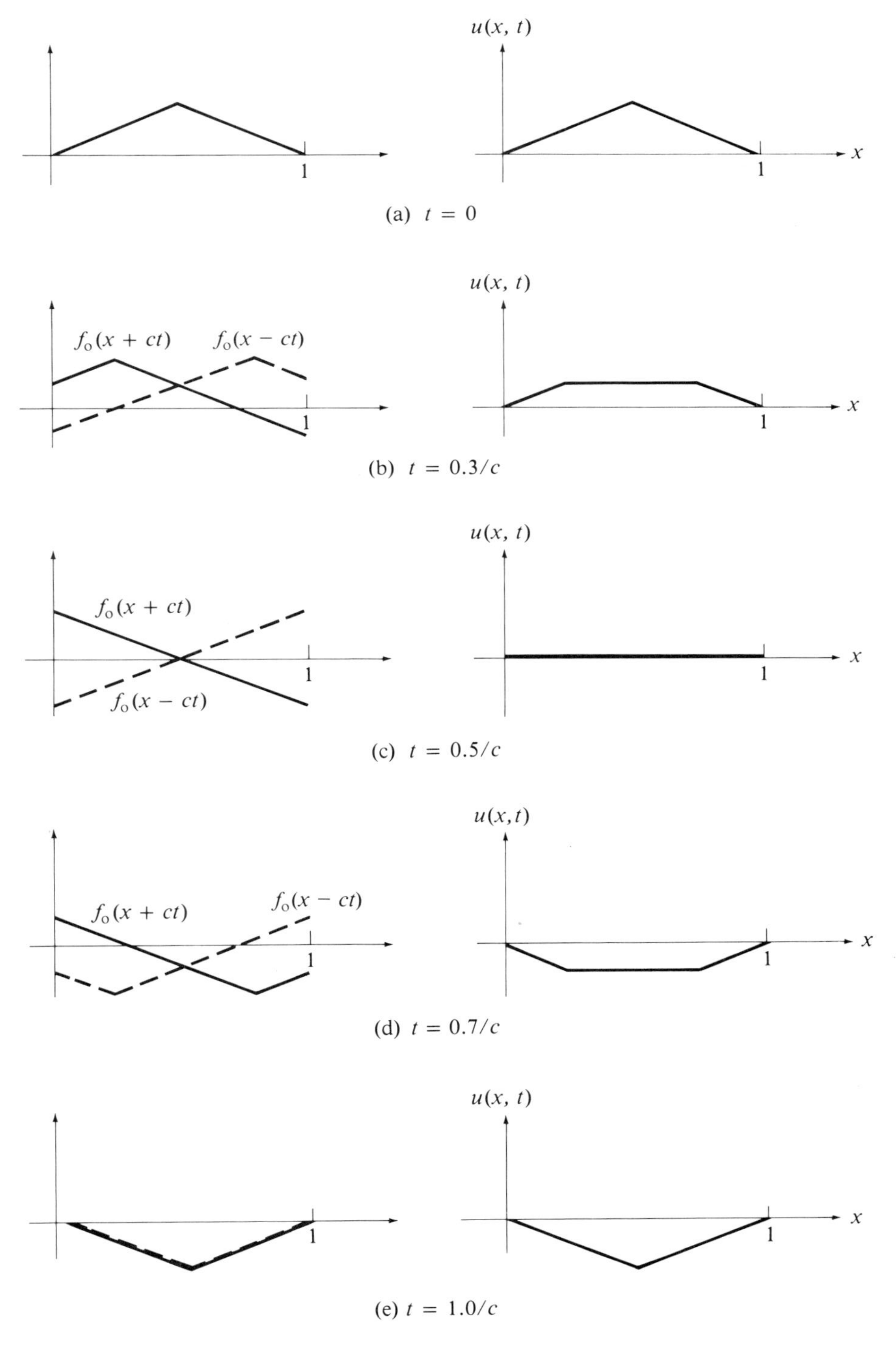

FIGURE 7.5

Displacement of string during first half-period

where a_n is defined by (20), is a formal solution of the boundary value problem

$$u_{xx} = c^{-2}u_{tt}, \qquad 0 < x < p, \qquad t > 0$$

$$\text{B.C.:} \qquad u(0, t) = 0, \qquad u(p, t) = 0 \tag{25}$$

$$\text{I.C.:} \qquad u(x, 0) = f(x), \qquad u_t(x, 0) = 0,$$

and the series

$$w(x, t) = \sum_{n=1}^{\infty} b_n \sin \frac{nc\pi t}{p} \sin \frac{n\pi x}{p}, \tag{26}$$

where b_n is given by (21), is the formal solution of

$$u_{xx} = c^{-2}u_{tt}, \qquad 0 < x < p, \qquad t > 0$$

$$\text{B.C.:} \qquad u(0, t) = 0, \qquad u(p, t) = 0 \tag{27}$$

$$\text{I.C.:} \qquad u(x, 0) = 0, \qquad u_t(x, 0) = g(x).$$

The sum $u(x, t) = v(x, t) + w(x, t)$ is, therefore, the formal solution of the boundary value problem described in (6)–(8).

We claim that the function $v(x, t)$, defined by the series (24), is a strict solution of the boundary value problem (25) provided that f, f', and f'' are continuous functions on $[0, p]$, and also that

$$f(0) = f''(0) = f(p) = f''(p) = 0. \tag{28}$$

To prove this claim, let us recall the Fourier series representation [see (18) and (20)]

$$f(x) = \sum_{n=1}^{\infty} a_n \sin \frac{n\pi x}{p}, \qquad 0 < x < p. \tag{29}$$

Under the hypothesis that f and f' are continuous in $[0, p]$ and that $f(0) = f(p) = 0$, it follows from Theorem 4.3 that the series (29) is uniformly convergent on every finite interval and converges to the odd periodic extension of f. Following the same procedure used in deriving (23), we find that (24) can be expressed in the form

$$v(x, t) = \frac{1}{2}[f_0(x + ct) + f_0(x - ct)], \tag{30}$$

where f_0 is the odd periodic extension of f. Clearly $v(x, t)$ is a continuous function. Moreover, at $x = 0$ and $x = p$, we find that

$$\begin{aligned} v(0, t) &= \frac{1}{2}[f_0(ct) + f_0(-ct)] \\ &= \frac{1}{2}[f_0(ct) - f_0(ct)] \\ &= 0 \end{aligned}$$

and

$$\begin{aligned} v(p, t) &= \frac{1}{2}[f_0(p + ct) + f_0(p - ct)] \\ &= \frac{1}{2}[f_0(p + ct) + f_0(-p - ct)] \\ &= \frac{1}{2}[f_0(p + ct) - f_0(p + ct)] \\ &= 0, \end{aligned}$$

and thus $v(x, t)$ satisfies both boundary conditions. To show that $v(x, t)$ also satisfies the initial conditions, we first observe that

$$\begin{aligned} v(x, 0) &= \frac{1}{2}[f_0(x) + f_0(x)] \\ &= f_0(x) \\ &= f(x), \qquad 0 < x < p. \end{aligned}$$

Then differentiating $v(x, t)$, we have

$$v_t(x, t) = \frac{1}{2}[cf_0'(x + ct) - cf_0'(x - ct)], \tag{31}$$

from which we deduce

$$v_t(x, 0) = \frac{c}{2}[f_0'(x) - f_0'(x)] = 0.$$

Our final task is to show that $v_{xx}(x, t)$ and $v_{tt}(x, t)$ are continuous functions and that $v(x, t)$ actually satisfies the wave equation. From (30) we see that

$$v_{xx}(x, t) = \frac{1}{2}[f_0''(x + ct) + f_0''(x - ct)], \tag{32}$$

$$v_{tt}(x, t) = \frac{1}{2}[c^2 f_0''(x + ct) + c^2 f_0''(x - ct)], \tag{33}$$

which are indeed continuous functions since we have assumed that f'' (and, hence, f_0'') is continuous. The proof is completed by observing that

$$\begin{aligned} v_{xx} - c^{-2}v_{tt} &= \frac{1}{2}[f_0''(x + ct) + f_0''(x - ct)] - \frac{1}{2}[f_0''(x + ct) + f_0''(x - ct)] \\ &= 0. \end{aligned}$$

In the same manner, it can be proven that the function $w(x, t)$, defined by the series (26), is a strict solution of (27) provided that g and g' are continuous functions on $[0, p]$, and also that

$$g(0) = g(p) = 0. \tag{34}$$

We leave the proof of this result to the reader (see Problem 14 in Exercises 7.2).

In some cases, it may turn out that we cannot obtain a strict solution of the problem because the initial data do not satisfy the above conditions. For instance, the function f defined by the triangular shape in Example 1 clearly does not have a derivative everywhere on the interval of interest. Thus, the series obtained for $u(x, t)$ is not a solution in the strict sense, but is what we sometimes call a *generalized solution*, or *weak solution*. Actually, this peculiar situation arises because of our mathematical model of the problem. It is physically impossible to deflect a string in the triangular shape assumed in Example 1. The true shape will always be a smooth curve, and thus differentiable. The generalized solution represents a reasonable tradeoff when one considers the complexities associated with solving a more accurate mathematical model.

REMARK: *One of the major differences between solutions of the wave equation and those of the heat equation is that the latter have negative exponential terms in the variable t that tend to zero quite rapidly for increasing n, whereas those of the former have oscillatory terms in t. Hence, series solutions $u(x, t)$ of the heat equation usually converge quite rapidly, as well as the corresponding series for $u_{xx}(x, t)$ and $u_t(x, t)$, since they also contain the fast decaying negative exponential terms. On the other hand, in the case of the wave equation we find, for example,*

$$u_{xx}(x, t) = -\sum_{n=1}^{\infty} \left(\frac{n\pi}{p}\right)^2 \left[a_n \cos \frac{nc\pi t}{p} + b_n \sin \frac{nc\pi t}{p}\right] \sin \frac{n\pi x}{p}.$$

Because of the factor n^2 in the numerator of this series it may not converge, even though the corresponding series for $u(x, t)$ may indeed converge. It is for this reason that stronger requirements must be placed on the initial data in problems involving the wave equation than those involving the heat equation. Finally, we remark that any discontinuities in the initial data in wave propagation problems will be preserved in the solution $u(x, t)$ for all time. In contrast, such discontinuities are instantly damped out in problems of heat conduction.

EXERCISES 7.2

In Problems 1–4, solve the wave equation subject to the prescribed boundary and initial conditions.

1. $u_{xx} = c^{-2}u_{tt}, \quad 0 < x < 1, \quad t > 0$
   B.C.: $u(0, t) = 0, \quad u(1, t) = 0$
   I.C.: $u(x, 0) = 0, \quad u_t(x, 0) = v_0 \quad$ (v_0 constant)

2. $u_{xx} = c^{-2}u_{tt}, \quad 0 < x < \pi, \quad t > 0$
   B.C.: $u_x(0, t) = 0, \quad u_x(\pi, t) = 0$
   I.C.: $u(x, 0) = 1 - 2\cos 3x, \quad u_t(x, 0) = 5\cos 2x$

3. $u_{xx} = c^{-2}u_{tt}, \quad 0 < x < 1, \quad t > 0$
   B.C.: $u(0, t) = 0, \quad u(1, t) = 0$
   I.C.: $u(x, 0) = x(1 - x), \quad u_t(x, 0) = 0$

4. $u_{xx} = c^{-2}u_{tt}$, $\quad 0 < x < p$, $\quad t > 0$

B.C.: $u(0, t) = 0$, $\quad u(p, t) = 0$

I.C.: $u(x, 0) = p \sinh x - x \sinh p$, $\quad u_t(x, 0) = 0$

5. Determine the relationship between the fundamental angular frequency of a vibrating string to its length p, the tension T in the string, and the mass m per unit length.

6. The initial displacement of a string of length 2 m is $0.1 \sin \pi x$. Assuming the string is released from rest, determine the maximum velocity in the string and state its location.

7. A string π meters long is stretched between fixed supports until the wave speed $c = 40$ m/s. If the string is given an initial velocity of $4 \sin x$ from its equilibrium position, calculate the maximum displacement and state its location.

8. A stretched string of unit length lies along the x-axis with ends fixed at $(0, 0)$ and $(1, 0)$. If the string is initially displaced into the curve $u(x, 0) = A \sin^3 \pi x$, where A is a small constant, and then let go from rest, show that subsequent displacements are given by

$$u(x, t) = \frac{3A}{4} \sin \pi x \cos c\pi t - \frac{A}{4} \sin 3\pi x \cos 3c\pi t.$$

9. A stretched string of length p is "plucked" at the point $x = b$ and its initial shape is described by

$$u(x, 0) = f(x) = \begin{cases} x/b, & 0 < x \leqslant b \\ (p - x)/(p - b), & b < x < p. \end{cases}$$

Assuming fixed supports at $(0, 0)$ and $(p, 0)$, and zero initial velocity, find the subsequent motion of the string.

★10. When a piano wire of length p is struck by a piano hammer near the point $x = b$, the initial velocity of the string is described by

$$u_t(x, 0) = g(x) = \begin{cases} v_0 \cos [\pi(x - b)/\epsilon], & |x - b| < \frac{1}{2}\epsilon \\ 0, & |x - b| \geqslant \frac{1}{2}\epsilon. \end{cases}$$

Assuming fixed ends and no initial displacement, find the subsequent motion of the piano wire. (Assume v_0, ϵ constant.)

★11. The air resistance encountered by a vibrating string is proportional to the velocity of the string. Such conditions lead to the boundary value problem described by

$$u_{tt} = c^2 u_{xx} - 2ku_t, \quad 0 < x < 1, \quad t > 0 \quad (0 < k < \pi c)$$

B.C.: $u(0, t) = 0$, $\quad u(1, t) = 0$

I.C.: $u(x, 0) = f(x)$, $\quad u_t(x, 0) = g(x)$.

Find a solution of the problem for the special case when $f(x) = A \sin \pi x$ (A constant) and $g(x) = 0$.

★12. Find a formal solution of

$$u_{xx} + 2u_x + u = u_{tt}, \quad 0 < x < \pi, \quad t > 0$$

B.C.: $u(0, t) = 0$, $\quad u(\pi, t) = 0$

I.C.: $u(x, 0) = e^{-x}$, $\quad u_t(x, 0) = 0$.

★13. If the mass density of a stretched string of unit length is not constant, i.e., if $\rho = \rho(x)$, and if the tension is given by $T = T(x)$, the governing PDE for the displacements of the string is

$$\frac{\partial}{\partial x}[T(x)u_x] = \rho(x)u_{tt}, \qquad 0 < x < 1, \qquad t > 0.$$

Show that separation of variables with $u(x, t) = X(x)W(t)$ leads to the Sturm-Liouville equation

$$\frac{d}{dx}[T(x)X'] + \lambda\rho(x)X = 0, \qquad 0 < x < 1.$$

★14. Prove that $w(x, t)$, defined by (26), is a strict solution of the boundary value problem (27) provided that g and g' are continuous functions on $[0, p]$, and further that $g(0) = g(p) = 0$.

7.3 FORCED MOTIONS OF A VIBRATING STRING

The motion of an elastic string under the influence of an external stimulus is called *forced*. It is characterized by the *nonhomogeneous equation*

$$u_{xx} = c^{-2}u_{tt} - q(x, t), \qquad 0 < x < p, \qquad t > 0, \tag{35}$$

where $q(x, t)$ is proportional to the external force [see Equation (5)]. The external force may be the result of some driving force exerted on the string, or simply a consequence of taking into account the effect of gravity acting on the string.

Let us assume the boundary conditions are general, but *homogeneous*, so that

$$\text{B.C.:} \qquad \begin{cases} B_1[u] \equiv a_{11}u(0, t) + a_{12}u_x(0, t) = 0, \\ B_2[u] \equiv a_{21}u(p, t) + a_{22}u_x(p, t) = 0, \end{cases} \qquad t > 0 \tag{36}$$

and the initial conditions are

$$\text{I.C.:} \qquad u(x, 0) = f(x), \qquad u_t(x, 0) = g(x), \qquad 0 < x < p. \tag{37}$$

Depending upon the nature of the forcing function $q(x, t)$, we might employ different methods of solution. We will first illustrate a method that is restricted to certain types of forcing functions, and then introduce a more general technique.

7.3.1 SINUSOIDAL FORCING FUNCTIONS

In some cases of interest the forcing function $q(x, t)$ may assume a simple functional form, such as a periodic function described by a sine or cosine function. Let us assume the problem we wish to solve is described by

$$\begin{gathered} u_{xx} = c^{-2}u_{tt} - P(x)\sin\omega t, \qquad 0 < x < p, \qquad t > 0 \\ \text{B.C.:} \qquad B_1[u] = 0, \qquad B_2[u] = 0, \qquad t > 0 \\ \text{I.C.:} \qquad u(x, 0) = f(x), \qquad u_t(x, 0) = g(x), \qquad 0 < x < p, \end{gathered} \tag{38}$$

where $P(x)$ is a known function. Problems of this nature can be solved by using techniques common to ordinary DEs. That is, we begin by assuming the solution consists of two parts,

$$u(x, t) = u_P(x, t) + u_H(x, t), \tag{39}$$

where $u_H(x, t)$, called the *homogeneous solution*, is a solution of

$$\begin{aligned} &u_{xx} = c^{-2}u_{tt}, \\ \text{B.C.:}\quad &B_1[u] = 0, \quad B_2[u] = 0, \end{aligned} \tag{40}$$

and $u_P(x, t)$ is any *particular solution* of

$$\begin{aligned} &u_{xx} = c^{-2}u_{tt} - P(x)\sin \omega t, \\ \text{B.C.:}\quad &B_1[u] = 0, \quad B_2[u] = 0. \end{aligned} \tag{41}$$

REMARK: *The subscript labels in (39) do not refer to partial differentiation in this instance. Since neither P nor H is one of the designated independent variables, such notation should not be a source of confusion.*

The problem described by (40) can be solved by separation of variables. By so doing, we find that

$$u_H(x, t) = \sum_{n=1}^{\infty} (a_n \cos k_n ct + b_n \sin k_n ct)\phi_n(x), \tag{42}$$

where $\lambda_n = k_n^2$ and $\phi_n(x)$ denote the eigenvalues and eigenfunctions of the related eigenvalue problem

$$X'' + \lambda X = 0, \quad B_1[X] = 0, \quad B_2[X] = 0. \tag{43}$$

REMARK: *At this point, the constants a_n and b_n in (42) remain undetermined since the initial conditions prescribed in (38) are not imposed until we have the complete solution (39).*

To find a particular solution of (41), we use a modification of a technique used for solving ODEs called the *method of undetermined coefficients.* We assume there exists a solution form

$$u_P(x, t) = Y(x)\sin \omega t + Z(x)\cos \omega t, \tag{44}$$

where $Y(x)$ and $Z(x)$ are the "undetermined coefficients." The direct substitution of (44) into the PDE in (41) leads to

$$Y'' \sin \omega t + Z'' \cos \omega t = \left[-\left(\frac{\omega}{c}\right)^2 Y - P(x)\right] \sin \omega t - \left(\frac{\omega}{c}\right)^2 Z \cos \omega t,$$

and by equating like coefficients of $\sin \omega t$ and $\cos \omega t$, we get the two ordinary DEs

$$Y'' + \left(\frac{\omega}{c}\right)^2 Y = -P(x) \tag{45}$$

and

$$Z'' + \left(\frac{\omega}{c}\right)^2 Z = 0. \tag{46}$$

Since any particular solution $u_P(x, t)$ suffices, it is convenient to take $Z(x) \equiv 0$. In this case, the homogeneous boundary conditions in (41) reduce to

$$B_1[Y] = 0, \qquad B_2[Y] = 0. \tag{47}$$

Solving (45) subject to (47) is a boundary value problem of the type discussed in Chapter 1. Once $Y(x)$ is found, we combine solutions to get

$$\begin{aligned} u(x, t) &= u_P(x, t) + u_H(x, t) \\ &= Y(x) \sin \omega t + \sum_{n=1}^{\infty} (a_n \cos k_n ct + b_n \sin k_n ct)\phi_n(x). \end{aligned} \tag{48}$$

The initial conditions prescribed in (38) are now imposed upon the solution (48), from which we find

$$f(x) = \sum_{n=1}^{\infty} a_n \phi_n(x) \tag{49}$$

and

$$g(x) = \omega Y(x) + \sum_{n=1}^{\infty} k_n c b_n \phi_n(x),$$

or, upon rearranging this last expression,

$$g(x) - \omega Y(x) = \sum_{n=1}^{\infty} k_n c b_n \phi_n(x). \tag{50}$$

Hence, we deduce that

$$a_n = \| \phi_n(x) \|^{-2} \int_0^p f(x)\phi_n(x)\, dx, \qquad n = 1, 2, 3, \ldots \tag{51}$$

and

$$k_n c b_n = \| \phi_n(x) \|^{-2} \int_0^p [g(x) - \omega Y(x)]\phi_n(x)\, dx, \qquad n = 1, 2, 3, \ldots, \tag{52}$$

where $\| \phi_n(x) \|^2$ is the normalization factor of the eigenfunctions.

EXAMPLE 2

A guy wire stretched tightly between fixed supports at $x = 0$ and $x = \pi$ is initially at rest until a gust of wind comes along. Assuming the wind can be modeled as a simple sinusoidal function of time (constant amplitude) applied normally to the wire, determine the subsequent motions of the wire.

SOLUTION The problem can be mathematically formulated by (P constant)

$$u_{xx} = c^{-2}u_{tt} - P \sin \omega t, \qquad 0 < x < \pi, \qquad t > 0$$

B.C.: $\quad u(0, t) = 0, \quad u(\pi, t) = 0, \quad t > 0$

I.C.: $\quad u(x, 0) = 0, \quad u_t(x, 0) = 0, \quad 0 < x < \pi,$

where $P \sin \omega t$ is proportional to the wind force. The associated eigenvalue problem is

$$X'' + \lambda X = 0, \qquad X(0) = 0, \qquad X(\pi) = 0,$$

with solution $\lambda_n = n^2$ and $\phi_n(x) = \sin nx$, $n = 1, 2, 3, \ldots$. Hence, the homogeneous solution is

$$u_H(x, t) = \sum_{n=1}^{\infty} (a_n \cos nct + b_n \sin nct) \sin nx.$$

To find the particular solution $u_P(x, t)$, we must first solve the boundary value problem

$$Y'' + \left(\frac{\omega}{c}\right)^2 Y = -P, \qquad Y(0) = 0, \qquad Y(\pi) = 0.$$

Writing $k = \omega/c$, the solution of this problem is readily found to be

$$Y(x) = \frac{P}{k^2}(\cos kx - 1) + \frac{P(1 - \cos k\pi) \sin kx}{k^2 \sin k\pi},$$

where k is restricted such that $k \neq 1, 2, 3, \ldots$. Upon combining $u_P(x, t)$ and $u_H(x, t)$, we have

$$u(x, t) = Y(x) \sin \omega t + \sum_{n=1}^{\infty} (a_n \cos nct + b_n \sin nct) \sin nx.$$

Finally, imposing the homogeneous initial conditions on $u(x, t)$, we find that

$$u(x, 0) = 0 = \sum_{n=1}^{\infty} a_n \sin nx$$

and

$$u_t(x, 0) = 0 = \omega Y(x) + \sum_{n=1}^{\infty} ncb_n \sin nx.$$

Thus, clearly $a_n = 0$ for all n, and*

$$ncb_n = -\frac{2\omega}{\pi}\int_0^{\pi} Y(x) \sin nx \, dx = \begin{cases} -\dfrac{4\omega P}{n\pi(n^2 - k^2)}, & n = 1, 3, 5, \ldots \\ 0, & n = 2, 4, 6, \ldots. \end{cases}$$

---

*This result for b_n is obtained only after considerable algebraic manipulation.

The solution we seek is therefore (now replacing k by ω/c)

$$u(x, t) = \frac{4Pc\omega}{\pi} \sum_{\substack{n=1 \\ (\text{odd})}}^{\infty} \frac{\sin nct \sin nx}{n^2(\omega^2 - n^2c^2)} + \frac{Pc^2 \sin \omega t}{\omega^2}[\cos(\omega x/c) - 1]$$

$$+ \frac{Pc^2 \sin(\omega x/c) \sin \omega t}{\omega^2 \sin(\omega\pi/c)}[1 - \cos(\omega\pi/c)],$$

where $\omega \neq nc$, $n = 1, 3, 5, \ldots$.

If the input frequency ω in Example 2 is one of the values

$$\omega_n = nc, \qquad n = 1, 3, 5, \ldots,$$

the amplitude of the resulting motion of the guy wire becomes unbounded, and *resonance* is said to occur. Of course, we recognize that in any physical problem the amplitude cannot be unbounded. Although not shown in our model, a certain amount of damping is always present. This has the effect of limiting the amplitude. But if the amplitude should become too large, the guy wire may snap. In our particular example, this can happen if the frequency of the wind is $\omega_1 = c$ or any odd harmonic of this frequency.

Resonance has played a deleterious role in the collapse of bridges. For example, on November 7, 1940, only four months after its grand opening, the Tacoma Narrows bridge at Puget Sound in the state of Washington collapsed from the large undulations caused by the wind blowing across the superstructure. Because of such possibilities, it is an important aspect of design that the natural frequency of the structure be different from the frequency of any probable forcing function.

7.3.2 EIGENFUNCTION EXPANSION METHOD

Another approach of solving the nonhomogeneous problem is called for when the forcing function $q(x, t)$ is of a more general nature. Again assuming the boundary conditions are *homogeneous,* we now consider the problem described by

$$u_{xx} = c^{-2}u_{tt} - q(x, t), \qquad 0 < x < p, \qquad t > 0$$

$$\text{B.C.:} \qquad B_1[u] = 0, \qquad B_2[u] = 0, \qquad t > 0 \tag{53}$$

$$\text{I.C.:} \qquad u(x, 0) = f(x), \qquad u_t(x, 0) = g(x), \qquad 0 < x < p.$$

Following the *generalized eigenfunction expansion method* used in Section 6.3.2, we assume that (53) has a solution of the form

$$u(x, t) = \sum_{n=1}^{\infty} E_n(t)\phi_n(x), \tag{54}$$

where the ϕ's are eigenfunctions of the associated eigenvalue problem

$$X'' + \lambda X = 0, \qquad B_1[X] = 0, \qquad B_2[X] = 0, \tag{55}$$

and the E's represent the unknowns. By writing the PDE in the form

$$c^2 q(x, t) = u_{tt} - c^2 u_{xx} \tag{56}$$

and substituting (54) directly into (56), we get

$$c^2 q(x, t) = \sum_{n=1}^{\infty} [E_n''(t)\phi_n(x) - c^2 E_n(t)\phi_n''(x)]$$

or, since $\phi_n''(x) = -\lambda_n \phi_n(x) = -k_n^2 \phi_n(x)$, we have

$$c^2 q(x, t) = \sum_{n=1}^{\infty} [E_n''(t) + c^2 k_n^2 E_n(t)]\phi_n(x). \tag{57}$$

If we think of t as fixed, then (57) is simply a generalized Fourier series of the function $c^2 q(x, t)$. The "coefficients" are therefore

$$\begin{aligned} E_n''(t) + c^2 k_n^2 E_n(t) &= c^2 \|\phi_n(x)\|^{-2} \int_0^p q(x, t)\phi_n(x)\, dx \\ &= c^2 Q_n(t), \qquad n = 1, 2, 3, \ldots . \end{aligned} \tag{58}$$

For each n, Equation (58) is a time-varying nonhomogeneous ODE whose general solution in terms of the *one-sided Green's function** can be expressed as (assuming $k_n \neq 0$ for all n)

$$\begin{aligned} E_n(t) = \frac{c}{k_n} \int_0^t \sin[k_n c(t - \tau)] Q_n(\tau)\, d\tau + a_n \cos k_n ct \\ + b_n \sin k_n ct, \qquad n = 1, 2, 3, \ldots , \end{aligned} \tag{59}$$

where a_n and b_n represent arbitrary constants. The substitution of (59) into (54) leads to the solution formula

$$\begin{aligned} u(x, t) = \sum_{n=1}^{\infty} \left\{ \frac{c}{k_n} \int_0^t \sin[k_n c(t - \tau)] Q_n(\tau)\, d\tau \right. \\ \left. + a_n \cos k_n ct + b_n \sin k_n ct \right\} \phi_n(x), \end{aligned} \tag{60}$$

where [from (58)]

$$Q_n(t) = \|\phi_n(x)\|^{-2} \int_0^p q(x, t)\phi_n(x)\, dx, \qquad n = 1, 2, 3, \ldots \tag{61}$$

and

$$\|\phi_n(x)\|^2 = \int_0^p [\phi_n(x)]^2\, dx, \qquad n = 1, 2, 3, \ldots . \tag{62}$$

---

* See Section 2.2.

Imposing the initial conditions in (53) upon the solution (60), we see that

$$u(x, 0) = f(x) = \sum_{n=1}^{\infty} a_n \phi_n(x) \tag{63}$$

and

$$u_t(x, 0) = g(x) = \sum_{n=1}^{\infty} k_n c b_n \phi_n(x), \tag{64}$$

from which we deduce

$$a_n = \| \phi_n(x) \|^{-2} \int_0^p f(x)\phi_n(x)\, dx, \qquad n = 1, 2, 3, \ldots \tag{65}$$

and

$$k_n c b_n = \| \phi_n(x) \|^{-2} \int_0^p g(x)\phi_n(x)\, dx, \qquad n = 1, 2, 3, \ldots . \tag{66}$$

REMARK: *As in Section 6.3.2 we remind the reader that if* $\lambda_0 = 0$ *and* $\phi_0(x)$ *form an eigenvalue–eigenfunction pair, the solution form (54) must be modified to read*

$$u(x, t) = E_0(t)\phi_0(x) + \sum_{n=1}^{\infty} E_n(t)\phi_n(x).$$

REMARK: *As a final comment we might point out that it is not necessary to resort to the Green's function method in solving (58). Any solution technique will work; however, this may lead to different expressions for the constants* a_n *and* b_n *than derived above.*

EXAMPLE 3

Solve Example 2 by the eigenfunction expansion method, i.e., solve

$$u_{xx} = c^{-2}u_{tt} - P \sin \omega t, \qquad 0 < x < \pi, \qquad t > 0$$

B.C.: $\quad u(0, t) = 0, \quad u(\pi, t) = 0, \quad t > 0$

I.C.: $\quad u(x, 0) = 0, \quad u_t(x, 0) = 0, \quad 0 < x < \pi.$

SOLUTION The associated eigenvalue problem

$$X'' + \lambda X = 0, \qquad X(0) = 0, \qquad X(\pi) = 0$$

has solutions

$$\lambda_n = n^2, \qquad \phi_n(x) = \sin nx, \qquad n = 1, 2, 3, \ldots .$$

Thus, by writing the PDE in the form

$$Pc^2 \sin \omega t = u_{tt} - c^2 u_{xx}$$

and substituting the derivatives of

$$u(x, t) = \sum_{n=1}^{\infty} E_n(t) \sin nx$$

into the PDE, we get (after simplification)

$$Pc^2 \sin \omega t = \sum_{n=1}^{\infty} [E_n''(t) + n^2c^2E_n(t)] \sin nx.$$

The unknown E's are, therefore, solutions of the DE

$$\begin{aligned} E_n''(t) + n^2c^2E_n(t) &= \frac{2Pc^2}{\pi} \sin \omega t \int_0^{\pi} \sin nx \, dx \\ &= \begin{cases} \dfrac{4Pc^2}{n\pi} \sin \omega t, & n = 1, 3, 5, \ldots \\ 0, & n = 2, 4, 6, \ldots, \end{cases} \end{aligned}$$

from which we deduce

$$E_n(t) = \begin{cases} \dfrac{4Pc}{\pi n^2} \displaystyle\int_0^t \sin[nc(t - \tau)] \sin \omega\tau \, d\tau + a_n \cos nct \\ \qquad\qquad + b_n \sin nct, & n = 1, 3, 5, \ldots \\ a_n \cos nct + b_n \sin nct, & n = 2, 4, 6, \ldots, \end{cases}$$

and, consequently,

$$\begin{aligned} u(x, t) = {} & \frac{4Pc}{\pi} \sum_{\substack{n=1 \\ (\text{odd})}}^{\infty} \frac{\sin nx}{n^2} \int_0^t \sin[nc(t - \tau)] \sin \omega\tau \, d\tau \\ & + \sum_{n=1}^{\infty} (a_n \cos nct + b_n \sin nct) \sin nx. \end{aligned}$$

Because the prescribed initial conditions are homogeneous, we find that $a_n = b_n = 0$, $n = 1, 2, 3, \ldots$, and the above solution reduces to (upon evaluating the integral)

$$u(x, t) = \frac{4Pc}{\pi} \sum_{\substack{n=1 \\ (\text{odd})}}^{\infty} \frac{\sin nx}{n^2(\omega^2 - n^2c^2)} (\omega \sin nct - nc \sin \omega t),$$

where it is assumed that $\omega \neq nc$ $(n = 1, 3, 5, \ldots)$.*

---

* Recall the discussion immediately following Example 2.

□ 7.3.3 NONHOMOGENEOUS BOUNDARY CONDITIONS

In the procedures for problems with nonhomogeneous PDE, we made the assumption that the boundary conditions were homogeneous. Although homogeneous boundary conditions are perhaps the most common type of boundary conditions associated with the wave equation, in some situations the prescribed conditions are *nonhomogeneous*. For example, the nonhomogeneous condition $u(0, t) = \alpha(t)$ in connection with the vibrating string problem physically describes a certain motion of the string at the endpoint $x = 0$. In studying the longitudinal displacements of an elastic bar the boundary condition $u_x(0, t) = \alpha(t)$ might represent a prescribed force in the direction of the u-axis. Other nonhomogeneous conditions are also physically possible, but in order to illustrate a procedure and retain some mathematical simplicity let us assume the problem we wish to solve is given by

$$u_{xx} = c^{-2}u_{tt} - q(x, t), \qquad 0 < x < p, \qquad t > 0$$

$$\text{B.C.:} \qquad B_1[u] = \alpha, \qquad B_2[u] = \beta, \qquad t > 0 \tag{67}$$

$$\text{I.C.:} \qquad u(x, 0) = f(x), \qquad u_t(x, 0) = g(x), \qquad 0 < x < p,$$

where α and β are *constants*.

Guided to some extent by a similar problem involving heat conduction, we choose to write the solution of (67) as the sum*

$$u(x, t) = S(x) + v(x, t), \tag{68}$$

where $S(x)$ is selected to satisfy the time-independent problem†

$$S'' = 0, \qquad B_1[S] = \alpha, \qquad B_2[S] = \beta, \tag{69}$$

and $v(x, t)$, being the difference between $u(x, t)$ and $S(x)$, must then satisfy the problem

$$v_{xx} = c^{-2}v_{tt} - q(x, t), \qquad 0 < x < p, \qquad t > 0$$

$$\text{B.C.:} \qquad B_1[v] = 0, \qquad B_2[v] = 0, \qquad t > 0 \tag{70}$$

$$\text{I.C.:} \qquad v(x, 0) = f(x) - S(x), \qquad v_t(x, 0) = g(x), \qquad 0 < x < p.$$

The boundary value problem (69) can readily be solved by methods discussed in Chapter 1 while (70) is of the variety treated in Section 7.3.2 (i.e., homogeneous boundary conditions).

Problems involving time-dependent boundary conditions are taken up in Exercises 7.3 (see Problems 12–14).

---

* By analogy with the method illustrated in Section 6.2.2, we can identify the function $S(x)$ with the *steady-state solution* and $v(x, t)$ with the *transient solution*. A major distinction in this case, however, is that $v(x, t)$ will not die out for large t. Thus, these solutions are not truly steady-state and transient solutions of the problem described by (67).

† In some cases there may not be a solution of (69). See the first footnote on p. 256 in connection with heat conduction problems.

EXERCISES 7.3

In Problems 1–3, solve by the method of Section 7.3.1.

1. $u_{xx} = c^{-2}u_{tt} - P\cos\omega t, \quad 0 < x < \pi, \quad t > 0 \quad (P,\ \omega \text{ constants})$
   B.C.: $u(0, t) = 0, \quad u(\pi, t) = 0$
   I.C.: $u(x, 0) = 0, \quad u_t(x, 0) = 0$

2. $u_{xx} = u_{tt} - 1, \quad 0 < x < 1, \quad t > 0$
   B.C.: $u_x(0, t) = 0, \quad u_x(1, t) = 0$
   I.C.: $u(x, 0) = 1, \quad u_t(x, 0) = 0$

3. $u_{xx} = u_{tt} - \sin x \cos t, \quad 0 < x < \dfrac{\pi}{2}, \quad t > 0$
   B.C.: $u(0, t) = 0, \quad u\left(\dfrac{\pi}{2}, t\right) = 0$
   I.C.: $u(x, 0) = 0, \quad u_t(x, 0) = 0$

In Problems 4–7, solve by the method of Section 7.3.2.

4. The wind is blowing over a suspension cable on a bridge that is 15 m long. The distributed force caused by the wind is approximated by the function $0.02\sin(21\pi t)$. Assume the tension in the cable is 40,000 N and the cable has a mass of 10 kg/m. Determine the subsequent motion. Will resonance occur? Explain.

5. Do Problem 1.

6. $u_{xx} = u_{tt} - 1, \quad 0 < x < 1, \quad t > 0$
   B.C.: $u(0, t) = 0, \quad u(1, t) = 0$
   I.C.: $u(x, 0) = x(1 - x), \quad u_t(x, 0) = 0$

7. $u_{xx} = u_{tt} - kx, \quad 0 < x < 1, \quad t > 0 \quad (k > 0)$
   B.C.: $u(0, t) = 0, \quad u(1, t) = 0$
   I.C.: $u(x, 0) = 0, \quad u_t = V_0 \quad \text{(constant)}$

In Problems 8–10, solve by the method of Section 7.3.3.

8. $u_{xx} = c^{-2}u_{tt}, \quad 0 < x < \pi, \quad t > 0$
   B.C.: $u(0, t) = 1, \quad u(\pi, t) = 3$
   I.C.: $u(x, 0) = 1, \quad u_t(x, 0) = 0$

9. $u_{xx} = u_{tt} - 10, \quad 0 < x < 1, \quad t > 0$
   B.C.: $u(0, t) = 2, \quad u_x(1, t) = 0$
   I.C.: $u(x, 0) = 2, \quad u_t(x, 0) = \sin\dfrac{3\pi x}{2}$

10. $u_{xx} = u_{tt} - x^2, \quad 0 < x < 1, \quad t > 0$
    B.C.: $u(0, t) = 0, \quad u(1, t) = 1$
    I.C.: $u(x, 0) = x, \quad u_t(x, 0) = 0$

11. A bar of length p with fixed endpoint $x = 0$ is initially at rest. If a constant force F_0 per unit area is applied longitudinally at the free end, the problem is characterized by

$$u_{xx} = c^{-2}u_{tt}, \qquad 0 < x < p, \qquad t > 0$$

B.C.: $u(0, t) = 0, \quad Eu_x(p, t) = F_0$

I.C.: $u(x, 0) = 0, \quad u_t(x, 0) = 0.$

Show that the longitudinal displacements are given by

$$u(x, t) = \frac{F_0}{E}\left\{x + \frac{8p}{\pi^2}\sum_{n=1}^{\infty}\frac{(-1)^n}{(2n-1)^2}\sin\left[\left(n - \frac{1}{2}\right)\frac{\pi x}{p}\right]\cos\left[\left(n - \frac{1}{2}\right)\frac{\pi ct}{p}\right]\right\}.$$

In Problems 12–14, use the method of Section 6.3.3 to find solutions.

★12. $u_{xx} = u_{tt}, \qquad 0 < x < 1, \qquad t > 0$

B.C.: $u(0, t) = 0, \quad u(1, t) = 1 + t$

I.C.: $u(x, 0) = x, \quad u_t(x, 0) = 0$

★13. $u_{xx} = u_{tt} - 1, \qquad 0 < x < 1, \qquad t > 0$

B.C.: $u(0, t) = \frac{1}{2}t^2, \quad u(1, t) = -\cos t$

I.C.: $u(x, 0) = -x, \quad u_t(x, 0) = 0$

★14. $u_{xx} = u_{tt}, \qquad 0 < x < 1, \qquad t > 0$

B.C.: $u(0, t) = \sin t, \quad u(1, t) = 0$

I.C.: $u(x, 0) = 0, \quad u_t(x, 0) = 1 - x$

★15. Show that the solution of

$$u_{xx} - u_{tt} = -\delta(x - s)\delta(t - \tau), \qquad 0 < x < \pi, \qquad t > 0$$

B.C.: $u(0, t) = 0, \quad u(\pi, t) = 0$

I.C.: $u(x, 0) = 0, \quad u_t(x, 0) = 0,$

where δ denotes the delta function (Section 2.3), leads to

$$u(x, t) = \begin{cases} 0, & t < \tau \\ \dfrac{2}{\pi}\displaystyle\sum_{n=1}^{\infty}\frac{\sin nx \sin ns}{n}\sin[n(t - \tau)], & t > \tau. \end{cases}$$

[This solution can be interpreted as a bilinear formula for a type of Green's function $g(x, s; t, \tau)$ for the wave equation.]

□ 7.4 INFINITE DOMAINS AND FOURIER INTEGRALS

In certain settings it is convenient to imagine our vibrating string to be of infinite extent. Our reasons for doing so are basically the same as those stated in Chapter 6 in connection with heat conduction in an infinite domain. Also, the technique of Fourier integrals applied to the wave equation is virtually the same as that for the heat equation. Hence, our treatment here of wave problems is intentionally terse.

7.4.1 A SEMI-INFINITE STRING

Let us consider the transverse deflections of a semi-infinite string which is given an initial deflection $f(x)$ and initial velocity $g(x)$. In the absence of any external forces, the problem is described by

$$u_{xx} = c^{-2}u_{tt}, \qquad 0 < x < \infty, \qquad t > 0$$

$$\text{B.C.:} \quad u(0, t) = 0, \quad u \text{ and } u_x \text{ finite as } x \to \infty, \quad t > 0 \tag{71}$$

$$\text{I.C.:} \quad u(x, 0) = f(x), \quad u_t(x, 0) = g(x), \quad 0 < x < \infty.$$

In this setting we assume that f and g are piecewise smooth and absolutely integrable.

By separating variables, we find

$$X'' + \lambda X = 0,$$

$$\text{B.C.:} \quad X(0) = 0, \quad X \text{ and } X' \text{ finite as } x \to \infty \tag{72}$$

and

$$W'' + \lambda c^2 W = 0. \tag{73}$$

Bounded solutions of (72) vanishing at $x = 0$ are given by

$$X(x) = \sin sx, \qquad s > 0 \tag{74}$$

($\lambda = s^2$) and the corresponding solutions of (73) are, therefore,

$$W(t) = A(s) \cos sct + B(s) \sin sct, \qquad s > 0 \tag{75}$$

where $A(s)$ and $B(s)$ are arbitrary functions. Combining solutions by the superposition principle leads to

$$u(x, t) = \int_0^\infty [A(s) \cos sct + B(s) \sin sct] \sin sx \, ds. \tag{76}$$

To determine $A(s)$ and $B(s)$, we first set $t = 0$ in (76) to get

$$u(x, 0) = f(x) = \int_0^\infty A(s) \sin sx \, ds, \tag{77}$$

from which we deduce

$$A(s) = \frac{2}{\pi} \int_0^\infty f(x) \sin sx \, dx \tag{78}$$

(see Section 4.7). Also, by differentiating (76), we find

$$u_t(x, 0) = g(x) = \int_0^\infty scB(s) \sin sx \, ds \tag{79}$$

where

$$scB(s) = \frac{2}{\pi} \int_0^\infty g(x) \sin sx \, dx. \tag{80}$$

EXERCISES 7.4

1. Solve the problem described by (71) when $f(x) = 0$ and $g(x) = e^{-x}$.

2. By using the identity

$$\sin A \cos B = \frac{1}{2}[\sin(A + B) + \sin(A - B)],$$

show for the special case when $g(x) \equiv 0$ that Equation (76) can be expressed in the form

$$u(x, t) = \frac{1}{2}[f_0(x + ct) + f_0(x - ct)],$$

where f_0 denotes the odd extension of f over the entire axis.

3. Find a solution formula similar to (76) for the vibrating string problem

$$u_{xx} = c^{-2}u_{tt}, \qquad 0 < x < \infty, \qquad t > 0$$

B.C.: $u_x(0, t) = 0, \quad u$ and u_x finite as $x \to \infty$.
I.C.: $u(x, 0) = f(x), \quad u_t(x, 0) = g(x)$.

4. Find a solution formula similar to (76) for the vibrating string problem on the infinite domain

$$u_{xx} = c^{-2}u_{tt}, \qquad -\infty < x < \infty, \qquad t > 0$$

B.C.: u and u_x finite as $|x| \to \infty$
I.C.: $u(x, 0) = f(x), \quad u_t(x, 0) = g(x)$.

5. If $g(x) \equiv 0$ in Problem 4, show that the solution so obtained can be put in the form

$$u(x, t) = \frac{1}{2}[f(x - ct) + f(x + ct)].$$

6. Solve Problem 4 when $f(x) = e^{-|x|}$ and $g(x) \equiv 0$.

7. Solve Problem 4 when $f(x) \equiv 0$ and $g(x) = e^{-|x|}$.

□ 7.5 METHOD OF INTEGRAL TRANSFORMS

The integral transforms of Laplace and Fourier can also be used to solve the wave equation on infinite domains. The technique is basically the same as for the heat equation.

7.5.1 LAPLACE TRANSFORM

Let $u(x, t)$ denote the transverse displacement of a semi-infinite stretched string, one end of which is fixed far out on the x-axis and the other end looped around the u-axis. The string is presumed to be at rest initially with the looped end later moved in some prescribed manner along the u-axis, i.e., $u(0, t) = f(t)$. Because the string is initially at rest, it follows that $f(0) = 0$. If no external forces are acting on the string, the above conditions are characterized by the boundary value problem

$$u_{xx} = c^{-2}u_{tt}, \qquad 0 < x < \infty, \qquad t > 0$$

$$\text{B.C.:} \qquad u(0, t) = f(t), \qquad u(x, 0) \to 0 \quad \text{as} \quad x \to \infty, \qquad t > 0 \tag{81}$$

$$\text{I.C.:} \qquad u(x, 0) = 0, \qquad u_t(x, 0) = 0, \qquad 0 < x < \infty.$$

If we let $\mathscr{L}\{u(x, t); t \to p\} = U(x, p)$ and $\mathscr{L}\{f(t); p\} = F(p)$, the transformed problem reads

$$U_{xx} - \left(\frac{p}{c}\right)^2 U = 0, \qquad 0 < x < \infty \tag{82}$$

$$\text{B.C.:} \qquad U(0, p) = F(p), \qquad U(x, p) \to 0 \quad \text{as} \quad x \to \infty.$$

Treating p as a positive constant, the general solution of (82) is

$$U(x, p) = A(p)e^{-(x/c)p} + B(p)e^{(x/c)p}, \tag{83}$$

where $A(p)$ and $B(p)$ are arbitrary functions of p. We choose $B(p) = 0$ so that $U(x, p)$ will vanish for large x, and hence $A(p) = F(p)$. Thus,

$$U(x, p) = F(p)e^{-(x/c)p}. \tag{84}$$

The *translation property* of the Laplace transform enables us to invert (84) directly, from which we get*

$$u(x, t) = f\left(t - \frac{x}{c}\right)h\left(t - \frac{x}{c}\right) = \begin{cases} 0, & t \leq x/c \\ f(t - x/c), & t > x/c. \end{cases} \tag{85}$$

Based on physical considerations, $f(t)$ must be a continuous function. Since $f(0) = 0$, it follows that our solution (85) is likewise a continuous function. The interpretation of this solution is that a point on the string x units from the origin remains at rest until time $t = x/c$, and then it executes the same motion as the loop at the u-axis (see Figure 7.6).

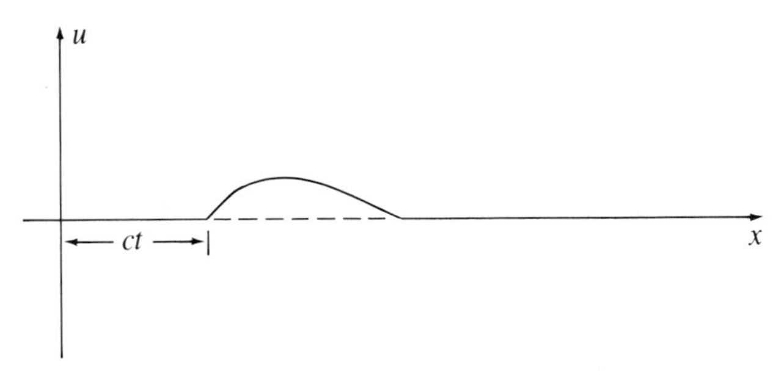

FIGURE 7.6

* See Problem 51 in Exercises 5.3.

7.5.2 FOURIER TRANSFORMS

To illustrate the use of the Fourier transform, we seek the transverse deflections of an infinite string which is given an initial displacement $f(x)$ and initial velocity $g(x)$. If no external forces are present, the problem is described by

$$
\begin{aligned}
&u_{xx} = c^{-2}u_{tt}, \qquad -\infty < x < \infty, \qquad t > 0 \\
\text{B.C.:} \quad &u(x, t) \to 0, \qquad u_x(x, t) \to 0 \quad \text{as} \quad |x| \to \infty, \qquad t > 0 \\
\text{I.C.:} \quad &u(x, 0) = f(x), \qquad u_t(x, 0) = g(x), \qquad -\infty < x < \infty.
\end{aligned}
\tag{86}
$$

The infinite extent on x suggests use of the *Fourier exponential transform*. Hence, by introducing $\mathscr{F}\{u(x, t); x \to s\} = U(s, t)$, $\mathscr{F}\{f(x); s\} = F(s)$, and $\mathscr{F}\{g(x); s\} = G(s)$, we get the transformed problem

$$
\begin{aligned}
&U_{tt} + c^2s^2U = 0, \qquad t > 0 \\
\text{I.C.:} \quad &U(s, 0) = F(s), \qquad U_t(s, 0) = G(s).
\end{aligned}
\tag{87}
$$

By standard solution techniques, we find

$$U(s, t) = F(s)\cos cst + \frac{G(s)}{cs}\sin cst, \tag{88}$$

the inverse transform of which leads to the integral representation

$$u(x, t) = \frac{1}{\sqrt{2\pi}}\int_{-\infty}^{\infty} e^{-isx}\left[F(s)\cos cst + \frac{G(s)}{cs}\sin cst\right] ds. \tag{89}$$

Although (89) is a formal solution of (86), an interesting result emerges if we choose to write the sine and cosine appearing in (89) in terms of complex exponentials by use of the Euler formulas

$$\cos x = \frac{1}{2}(e^{ix} + e^{-ix}), \qquad \sin x = \frac{1}{2i}(e^{ix} - e^{-ix}).$$

Making the appropriate substitutions, we get

$$
\begin{aligned}
u(x, t) = {} & \frac{1}{2\sqrt{2\pi}}\int_{-\infty}^{\infty} [e^{-is(x-ct)} + e^{-is(x+ct)}]F(s)\, ds \\
& + \frac{1}{2\sqrt{2\pi}}\int_{-\infty}^{\infty} [e^{-is(x-ct)} - e^{-is(x+ct)}]\frac{G(s)}{ics}\, ds.
\end{aligned}
\tag{90}
$$

By recalling the inverse transform relation

$$f(x) = \frac{1}{\sqrt{2\pi}}\int_{-\infty}^{\infty} e^{-isx}F(s)\, ds,$$

the first integral in (90) is recognized as

$$\frac{1}{2\sqrt{2\pi}}\int_{-\infty}^{\infty} [e^{-is(x-ct)} + e^{-is(x+ct)}]F(s)\, ds = \frac{1}{2}[f(x - ct) + f(x + ct)]. \tag{91}$$

The second integral can be similarly identified if we start with the inverse transform relation

$$g(z) = \frac{1}{\sqrt{2\pi}} \int_{-\infty}^{\infty} e^{-isz} G(s)\, ds$$

and integrate both sides from $x - ct$ to $x + ct$. Hence,

$$\int_{x-ct}^{x+ct} g(z)\, dz = \frac{1}{\sqrt{2\pi}} \int_{-\infty}^{\infty} [e^{-is(x-ct)} - e^{-is(x+ct)}] \frac{G(s)}{ics}\, ds, \tag{92}$$

and we deduce that (90) is equivalent to

$$u(x, t) = \frac{1}{2}[f(x - ct) + f(x + ct)] + \frac{1}{2c} \int_{x-ct}^{x+ct} g(z)\, dz, \tag{93}$$

which is *d'Alembert's solution* of the wave equation once again (see Section 3.2.2).

REMARK: *Generally speaking, because of d'Alembert's solution we do not use integral transforms to solve the one-dimensional wave equation on infinite domains. They are introduced here primarily as an alternate and novel way of solving the wave equation. However, integral transform methods can be quite useful in solving the wave equation in higher dimensions where a d'Alembert type of solution does not exist, or in solving related problems of vibration concerning elastic bars where the governing PDE is fourth order (see Problems 14 and 15), and so on.*

EXERCISES 7.5

In Problems 1–9, use the Laplace transform to solve the problem.

1. Show that

$$u(x, t) = (t - x)h(t - x),$$

where h is the Heaviside unit function, is the solution of

$$u_{xx} = u_{tt}, \qquad 0 < x < \infty, \qquad t > 0$$

B.C.: $u_x(0, t) = -1, \quad u(x, t) \to 0 \text{ as } x \to \infty$

I.C.: $u(x, 0) = 0, \quad u_t(x, 0) = 0.$

2. Given the boundary value problem

$$u_{xx} = u_{tt}, \qquad 0 < x < \infty, \qquad t > 0$$

B.C.: $u_x(0, t) = f(t), \quad u(x, t) \to 0 \text{ as } x \to \infty$

I.C.: $u(x, 0) = 0, \quad u_t(x, 0) = 0,$

(a) show that $u(x, t) = g(t - x)$, where

$$g(z) = \begin{cases} -\int_0^z f(\tau)\, d\tau, & z \geq 0 \\ 0, & z < 0. \end{cases}$$

(b) Determine $u(0, t)$.

3. Solve Problem 2 when
   (a) $f(t) = t$.
   (b) $f(t) = A \sin \omega t$, $(A, \omega$ constants).

4. Consider the motions of a string fastened at the origin but whose far end is looped around a frictionless peg that exerts no vertical force on the loop. The string is initially supported at rest along the x-axis and is released at time $t = 0$, moving downward under the action of gravity. Determine the subsequent displacements given that the problem is characterized by

$$c^2 u_{xx} = u_{tt} - g, \qquad 0 < x < \infty, \qquad t > 0 \qquad (g \text{ constant})$$

B.C.: $u(0, t) = 0, \quad u(x, t) \to 0$ as $x \to \infty$
I.C.: $u(x, 0) = 0, \quad u_t(x, 0) = 0.$

5. Show that the boundary value problem

$$u_{xx} = u_{tt}, \qquad 0 < x < \infty, \qquad t > 0$$

B.C.: $u_x(0, t) = 0, \quad u(x, t) \to 0$ as $x \to \infty$
I.C.: $u(x, 0) = e^{-x}, \quad u_t(x, 0) = 0,$

has the solution

$$u(x, t) = \begin{cases} e^{-t} \cosh x, & x < t \\ e^{-x} \cosh t, & x > t. \end{cases}$$

6. Show that the boundary value problem

$$u_{xx} = c^{-2} u_{tt} - A \sin \pi x, \qquad 0 < x < 1, \qquad t > 0$$

B.C.: $u(0, t) = 0, \quad u(1, t) = 0$
I.C.: $u(x, 0) = 0, \quad u_t(x, 0) = 0,$

has the solution (A constant)

$$u(x, t) = \frac{A}{\pi^2}(1 - \cos \pi ct) \sin \pi x.$$

7. $u_{xx} = c^{-2} u_{tt}, \qquad 0 < x < 1, \qquad t > 0$

B.C.: $u(0, t) = 0, \quad u(1, t) = 1$
I.C.: $u(x, 0) = 0, \quad u_t(x, 0) = 0.$

Hint: $\mathscr{L}^{-1}\left\{\frac{\sinh xp}{p \sinh ap}; p \to t\right\} = \frac{x}{a} + \frac{2}{\pi} \sum_{n=1}^{\infty} \frac{(-1)^n}{n} \sin(n\pi x/a) \cos(n\pi t/a).$

8. $u_{xx} = c^{-2} u_{tt}, \qquad 0 < x < 1, \qquad t > 0$

B.C.: $u_x(0, t) = 0, \quad u(1, t) = 1$
I.C.: $u(x, 0) = 0, \quad u_t(x, 0) = 0$

Hint: $\mathscr{L}^{-1}\left\{\dfrac{\cosh xp}{p\cosh ap}; p\to t\right\}$

$$= 1 + \frac{4}{\pi}\sum_{n=1}^{\infty}\frac{(-1)^n}{2n-1}\cos\left[\left(n-\frac{1}{2}\right)\frac{\pi x}{a}\right]\cos\left[\left(n-\frac{1}{2}\right)\frac{\pi t}{a}\right]$$

★9. Given the nonhomogeneous boundary value problem

$$u_{xx} = u_{tt} - q(x, t), \qquad 0 < x < 1, \qquad t > 0$$

B.C.: $u(0, t) = 0, \qquad u(1, t) = 0$

I.C.: $u(x, 0) = 0, \qquad u_t(x, 0) = 0,$

(a) show that the solution of the transformed problem is

$$U(x, p) = \int_0^1 G(x, s; p)Q(s, p)\, ds,$$

where $Q(x, p)$ is the Laplace transform of $q(x, t)$ and $G(x, s; p)$ is the Green's function (see Section 2.4) defined by

$$G(x, s; p) = \begin{cases} \dfrac{\sinh px \sinh p(1-x)}{p \sinh p}, & 0 < s \leqslant x \\ \dfrac{\sinh ps \sinh p(1-s)}{p \sinh p}, & x < s < 1. \end{cases}$$

(b) Express the Green's function in (a) in a bilinear formula and use the convolution theorem of Laplace to deduce that

$$u(x, t) = \frac{1}{\pi}\sum_{n=1}^{\infty}\frac{1}{n}\int_0^t \sin[n\pi(t-\tau)]Q_n(\tau)\, d\tau,$$

where

$$Q_n(t) = 2\int_0^1 q(x, t)\sin(n\pi x)\, dx.$$

10. Use the Fourier sine transform to find an integral solution of

$$u_{xx} = c^{-2}u_{tt}, \qquad 0 < x < \infty, \qquad t > 0$$

B.C.: $u(0, t) = 0, \qquad u(x, t) \to 0 \;\text{ as }\; x \to \infty$

I.C.: $u(x, 0) = f(x), \qquad u_t(x, 0) = g(x).$

11. Use the Fourier cosine transform to find an integral solution of

$$u_{xx} = c^{-2}u_{tt}, \qquad 0 < x < \infty, \qquad t > 0$$

B.C.: $u_x(0, t) = 0, \qquad u(x, t) \to 0 \;\text{ as }\; x \to \infty$

I.C.: $u(x, 0) = f(x), \qquad u_t(x, 0) = g(x).$

12. Solve Problem 1 by the Fourier cosine transform.

Hint: $\mathscr{F}_C\{(1-x)h(1-x); s\} = \sqrt{\dfrac{2}{\pi}}\, s^{-2}(1-\cos s)$

13. Using (89), find an integral solution of the problem described by (86) when $f(x) \equiv 0$ and $g(x) = e^{-|x|}$.

★14. The free vibrations of an infinite elastic bar initially distorted in the shape $f(x)$ with zero velocity are governed by

$$u_{xxxx} + u_{tt} = 0, \qquad -\infty < x < \infty, \qquad t > 0$$

B.C.: $u(x, t) \to 0, \quad u_x(x, t) \to 0 \quad \text{as } |x| \to \infty$

I.C.: $u(x, 0) = f(x), \quad u_t(x, 0) = 0.$

If $F(s)$ denotes the Fourier transform of $f(x)$, show that

$$u(x, t) = \frac{1}{\sqrt{2\pi}} \int_{-\infty}^{\infty} e^{-isx} F(s) \cos(ts^2)\, ds.$$

★15. Using the convolution theorem of Fourier,

(a) show that the solution of Problem 14 takes the form

$$u(x, t) = \frac{1}{2\sqrt{2\pi t}} \int_{-\infty}^{\infty} f(x - \xi)[\cos(\xi^2/4t) + \sin(\xi^2/4t)]\, d\xi.$$

Hint: Use Problem 15 in Exercises 5.5.

(b) By making the change of variable $\eta = x - \xi$ in (a), show that

$$u(x, t) = \frac{1}{2\sqrt{\pi t}} \int_{-\infty}^{\infty} f(\eta) \sin\left[\frac{(x - \xi)^2}{4t} + \frac{1}{4}\pi\right] d\eta.$$

★16. If $f(x) = e^{-x^2/4}$ in Problem 14,

(a) show that

$$u(x, t) = \frac{1}{2}\left[\frac{e^{-x^2/4(1+it)}}{(1 + it)^{1/2}} + \frac{e^{-x^2/4(1-it)}}{(1 - it)^{1/2}}\right].$$

(b) By setting $1 + it = Re^{i\phi}$, where

$$R = (1 + t^2)^{1/2}, \qquad \tan \phi = t,$$

show that the solution in (a) becomes

$$u(x, t) = R^{-1/2} e^{-x^2(\cos\phi)/4R} \cos\left(\frac{x^2 \sin \phi}{4R} - \frac{1}{2}\phi\right).$$

□ 7.6 NUMERICAL METHOD: FINITE DIFFERENCES

In this section we will extend the method of finite differences introduced in Section 6.6 for the heat equation to similar problems involving the one-dimensional wave equation.

If we assume the problem in Section 7.2 has been made nondimensional, the simple vibrating-string problem will be characterized by

$$u_{xx} = u_{tt}, \qquad 0 < x < 1, \qquad t > 0$$

$$\text{B.C.:} \quad u(0, t) = 0, \quad u(1, t) = 0, \quad t > 0 \tag{94}$$

$$\text{I.C.:} \quad u(x, 0) = f(x), \quad u_t(x, 0) = g(x), \quad 0 < x < 1.$$

Because both spatial and time derivatives in the wave equation are of second order, we can approximate both derivatives by central differences, which gives us

$$u_{xx} \simeq \frac{1}{h^2}(u_{i+1,j} - 2u_{i,j} + u_{i-1,j})$$

and

$$u_{tt} \simeq \frac{1}{\tau^2}(u_{i,j+1} - 2u_{i,j} + u_{i,j-1}).$$

As before, we are using the notation $u_{i,j} = u(x_i, t_j)$, $h = x_{i+1} - x_i$, and $\tau = t_{j+1} - t_j$. The complete replacement equations for (94) are

$$\begin{aligned} &u_{i,j+1} = \rho^2 u_{i-1,j} + 2(1-\rho^2)u_{i,j} + \rho^2 u_{i+1,j} - u_{i,j-1}; \\ &\qquad\qquad i = 1, 2, \ldots, n-1; \qquad j = 1, 2, 3, \ldots \\ &u_{0,j} = 0, \qquad u_{n,j} = 0; \qquad j = 1, 2, 3, \ldots \\ &u_{i,0} = f_i, \qquad u_{i,1} - u_{i,0} = \tau g_i; \qquad i = 0, 1, \ldots, n \end{aligned} \tag{95}$$

where $\rho = \tau/h$ and where we are using a forward difference to approximate the initial velocity. Here we find that in order to obtain stable solutions we must restrict $\rho \leqslant 1$. This means that the time step cannot exceed the space step.

EXAMPLE 4

Use finite differences to solve

$$u_{xx} = u_{tt}, \qquad 0 < x < 1, \qquad t > 0$$

B.C.: $\quad u(0, t) = 0, \qquad u(1, t) = 0, \qquad t > 0$

I.C.: $\quad u(x, 0) = x(1 - x), \qquad u_t(x, 0) = 0, \qquad 0 < x < 1.$

SOLUTION By selecting $h = \tau = 1/4$, the initial conditions become

$$u_{i,0} = \frac{i}{4}\left(1 - \frac{i}{4}\right); \qquad i = 0, 1, 2, 3, 4$$

$$u_{i,1} = u_{i,0} = \frac{i}{4}\left(1 - \frac{i}{4}\right); \qquad i = 0, 1, 2, 3, 4$$

and the first equation in (95) takes the form

$$u_{1,j+1} = u_{0,j} + u_{2,j} - u_{1,j-1}$$

$$u_{2,j+1} = u_{1,j} + u_{3,j} - u_{2,j-1}$$

$$u_{3,j+1} = u_{2,j} + u_{4,j} - u_{3,j-1}$$

TABLE 7.1 Numerical Solution of Example 4.

t \ x	0	$\frac{1}{4}$	$\frac{1}{2}$	$\frac{3}{4}$	1
0	0	0.1875	0.2500	0.1875	0
$\frac{1}{4}$	0	0.1875	0.2500	0.1875	0
$\frac{1}{2}$	0	0.0625	0.1250	0.0625	0
$\frac{3}{4}$	0	−0.0625	−0.1250	−0.0625	0
1	0	−0.1875	−0.2500	−0.1875	0

where $j = 1, 2, 3, \ldots$. Thus, using the boundary values $u_{0,j} = u_{4,j} = 0$ and successively substituting $j = 1, 2, 3, \ldots$ into the above equations, we obtain the values listed in Table 7.1.

If the initial data do not involve discontinuities, the method of finite differences shown here is satisfactory for solving the one-dimensional wave equation. However, if discontinuities are present in the initial data, it is usually best to use the *method of characteristics* whereby the problem is reduced to numerically solving an ODE.* The reason for this is that discontinuities in the wave equation are propagated into the xt-plane along the characteristics and are difficult to deal with along any grid other than a grid of characteristics. This situation is quite different from that involving the heat equation where discontinuities are smoothed out immediately for increasing time.

EXERCISES 7.6

In Problems 1–4, use finite differences with $h = \tau = 1/4$ to solve the given wave equation. Construct a table like Table 7.1 for your answer.

1. $u_{xx} = u_{tt}, \quad 0 < x < 1, \quad t > 0$
   B.C.: $u(0, t) = 0, \quad u(1, t) = 0$
   I.C.: $u(x, 0) = \sin \pi x, \quad u_t(x, 0) = 0$

2. $u_{xx} = u_{tt}, \quad 0 < x < 1, \quad t > 0$
   B.C.: $u(0, t) = 0, \quad u(1, t) = 0$
   I.C.: $u(x, 0) = 0, \quad u_t(x, 0) = \sin \pi x$

*For a discussion of the method of characteristics, see G. D. Smith, *Numerical Solution of Partial Differential Equations*, 2nd ed., Clarendon Press, Oxford (1978).

3. $u_{xx} = u_{tt}, \qquad 0 < x < 1, \qquad t > 0$
   B.C.: $u(0, t) = 0, \qquad u(1, t) = 0$
   I.C.: $u(x, 0) = 0, \qquad u_t(x, 0) = 1$

4. $u_{xx} = u_{tt}, \qquad 0 < x < 1, \qquad t > 0$
   B.C.: $u(0, t) = \sin t, \qquad u(1, t) = 0$
   I.C.: $u(x, 0) = 0, \qquad u_t(x, 0) = 0$

CHAPTER 8

THE POTENTIAL EQUATION

Historically, the name *boundary value problem* was attributed to only problems for which the PDE was of the elliptic variety. The first boundary value problems to be studied in detail involved the *potential equation* (widely known as *Laplace's equation*) and are now referred to as *boundary value problems of the first and second kinds*. Today we use the term boundary value problem in a much wider sense. For example, problems associated with hyperbolic and parabolic PDEs were originally called *initial-boundary value problems,* since they involve both initial and boundary conditions. Today, however, it is common to include these PDEs under the general title of boundary value problems.

In Section 8.2 we define the two most important boundary value problems involving the potential equation. These are known as the *Dirichlet problem* (boundary value problem of the first kind) and the *Neumann problem* (boundary value problem of the second kind). Some simple examples of these problems are discussed in terms of rectangular domains. In Section 8.3 we will consider potential problems in circular domains.

The continuous solutions of the potential equation are called *harmonic functions*. There are many salient properties associated with this class of functions, some of which are discussed in Section 8.4. In particular, we discuss the *uniqueness properties* of the Dirichlet and Neumann problems and state the important *maximum-minimum principle*.

The methods of *Fourier integrals* and *Fourier transforms* are used in Section 8.5 to investigate problems formulated on *unbounded domains*. The Dirichlet problem for the half-plane is solved and its solution is put in a form known as the *Poisson integral formula for the half-plane*.

In the final section we look briefly at two *approximation methods* for solving potential problems. The first method is a generalization of the *weighted residual methods* discussed in Chapter 1, and the second is the *finite-difference method* that was used in Chapter 6 to solve the heat equation.

8.1 INTRODUCTION

Perhaps the single most important PDE in mathematical physics is the simplest example of an elliptic equation called the *potential equation,* or *Laplace's equation*. In two and three dimensions, respectively, we have the rectangular coordinate representations

$$u_{xx} + u_{yy} = 0, \tag{1}$$

$$u_{xx} + u_{yy} + u_{zz} = 0. \tag{2}$$

In general we write the potential equation as

$$\nabla^2 u = 0, \tag{3}$$

regardless of the coordinate system or number of dimensions.

The potential equation arises in steady-state heat conduction problems involving homogeneous solids.* This same equation is satisfied by the gravitational potential in free space, the electrostatic potential in a uniform dielectric, the magnetic potential in free space, the electric potential in the theory of steady flow of currents in solid conductors, and the velocity potential of inviscid, irrotational fluids. The mathematical formulation of all potential problems is essentially the same despite the physical differences of the applications. Because of this, all solutions of the potential equation are collectively called *potential functions,* and the study of the many properties associated with these functions forms that branch of mathematics known as *potential theory*.†

One of the most fundamental properties of the continuous solutions of the potential equation is smoothness. This property is a consequence of the fact that the equation describes "steady states." However, not all solutions are continuous. For example, the function $u(x, y) = \log(x^2 + y^2)$ satisfies (1) but is discontinuous at (0, 0). The continuous solutions of the potential equation that also have continuous second partial derivatives in some domain R are commonly called *harmonic functions*.

8.2 DIRICHLET AND NEUMANN PROBLEMS

A properly-posed problem involving the potential equation consists of finding a harmonic function in a region R subject to a *single* boundary condition. In this chapter we will restrict our attention to two-dimensional domains for which Equation (1), or an equivalent representation in polar coordinates, is the governing PDE.

The most common boundary conditions fall mainly into two categories, giving us two primary types of boundary value problems. If R denotes a region in the plane and C its boundary curve, then one type of problem is characterized by

$$\begin{aligned} \nabla^2 u &= 0 \quad \text{in} \quad R \\ u &= f \quad \text{on} \quad C, \end{aligned} \tag{4}$$

*The heat equation $\nabla^2 u = a^{-2}u_t$ reduces to the potential equation when $u_t = 0$, i.e., when u is independent of t.

†For example, see O. D. Kellog, *Foundations of Potential Theory* (New York: Dover, 1953).

which is called a *Dirichlet problem* or *boundary value problem of the first kind.* In this problem we specify the value of u at each point of the boundary. An example of a Dirichlet problem is to find the steady-state temperature distribution in a region R given that the temperature is known everywhere on the boundary of R. We refer to

$$\nabla^2 u = 0 \quad \text{in} \quad R \qquad (5)$$
$$\frac{\partial u}{\partial n} = f \quad \text{on} \quad C$$

as a *Neumann problem* or *boundary value problem of the second kind.* The derivative $\partial u/\partial n$ is called the *normal derivative* of u and is positive in the direction of the outward normal to the boundary curve C.* In steady-state temperature problems, the normal derivative specifies the heat flow across the boundary of R. There is also a third boundary value problem, called *Robin's problem,* in which the boundary condition is a linear combination of u and its normal derivative. We will not, however, give separate treatment of it. In most potential problems the region R is a bounded region, and the problem is then called an *interior problem.* An *exterior problem* arises when the region R is unbounded but contains a hole.

8.2.1 RECTANGULAR DOMAINS

To begin, we wish to consider potential problems in rectangular domains. A simple example of this is formulated by the *Dirichlet problem*

$$u_{xx} + u_{yy} = 0, \qquad 0 < x < a, \qquad 0 < y < b$$

$$\text{B.C.:} \quad \begin{cases} u(0, y) = 0, & u(a, y) = 0 \\ u(x, 0) = 0, & u(x, b) = f(x) \end{cases} \qquad (6)$$

(see Figure 8.1).

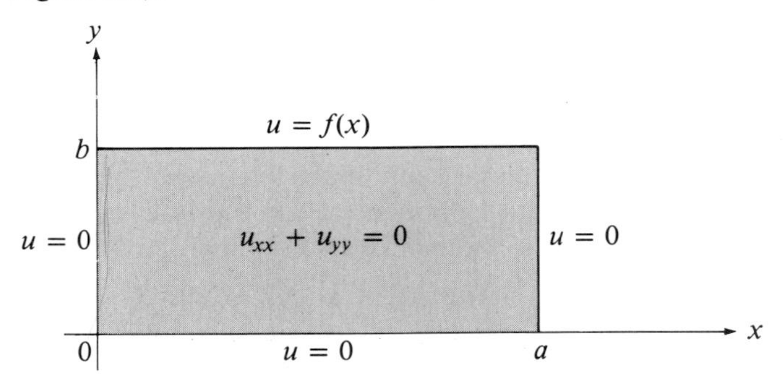

FIGURE 8.1

Dirichlet problem for a rectangle

*Recall that $\partial u/\partial n = \nabla u \cdot \vec{n}$, where $\vec{n}$ is the outward unit normal to C.

A problem like (6) would physically arise if three edges of a thin rectangular isotropic* plate with insulated flat surfaces were maintained at 0°C and the fourth edge was maintained at temperture f until steady-state conditions prevailed throughout the plate. The solution $u(x, y)$ is the *steady-state temperature distribution* interior to the plate in this case. The same mathematical problem describes the steady-state temperature distribution in a long rectangular bar bounded by the planes $x = 0$, $x = a$, $y = 0$, $y = b$, whose temperature variation in the z-direction may be neglected. Other physical situations also lead to the same mathematical problem.

Our solution technique is basically the same as that for the heat and wave equations. That is, we seek solutions of (6) in the form $u(x, y) = X(x)Y(y)$, which, when substituted into the potential equation with variables separated, leads to

$$X'' + \lambda X = 0, \tag{7}$$

$$Y'' - \lambda Y = 0, \tag{8}$$

where λ is the separation constant. Since the first three boundary conditions in (6) are homogeneous, they reduce immediately to

$$X(0) = 0, \qquad X(a) = 0, \qquad Y(0) = 0. \tag{9}$$

The fourth condition, which is nonhomogeneous, must be handled separately.

The solution of (7) when subject to the first two boundary conditions in (9) is

$$\lambda_n = \frac{n^2\pi^2}{a^2}, \qquad X_n(x) = \sin\frac{n\pi x}{a}, \qquad n = 1, 2, 3, \ldots. \tag{10}$$

Corresponding to these values of λ, the solutions of (8) satisfying the remaining condition in (9) become

$$Y_n(y) = \sinh\frac{n\pi y}{a}, \qquad n = 1, 2, 3, \ldots. \tag{11}$$

Hence, for any choice of constants c_n, $n = 1, 2, 3, \ldots$, the function

$$u(x, y) = \sum_{n=1}^{\infty} c_n \sin\frac{n\pi x}{a}\sinh\frac{n\pi y}{a} \tag{12}$$

satisfies the potential equation and the three homogeneous boundary conditions prescribed in (6). The remaining task, therefore, is to determine the constants c_n in such a way that the fourth boundary condition is satisfied. We do this by setting $y = b$ in (12), which yields

$$u(x, b) = f(x) = \sum_{n=1}^{\infty} c_n \sinh\frac{n\pi b}{a}\sin\frac{n\pi x}{a}, \qquad 0 < x < a. \tag{13}$$

* A body is called *isotropic* if the thermal conductivity at each point in the body is independent of the direction of heat flow through the point.

From the theory of *Fourier series* it follows that

$$c_n \sinh \frac{n\pi b}{a} = \frac{2}{a}\int_0^a f(x) \sin \frac{n\pi x}{a}\, dx, \qquad n = 1, 2, 3, \ldots, \tag{14}$$

and the problem is formally solved.

EXAMPLE 1

Solve the Dirichlet problem

$$u_{xx} + u_{yy} = 0, \qquad 0 < x < 1, \qquad 0 < y < 1$$

$$\text{B.C.:} \quad \begin{cases} u(0, y) = 0, & u(1, y) = 0 \\ u(x, 0) = 0, & u(x, 1) = T_0 x(1 - x), \qquad T_0 \text{ constant} \end{cases}$$

SOLUTION By separating variables, we obtain

$$X'' + \lambda X = 0, \qquad X(0) = 0, \qquad X(1) = 0$$

$$Y'' - \lambda Y = 0, \qquad Y(0) = 0.$$

The solution for X leads to

$$\lambda_n = n^2\pi^2, \qquad X_n(x) = \sin n\pi x, \qquad n = 1, 2, 3, \ldots,$$

and for these values of λ, we find that

$$Y_n(y) = \sinh n\pi y, \qquad n = 1, 2, 3, \ldots.$$

The superposition principle then yields the result

$$u(x, y) = \sum_{n=1}^{\infty} c_n \sinh n\pi y \sin n\pi x.$$

Finally, by imposing the remaining boundary condition, we get

$$u(x, 1) = T_0 x(1 - x) = \sum_{n=1}^{\infty} c_n \sinh n\pi \sin n\pi x,$$

and thus

$$\begin{aligned} c_n \sinh n\pi &= 2T_0 \int_0^1 x(1 - x) \sin n\pi x\, dx \\ &= \frac{4T_0}{n^3\pi^3}[1 - (-1)^n] \\ &= \begin{cases} \dfrac{8T_0}{n^3\pi^3}, & n = 1, 3, 5, \ldots \\ 0, & n = 2, 4, 6, \ldots. \end{cases} \end{aligned}$$

Upon changing the index, the solution we seek then takes the form

$$u(x, y) = \frac{8T_0}{\pi^3} \sum_{n=1}^{\infty} \frac{\sinh[(2n-1)\pi y] \sin[(2n-1)\pi x]}{(2n-1)^3 \sinh[(2n-1)\pi]}.$$

A more realistic situation occurs when the temperature is prescribed by nonzero values along all four edges of the plate, rather than along just one edge as in (6). To solve this more general problem we simply superimpose two solutions, each of which corresponds to a problem in which temperatures of 0°C are prescribed along parallel sides of the plate (see Problems 5–8 in Exercises 8.2).

The Neumann problem for the rectangle is similarly solved, as are problems featuring a mix of Dirichlet and Neumann conditions. The following example features a mix of boundary conditions, while others are taken up in Exercises 8.2.

EXAMPLE 2

Solve

$$u_{xx} + u_{yy} = 0, \quad 0 < x < \pi, \quad 0 < y < 1$$

$$\text{B.C.:} \quad \begin{cases} u_x(0, y) = 0, & u_x(\pi, y) = 0 \\ u(x, 0) = T_0 \cos x, & u(x, 1) = T_0 \cos^2 x. \end{cases}$$

SOLUTION Physically, this problem corresponds to a steady-state temperature distribution problem for a rectangular plate with temperatures prescribed along its edges $y = 0$ and $y = 1$, but whose edges $x = 0$ and $x = \pi$ are insulated.

Proceeding with separation of variables leads to

$$X'' + \lambda X = 0, \quad X'(0) = 0, \quad X'(\pi) = 0$$

$$Y'' - \lambda Y = 0.$$

The problem for X is one of our standard eigenvalue problems for which

$$\lambda_0 = 0, \quad X_0(x) = 1,$$

$$\lambda_n = n^2, \quad X_n(x) = \cos nx, \quad n = 1, 2, 3, \ldots.$$

Thus the second equation above yields the general solution

$$Y_n(y) = \begin{cases} a_0 + b_0 y, & n = 0 \\ a_n \cosh ny + b_n \sinh ny, & n = 1, 2, 3, \ldots. \end{cases}$$

By combining solutions via the superposition principle, we obtain

$$u(x, y) = a_0 + b_0 y + \sum_{n=1}^{\infty} (a_n \cosh ny + b_n \sinh ny) \cos nx.$$

The boundary condition at $y = 0$ requires that

$$u(x, 0) = T_0 \cos x = a_0 + \sum_{n=1}^{\infty} a_n \cos nx.$$

Because the boundary condition contains an eigenfunction, we can solve for the unknown constants by simply matching coefficients of like terms on each side of the equation. Hence, by comparing like terms, we deduce that $a_0 = 0$, $a_1 = T_0$, and $a_n = 0$, $n = 2, 3, 4, \ldots$. Therefore, our solution becomes

$$u(x, y) = b_0 y + T_0 \cosh y \cos x + \sum_{n=1}^{\infty} b_n \sinh ny \cos nx.$$

By imposing the remaining boundary condition at $y = 1$, we find

$$T_0 \cos^2 x = b_0 + T_0 \cosh 1 \cos x + \sum_{n=1}^{\infty} b_n \sinh n \cos nx,$$

or upon rearranging terms and writing $\cos^2 x = (1 + \cos 2x)/2$,

$$\frac{1}{2} T_0(1 + \cos 2x) - T_0 \cosh 1 \cos x = b_0 + \sum_{n=1}^{\infty} b_n \sinh n \cos nx.$$

Matching up coefficients of like terms once again, we see that

$$b_0 = \frac{1}{2} T_0, \qquad b_1 = -\frac{T_0 \cosh 1}{\sinh 1}, \qquad b_2 = \frac{T_0}{2 \sinh 2},$$

and $b_n = 0$, $n = 3, 4, 5, \ldots$; thus

$$u(x, y) = \frac{1}{2} T_0 y + T_0\left(\cosh y - \frac{\cosh 1 \sinh y}{\sinh 1}\right) \cos x + T_0 \frac{\sinh 2y}{2 \sinh 2} \cos 2x.$$

Finally, with the aid of the identity

$$\sinh(A - B) = \sinh A \cosh B - \cosh A \sinh B,$$

our solution can be expressed in the more compact form

$$u(x, y) = T_0\left[\frac{1}{2} y + \frac{\sinh(1 - y)}{\sinh 1} \cos x + \frac{\sinh 2y}{2 \sinh 2} \cos 2x\right].$$

Sometimes when formulating math models it happens that we prescribe boundary conditions along adjacent edges of the rectangular domain that lead to different values at the common corner. In such cases, the solution obtained will not be valid in the immediate vicinity of the corner (see also the comments at the end of Section 6.2.1).

EXERCISES 8.2

In Problems 1–4, solve the given potential problem.

1. $u_{xx} + u_{yy} = 0, \quad 0 < x < \pi, \quad 0 < y < 1$
   B.C.: $u(0, y) = 0, \quad u(\pi, y) = 0, \quad u(x, 0) = 0, \quad u(x, 1) = \sin x$

2. $u_{xx} + u_{yy} = 0, \quad 0 < x < \pi, \quad 0 < y < \pi$
   B.C.: $u(0, y) = 0, \quad u(\pi, y) = 1, \quad u(x, 0) = 0, \quad u(x, \pi) = 0$

3. $u_{xx} + u_{yy} = 0, \quad 0 < x < 1, \quad 0 < y < 1$
   B.C.: $u_x(0, y) = 0, \quad u_x(1, y) = 0, \quad u(x, 0) = x^2, \quad u_y(x, 1) = 0$

4. $u_{xx} + u_{yy} = 0, \quad 0 < x < 1, \quad 0 < y < 1$
   B.C.: $u(0, y) = 0, \quad u(1, y) = 0, \quad u_y(x, 0) = 0, \quad u_y(x, 1) = x(1 - x)$

5. Show that $u(x, y) = v(x, y) + w(x, y)$ is a solution of the potential problem

$$u_{xx} + u_{yy} = 0, \qquad 0 < x < a, \qquad 0 < y < b$$

$$\text{B.C.:} \quad \begin{cases} u(x, 0) = f_1(x), & u(x, b) = f_2(x) \\ u(0, y) = g_1(y), & u(a, y) = g_2(y), \end{cases}$$

given that $v(x, y)$ and $w(x, y)$ are solutions, respectively, of

$$v_{xx} + v_{yy} = 0, \qquad 0 < x < a, \qquad 0 < y < b$$

$$\text{B.C.:} \quad \begin{cases} v(x, 0) = f_1(x), & v(x, b) = f_2(x) \\ v(0, y) = 0, & v(a, y) = 0 \end{cases}$$

and

$$w_{xx} + w_{yy} = 0, \qquad 0 < x < a, \qquad 0 < y < b$$

$$\text{B.C.:} \quad \begin{cases} w(x, 0) = 0, & w(x, b) = 0 \\ w(0, y) = g_1(y), & w(a, y) = g_2(y). \end{cases}$$

6. Show that solutions of the potential problem for $v(x, y)$ in Problem 5 are of the form

$$v(x, y) = \sum_{n=1}^{\infty} \left(A_n \cosh \frac{n\pi y}{a} + B_n \sinh \frac{n\pi y}{a} \right) \sin \frac{n\pi x}{a}.$$

Find formal expressions for the constants A_n and B_n.

7. Show that the solution in Problem 6 can also be expressed as

$$v(x, y) = \sum_{n=1}^{\infty} \left[\alpha_n \sinh \frac{n\pi y}{a} + \beta_n \sinh \frac{n\pi(b - y)}{a} \right] \sin \frac{n\pi x}{a}.$$

Find formal expressions for the constants α_n and β_n.

★8. Solve

$$u_{xx} + u_{yy} = 0, \qquad 0 < x < \pi, \qquad 0 < y < \pi$$

$$\text{B.C.:} \quad \begin{cases} u(x, 0) = 0, & u(x, \pi) = \begin{cases} x, & 0 < x < \pi/2 \\ \pi - x, & \pi/2 < x < \pi \end{cases} \\ u(0, y) = \begin{cases} y, & 0 < y < \pi/2 \\ \pi - y, & \pi/2 < y < \pi, \end{cases} & u(\pi, y) = 0 \end{cases}$$

Hint: See Problems 5–7.

9. Determine the steady-state temperature distribution in a square plate of side 2 m if three sides are maintained at 100°C and the remaining side (at $y = 2$) is held at 200°C. Assume the two flat surfaces are insulated.
Hint: Split the problem into two problems, one of which has temperature 100°C on all four sides.

10. A square plate has its faces and edges $x = 0$ and $x = \pi$ insulated. The edges $y = 0$ and $y = \pi$ are kept at zero temperature and $f(x)$, respectively.
(a) Show that a formal solution is given by

$$u(x, y) = \frac{1}{2\pi} a_0 y + \sum_{n=1}^{\infty} a_n \frac{\sinh ny}{\sinh n\pi} \cos nx.$$

(b) Find the solution when $f(x) = T_0$ (constant).

★11. Find a formal solution of

$$u_{xx} + u_{yy} = 0, \qquad 0 < x < 1, \qquad 0 < y < 2$$

$$\text{B.C.:} \quad \begin{cases} u_x(0, y) = 0, & u_x(1, y) + u(1, y) = 0 \\ u(x, 0) = 0, & u(x, 2) = 1 \end{cases}$$

★12. Find a formal solution of

$$u_{xx} + u_{yy} = 0, \qquad 0 < x < \pi, \qquad 0 < y < \pi$$

$$\text{B.C.:} \quad \begin{cases} u_x(0, y) = 0, & u(\pi, y) = 1 \\ u(x, 0) = 1, & u(x, \pi) = 1 \end{cases}$$

8.3 CIRCULAR DOMAINS: POLAR COORDINATES

For problems involving circular-shaped domains it is convenient to formulate our mathematical descriptions in terms of polar coordinates. This means we need to find a polar representation of the potential equation (3).

In two-dimensional rectangular coordinates (x, y) the Laplacian has been defined by

$$\nabla^2 u = u_{xx} + u_{yy}. \tag{15}$$

To find a comparable expression in polar coordinates (r, θ), we simply apply the chain rule to (15) using the equations of transformation

$$x = r \cos \theta, \tag{16a}$$
$$y = r \sin \theta,$$

and inverse transformation

$$r = \sqrt{x^2 + y^2}, \quad \theta = \tan^{-1}\left(\frac{y}{x}\right). \tag{16b}$$

From the chain rule, we immediately have

$$u_x = u_r r_x + u_\theta \theta_x,$$

where

$$r_x = \frac{x}{\sqrt{x^2 + y^2}} = \frac{x}{r},$$

$$\theta_x = \frac{-y/x^2}{1 + (y/x)^2} = -\frac{y}{r^2}.$$

Thus,

$$u_x = \frac{x}{r} u_r - \frac{y}{r^2} u_\theta, \tag{17}$$

and differentiating both sides of this expression with respect to x yields

$$\begin{aligned} u_{xx} &= \frac{\partial}{\partial x}\left(\frac{x}{r} u_r - \frac{y}{r^2} u_\theta\right) \\ &= \frac{x}{r}(u_{rr} r_x + u_{r\theta}\theta_x) - \left(\frac{r - x r_x}{r^2}\right) u_r - \frac{y}{r^2}(u_{r\theta} r_x + u_{\theta\theta}\theta_x) + \frac{2y}{r^3} r_x u_\theta, \end{aligned}$$

or

$$u_{xx} = \frac{x^2}{r^2} u_{rr} - \frac{2xy}{r^3} u_{r\theta} + \frac{y^2}{r^4} u_{\theta\theta} + \frac{y^2}{r^3} u_r + \frac{2xy}{r^4} u_\theta. \tag{18}$$

In a similar fashion it follows that

$$u_{yy} = \frac{y^2}{r^2} u_{rr} + \frac{2xy}{r^3} u_{r\theta} + \frac{x^2}{r^4} u_{\theta\theta} + \frac{x^2}{r^3} u_r - \frac{2xy}{r^4} u_\theta. \tag{19}$$

By combining the results of $u_{xx} + u_{yy}$, we deduce that the Laplacian in polar coordinates is

$$\nabla^2 u = u_{rr} + \frac{1}{r} u_r + \frac{1}{r^2} u_{\theta\theta}. \tag{20}$$

8.3.1 DIRICHLET PROBLEM FOR A DISK

Let us consider a steady-state heat conduction problem in a flat plate in the shape of a circular disk with boundary curve $x^2 + y^2 = \rho^2$ (see Figure 8.2). We assume that the plate is isotropic, that the flat surfaces are insulated, and that the temperature is known everywhere on the circular boundary. The temperature inside the disk is then a solution of the *Dirichlet problem* (in polar coordinates)

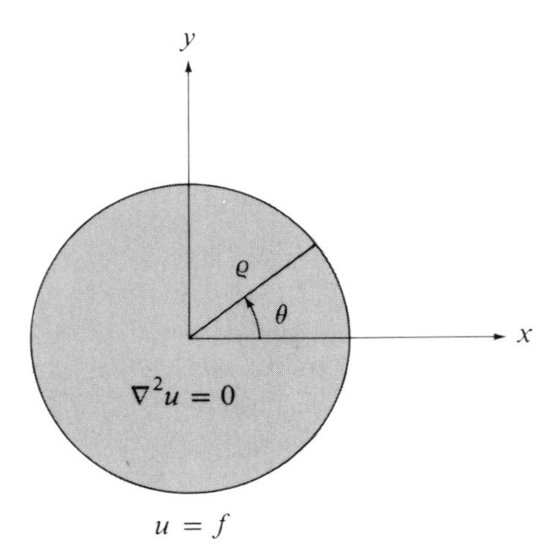

FIGURE 8.2

Dirichlet problem for a circle

$$u_{rr} + \frac{1}{r} u_r + \frac{1}{r^2} u_{\theta\theta} = 0, \qquad 0 < r < \rho, \qquad -\pi < \theta < \pi \tag{21}$$

B.C.: $$u(\rho, \theta) = f(\theta), \qquad -\pi < \theta < \pi.$$

Before we attempt to solve (21), some observations about the solution are in order. First, we need to recognize that $r = 0$ is a "mathematical boundary" of the problem, although it is clearly not a physical boundary. Moreover, $r = 0$ is a *singular point* of the PDE in (21), and this means that we should impose the implicit boundary condition

B.C.: $$u(r, \theta) \quad \text{finite as} \quad r \to 0^+ \tag{22}$$

(see Section 1.6.4).* Second, in order to allow θ to assume any value rather than be restricted to the interval $-\pi < \theta \leq \pi$, we must require that $f(\theta)$ be a periodic function with period 2π. Thus, the solution $u(r, \theta)$ must exhibit the same periodicity as $f(\theta)$ in order to preserve its single-valuedness. [We would expect this condition from a purely physical point of view since the temperature at a point (r, θ) should not vary if we describe the same point by the coordinates $(r, \theta + 2\pi)$.] Therefore, we require that

B.C.: $$u(r, -\pi) = u(r, \pi), \qquad u_\theta(r, -\pi) = u_\theta(r, \pi), \tag{23}$$

which are actually continuity requirements along the slit $\theta = \pi$.

* Although we also want $u_r(r, \theta)$ and $u_{rr}(r, \theta)$ bounded at $r = 0$, it can be shown that condition (22) is sufficient for this to happen.

Separation of variables with $u(r, \theta) = R(r)H(\theta)$ leads to

$$H'' + \lambda H = 0, \qquad H(-\pi) = H(\pi), \qquad H'(-\pi) = H'(\pi), \tag{24}$$

$$r^2R'' + rR' - \lambda R = 0, \qquad R(r) \quad \text{finite as} \quad r \to 0^+, \tag{25}$$

where we have incorporated conditions (22) and (23). The solution of (24) demands that $\lambda = \lambda_n = n^2$ ($n = 0, 1, 2, \ldots$), and hence*

$$H_n(\theta) = \begin{cases} \frac{1}{2} a_0, & n = 0 \\ a_n \cos n\theta + b_n \sin n\theta, & n = 1, 2, 3, \ldots, \end{cases} \tag{26}$$

with the constant $\frac{1}{2}a_0$ selected for conventional reasons. Equation (25) is a *Cauchy-Euler equation* whose general solutions for $\lambda = n^2$ are

$$R_n(r) = \begin{cases} C_1 + C_2 \log r, & n = 0 \\ C_3 r^n + C_4 r^{-n}, & n = 1, 2, 3, \ldots. \end{cases} \tag{27}$$

However, we must set $C_2 = C_4 = 0$ since the terms $\log r$ and r^{-n} would lead to infinite discontinuities (and therefore infinite temperatures) at the center of the disk $r = 0$.

Collecting the solutions (26) and (27), with $C_1 = 1$ and $C_3 = 1/\rho^n$ for mathematical convenience, we find that the superposition principle yields the formal solution

$$u(r, \theta) = \frac{1}{2} a_0 + \sum_{n=1}^{\infty} \left(\frac{r}{\rho}\right)^n (a_n \cos n\theta + b_n \sin n\theta). \tag{28}$$

At $r = \rho$, the boundary condition in (21) leads to

$$f(\theta) = \frac{1}{2} a_0 + \sum_{n=1}^{\infty} (a_n \cos n\theta + b_n \sin n\theta), \tag{29}$$

which is a *full trigonometric series*.† Hence, we deduce that

$$a_n = \frac{1}{\pi} \int_{-\pi}^{\pi} f(\theta) \cos n\theta \, d\theta, \qquad n = 0, 1, 2, \ldots \tag{30}$$

and

$$b_n = \frac{1}{\pi} \int_{-\pi}^{\pi} f(\theta) \sin n\theta \, d\theta, \qquad n = 1, 2, 3, \ldots. \tag{31}$$

Based upon our formal solution of the Dirichlet problem for a disk, we can make certain observations. For example, by setting $r = 0$ in (28), we obtain the result

---

* See Example 21 in Section 1.6.3.

† By *full* we mean not simply a cosine or sine series as has generally been the case.

$$u(0, \theta) = \frac{1}{2} a_0 = \frac{1}{2\pi} \int_{-\pi}^{\pi} f(\theta)\, d\theta. \tag{32}$$

Here we see that the temperature at the center of the disk is simply the average of the temperature on the boundary of the disk. A similar result holds for any fixed radius from the center of the plate. That is, it can be shown that

$$u(0, \theta) = \frac{1}{2\pi} \int_{-\pi}^{\pi} u(r, \theta)\, d\theta \tag{33}$$

for *any value* of r between 0 and ρ. This characteristic of the solution is called the *mean value property* of harmonic functions and is very important in approximation methods involving the potential equation.

As a final comment here, we point out that the series solution (28) can be put in the form of an integral by replacing the coefficients a_0, a_n, and b_n with their integral representations. The result is *Poisson's integral formula for a circle*

$$u(r, \theta) = \frac{\rho^2 - r^2}{2\pi} \int_{-\pi}^{\pi} \frac{f(x)}{\rho^2 - 2\rho r \cos(x - \theta) + r^2}\, dx \tag{34}$$

(see Problem 11 in Exercises 8.3). This form of the solution can be quite useful for developing certain properties of harmonic functions.*

EXAMPLE 3

Solve the Dirichlet problem

$$u_{rr} + \frac{1}{r} u_r + \frac{1}{r^2} u_{\theta\theta} = 0, \qquad 0 < r < 1, \qquad -\pi < \theta < \pi$$

B.C.:
$$u(1, \theta) = \begin{cases} 0, & -\pi < \theta < 0 \\ T_0, & 0 < \theta < \pi, \end{cases}$$

where T_0 is constant (see Figure 8.3).

SOLUTION We can physically interpret this problem as a steady-state temperature distribution problem in a unit disk, where half the boundary is kept at 0°C while the other half is maintained at temperature T_0.

The formal solution is given by (28), where $\rho = 1$ and

$$a_0 = \frac{1}{\pi} \int_0^{\pi} T_0\, d\theta = T_0,$$

$$a_n = \frac{1}{\pi} \int_0^{\pi} T_0 \cos n\theta\, d\theta = 0, \qquad n = 1, 2, 3, \ldots,$$

---

*For example, see Section 11.6 in P. W. Berg and J. L. McGregor, *Elementary Partial Differential Equations*, (San Francisco: Holden–Day, 1966).

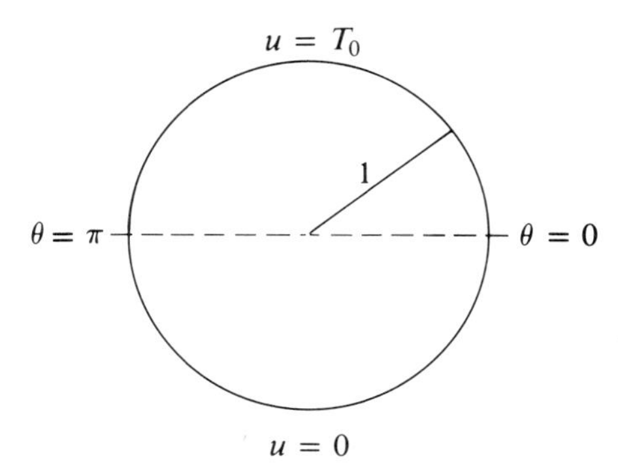

FIGURE 8.3

and

$$b_n = \frac{1}{\pi} \int_0^{\pi} T_0 \sin n\theta \, d\theta = \frac{T_0}{n\pi} [1 - (-1)^n], \qquad n = 1, 2, 3, \ldots .$$

Following a change in index we obtain*

$$u(r, \theta) = \frac{1}{2} T_0 + \frac{2T_0}{\pi} \sum_{n=1}^{\infty} \frac{r^{2n-1}}{2n-1} \sin[(2n-1)\theta].$$

The solution (28) gives the temperature distribution *interior* to a circular disk. A related problem would be to find the temperature distribution *exterior* to the disk. This latter problem might correspond to finding the temperature distribution in a large plate (occupying the entire xy-plane) that has a hole of radius ρ, for which the temperature around the boundary of the hole is known. The problem is characterized by

$$u_{rr} + \frac{1}{r} u_r + \frac{1}{r^2} u_{\theta\theta} = 0, \qquad \rho < r < \infty, \qquad -\pi < \theta < \pi \tag{35}$$

$$\text{B.C.:} \qquad u(\rho, \theta) = f(\theta), \qquad -\pi < \theta < \pi,$$

which is the same as (21) except for the range on r. The primary difference in solving this problem as compared with solving (21) is that we no longer need to require u bounded at $r = 0$, but instead require that u remain bounded as $r \to \infty$. Thus, this time we set $C_2 = C_3 = 0$ in the solution set (27), and select $C_1 = 1$ and $C_4 = \rho^n$. The superposition principle then yields the result

$$u(r, \theta) = \frac{1}{2} a_0 + \sum_{n=1}^{\infty} \left(\frac{\rho}{r}\right)^n (a_n \cos n\theta + b_n \sin n\theta), \tag{36}$$

where the constants a_0, a_n, and b_n are still determined by (30) and (31).

*Note that the solution is $\frac{1}{2}T_0$ along the diameter defined by $\theta = 0$ and $\theta = \pi$. This is simply the average value of the two temperatures prescribed on the boundary of the disk.

8.3.2 NEUMANN PROBLEM FOR A DISK

The normal derivative for a circular domain is in the radial direction. Thus, for a *Neumann problem* on a circular disk we must solve

$$u_{rr} + \frac{1}{r} u_r + \frac{1}{r^2} u_{\theta\theta} = 0, \qquad 0 < r < \rho, \qquad -\pi < \theta < \pi \tag{37}$$

B.C.: $$u_r(\rho, \theta) = f(\theta), \qquad -\pi < \theta < \pi.$$

Of course, the solution of (37) must also satisfy the *implicit conditions* (22) and (23).

The solution of the potential equation satisfying (22) and (23) is that given by (28). If we impose the boundary condition in (37) upon this solution, we obtain

$$f(\theta) = \frac{1}{\rho} \sum_{n=1}^{\infty} n(a_n \cos n\theta + b_n \sin n\theta), \tag{38}$$

where

$$na_n = \frac{\rho}{\pi} \int_{-\pi}^{\pi} f(\theta) \cos n\theta \, d\theta, \qquad n = 1, 2, 3, \ldots \tag{39}$$

and

$$nb_n = \frac{\rho}{\pi} \int_{-\pi}^{\pi} f(\theta) \sin n\theta \, d\theta, \qquad n = 1, 2, 3, \ldots. \tag{40}$$

Hence, the Neumann problem (37) has the solution (28) where a_n and b_n are given by (39) and (40). Notice, however, that the constant $\frac{1}{2}a_0$ that is missing in (38) remains indeterminate in the solution (28). Oddly enough, this is a basic characteristic of the solution of Neumann problems (see Section 8.4.3).

8.3.3 POTENTIAL PROBLEMS IN A CIRCULAR ANNULUS

A circular annulus is formed by two concentric circles. Suppose the temperature distributions along the inner and outer radii of a circular annular plate are maintained at $f(\theta)$ and $g(\theta)$, respectively, until steady-state conditions prevail.* Assuming the flat surfaces of the annular plate are insulated, the steady-state temperature distribution at internal points of the annular plate is then a solution of the Dirichlet problem

$$u_{rr} + \frac{1}{r} u_r + \frac{1}{r^2} u_{\theta\theta} = 0, \qquad a < r < b, \qquad -\pi < \theta < \pi \tag{41}$$

B.C.: $$u(a, \theta) = f(\theta), \qquad u(b, \theta) = g(\theta), \qquad -\pi < \theta < \pi$$

(see Figure 8.4).

---

* We assume the plate is isotropic.

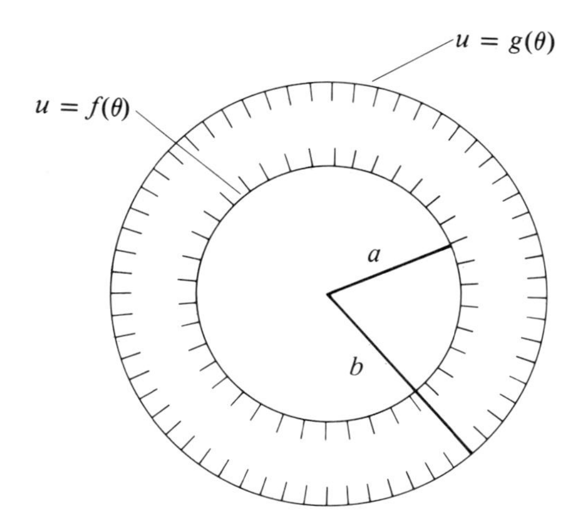

FIGURE 8.4

Circular annulus

The separation of variables technique with $u(r, \theta) = R(r)H(\theta)$ will once again lead to the solution forms

$$H_n(\theta) = \begin{cases} \frac{1}{2} a_0, & n = 0 \\ a_n \cos n\theta + b_n \sin n\theta, & n = 1, 2, 3, \ldots \end{cases} \tag{42}$$

and

$$R_n(r) = \begin{cases} C_1 + C_2 \log r, & n = 0 \\ C_3 r^n + C_4 r^{-n}, & n = 1, 2, 3, \ldots \end{cases} \tag{43}$$

[see (26) and (27)]. However, here we are not concerned with bounded solutions at $r = 0$ or $r \to \infty$ since these points are outside the domain of interest. The solution we seek is, therefore, a linear combination of the functions

$$1, \qquad \log r$$

when $n = 0$. For $n \neq 0$, the solution is a linear combination of the four functions

$$\left.\begin{matrix} r^n \cos n\theta, & r^{-n} \cos n\theta, \\ r^n \sin n\theta, & r^{-n} \sin n\theta, \end{matrix}\right\} \quad n = 1, 2, 3, \ldots.$$

After redefining all constants, the superposition principle yields

$$u(r, \theta) = \frac{1}{2}(A_0 + B_0 \log r) + \sum_{n=1}^{\infty} [(A_n r^n + B_n r^{-n}) \cos n\theta + (C_n r^n + D_n r^{-n}) \sin n\theta]. \tag{44}$$

By applying the boundary conditions in (41) to the solution (44), we obtain the relations

$$f(\theta) = \frac{1}{2}(A_0 + B_0 \log a) + \sum_{n=1}^{\infty} [(A_n a^n + B_n a^{-n}) \cos n\theta + (C_n a^n + D_n a^{-n}) \sin n\theta], \tag{45}$$

$$g(\theta) = \frac{1}{2}(A_0 + B_0 \log b) + \sum_{n=1}^{\infty} [(A_n b^n + B_n b^{-n}) \cos n\theta + (C_n b^n + D_n b^{-n}) \sin n\theta]. \tag{46}$$

The Fourier coefficients are determined in pairs by

$$\left.\begin{aligned} A_0 + B_0 \log a &= \frac{1}{\pi}\int_{-\pi}^{\pi} f(\theta)\, d\theta \\ A_0 + B_0 \log b &= \frac{1}{\pi}\int_{-\pi}^{\pi} g(\theta)\, d\theta \end{aligned}\right\}, \tag{47}$$

$$\left.\begin{aligned} A_n a^n + B_n a^{-n} &= \frac{1}{\pi}\int_{-\pi}^{\pi} f(\theta) \cos n\theta\, d\theta \\ A_n b^n + B_n b^{-n} &= \frac{1}{\pi}\int_{-\pi}^{\pi} g(\theta) \cos n\theta\, d\theta \end{aligned}\right\}, \quad n = 1, 2, 3, \ldots, \tag{48}$$

$$\left.\begin{aligned} C_n a^n + D_n a^{-n} &= \frac{1}{\pi}\int_{-\pi}^{\pi} f(\theta) \sin n\theta\, d\theta \\ C_n b^n + D_n b^{-n} &= \frac{1}{\pi}\int_{-\pi}^{\pi} g(\theta) \sin n\theta\, d\theta \end{aligned}\right\}, \quad n = 1, 2, 3, \ldots. \tag{49}$$

Equations (47)–(49) must be solved in simultaneous pairs as grouped. For instance, if α and β denote the values on the right-hand sides in (47), then

$$A_0 + B_0 \log a = \alpha,$$

$$A_0 + B_0 \log b = \beta,$$

the simultaneous solution of which yields

$$A_0 = \frac{\log(b^\alpha/a^\beta)}{\log(b/a)}, \qquad B_0 = \frac{\alpha - \beta}{\log(b/a)}. \tag{50}$$

The other Fourier coefficients can be determined in a similar manner.

EXAMPLE 4

Find the steady-state temperature distribution inside a circular annular plate whose outer radius $r = 2$ is insulated and whose inner radius $r = 1$ is maintained at temperatures described by $\sin^2 \theta$.

SOLUTION The problem is characterized by

$$u_{rr} + \frac{1}{r} u_r + \frac{1}{r^2} u_{\theta\theta} = 0, \qquad 1 < r < 2, \qquad -\pi < \theta < \pi$$

B.C.: $\quad u(1, \theta) = \sin^2 \theta, \qquad u_r(2, \theta) = 0, \qquad -\pi < \theta < \pi.$

The solution is given by (44), where the above boundary conditions demand that

$$\sin^2 \theta = \frac{1}{2} A_0 + \sum_{n=1}^{\infty} [(A_n + B_n) \cos n\theta + (C_n + D_n) \sin n\theta],$$

$$0 = \frac{1}{4} B_0 + \sum_{n=1}^{\infty} n[(A_n 2^{n-1} - B_n 2^{-n-1}) \cos n\theta + (C_n 2^{n-1} - D_n 2^{-n-1}) \sin n\theta].$$

Recognizing that $\sin^2 \theta = \frac{1}{2}(1 - \cos 2\theta)$, we can simply equate coefficients of like terms to obtain the sets of equations

$$\left.\begin{aligned} A_0 &= 1 \\ B_0 &= 0 \end{aligned}\right\},$$

$$\left.\begin{aligned} A_2 + B_2 &= -\frac{1}{2} \\ 2A_2 - \frac{1}{8} B_2 &= 0 \end{aligned}\right\},$$

$$\left.\begin{aligned} &A_n + B_n = 0 \\ &2^{n-1} A_n - 2^{-n-1} B_n = 0 \end{aligned}\right\}, \qquad n = 1, 3, 4, 5, \ldots,$$

$$\left.\begin{aligned} &C_n + D_n = 0 \\ &2^{n-1} C_n + 2^{-n-1} D_n = 0 \end{aligned}\right\}, \qquad n = 1, 2, 3, \ldots.$$

Solving, we find $A_0 = 1$, $B_0 = 0$, $A_2 = -1/34$, $B_2 = -8/17$, and all other Fourier coefficients are zero. Hence, our solution is simply

$$u(r, \theta) = \frac{1}{2} - \frac{1}{17}\left(\frac{r^2}{2} + \frac{8}{r^2}\right) \cos 2\theta.$$

EXERCISES 8.3

In Problems 1–3, solve the given potential problem.

1. $u_{rr} + \frac{1}{r} u_r + \frac{1}{r^2} u_{\theta\theta} = 0, \qquad 0 < r < \rho, \qquad -\pi < \theta < \pi$

   B.C.: $u(\rho, \theta) = \cos^2 \theta, \qquad -\pi < \theta < \pi$

2. $u_{rr} + \frac{1}{r} u_r + \frac{1}{r^2} u_{\theta\theta} = 0, \qquad 0 < r < \rho, \qquad -\pi < \theta < \pi$

   B.C.: $u(\rho, \theta) = |\theta|, \qquad -\pi < \theta < \pi$

3. $u_{rr} + \frac{1}{r} u_r + \frac{1}{r^2} u_{\theta\theta} = 0, \qquad 0 < r < 1$

B.C.: $u_r(1, \theta) = \begin{cases} -1, & -\pi < \theta < 0 \\ 1, & 0 < \theta < \pi \end{cases}$

4. Solve the potential problem in the quadrant

$$u_{rr} + \frac{1}{r} u_r + \frac{1}{r^2} u_{\theta\theta} = 0, \qquad 0 < r < 1, \qquad 0 < \theta < \frac{\pi}{2},$$

when the boundary conditions are given by

(a) $u(r, 0) = 0, \qquad u\left(r, \frac{\pi}{2}\right) = 0, \qquad u(1, \theta) = T_0$ (constant).

(b) $u(r, 0) = 0, \qquad u\left(r, \frac{\pi}{2}\right) = 0, \qquad u(1, \theta) = \theta\left(\frac{\pi}{2} - \theta\right).$

5. Determine the steady-state temperature distribution in the semi-circular disk bounded by $y = \sqrt{1 - x^2}$, $y = 0$, with 0°C prescribed along the diameter $y = 0$, while along the circumference the temperature is prescribed by
(a) $u(1, \theta) = \sin \theta$.
(b) $u(1, \theta) = T_0$ (constant).

6. Find a formal solution of the *exterior* Dirichlet problem

$$u_{rr} + \frac{1}{r} u_r + \frac{1}{r^2} u_{\theta\theta} = 0, \qquad \rho < r < \infty, \qquad -\pi < \theta < \pi$$

B.C.: $u(\rho, \theta) = |\theta|, \qquad -\pi < \theta < \pi.$

★7. Show that the potential problem

$$u_{rr} + \frac{1}{r} u_r + \frac{1}{r^2} u_{\theta\theta} = 0, \qquad 0 < r < 1, \qquad 0 < \theta < \pi$$

B.C.: $u(r, 0) = 0, \qquad u(r, \pi) = T_0, \qquad u(1, \theta) = T_0,$

has the solution

$$u(r, \theta) = \frac{T_0}{\pi}\left(\theta + 2 \sum_{n=1}^{\infty} \frac{r^n}{n} \sin n\theta\right).$$

8. Determine the steady-state temperature distribution in a circular annulus $10 < r < 20$, where the boundary conditions are

$$u(10, \theta) = 15 \cos \theta, \qquad u(20, \theta) = 30 \sin \theta.$$

9. Determine the steady-state temperature distribution in a circular annulus $a < r < b$, where the boundary conditions are (T_1, T_2 constant)

$$u(a, \theta) = T_1, \qquad u(b, \theta) = T_2.$$

★10. A thin metal plate lies in the plane $z = 0$ and is bounded by $r = a$, $r = b$ $(a < b)$, $\theta = 0$, and $\theta = \pi/2$. The boundary $r = a$ is maintained at temperature $\theta[(\pi/2) - \theta]$, and the other boundaries are kept at 0°C. Show that the steady-state solution is

$$u(r, \theta) = \frac{2}{\pi} \sum_{n=0}^{\infty} \left[\frac{(r/b)^{4n+2} - (b/r)^{4n+2}}{(a/b)^{4n+2} - (b/a)^{4n+2}}\right] \frac{\sin(4n + 2)\theta}{(2n + 1)^3}.$$

★11. By substituting the integral representations (30) and (31) into the solution formula (28), with x as the dummy integration variable,

(a) show that

$$u(r, \theta) = \frac{1}{2\pi} \int_{-\pi}^{\pi} f(x)\left[1 + 2\sum_{n=1}^{\infty} \left(\frac{r}{\rho}\right)^n \cos n(x - \theta)\right] dx.$$

(b) Verify the identity

$$1 + 2\sum_{n=1}^{\infty} R^n \cos nz = \frac{1 - R^2}{1 - 2R\cos z + R^2}, \qquad |R| < 1.$$

Hint: Use $\cos x = \frac{1}{2}(e^{ix} + e^{-ix})$ and $\sum_{n=0}^{\infty} x^n = 1/(1 - x)$ for $|x| < 1$.

(c) From (a) and (b), derive Poisson's integral formula (34).

★12. Along the circumference of the unit circle $r = 1$, a solution of the potential equation is required to take on the value T_0 when $0 < \theta < \pi$ and the value zero elsewhere. Use the Poisson integral formula to show that (see also Example 3)

$$u(r, \theta) = T_0 - \frac{T_0}{\pi} \arctan\left(\frac{2r \sin \theta}{1 - r^2}\right).$$

Hint: Make the substitution $x - \theta = \tan \frac{1}{2}z$.

13. Verify that (20) can also be written in the equivalent form

$$\nabla^2 u = \frac{1}{r}\frac{\partial}{\partial r}(ru_r) + \frac{1}{r^2} u_{\theta\theta}.$$

14. Find the most general solution of $\nabla^2 u = 0$ in polar coordinates that is independent of θ.
Hint: See Problem 13.

★15. If $u(r, \theta)$ satisfies $\nabla^2 u = 0$, show that $v(r, \theta) = u\left(\frac{1}{r}, \theta\right)$ satisfies $\nabla^2 v = 0$.

8.4 PROPERTIES OF HARMONIC FUNCTIONS

The study of harmonic functions is an important subject in its own right. There are many important properties associated with these functions, the proofs of which generally involve various specializations of *Green's theorem in the plane*. In symbols, Green's theorem in the plane takes the form*

$$\iint_R \left(\frac{\partial Q}{\partial x} - \frac{\partial P}{\partial y}\right) dx\, dy = \oint_C P(x, y)\, dx + Q(x, y)\, dy, \tag{51}$$

*For a proof of Green's theorem in the plane, see pp. 198–199 in H. F. Davis and A. D. Snider, *Introduction to Vector Analysis*, 4th ed., (Boston: Allyn & Bacon, 1979).

where the partial derivatives of $P(x, y)$ and $Q(x, y)$ are assumed to be continuous everywhere within the bounded, closed, regular region R whose boundary is a closed, regular curve C. (Note: The orientation of C is counterclockwise and $\oint$ denotes a closed path line integral.)

The specialized forms of (51) that concern us are widely known as Green's identities. In particular, *Green's first and second identities,* respectively, are given by

$$\iint_R (v\nabla^2 u + \nabla v \cdot \nabla u)\, dx\, dy = \oint_C v \frac{\partial u}{\partial n}\, ds \tag{52}$$

and

$$\iint_R (v\nabla^2 u - u\nabla^2 v)\, dx\, dy = \oint_C \left(v \frac{\partial u}{\partial n} - u \frac{\partial v}{\partial n} \right) ds, \tag{53}$$

where u and v are differentiable functions throughout the region R. The expression $\nabla v \cdot \nabla u$ is simply the dot or scalar product of vector analysis between two gradients, and $\partial u/\partial n$ and $\partial v/\partial n$ are normal derivatives. The derivations of (52) and (53) from (51) are found in the exercises (see Problems 1–3 in Exercises 8.4).

Although we have stated Green's identities only for two-dimensional regions, analogous forms of these identities exist for three-dimensional regions.* In this regard, the results about to be proven based upon Green's identities can be extended to harmonic functions in three-dimensional regions.

8.4.1 UNIQUENESS OF THE DIRICHLET PROBLEM

In Sections 8.2 and 8.3 we derived formal solutions of certain Dirichlet problems in both rectangular and circular domains. Using Green's identities, it is relatively simple to show that such solutions are *unique*. We start with the following lemma.†

LEMMA 8.1

If u is a harmonic function in a region R and $u = 0$ at all points on the boundary C of R, then $u = 0$ everywhere inside R.

---

* In three-dimensional regions, Green's identities are actually alternate forms of the *divergence theorem of Gauss*.

† Throughout our discussion we will assume that the region R is the kind of region for which Green's identities are valid.

PROOF Let $v = u$ in Green's first identity (52), which yields

$$\iint_R (u\nabla^2 u + \|\nabla u\|^2)\,dx\,dy = \oint_C u\,\frac{\partial u}{\partial n}\,ds.$$

Because u is harmonic in R, it follows that $\nabla^2 u = 0$ in R. Also, because $u = 0$ on the boundary curve C, the line integral on the right vanishes. Hence, Green's first identity reduces to

$$\iint_R \|\nabla u\|^2\,dx\,dy = 0.$$

The term $\|\nabla u\|^2$ is never negative, and so the preceding relation is possible only if ∇u is the zero vector. Therefore, u itself is a constant, and since $u = 0$ on C, it must assume this same value everywhere in R.

A more general form of Lemma 8.1 is contained in the following theorem.

THEOREM 8.1

If u_1 and u_2 are both harmonic functions in a region R and $u_1 = u_2$ everywhere on the boundary C of R, then $u_1 = u_2$ everywhere in R.

PROOF To begin, let $u = u_1 - u_2$. From the linearity property it follows that $\nabla^2 u = \nabla^2 u_1 - \nabla^2 u_2 = 0$, and thus u is harmonic. Also, $u = 0$ at all points on the boundary C. Consequently, by Lemma 8.1 we conclude that $u = 0$ inside R, which in turn implies $u_1 = u_2$ inside R.

Theorem 8.1 can be rephrased as saying—*a function harmonic in a region R is uniquely determined by its values on the boundary of R*. In other words, the Dirichlet problem has a *unique solution* provided that a solution *exists*. Proving that the Dirichlet problem has a solution is more difficult to resolve and will not be treated here.*

8.4.2 MAXIMUM–MINIMUM PRINCIPLE AND STABILITY

To show that the Dirichlet problem is properly posed, we must also verify that the solution is *stable*. This means that the solution depends continuously on the boundary values.† To prove this important result, we need the follow-

* For a proof of the existence theorem, see Section 31 in I. G. Petrovsky, *Lectures on Partial Differential Equations*, (New York: Interscience, 1954).

† Recall the discussion in Section 3.1.1.

ing theorem concerning the maximum and minimum values of a harmonic function.

THEOREM 8.2

Maximum-minimum principle. If u is a harmonic function in a region R, then both the maximum and minimum values of u are attained on the boundary C of R.

Theorem 8.2, which is of great importance in the study of potential theory, is a consequence of the *mean value property* of harmonic functions [e.g., see Equation (33)]. Although we won't provide a rigorous proof of the theorem, we can provide an intuitive proof. Suppose we think of u as the steady-state temperature distribution in a region R, such as a metal plate. The temperature at any one interior point cannot exceed all other nearby points, for if it did, heat would flow from the hot points to the cooler points. But then the temperature would be changing in time, and this is a contradiction that steady-state conditions prevail. Also, an interior maximum or minimum temperature could never be the average of its neighbors as required by the mean value property.

THEOREM 8.3

If the Dirichlet problem in a region R has a solution, it is stable.

PROOF Let u_1 and u_2 be two solutions of the Dirichlet problem, corresponding to boundary values described by f_1 and f_2, respectively. Suppose further that for some given $\epsilon > 0$, we have

$$|f_1 - f_2| < \epsilon \quad \text{on} \quad C.$$

Since $u_1 = f_1$ and $u_2 = f_2$ on C, it is also true that

$$|u_1 - u_2| < \epsilon \quad \text{on} \quad C.$$

From the maximum-minimum principle (Theorem 8.2), both u_1 and u_2 assume their maximum and minimum values on C. Hence, we can write

$$|\max u_1 - \min u_2| < \epsilon \quad \text{on} \quad C,$$

and therefore $|u_1 - u_2| < \epsilon$ everywhere in R. But this condition is precisely what we mean by continuous dependence upon the boundary data, or stability.

8.4.3 UNIQUENESS OF THE NEUMANN PROBLEM

If the normal derivative of a harmonic function vanishes everywhere on the boundary C of a region R, that function is necessarily a constant. Moreover, if u_1 and u_2 are both harmonic functions in R and their normal derivatives are equal everywhere on the boundary curve C, then u_1 and u_2 differ by, at most, a constant within R. The proofs of these statements are analogous to the proofs of Lemma 8.1 and Theorem 8.1, and are left to Problems 4 and 5 in Exercises 8.4.

These remarks about harmonic functions imply that the *Neumann problem has a unique solution only to within an additive constant*. That is, if u is a solution of the Neumann problem, so is $u + K$, where K is any constant. However, like the Dirichlet uniqueness theorem, no assertion is made concerning the *existence* of a solution. In fact, the Neumann problem does not always have a solution.

For the Neumann problem to have a solution, a certain *compatibility condition* must be satisfied. To derive this condition, we set $v = 1$ in Green's first identity (52) to find

$$\iint_R \nabla^2 u \, dx \, dy = \oint_C \frac{\partial u}{\partial n} \, ds. \tag{54}$$

If u is harmonic, then $\nabla^2 u = 0$ and this condition leads to the following theorem.

THEOREM 8.4

Compatibility condition. The Neumann problem

$$\nabla^2 u = 0 \quad \text{in} \quad R, \qquad \frac{\partial u}{\partial n} = f \quad \text{on} \quad C,$$

has a solution only if

$$\oint_C f \, ds = 0.$$

The compatibility condition can also be argued from purely physical considerations. Suppose u denotes the steady-state temperature in a flat plate. Then $\partial u / \partial n = f$ denotes the flux (heat loss across the boundary), and the net flux must remain zero if there are no heat sources or sinks within R. That is, heat cannot be created or absorbed within R.

EXERCISES 8.4

1. Green's theorem in the plane is mathematically described by

$$\iint_R \left(\frac{\partial Q}{\partial x} - \frac{\partial P}{\partial y} \right) dx \, dy = \oint_C P(x, y) \, dx + Q(x, y) \, dy.$$

By defining the vector function $\vec{F} = \langle Q, -P \rangle$, show that Green's theorem can also be written in the form

$$\iint_R (\nabla \cdot \vec{F})\, dx\, dy = \oint_C (\vec{F} \cdot \vec{n})\, ds,$$

where $\vec{n} = \left\langle \dfrac{dy}{ds}, -\dfrac{dx}{ds} \right\rangle$ is the outward unit normal to C.

2. Derive Green's first identity (52) by setting $\vec{F} = v\nabla u$ in the result of Problem 1 and using the vector identities

$$\nabla \cdot v\nabla u = v\nabla^2 u + \nabla v \cdot \nabla u,$$

$$\nabla u \cdot \vec{n} = \frac{\partial u}{\partial n}.$$

3. Derive Green's second identity (53) by first interchanging the roles of u and v in Green's first identity (52) and then subtracting the result from (52).

4. If u is a harmonic function in a region R and $\partial u/\partial n = 0$ everywhere on the boundary curve C of R, show that u is a constant inside R.

5. If u_1 and u_2 are both harmonic functions in a region R and $\partial u_1/\partial n = \partial u_2/\partial n$ everywhere on the boundary curve C of R, show that the difference $u_1 - u_2$ is a constant inside R.
Hint: Use Problem 4.

6. Prove uniqueness of *Robin's problem*

$$\nabla^2 u = 0 \quad \text{in} \quad R,$$

$$\frac{\partial u}{\partial n} + hu = f \quad \text{on} \quad C \qquad (h > 0).$$

Hint: Show that $\displaystyle\iint_R \|\nabla u\|^2\, dx\, dy = -h \oint_C u^2\, ds.$

★7. If $k < 0$, prove uniqueness of

$$\nabla^2 u + ku = -q \quad \text{in} \quad R,$$

$$u = f \quad \text{on} \quad C.$$

★8. If $k < 0$, prove uniqueness of

$$\nabla^2 u + ku = -q \quad \text{in} \quad R,$$

$$\frac{\partial u}{\partial n} = f \quad \text{on} \quad C.$$

9. Verify the Neumann compatibility condition (Theorem 8.4) for Problem 3 in Exercises 8.3.

10. Find a compatibility condition like Theorem 8.4 for

$$\nabla^2 u = -q \quad \text{in} \quad R,$$

$$\frac{\partial u}{\partial n} = f \quad \text{on} \quad C.$$

11. Demonstrate that the following problem has no solution. Can you explain why?

$$u_{xx} + u_{yy} = 0, \qquad 0 < x < \pi, \qquad 0 < y < \pi$$

$$\text{B.C.:} \quad \begin{cases} u_x(0, y) = y, & u_x(\pi, y) = 1 \\ u_y(x, 0) = 0, & u_y(x, \pi) = 0 \end{cases}$$

12. If the boundary condition $u_x(\pi, y) = 1$ in Problem 11 is replaced by $u_x(\pi, y) = \pi/2$, show that the resulting problem would have a solution. Is the solution unique?

13. Determine the conditions on the function $f(x)$ so that the following Neumann problem has a solution:

$$u_{xx} + u_{yy} = 0, \qquad 0 < x < 1, \qquad 0 < y < 1$$

$$\text{B.C.:} \quad \begin{cases} u_x(0, y) = 0, & u_x(1, y) = 0 \\ u_y(x, 0) = f(x), & u_y(x, 1) = 0 \end{cases}$$

★14. By defining a Green's function $g(x, s; y, t)$ such that

$$\nabla^2 g = -\delta(x - s)\delta(y - t) \quad \text{in} \quad R,$$

$$g = 0 \quad \text{on} \quad C,$$

where δ is the delta function (Section 2.3), show that the Dirichlet problem for the same region R has the formal solution

$$u = -\oint_C f \frac{\partial g}{\partial n}\, ds$$

(assume $u = f$ on C).
Hint: Set $v = g$ in Green's second identity (53).

★15. By using the Green's function defined in Problem 14, show that the nonhomogeneous problem*

$$\nabla^2 u = -F \quad \text{in} \quad R,$$

$$u = f \quad \text{on} \quad C,$$

has the formal solution

$$u = \iint_R gF\, dx\, dy - \oint_C f \frac{\partial g}{\partial n}\, ds.$$

★16. By defining a Green's function $g(x, s; y, t)$ such that

$$\nabla^2 g = -\delta(x - s)\delta(y - t) \quad \text{in} \quad R,$$

$$\frac{\partial g}{\partial n} = 0 \quad \text{on} \quad C,$$

where δ is the delta function (Section 2.3), show that the Neumann problem for the same region R has the formal solution

---

* The nonhomogeneous equation $\nabla^2 u = -F$ is known as *Poisson's equation* (see Section 10.6).

$$u = \oint_C fg \, ds$$

(assume $\partial u/\partial n = f$ on C).
Hint: Set $v = g$ in Green's second identity (53).

★17. By using the Green's function defined in Problem 16, show that the nonhomogeneous problem

$$\nabla^2 u = -F \quad \text{in} \quad R,$$

$$\frac{\partial u}{\partial n} = f \quad \text{on} \quad C,$$

has the formal solution

$$u = \iint_R gF \, dx \, dy + \oint_C fg \, ds.$$

8.5 UNBOUNDED DOMAINS

Problems involving Laplace's equation on infinite domains can often be handled by either *Fourier integrals* or *Fourier transforms*. Generally speaking, the Laplace transform is not an appropriate tool for problems of this nature.*

8.5.1 METHOD OF FOURIER INTEGRALS

Suppose we wish to find the steady-state temperature distribution in a very large rectangular plate when the temperature is prescribed by the function f along one edge of the plate and tends to zero along each of the other edges. Such a problem might be mathematically modeled by

$$u_{xx} + u_{yy} = 0, \qquad -\infty < x < \infty, \qquad 0 < y < \infty$$

$$\text{B.C.:} \quad \begin{cases} u(x, 0) = f(x), & -\infty < x < \infty \\ u(x, y) \to 0 \quad \text{as} & (x^2 + y^2) \to \infty \end{cases} \tag{55}$$

(see Figure 8.5).

* For example, consider the Laplace transform for $x > 0$,

$$\mathcal{L}\{u_{xx}(x, t); x \to p\} = p^2 U(p, t) - pu(0, t) - u_x(0, t).$$

To use this expression would require knowledge of both u and u_x at the boundary $x = 0$. Yet, prescribing both u and u_x on the boundary would generally lead to an *ill-posed problem* (i.e., not properly posed). Even if either u or u_x is left undetermined until later in the problem, the problem would almost surely become unwieldy. For this reason, the transforms of Fourier are better suited for solving the potential equation.

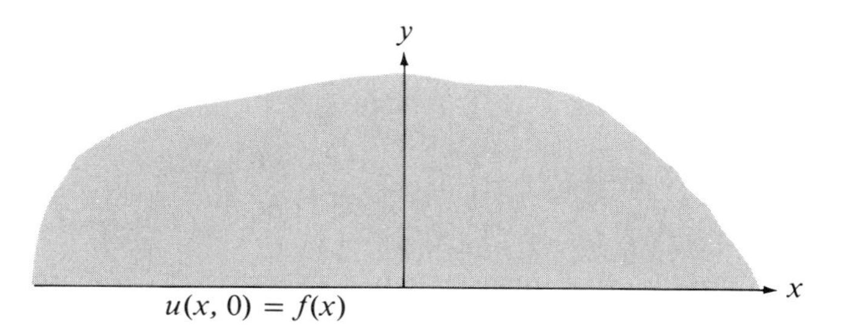

FIGURE 8.5

Dirichlet problem for a half-plane

By separating variables with $u(x, y) = X(x)Y(y)$, we obtain

$$X'' + \lambda X = 0, \qquad -\infty < x < \infty, \tag{56}$$

$$Y'' - \lambda Y = 0, \qquad 0 < y < \infty. \tag{57}$$

In order to satisfy the limiting boundary condition in (55), and eventually satisfy the finite condition, we must set $\lambda = s^2$ ($s > 0$), which leads to the solution forms

$$X(x) = A(s) \cos sx + B(s) \sin sx$$

and

$$Y(y) = D(s)e^{-sy} + E(s)e^{sy}.$$

For $s > 0$, the function e^{sy} does not tend to zero as $y \to \infty$. Therefore, we choose $E(s) = 0$ and $D(s) = 1$ (for convenience), and form the solution

$$u(x, y) = \int_0^\infty [A(s) \cos sx + B(s) \sin sx]e^{-sy}\, ds. \tag{58}$$

Our remaining task is to satisfy the first boundary condition in (55). This we do by setting $y = 0$ in (58), which yields

$$f(x) = \int_0^\infty [A(s) \cos sx + B(s) \sin sx]\, ds.$$

Then we deduce that

$$A(s) = \frac{1}{\pi} \int_{-\infty}^\infty f(x) \cos sx\, dx \tag{59}$$

and

$$B(s) = \frac{1}{\pi} \int_{-\infty}^\infty f(x) \sin sx\, dx, \tag{60}$$

and the problem is solved.

EXAMPLE 5

Solve the problem described by (55) when

$$f(x) = \begin{cases} T_0, & |x| < b \\ 0, & |x| > b. \end{cases}$$

SOLUTION The solution of this problem is given by (58)–(60), where

$$A(s) = \frac{T_0}{\pi} \int_{-b}^{b} \cos sx \, dx = \frac{2T_0}{\pi} \frac{\sin bs}{s}$$

and

$$B(s) = \frac{T_0}{\pi} \int_{-b}^{b} \sin sx \, dx = 0.$$

Therefore, we have

$$u(x, y) = \frac{2T_0}{\pi} \int_0^{\infty} e^{-sy} \left(\frac{\sin bs}{s} \right) \cos sx \, ds.$$

While we have provided a formal solution to the problem in Example 5, the resulting integral is difficult to evaluate by elementary techniques. An alternate solution form that is often more useful than (58) for extracting information about problems of this nature is developed in Section 8.5.2.

8.5.2 POISSON INTEGRAL FORMULA FOR THE HALF-PLANE

By rewriting (59) and (60) using the dummy variable t instead of x, and then substituting these integral expressions into (58), we find

$$u(x, y) = \int_0^{\infty} \left[\frac{1}{\pi} \int_{-\infty}^{\infty} f(t) \cos st \cos sx \, dt \right.$$
$$\left. + \frac{1}{\pi} \int_{-\infty}^{\infty} f(t) \sin st \sin sx \, dt \right] e^{-sy} \, ds. \tag{61}$$

Next, interchanging the order of integration in (61) and using the trigonometric identity

$$\cos A \cos B + \sin A \sin B = \cos(A - B)$$

leads to

$$u(x, y) = \frac{1}{\pi} \int_{-\infty}^{\infty} f(t) \int_0^{\infty} \cos[s(t - x)] e^{-sy} \, ds \, dt. \tag{62}$$

Finally, upon integrating the inner integral in (62), we obtain the important result

$$u(x, y) = \frac{y}{\pi} \int_{-\infty}^{\infty} \frac{f(t)\, dt}{(t - x)^2 + y^2}, \qquad y > 0, \tag{63}$$

called the *Poisson integral formula for the half-plane*. It is comparable to the integral formula (34) in Section 8.3.1 for the circle. In addition to representing solutions for all Dirichlet problems in the upper half-plane, (63) is a useful result for deriving properties of harmonic functions in this domain.

EXAMPLE 6

Solve Example 5 using Poisson's integral formula.

SOLUTION By substituting f directly into the Poisson integral formula (63), we obtain the solution

$$\begin{aligned} u(x, y) &= \frac{yT_0}{\pi} \int_{-b}^{b} \frac{dt}{(t - x)^2 + y^2} \\ &= \frac{T_0}{\pi} \left[\tan^{-1}\left(\frac{x + b}{y}\right) - \tan^{-1}\left(\frac{x - b}{y}\right) \right]. \end{aligned}$$

Using the trigonometric identity

$$\tan(A - B) = \frac{\tan A - \tan B}{1 + \tan A \tan B},$$

we can express our solution in the more convenient form

$$u(x, y) = \frac{T_0}{\pi} \tan^{-1}\left(\frac{2by}{x^2 + y^2 - b^2}\right).$$

Curves in the upper half-plane for which the steady-state temperature is constant are called *isotherms*. For our particular problem, these curves are defined by the family of circular arcs (C constant)

$$x^2 + y^2 - Cy = b^2,$$

which have centers on the y-axis and endpoints on the x-axis at $x = \pm b$ (see Figure 8.6). The form of our solution here made it easy to identify these curves. This would not have been the case had we used the solution form in Example 5.

□ 8.5.3 FOURIER TRANSFORM METHOD

To illustrate the use of Fourier transforms in solving Laplace's equation, we consider the problem of determining the steady-state temperature distribution in a thick slab in which one of the infinite faces of the slab is subjected to a prescribed temperature distribution f and the other two faces are insulated against the flow of heat. The problem is formulated by

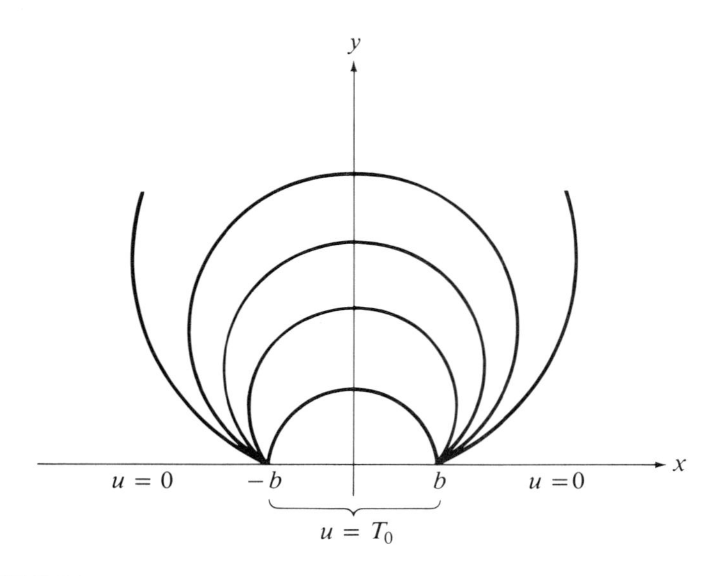

FIGURE 8.6

Family of isotherms

$$u_{xx} + u_{yy} = 0, \qquad 0 < x < \infty, \qquad 0 < y < a$$

$$\text{B.C.:} \quad \begin{cases} u(x, 0) = f(x), & u_y(x, a) = 0, \quad 0 < x < \infty \\ u_x(0, y) = 0, & 0 < y < a \end{cases} \tag{64}$$

(see Figure 8.7).

Because the independent variable x is defined on a semi-infinite domain and the boundary condition $u_x(0, y)$ is prescribed, we find the *Fourier cosine transform* to be the appropriate tool in this case. Thus, we introduce

$$\mathcal{F}_C\{u(x, y); x \to s\} = U(s, y), \tag{65}$$

and

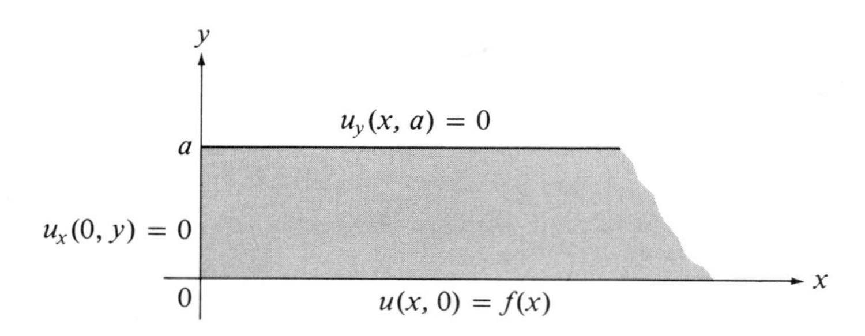

FIGURE 8.7

Semi-infinite slab

$$\mathscr{F}_C\{u_{xx}(x, y); x \to s\} = -s^2U(s, y) - \sqrt{\frac{2}{\pi}}\, u_x(0, y)$$
$$= -s^2U(s, y), \tag{66}$$

from which we generate the transformed problem

$$U_{yy} - s^2U = 0, \qquad 0 < y < a$$
$$\text{B.C.:} \quad U(s, 0) = F(s), \qquad U_y(s, a) = 0, \tag{67}$$

where $F(s)$ is the cosine transform of $f(x)$. The general solution of (67) is

$$U(s, y) = A(s) \cosh sy + B(s) \sinh sy. \tag{68}$$

Imposing the boundary conditions, we find

$$U(s, 0) = A(s) = F(s),$$
$$U_y(s, a) = sA(s) \sinh sa + sB(s) \cosh sa = 0,$$

the simultaneous solution of which leads to the solution form

$$U(s, y) = F(s) \frac{\cosh(a - y)s}{\cosh as}. \tag{69}$$

Finally, if we let*

$$g(x, y) = \mathscr{F}_C^{-1}\left\{\frac{\cosh(a - y)s}{\cosh as}; s \to x\right\}, \tag{70}$$

then through use of the convolution theorem, the inversion of (69) yields the solution formula

$$u(x, y) = \frac{1}{\sqrt{2\pi}} \int_0^\infty f(\xi)[g(x + \xi, y) + g(|x - \xi|, y)]\, d\xi \tag{71}$$

(see Problem 35 in Exercises 5.5).

EXERCISES 8.5

1. Use the solution formula (58) to solve (55) when

$$f(x) = \begin{cases} 1 - x^2, & |x| < 1 \\ 0, & |x| > 1. \end{cases}$$

2. Use the solution formula (58) to solve (55) when

---

*The inverse transform (70) must be evaluated by complex variable methods, which yields the result

$$g(x, y) = \frac{\sin(\pi y/2a) \cosh(\pi x/2a)}{\cosh(\pi x/a) - \cos(\pi y/a)}.$$

$$f(x) = \begin{cases} 0, & x < 0 \\ e^{-x}, & x > 0. \end{cases}$$

In Problems 3–5, use the prescribed function f and the Poisson integral formula (63) to find an explicit solution of (55).

3. $f(x) = \begin{cases} 0, & x < 0 \\ 1, & x > 0 \end{cases}$

4. $f(x) = \begin{cases} 1, & x < 0 \\ 0, & x > 0 \end{cases}$

5. $f(x) = T_0, \quad -\infty < x < \infty \quad$ (T_0 constant)

6. Given the Neumann problem for the half-plane

$$u_{xx} + u_{yy} = 0, \qquad -\infty < x < \infty, \qquad y > 0$$

$$\text{B.C.:} \quad \begin{cases} u_y(x, 0) = f(x), & -\infty < x < \infty \\ u(x, y) \to 0, \quad u_y(x, y) \to 0 & \text{as} \quad (x^2 + y^2) \to \infty. \end{cases}$$

(a) show that $v(x, y) = u_y(x, y)$ satisfies the Dirichlet problem

$$v_{xx} + v_{yy} = 0, \qquad -\infty < x < \infty, \qquad y > 0$$

$$\text{B.C.:} \quad \begin{cases} v(x, 0) = f(x), & -\infty < x < \infty \\ v(x, y) \to 0 \quad \text{as} & (x^2 + y^2) \to \infty. \end{cases}$$

(b) Using the Poisson integral formula (63) to solve (a), show that the solution of the Neumann problem is

$$u(x, y) = \frac{1}{2\pi} \int_{-\infty}^{\infty} f(t) \log[(t - x)^2 + y^2] \, dt + K,$$

where K is an arbitrary constant.

7. Use the Fourier transform to solve the Dirichlet problem for the half-plane described by (55).
Hint: Use the convolution theorem.

8. Use the Fourier sine transform to show that

$$u_{xx} + u_{yy} = 0, \qquad 0 < x < \infty, \qquad y > 0$$

$$\text{B.C.:} \quad \begin{cases} u(0, y) = 0, & u(x, 0) = f(x), \\ u(x, y) \to 0 \quad \text{as} & (x^2 + y^2) \to \infty \end{cases}$$

has the formal solution

$$u(x, y) = \frac{y}{\pi} \int_0^{\infty} f(t) \left[\frac{1}{(t - x)^2 + y^2} - \frac{1}{(t + x)^2 + y^2} \right] dt.$$

9. When $f(x) = 1$, show that the solution in Problem 8 reduces to

$$u(x, y) = \frac{2}{\pi} \arctan \frac{x}{y}.$$

10. Use the Fourier cosine transform to solve

$$u_{xx} + u_{yy} = 0, \qquad 0 < x < \infty, \qquad y > 0$$

$$\text{B.C.:} \quad \begin{cases} u_x(0, y) = 0, & u(x, 0) = f(x) \\ u(x, y) \to 0 \quad \text{as} & (x^2 + y^2) \to \infty, \end{cases}$$

and show that the solution has a form similar to that in Problem 8.

11. Solve Problem 10 when

$$f(x) = \begin{cases} 1, & 0 < x < c \\ 0, & c < x < \infty. \end{cases}$$

★12. Given the nonhomogeneous boundary value problem

$$u_{xx} + u_{yy} = -xe^{-x^2}, \qquad -\infty < x < \infty, \qquad y > 0$$

B.C.: $u(x, 0) = 0, \qquad u(x, y)$ finite as $y \to \infty$,

(a) show that the transformed problem via the Fourier transform has the solution

$$U(s, y) = (1 - e^{-|s|y}) \frac{i}{2\sqrt{2}s} e^{-s^2/4}.$$

(b) Taking the inverse Fourier transform of (a), deduce that

$$u(x, y) = \frac{1}{2\pi} \int_0^\infty (1 - e^{-sy}) \frac{\sin xs}{s} e^{-s^2/4}\, ds.$$

Hint: $\mathscr{F}\{xe^{-x^2/2}; s\} = ise^{-s^2/2}$.

□ 8.6 APPROXIMATION METHODS

It is either convenient or necessary to solve some problems involving the potential equation by an approximation or numerical method. We do this for basically the same reason as discussed in Section 6.6 in connection with heat conduction problems. In this section we will briefly discuss two methods of approximation. The first is a generalization of the *weighted residual methods* introduced in Section 1.7, which leads to an analytical approximation to the true solution, while the second technique is the *finite-difference numerical method* introduced in Section 6.6.

8.6.1 WEIGHTED RESIDUALS

The weighted residual methods discussed in Section 1.7 for ODEs can be extended to certain higher-dimensional potential problems. For the sake of illustration, we will select a rectangular domain and consider the particular Dirichlet problem

$$u_{xx} + u_{yy} = 0, \qquad 0 < x < a, \qquad 0 < y < b$$

$$\text{B.C.:} \qquad \begin{cases} u(0, y) = 0, & u(a, y) = 0 \\ u(x, 0) = 0, & u(x, b) = f(x). \end{cases} \tag{72}$$

The method of weighted residuals is a technique whereby we formulate a *trial solution* which satisfies exactly the prescribed boundary conditions but only approximately satisfies the PDE. Using a one-term trial solution, we will approximate the solution of (72) by

$$\psi(x, y) = c_1(y)f(x), \tag{73}$$

where $c_1(y)$ is an unknown function. In order that $\psi(x, y)$ satisfy the prescribed boundary conditions, we will restrict our choice of $f(x)$ to only those functions for which $f(0) = f(a) = 0$. In addition, the unknown function $c_1(y)$ must satisfy the conditions

$$\psi(x, 0) = c_1(0)f(x) = 0,$$

$$\psi(x, b) = c_1(b)f(x) = f(x),$$

which require that

$$c_1(0) = 0, \qquad c_1(b) = 1. \tag{74}$$

Next, we define the *error residual* by

$$E_1(x, y) = \psi_{xx} + \psi_{yy} = c_1(y)f''(x) + c_1''(y)f(x), \tag{75}$$

which we wish to minimize as a function of x. There are several ways in which this can be accomplished, but we will discuss only the *method of collocation* and *Galerkin's method*. The collocation method minimizes (75) by setting $E_1(x, y) = 0$ for some value of x, usually at $x = a/2$. This action leads to a second-order ODE in $c_1(y)$, which is then forced to satisfy the boundary conditions (74). Galerkin's method leads to a similar boundary value problem for $c_1(y)$ by setting

$$\int_0^a E_1(x, y)f(x)\,dx = 0. \tag{76}$$

EXAMPLE 7

Find an approximate solution of the Dirichlet problem in Example 1, i.e.,

$$u_{xx} + u_{yy} = 0, \qquad 0 < x < 1, \qquad 0 < y < 1$$

$$\text{B.C.:} \qquad \begin{cases} u(0, y) = 0, & u(1, y) = 0 \\ u(x, 0) = 0, & u(x, 1) = T_0x(1 - x). \end{cases}$$

SOLUTION Our trial solution has the form

$$\psi(x, y) = T_0c_1(y)x(1 - x), \qquad c_1(0) = 0, \qquad c_1(1) = 1,$$

which leads to the error residual

$$E_1(x, y) = T_0[c_1''(y)x(1 - x) - 2c_1(y)].$$

Employing first the collocation method, we set $E_1(x, y) = 0$ at $x = 1/2$ and obtain the boundary value problem

$$c_1''(y) - 8c_1(y) = 0, \qquad c_1(0) = 0, \qquad c_1(1) = 1.$$

The solution of this problem is

$$c_1(y) = \frac{\sinh 2\sqrt{2}y}{\sinh 2\sqrt{2}},$$

and thus our approximate solution of the Dirichlet problem is

$$\psi(x, y) = T_0 x(1 - x) \frac{\sinh 2\sqrt{2}y}{\sinh 2\sqrt{2}} \qquad \text{(collocation)}.$$

For Galerkin's method, we set

$$\int_0^1 E_1(x, y)x(1 - x)\, dx = 0,$$

which yields

$$c_1''(y) - 10c_1(y) = 0, \qquad c_1(0) = 0, \qquad c_1(1) = 1.$$

Therefore

$$c_1(y) = \frac{\sinh \sqrt{10}y}{\sinh \sqrt{10}},$$

from which it follows that

$$\psi(x, y) = T_0 x(1 - x) \frac{\sinh \sqrt{10}y}{\sinh \sqrt{10}} \qquad \text{(Galerkin)}.$$

As a means of comparison of our solutions, we note that the first two terms of the series solution in Example 1 yield

$$u(x, y) \simeq \frac{8T_0}{\pi^3}\left[\frac{\sinh \pi y \sin \pi x}{\sinh \pi} + \frac{\sinh 3\pi y \sin 3\pi x}{27 \sinh 3\pi}\right].$$

Evaluating all three solutions at the center of the plate $(\frac{1}{2}, \frac{1}{2})$, we have

$$\psi\left(\frac{1}{2}, \frac{1}{2}\right) = 0.0574T_0 \qquad \text{(collocation)}$$

$$\psi\left(\frac{1}{2}, \frac{1}{2}\right) = 0.0493T_0 \qquad \text{(Galerkin)}$$

$$u\left(\frac{1}{2}, \frac{1}{2}\right) \simeq 0.0506T_0 \qquad \text{(first two terms of series)}.$$

As discussed in Section 1.7, the Galerkin method is generally more accurate than the collocation method. Greater accuracy can be achieved in *both* techniques, however, if additional terms are included in the trial solution. Adding more terms to the trial solution here would necessitate solving simultaneous ODEs for the unknown functions $c_1(y)$, $c_2(y)$, and so on. If this is necessary to achieve the desired accuracy, it is probably best to abandon the method of weighted residuals and use a numerical technique like finite differences to solve the problem. Generally speaking, the weighted residual method is used only in those problems for which a rough approximation is satisfactory. The great appeal for such methods is that they can normally be applied quite easily, and in some cases they can be very accurate with only one term.

REMARK: *Although we have illustrated the weighted residual methods for only a Dirichlet problem, they can also be used for Neumann problems and those featuring a mix of Dirichlet and Neumann boundary conditions. Some of these other kinds of problems are taken up in the exercises. However, we should point out that weighted residual methods are normally restricted to regular-shaped domains such as rectangles or circles. For irregular-shaped domains, one must usually resort to a numerical procedure such as finite differences.*

8.6.2 FINITE DIFFERENCES

The *finite-difference numerical method* introduced in Section 6.6.1 can also be used to solve potential problems in two dimensions. For illustrative purposes, we will consider a simple Dirichlet problem for a square, which is characterized by

$$u_{xx} + u_{yy} = 0, \quad 0 < x < \frac{1}{2}, \quad 0 < y < \frac{1}{2}$$

$$\text{B.C.:} \quad \begin{cases} u(0, y) = 0, & u\left(\frac{1}{2}, y\right) = 200y \\ u(x, 0) = 0, & u\left(x, \frac{1}{2}\right) = 200x. \end{cases} \tag{77}$$

By recalling the central difference relation [see Equation (159) in Section 6.6]

$$f_i'' \simeq \frac{1}{h^2}(f_{i+1} - 2f_i + f_{i-1}), \tag{78}$$

we make the approximations

$$u_{xx} \simeq \frac{1}{h^2}(u_{i+1,j} - 2u_{i,j} + u_{i-1,j}) \tag{79}$$

and

$$u_{yy} \simeq \frac{1}{h^2}(u_{i,j+1} - 2u_{i,j} + u_{i,j-1}), \tag{80}$$

where $u_{i,j} = u(x_i, y_j)$, $i = 0, 1, 2, \ldots, n$ and $j = 0, 1, 2, \ldots, n$. We choose the same increment h here in both the x and y directions because the region of interest is a square. The next step is to choose a number n that defines the mesh size h. For $n = 4$, we find that $h = 1/8$, and by partitioning the square accordingly we obtain the mesh shown in Figure 8.8.

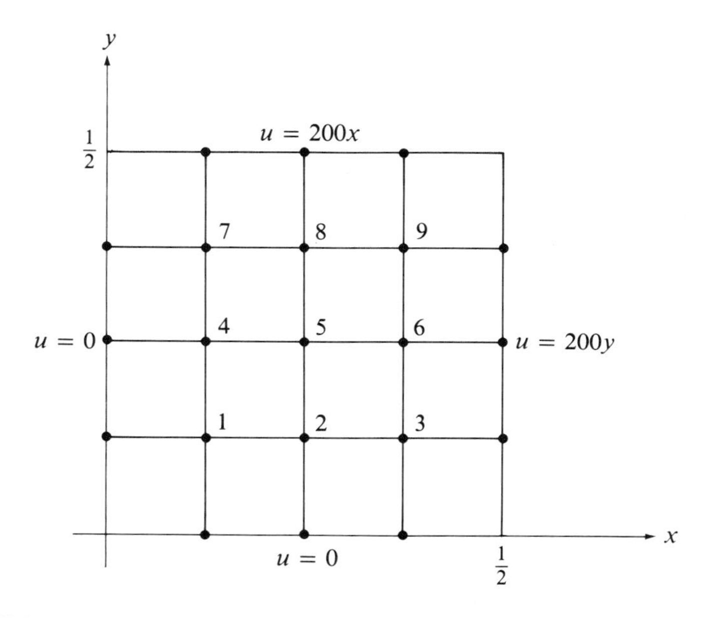

FIGURE 8.8

The potential equation and boundary conditions now take the form*

$$4u_{i,j} - [u_{i+1,j} + u_{i-1,j} + u_{i,j+1} + u_{i,j-1}] = 0; \qquad i = 1, 2, 3; \qquad j = 1, 2, 3$$

$$u_{0,j} = 0, \qquad u_{4,j} = 25j; \qquad j = 0, 1, 2, 3, 4 \tag{81}$$

$$u_{i,0} = 0, \qquad u_{i,4} = 25i; \qquad i = 0, 1, 2, 3, 4.$$

Expressing (81) in terms of the relabeled grid points in Figure 8.8, we obtain the system of equations:

* By solving the first equation in (81) for $u_{i,j}$, we find that it is equal to the average value of its four neighbors $u_{i+1,j}$, $u_{i-1,j}$, $u_{i,j+1}$, and $u_{i,j-1}$. This is simply the finite difference analog of the mean-value property given by (33) in Section 8.3.1.

$$
\begin{aligned}
4u_1 - u_2 - u_4 &= u_{1,0} + u_{0,1} \\
4u_2 - u_3 - u_1 - u_5 &= u_{2,0} \\
4u_3 - u_2 - u_6 &= u_{3,0} + u_{4,1} \\
4u_4 - u_5 - u_1 - u_7 &= u_{0,2} \\
4u_5 - u_6 - u_4 - u_2 - u_8 &= 0 \\
4u_6 - u_5 - u_3 - u_9 &= u_{4,2} \\
4u_7 - u_8 - u_4 &= u_{0,3} + u_{1,4} \\
4u_8 - u_9 - u_7 - u_5 &= u_{2,4} \\
4u_9 - u_8 - u_6 &= u_{3,4} + u_{4,3},
\end{aligned}
\tag{82}
$$

where the right-hand sides of the equations are obtained from the boundary conditions in (81). By evaluating these terms on the right-hand side, we get

$$
\begin{aligned}
&u_{1,0} = u_{2,0} = u_{3,0} = 0 \\
&u_{0,1} = u_{0,2} = u_{0,3} = 0 \\
&u_{1,4} = 25, \qquad u_{2,4} = 50, \qquad u_{3,4} = 75 \\
&u_{4,1} = 25, \qquad u_{4,2} = 50, \qquad u_{4,3} = 75.
\end{aligned}
\tag{83}
$$

The values of $u_1, u_2, \ldots, u_9$ are now found by solving the system of equations (82) with values for the right-hand sides given by (83). By using an elimination method, the values of u are those shown on the grid in Figure 8.9. Observe the symmetry in the solution, which should be incorporated into the solution process to reduce (82) to six equations and six unknowns.*

The mesh size $h = 1/4$ for the example provided here is much too coarse for most problems. A more realistic mesh size might be $h = 1/10$, which would lead to 81 linear algebraic equations with 81 unknowns which are reducible to a smaller number only if the problem exhibits some form of symmetry. For other kinds of boundary conditions, or domains other than rectangular, it may be necessary to consider even smaller mesh sizes which might produce several thousand equations to solve simultaneously. Even for today's computers, the solution of such large systems of equations by an elimination method may not be practical because of storage requirements, accumulation of roundoff errors, and time. In such cases it is almost always best to turn to an iterative technique, such as *Gauss-Seidel* or *Jacobi iteration,* for solving these large systems of equations. An iterative method starts with an initial approximation to the solution of the linear system of equations and then generates a sequence of solutions that, if it is stable, will converge to

* Note that $u_2 = u_4$, $u_3 = u_7$, and $u_6 = u_8$. Based on the boundary conditions, these symmetries should be anticipated in advance.

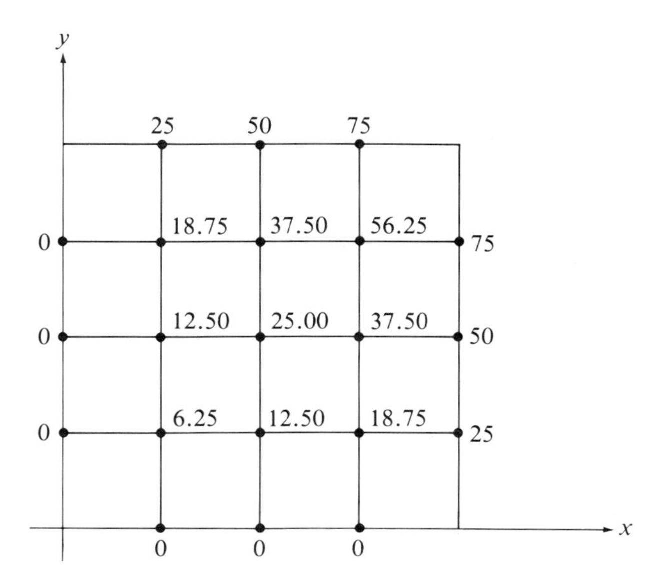

FIGURE 8.9

Numerical solution of (77)

the exact solution. Because only a finite number of iterations are possible, the solution obtained by an iterative method only approximates the solution of the linear system of algebraic equations (which themselves approximate the given boundary value problem). The number of iterations required for a prescribed accuracy is dependent upon the initial approximation. A good initial approximation is rewarded by fewer iterations to achieve the desired accuracy.* An important advantage of iterative methods over elimination methods is that the former suppress arithmetic errors that may be introduced at any earlier step of the solution process.

EXERCISES 8.6

In Problems 1–4, use the collocation method to find a one-term approximation to the solution.

1. $u_{xx} + u_{yy} = 0, \quad 0 < x < \pi, \quad 0 < y < 1$
   B.C.: $u(0, y) = 0, \quad u(\pi, y) = 0, \quad u(x, 0) = 0, \quad u(x, 1) = \sin x$
   Compare your answer with the exact solution.

* In practice, it is common to initially assume the solution is zero and then proceed with the iteration scheme.

2. $u_{xx} + u_{yy} = 0, \quad 0 < x < 1, \quad 0 < y < 1$
   B.C.: $u(0, y) = 0, \quad u(1, y) = 0, \quad u(x, 0) = x(1 - x), \quad u(x, 1) = 0$
   Hint: Choose $c_1(0) = 1$ and $c_1(1) = 0$.

3. $u_{xx} + u_{yy} = 0, \quad 0 < x < \pi, \quad 0 < y < 1$
   B.C.: $u_x(0, y) = 0, \quad u_x(\pi, y) = 0, \quad u(x, 0) = \cos x, \quad u(x, 1) = 0$

4. $u_{xx} + u_{yy} = 0, \quad 0 < x < \pi, \quad 0 < y < 1$
   B.C.: $u_x(0, y) = 0, \quad u_x(\pi, y) = 0, \quad u(x, 0) = 0, \quad u(x, 1) = \cos^2 x$

In Problems 5–8, use Galerkin's method to find a one-term approximation to the solution.

5. Do Problem 1.
6. Do Problem 2.
7. Do Problem 3.
8. Do Problem 4.

9. For the half-strip potential problem

$$u_{xx} + u_{yy} = 0, \quad 0 < x < 1, \quad y > 0$$

$$\text{B.C.:} \quad \begin{cases} u(x, 0) = x(1 - x), & u(x, y) \text{ finite as } y \to \infty, \quad 0 < x < 1 \\ u(0, y) = 0, & u(1, y) = 0, \quad y > 0, \end{cases}$$

   assume a trial solution of the form $\psi(x, y) = x(1 - x)c_1(y)$ and solve
   (a) by the collocation method.
   (b) by the Galerkin method.

★10. A square membrane under constant tension on all sides and supporting a uniformly distributed load leads to the *nonhomogeneous* PDE* and boundary conditions

$$u_{xx} + u_{yy} = -1, \quad 0 < x < 1, \quad 0 < y < 1$$

$$\text{B.C.:} \quad u(0, y) = 0, \quad u(1, y) = 0, \quad u(x, 0) = 0, \quad u(x, 1) = 0.$$

   Assuming a trial solution $\psi(x, y) = x(1 - x)c_1(y)$, find the maximum displacement of the membrane by using
   (a) the collocation method.
   (b) the Galerkin method.

In Problems 11–14, use the finite difference technique with $n = 4$ to find numerical solutions. Use symmetry to reduce the number of unknowns.

11. $u_{xx} + u_{yy} = 0, \quad 0 < x < 1, \quad 0 < y < 1$

$$\text{B.C.:} \quad \begin{cases} u(0, y) = 0, \quad u(1, y) = 0 \\ u(x, 0) = u(x, 1) = \begin{cases} 2x, & 0 < x < 1/2 \\ 2(1 - x), & 1/2 < x < 1 \end{cases} \end{cases}$$

   Hint: Using symmetry, there are only *four* unknowns.

---

* This is called *Poisson's equation* (see Section 10.6).

12. $u_{xx} + u_{yy} = -1, \quad 0 < x < 1, \quad 0 < y < 1$

$$\text{B.C.:} \begin{cases} u(0, y) = 0, & u(1, y) = 0 \\ u(x, 0) = 0, & u(x, 1) = 0 \end{cases}$$

13. $u_{xx} + u_{yy} = 0, \quad 0 < x < 1, \quad 0 < y < 1$

$$\text{B.C.:} \begin{cases} u(0, y) = 0, & u(1, y) = 0 \\ u(x, 0) = 0, & u(x, 1) = 100 \end{cases}$$

14. $u_{xx} + u_{yy} = 0, \quad 0 < x < 1, \quad 0 < y < 1$

$$\text{B.C.:} \begin{cases} u(0, y) = y, & u(1, y) = 0 \\ u(x, 0) = x, & u(x, 1) = 0 \end{cases}$$

★15. Find the finite difference representation of the following mixed potential problem for $n = 4$ in both the x and y directions, but do not solve.

$$u_{xx} + u_{yy} = 0, \quad 0 < x < \pi, \quad 0 < y < 1$$

$$\text{B.C.:} \begin{cases} u_x(0, y) = 0, & u_x(\pi, y) = 0 \\ u(x, 0) = \cos x, & u(x, 1) = \cos^2 x. \end{cases}$$

★16. By using the method of weighted residuals to solve (77), let

$$\psi(x, y) = c_1(y)200x.$$

(a) What is $c_1(y)$ so that all boundary conditions are satisfied?
(b) What is $E_1(x, y)$ in this case?
(c) Using the result of part (a), compare the values predicted by $\psi(x, y)$ with those given in Figure 8.9.

BIBLIOGRAPHY

Ames, W. F. *Numerical Methods for Partial Differential Equations*. New York: Barnes & Noble, 1969.

Bateman, H. *Partial Differential Equations of Mathematical Physics*. London: Cambridge University Press, 1959.

Berg, P. W., and J. L. McGregor. *Elementary Partial Differential Equations*. San Francisco: Holden-Day, 1966.

Butkov, E. *Mathematical Physics*. Reading, Mass.: Addison-Wesley, 1968.

Carslaw, H. S. *Introduction to the Theory of Fourier's Series and Integrals*. New York: Dover, 1950.

Chester, C. R. *Techniques in Partial Differential Equations*. New York: McGraw-Hill, 1971.

Churchill, R. V., and J. W. Brown. *Fourier Series and Boundary Value Problems*. 3rd ed. New York: McGraw-Hill, 1972.

Dennemeyer, R. *Introduction to Partial Differential Equations and Boundary Value Problems*. New York: McGraw-Hill, 1968.

Duff, G. F. D., and D. Naylor. *Differential Equations of Applied Mathematics*. New York: John Wiley & Sons, 1966.

Haberman, R. *Elementary Applied Partial Differential Equations*. Englewood Cliffs, N.J.: Prentice-Hall, 1983.

Kreyszig, E. *Advanced Engineering Mathematics*. 4th ed. New York: John Wiley & Sons, 1979.

Powers, D. *Boundary Value Problems*. 2nd ed. New York: Academic Press, 1979.

Rainville, E. D. *The Laplace Transform*. New York: Macmillan, 1963.

Sagan, H. *Boundary and Eigenvalue Problems in Mathematical Physics*. New York: John Wiley & Sons, 1961.

Smith, G. D. *Numerical Solution of Partial Differential Equations*. 2nd ed. Oxford: Clarendon Press, 1978.

Sneddon, I. N. *Elements of Partial Differential Equations*. New York: McGraw-Hill, 1957.

———. *The Use of Integral Transforms*. New York: McGraw-Hill, 1972.

Stakgold, I. *Boundary Value Problems of Mathematical Physics*. 2 vols. New York: Macmillan, 1967–1968.

Tolstov, G. P. *Fourier Series*. Englewood Cliffs, N.J.: Prentice-Hall, 1962.

Webster, A. G. *Partial Differential Equations of Mathematical Physics*. New York: Dover, 1955.

Weinberger, H. F. *A First Course in Partial Differential Equations*. New York: Blaisdell, 1965.

PART III

SPECIAL FUNCTIONS AND PARTIAL DIFFERENTIAL EQUATIONS

CHAPTER 9

SPECIAL FUNCTIONS

Most of the functions encountered in introductory analysis belong to the class of *elementary functions*. This class is composed of polynomials, rational functions, transcendental functions (trigonometric, inverse trigonometric, exponential, logarithmic, and so on), and functions constructed by combining two or more of these functions through addition, subtraction, multiplication, division, or composition. Beyond these lies a class of *special functions,* which are important in a variety of engineering and physics applications. For the problems that interest us, they arise mostly as *power series solutions* of certain variable–coefficient DEs.

A singular Sturm-Liouville system involving *Legendre's equation* is discussed in Section 9.2. The eigenfunctions of this system are polynomials known as the *Legendre polynomials*. We also briefly discuss *Legendre functions of the second kind* that arise as a second solution of Legendre's equation.

In Section 9.3 we introduce the *gamma function* by means of a nonelementary integral. Much of the usefulness of the gamma function stems from its role in defining other special functions, such as the *Bessel functions* that are studied in Section 9.4. Eigenvalue problems involving Bessel functions are discussed in Section 9.5, while in Section 9.6 we briefly look at *modified Bessel functions*.

Generalized Fourier series (Chapter 4) are introduced again in Section 9.7 in connection with Legendre polynomials and Bessel functions.

9.1 INTRODUCTION

If the material properties such as conductivity and mass density are varying quantities in a heat conduction problem, the governing heat equation takes on the more general form

$$\frac{\partial}{\partial x}[p(x)u_x] = r(x)u_t, \tag{1}$$

where $p(x)$ and $r(x)$ are positive known functions. Upon separating the variables with $u(x, t) = X(x)W(t)$, the resulting eigenequation becomes the *variable coefficient* DE

$$\frac{d}{dx}[p(x)X'] + \lambda r(x)X = 0, \tag{2}$$

which is a specialized form of the Sturm-Liouville equation studied in Section 1.6. This same equation also occurs in certain wave propagation problems involving non-constant parameters. Similar forms of variable coefficient DEs arise in solving the heat, wave, and potential equations in geometries other than rectangular (see Chapter 10). With the exception of the Cauchy-Euler equation, we have not thus far addressed this class of equations.

The solution of a linear, homogeneous DE with constant coefficients can always be expressed in terms of *elementary functions,* such as exponentials or sines and cosines. A variable coefficient DE, however, rarely yields explicit solutions in terms of elementary functions, except for Cauchy-Euler equations. Therefore, in attempting to solve such equations it is usually fruitless to assume some general solution form as a combination of elementary functions wherein only certain parameters need to be identified. Rather, we almost always resort to a power series method* in solving variable coefficients DEs, and in so doing, we use the resulting power series to define "new functions." Some of these DEs occur so often in practice that their solutions, called *special functions,* have been singled out and studied in great detail. We will consider two such DEs and some related equations in the present chapter.

9.2 LEGENDRE'S EQUATION

In solving certain potential problems displaying spherical symmetry, it is common for the separation of variables method to lead to the eigenequation

$$(1 - x^2)y'' - 2xy' + \lambda y = 0, \qquad -1 < x < 1, \tag{3}$$

called *Legendre's equation* after its discoverer.† By writing (3) in the self-adjoint form

$$\frac{d}{dx}[(1 - x^2)y'] + \lambda y = 0$$

and comparing this expression with the general self-adjoint form (Section 1.6)

$$\frac{d}{dx}[p(x)y'] + [q(x) + \lambda r(x)]y = 0,$$

we can readily identify the functions

---

*For a review of the power series method, see Chapter 9 in L. C. Andrews, *Ordinary Differential Equations with Applications,* (Glenview, Ill.: Scott, Foresman & Co., 1982).

†ADRIEN MARIE LEGENDRE (1752–1833) was a French mathematician who spent over 40 years of his life extending the theory of elliptic integrals. He also studied mechanics and astronomy.

$$p(x) = 1 - x^2, \qquad q(x) = 0, \qquad r(x) = 1.$$

Because $p(-1) = p(1) = 0$, we see that Legendre's equation has singularities at both endpoints $x = \pm 1$. Thus, in order to be assured of generating a set of orthogonal eigenfunctions of (3), we must impose the special boundary conditions

$$y(x), \quad y'(x) \quad \text{finite as} \quad x \to -1^+, \quad x \to 1^- \tag{4}$$

(see Section 1.6.4). It turns out that such conditions are consistent with physical considerations in connection with Legendre's equation.

To find a general solution of (3), we begin by assuming there is a power series solution of the form*

$$y = \sum_{k=0}^{\infty} c_k x^k \tag{5}$$

where the c's are constants to be determined. The termwise differentiation of (5) gives us the additional series

$$y' = \sum_{k=0}^{\infty} kc_k x^{k-1}, \qquad y'' = \sum_{k=0}^{\infty} k(k-1)c_k x^{k-2}. \tag{6}$$

By substituting the series (5) and (6) into Legendre's equation (3), we obtain

$$(1 - x^2)\sum_{k=0}^{\infty} k(k-1)c_k x^{k-2} - 2x\sum_{k=0}^{\infty} kc_k x^{k-1} + \lambda\sum_{k=0}^{\infty} c_k x^k = 0,$$

and by carrying out the multiplications indicated above we get

$$\sum_{k=0}^{\infty} k(k-1)c_k x^{k-2} - \sum_{k=0}^{\infty} k(k-1)c_k x^k - 2\sum_{k=0}^{\infty} kc_k x^k + \lambda\sum_{k=0}^{\infty} c_k x^k = 0.$$

If we replace the index k by $k - 2$ in the last three summations, adjust the summands accordingly, and combine the results of this action, we find

$$0 \cdot c_0 x^{-2} + 0 \cdot c_1 x^{-1} + \sum_{k=2}^{\infty} [k(k-1)c_k$$
$$- (k-2)(k-3)c_{k-2} - 2(k-2)c_{k-2} + \lambda c_{k-2}]x^{k-2} = 0.$$

Consequently, both c_0 and c_1 are arbitrary, and the remaining constants can be determined by equating to zero the coefficient of x^{k-2}; i.e.,

$$k(k-1)c_k - [(k-2)(k-1) - \lambda]c_{k-2} = 0, \qquad k = 2, 3, 4, \ldots,$$

which is called the *recurrence formula*. After making the substitution $\lambda = n(n + 1)$,† and rearranging the terms, we can write the recurrence relation in the more useful form

---

* Note that $x = 0$ is an ordinary point of the equation.

† Experience has shown that writing λ in this way simplifies the algebra.

$$c_k = \frac{(k-n-2)(k+n-1)}{k(k-1)} c_{k-2}, \qquad k = 2, 3, 4, \ldots . \tag{7}$$

This recurrence formula gives each coefficient in terms of the second one preceding it, except for c_0 and c_1 which are arbitrary. We find successively,

$$c_2 = -\frac{n(n+1)}{2!} c_0,$$

$$c_3 = -\frac{(n-1)(n+2)}{3!} c_1,$$

$$c_4 = -\frac{(n-2)(n+3)}{4 \cdot 3} c_2 = \frac{(n-2)n(n+1)(n+3)}{4!} c_0,$$

$$c_5 = -\frac{(n-3)(n+4)}{5 \cdot 4} c_3 = \frac{(n-3)(n-1)(n+2)(n+4)}{5!} c_1,$$

and so forth. By inserting these coefficients back into (5), we obtain the general solution

$$y = c_0 y_1(x) + c_1 y_2(x), \tag{8}$$

where

$$y_1(x) = 1 - \frac{n(n+1)}{2!} x^2 + \frac{(n-2)n(n+1)(n+3)}{4!} x^4 - \ldots , \tag{9a}$$

and

$$y_2(x) = x - \frac{(n-1)(n+2)}{3!} x^3 + \frac{(n-3)(n-1)(n+2)(n+4)}{5!} x^5 - \ldots . \tag{9b}$$

It can be shown that for general values of n, both y_1 and y_2 are unbounded at $x = \pm 1$,* and hence the boundary conditions (4) cannot be satisfied. However, if $n = 0, 1, 2, \ldots$, then either y_1 or y_2 will reduce to a polynomial of degree n, which is of course bounded at $x = \pm 1$. If n is even, y_1 reduces to a polynomial, and if n is odd, y_2 is a polynomial. That multiple of the polynomial of degree n that has the value unity when $x = 1$ is called the nth *Legendre polynomial* and is denoted by the symbol $P_n(x)$. Thus, Legendre's equation (3) subject to the boundary conditions (4) has the set of eigenvalues and eigenfunctions

$$\lambda_n = n(n+1), \qquad \phi_n(x) = P_n(x), \qquad n = 0, 1, 2, \ldots \tag{10}$$

* See Appendix 1 in W. W. Bell, *Special Functions for Scientists and Engineers,* (London: Van Nostrand, 1968).

where

$$P_n(x) = \begin{cases} y_1(x)/y_1(1), & n \text{ even} \\ y_2(x)/y_2(1), & n \text{ odd.} \end{cases} \tag{11}$$

The first few Legendre polynomials obtained from Equation (11) are readily found to be

$$P_0(x) = 1,$$

$$P_1(x) = x,$$

$$P_2(x) = \frac{1}{2}(3x^2 - 1),$$

$$P_3(x) = \frac{1}{2}(5x^3 - 3x),$$

$$P_4(x) = \frac{1}{8}(35x^4 - 30x^2 + 3),$$

$$P_5(x) = \frac{1}{8}(63x^5 - 70x^3 + 15x),$$

while in general it can be shown that*

$$P_n(x) = \sum_{k=0}^{[n/2]} \frac{(-1)^k (2n-2k)!\, x^{n-2k}}{2^n k!(n-k)!(n-2k)!}, \qquad n = 0, 1, 2, \ldots, \tag{12}$$

where

$$[n/2] = \begin{cases} n/2, & n \text{ even} \\ (n-1)/2, & n \text{ odd.} \end{cases} \tag{13}$$

Another formula commonly used to generate these polynomials is *Rodrigues' formula*†

$$P_n(x) = \frac{1}{2^n n!} \frac{d^n}{dx^n}[(x^2 - 1)^n], \qquad n = 0, 1, 2, \ldots \tag{14}$$

(see Problems 1 and 4 in Exercises 9.2). The first few Legendre polynomials are sketched over the interval $[-1, 1]$ in Figure 9.1.

9.2.1 LEGENDRE FUNCTIONS OF THE SECOND KIND

The function $P_n(x)$ represents only one solution of Legendre's equation (3), which corresponds to either y_1 or y_2 depending upon whether n is even or

---

* See Chapter 4 in L. C. Andrews, *Special Functions for Engineers and Applied Mathematicians*, (New York: Macmillan, 1985).

† Named after OLINDE RODRIGUES (1794–1851).

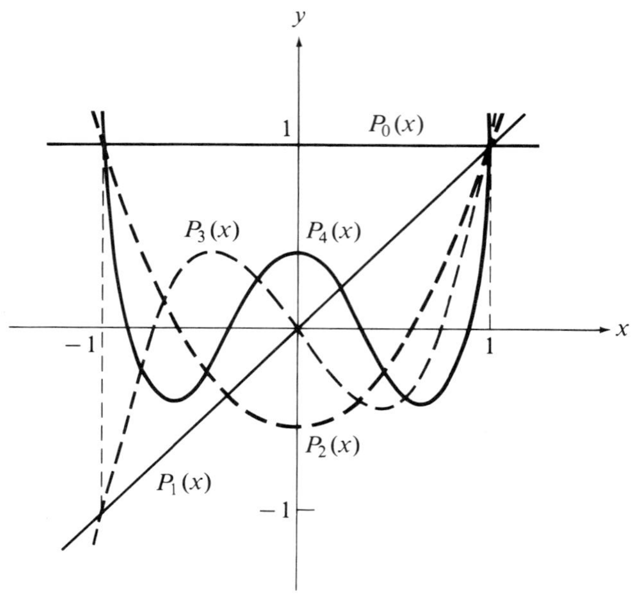

FIGURE 9.1

Graph of $P_n(x)$, $n = 0, 1, 2, 3, 4$

odd. The remaining infinite series in each case defines a second linearly independent solution, multiples of which are called *Legendre functions of the second kind*. We define such functions by

$$Q_n(x) = \begin{cases} y_1(1)y_2(x), & n \text{ even} \\ -y_2(1)y_1(x), & n \text{ odd} \end{cases} \tag{15}$$

(see also Problem 14 in Exercises 9.2). Thus, for $\lambda = n(n+1)$, $n = 0, 1, 2, \ldots$, we say that Legendre's equation (3) has the general solution

$$y = C_1P_n(x) + C_2Q_n(x), \qquad n = 0, 1, 2, \ldots. \tag{16}$$

Specifically, for $n = 0$ we find that $y_1(x) = 1$ and

$$y_2(x) = x + \frac{x^3}{3} + \frac{x^5}{5} + \frac{x^7}{7} + \cdots = \frac{1}{2}\log\frac{1+x}{1-x}, \qquad |x| < 1$$

so that using (15), we deduce that

$$Q_0(x) = \frac{1}{2}\log\frac{1+x}{1-x}. \tag{17}$$

By some involved manipulation, it can also be shown that

$$Q_1(x) = xQ_0(x) - 1,$$

$$Q_2(x) = P_2(x)Q_0(x) - \frac{3}{2}x,$$

$$Q_3(x) = P_3(x)Q_0(x) - \frac{5}{2}x^2 + \frac{2}{3}.$$

In general,

$$Q_n(x) = P_n(x)Q_0(x) - \sum_{k=0}^{[(n-1)/2]} \frac{(2n-4k-1)}{(2k+1)(n-k)} P_{n-2k-1}(x) \tag{18}$$

$(n = 1, 2, 3, \ldots)$.

Note that all $Q_n(x)$, $n = 0, 1, 2, \ldots$, have singularities at $x = \pm 1$ due to the presence of the logarithm term in $Q_0(x)$. Hence, because the Legendre functions of the second kind do not satisfy the boundary conditions (4), they cannot be eigenfunctions of Legendre's equation. We conclude, therefore, that there are no further eigenvalues and eigenfunctions of (3) and (4) other than those given by (10).

9.2.2 SPECIAL PROPERTIES OF $P_n(x)$

The Legendre polynomials are rich in recurrence relations and basic identities. For reference purposes, we present a list of some of the most useful of these relations, a few of which are taken up in the exercises.*

Basic Identities for $P_n(x)$

(L1): $$P_n(x) = \sum_{m=0}^{[n/2]} \frac{(-1)^m(2n-2m)!\,x^{n-2m}}{2^n m!(n-m)!(n-2m)!}, \qquad n = 0, 1, 2, \ldots$$

(L2): $$P_n(x) = \frac{1}{2^n n!}\frac{d^n}{dx^n}[(x^2-1)^n], \qquad n = 0, 1, 2, \ldots$$

(L3): $P_n(1) = 1, \qquad P_n(-1) = (-1)^n, \qquad n = 0, 1, 2, \ldots$

(L4): $P_n'(1) = \frac{1}{2}n(n+1), \qquad P_n'(-1) = (-1)^{n-1}\frac{1}{2}n(n+1),$
$n = 0, 1, 2, \ldots$

(L5): $$P_{2n}(0) = \frac{(-1)^n(2n)!}{2^{2n}(n!)^2}, \qquad P_{2n+1}(0) = 0, \qquad n = 0, 1, 2, \ldots$$

(L6): $(n+1)P_{n+1}(x) - (2n+1)xP_n(x) + nP_{n-1}(x) = 0, \qquad n = 1, 2, 3, \ldots$

(L7): $P_{n+1}'(x) - 2xP_n'(x) + P_{n-1}'(x) - P_n(x) = 0, \qquad n = 1, 2, 3, \ldots$

(L8): $P_{n+1}'(x) - xP_n'(x) - (n+1)P_n(x) = 0, \qquad n = 0, 1, 2, \ldots$

(L9): $xP_n'(x) - P_{n-1}'(x) - nP_n(x) = 0, \qquad n = 1, 2, 3, \ldots$

---

* For more details on the derivation of these properties, see Chapter 4 in L. C. Andrews, *Special Functions for Engineers and Applied Mathematicians,* (New York: Macmillan, 1985).

(L10): $P'_{n+1}(x) - P'_{n-1}(x) = (2n + 1)P_n(x), \qquad n = 1, 2, 3, \ldots$

(L11): $|P_n(x)| \leq 1, \qquad |x| \leq 1, \qquad n = 0, 1, 2, \ldots$

EXERCISES 9.2

1. Determine the Legendre polynomials $P_1(x)$, $P_2(x)$, $P_3(x)$, and $P_4(x)$
   (a) using (12).
   (b) using (14).

2. Show that $P_n(-x) = (-1)^n P_n(x), \quad n = 0, 1, 2, \ldots.$

3. Determine the subset of eigenvalues $\lambda_n = n(n + 1)$, $n = 0, 1, 2, \ldots$, and corresponding eigenfunctions for each of the following problems:
   (a) $(1 - x^2)y'' - 2xy' + \lambda y = 0, \quad y(0) = 0, \quad y(1)$ finite
   (b) $(1 - x^2)y'' - 2xy' + \lambda y = 0, \quad y'(0) = 0, \quad y(1)$ finite

4. Establish the formula

$$\frac{d^n}{dx^n}(x^m) = \begin{cases} \dfrac{m!}{(m-n)!}\, x^{m-n}, & n \leq m \\ 0, & n > m \end{cases}$$

   and use it to verify Rodrigues' formula (14).

   Hint: $(x^2 - 1)^n = \sum_{k=0}^{n} \dfrac{(-1)^n n!}{k!(n-k)!}\, x^{2n-2k}.$

5. Make the change of variable $x = \cos\phi$ in the following DE and show that it reduces to Legendre's equation:

$$\frac{1}{\sin\phi}\frac{d}{d\phi}\left(\sin\phi\,\frac{dy}{d\phi}\right) + n(n + 1)y = 0$$

   Hint: First show that $\dfrac{d}{d\phi} = -\sin\phi\,\dfrac{d}{dx}.$

★6. If the tension in a rotating string (Section 1.5.2) is a function of x given by $T(x) = 1 - x^2$, the resulting DE governing the motions of the string supporting a distributed load $q(x)$ is

$$\frac{d}{dx}[(1 - x^2)y'] + \rho\omega^2 y = q(x), \qquad -1 < x < 1$$

$$y(x),\ y'(x) \ \text{ finite as } \ x \to -1^+, \qquad x \to 1^-.$$

   Assuming the string is *not* rotating at one of the critical speeds ω_n, determine the displacements when
   (a) $q(x) = 1.$
   (b) $q(x) = x.$
   Hint: Use the eigenfunction expansion method of Section 4.6.

★7. Solve Problem 6 when the string is rotating at the second critical speed ω_2.

8. Without the use of Theorem 1.4, show directly that

$$\int_{-1}^{1} P_n(x)P_k(x)\,dx = 0, \qquad k \neq n.$$

Hint: Note that $P_n(x)$ and $P_k(x)$ satisfy, respectively,

$$\frac{d}{dx}[(1-x^2)P_n'(x)] + n(n+1)P_n(x) = 0,$$

$$\frac{d}{dx}[(1-x^2)P_k'(x)] + k(k+1)P_k(x) = 0.$$

Multiply the first DE by $P_k(x)$ and the second by $P_n(x)$, subtract the results, and integrate from -1 to 1.

9. Use the orthogonality property in Problem 8 to explain why

(a) $\displaystyle\int_{-1}^{1} P_n(x)\,dx = 0, \qquad n = 1, 2, 3, \ldots.$

(b) $\displaystyle\int_{-1}^{1} (ax+b)P_n(x)\,dx = 0, \qquad n = 2, 3, 4, \ldots \qquad (a, b \text{ constant}).$

In Problems 10–13, find a general solution in terms of $P_n(x)$ and $Q_n(x)$.

10. $(1-x^2)y'' - 2xy' = 0$
11. $(1-x^2)y'' - 2xy' + 2y = 0$
12. $(1-x^2)y'' - 2xy' + 12y = 0$
13. $(1-x^2)y'' - 2xy' + 30y = 0$
14. For appropriate values of A_n and B_n the Legendre functions of the second kind can be defined by ($n = 0, 1, 2, \ldots$)

$$Q_n(x) = A_nP_n(x) + B_nP_n(x)\int \frac{dx}{(1-x^2)[P_n(x)]^2}.$$

Using this relation, verify that
(a) if $A_0 = 0$ and $B_0 = 1$, then

$$Q_0(x) = \frac{1}{2}\log\frac{1+x}{1-x}.$$

(b) if $A_1 = 0$ and $B_1 = 1$, then

$$Q_1(x) = xQ_0(x) - 1.$$

15. Verify directly that the Wronskian of $P_0(x)$ and $Q_0(x)$ yields

$$W(P_0, Q_0)(x) = \frac{1}{1-x^2}.$$

16. For arbitrary values of A_n and B_n ($B_n \neq 0$), use the result of Problem 14 to deduce that the Wronskian of $P_n(x)$ and $Q_n(x)$ yields

$$W(P_n, Q_n)(x) = \frac{B_n}{1-x^2}, \qquad n = 0, 1, 2, \ldots.$$

★17. Construct a Green's function for the boundary value problem

$$\frac{d}{dx}[(1-x^2)y'] + 6y = f(x), \qquad y(0) = 0, \qquad y(1) \text{ finite.}$$

Hint: Use Problem 16.

★18. For $n = 0, 1, 2, \ldots$, show that

(a) $P_n'(1) = \frac{1}{2}n(n+1)$.

(b) $P_n'(-1) = (-1)^{n-1}\frac{1}{2}n(n+1)$.

★19. Starting with the binomial series*

$$(1-u)^{-1/2} = \sum_{n=0}^{\infty}\binom{-1/2}{n}(-1)^n u^n, \qquad |u| < 1,$$

(a) set $u = t(2x - t)$ and deduce that

$$(1 - 2xt + t^2)^{-1/2} = \sum_{n=0}^{\infty}\sum_{k=0}^{n}\binom{-1/2}{n}\binom{n}{k}(-1)^{n+k}t^{n+k}.$$

(b) Make the change of index $n \to (n - k)$ in (a) and show that

$$(1 - 2xt + t^2)^{-1/2} = \sum_{n=0}^{\infty}\left\{\sum_{k=0}^{[n/2]}\binom{-1/2}{n-k}\binom{n-k}{k}(-1)^n(2x)^{n-2k}\right\}t^n.$$

(c) Finally, verify that the inner series in (b) is the nth Legendre polynomial equivalent to (12), and hence deduce the *generating function* relation

$$(1 - 2xt + t^2)^{-1/2} = \sum_{n=0}^{\infty}P_n(x)t^n.$$

20. Given the generating function $w(x, t) = (1 - 2xt + t^2)^{-1/2}$ from Problem 19,
(a) show that it satisfies the identity

$$(1 - 2xt + t^2)\frac{\partial w}{\partial t} + (t - x)w = 0.$$

(b) Substitute the series $\sum_{n=0}^{\infty}P_n(x)t^n$ into the above identity for $w(x, t)$ and show that it reduces to

$$\sum_{n=2}^{\infty}[nP_n(x) - (2n-1)xP_{n-1}(x) + (n-1)P_{n-2}(x)]t^{n-1} = 0.$$

---

* Recall from calculus that $(1 + x)^a = \sum_{n=0}^{\infty}\binom{a}{n}x^n, |x| < 1$, where $\binom{a}{n}$ is the binomial coefficient defined by

$$\binom{a}{0} = 1, \qquad \binom{a}{n} = \frac{a(a-1)(a-2)\cdots(a-n+1)}{n!}, \qquad n = 1, 2, 3, \ldots.$$

(c) From (b), deduce the *recurrence formula* ($n = 1, 2, 3, \ldots$)

$$(n+1)P_{n+1}(x) - (2n+1)xP_n(x) + nP_{n-1}(x) = 0.$$

21. Given the generating function $w(x, t) = (1 - 2xt + t^2)^{-1/2}$ from Problem 19,
(a) show that it satisfies the identity

$$(1 - 2xt + t^2)\frac{\partial w}{\partial x} - tw = 0.$$

(b) Substitute the series $\sum_{n=0}^{\infty} P_n(x)t^n$ into the above identity for $w(x, t)$ and show that it reduces to

$$\sum_{n=2}^{\infty} [P_n'(x) - 2xP_{n-1}'(x) + P_{n-2}'(x) - P_{n-1}(x)]t^n = 0.$$

(c) From (b), deduce the relation ($n = 1, 2, 3, \ldots$)

$$P_{n+1}'(x) - 2xP_n'(x) + P_{n-1}'(x) - P_n(x) = 0.$$

22. Using Problem 20(c) and Problem 21(c), derive the following relations:
(a) $P_{n+1}'(x) - xP_n'(x) - (n+1)P_n(x) = 0$
(b) $xP_n'(x) - P_{n-1}'(x) - nP_n(x) = 0$
(c) $P_{n+1}'(x) - P_{n-1}'(x) = (2n+1)P_n(x)$
(d) $(1 - x^2)P_n'(x) = nP_{n-1}(x) - nxP_n(x)$

★23. *Associated Legendre functions.* Make the substitution $y = u(x)(1 - x^2)^{m/2}$ in the DE (called the *associated Legendre equation*)

$$(1 - x^2)y'' - 2xy' + \left[n(n+1) - \frac{m^2}{1 - x^2}\right]y = 0$$

(a) and show that $u(x)$ satisfies

$$(1 - x^2)u'' - 2(m+1)xu' + [n(n+1) - m(m+1)]u = 0.$$

(b) Verify that

$$u = \frac{d^m}{dx^m} P_n(x), \qquad m = 1, 2, 3, \ldots$$

is a solution of the DE in (a). The corresponding function $y(x)$ is denoted by $P_n^m(x)$ and called the *associated Legendre function*. Thus,

$$P_n^m(x) = (1 - x^2)^{m/2}\frac{d^m}{dx^m} P_n(x),$$

which plays an important role in quantum mechanics, is a solution of the original DE in y.

★24. *Laguerre polynomials.* By assuming a power series solution of the form

$$y = \sum_{k=0}^{\infty} c_k x^k,$$

show that for $c_0 = 1$, one solution of *Laguerre's equation*

$$xy'' + (1 - x)y' + ny = 0$$

becomes*

$$y = L_n(x) = \sum_{k=0}^{n} \frac{(-1)^k n! x^k}{(n-k)!(k!)^2}.$$

★25. *Hermite Polynomials.* The Hermite polynomials† are defined by

$$H_n(x) = \sum_{k=0}^{[n/2]} \frac{(-1)^k n!(2x)^{n-2k}}{k!(n-2k)!}, \qquad n = 0, 1, 2, \ldots.$$

(a) Show that $y = H_n(x)$ satisfies the DE

$$y'' - 2xy' + 2ny = 0.$$

(b) If $u = H_n(x)e^{-x^2/2}$ is a solution of *Weber's equation*

$$u'' + (\lambda - x^2)u = 0$$

deduce that $\lambda = 2n - 1, n = 0, 1, 2, \ldots$.

9.3 GAMMA FUNCTION

Special functions occur not only in the process of solving DEs, but in many cases arise out of evaluating a nonelementary integral. For example, in earlier chapters (Chapters 5 and 6) we encountered the error function erf(t) and complementary error function erfc(t). Here now we consider another important example, called the *gamma function,* which we define by

$$\Gamma(x) = \int_0^\infty e^{-t} t^{x-1}\, dt, \qquad x > 0. \tag{19}$$

The value of the integral depends only on the value of x, which is restricted to positive values. It is an improper integral due to the infinite limit of integration. Also, for $0 < x < 1$, the factor t^{x-1} becomes infinite at $t = 0$. Nonetheless, it can be shown that the integral (19) converges (uniformly) for all $a \leqslant x \leqslant b$, where $0 < a < b < \infty$.

If we set $x = 1$ in (19), we find that

$$\Gamma(1) = \int_0^\infty e^{-t}\, dt = -e^{-t}\Big|_0^\infty, \ddagger$$

or

$$\Gamma(1) = 1. \tag{20}$$

---

* The polynomial $L_n(x)$, called a *Laguerre polynomial,* is named after the French mathematician EDMOND LAGUERRE (1834–1886).

† Named after the French mathematician CHARLES HERMITE (1822–1901).

‡ For abbreviation, $\Big|_0^\infty$ will denote $\lim_{b\to\infty} (\quad)\Big|_0^b$.

Other values of the gamma function are not so easily obtained, but the substitution of $(x + 1)$ for x in (19), combined with an integration by parts, leads to

$$\Gamma(x+1) = \int_0^\infty e^{-t} t^x \, dt$$

$$= -e^{-t} t^x \Big|_0^\infty + x \int_0^\infty e^{-t} t^{x-1} \, dt,$$

from which we deduce the *recurrence formula*

$$\Gamma(x+1) = x\Gamma(x). \tag{21}$$

A direct connection between the gamma function and factorials can now be obtained from (20) and (21). That is, if we combine these relations, we see that

$$\begin{aligned} \Gamma(2) &= 1 \cdot \Gamma(1) = 1, \\ \Gamma(3) &= 2 \cdot \Gamma(2) = 2 \cdot 1 = 2!, \\ \Gamma(4) &= 3 \cdot \Gamma(3) = 3 \cdot 2! = 3!, \\ &\ldots\ldots\ldots\ldots\ldots\ldots\ldots, \end{aligned}$$

and through mathematical induction it can be verified that

$$\Gamma(n+1) = n!, \qquad n = 0, 1, 2, \ldots. \tag{22}$$

This last relation is interesting because it illustrates that the gamma function is a generalization of $n!$ from the domain of positive integers to the domain of positive numbers (and most negative numbers as we will soon show). Also, it confirms a result that beginning algebra students find puzzling to understand, namely, $0! = 1$.

It follows from the uniform convergence of the integral (19) that $\Gamma(x)$ is a continuous function for all $x > 0$.* To investigate the behavior of x in the vicinity of zero, we rewrite the recurrence formula (21) in the form

$$\Gamma(x) = \frac{\Gamma(x+1)}{x}, \qquad x \neq 0, \tag{23}$$

from which we deduce

$$\lim_{x \to 0^+} \Gamma(x) = \lim_{x \to 0^+} \frac{\Gamma(x+1)}{x} = +\infty. \tag{24}$$

---

* A basic theorem of advanced calculus states that $F(x) = \int_a^b f(x, t) \, dt$ is a continuous function on any interval for which the integral converges uniformly.

We can also use (23) to extend the definition of the gamma function to negative values of x. That is, since the right-hand side of (23) is defined for all $x > -1$, $x \neq 0$, it follows that the left-hand side is similarly defined. Hence, in particular we obtain the limiting values

$$\lim_{x \to 0^-} \Gamma(x) = -\infty, \qquad \lim_{x \to -1^+} \Gamma(x) = -\infty. \tag{25}$$

Replacing x with $(x + 1)$ in (23) leads to

$$\Gamma(x + 1) = \frac{\Gamma(x + 2)}{x + 1},$$

and thus we extend the definition to $x > -2$ by writing (23) as

$$\Gamma(x) = \frac{\Gamma(x + 2)}{x(x + 1)}, \qquad x \neq 0, -1. \tag{26}$$

Limiting values this time reveal that

$$\lim_{x \to -1^-} \Gamma(x) = +\infty, \qquad \lim_{x \to -2^+} \Gamma(x) = +\infty. \tag{27}$$

Continuing this process, we finally deduce that for integer k,

$$\Gamma(x) = \frac{\Gamma(x + k)}{x(x + 1)(x + 2) \cdots (x + k - 1)}, \qquad x \neq 0, -1, \ldots, -k + 1. \tag{28}$$

Equation (28) can be used to define the gamma function for $-k < x < 0$, except as noted. As a consequence of this result, we have the following theorem.

THEOREM 9.1

> If $n = 0, 1, 2, \ldots$, then $|\Gamma(-n)| = \infty$, or equivalently,
>
> $$\frac{1}{\Gamma(-n)} = 0.$$

The graph of the gamma function for both positive and negative values of x is shown in Figure 9.2. Values of $\Gamma(x)$ are commonly tabulated for the interval $1 \leqslant x \leqslant 2$, and other values of $\Gamma(x)$ can be generated through use of the recurrence formulas.

□ 9.3.1 ADDITIONAL PROPERTIES

The gamma function is important in the definition of other special functions like Bessel functions (see Sections 9.4 and 9.6). Another way in which it proves useful is in the evaluation of certain integrals. That is, when it appears in applications it frequently has the form suggested by (19) or some variation of it. For example, if we set $t = u^2$ in (19), we find

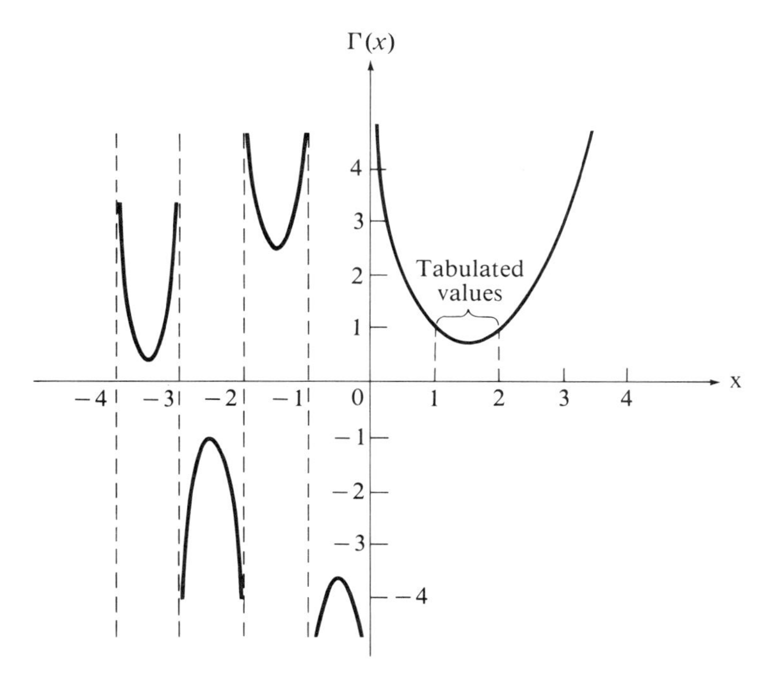

FIGURE 9.2

Graph of the gamma function

$$\Gamma(x) = 2\int_0^\infty e^{-u^2}u^{2x-1}\,du, \qquad x > 0, \tag{29}$$

whereas the substitution $t = \log(1/u)$ yields the integral form

$$\Gamma(x) = \int_0^1 \left(\log\frac{1}{u}\right)^{x-1} du, \qquad x > 0. \tag{30}$$

EXAMPLE 1

Evaluate $I = \int_0^\infty e^{-x^3}x^4\,dx$ in terms of the gamma function.

SOLUTION By making the substitution $t = x^3$, we find

$$I = \frac{1}{3}\int_0^\infty e^{-t}t^{2/3}\,dt,$$

and by comparison with Equation (19), we deduce that

$$I = \frac{1}{3}\Gamma\left(\frac{5}{3}\right).$$

EXAMPLE 2

Show that $\Gamma(1/2) = \sqrt{\pi}$.

SOLUTION Referring to Equation (29), we write

$$\Gamma\left(\frac{1}{2}\right) = 2\int_0^\infty e^{-u^2}\,du.$$

To evaluate this integral, we first consider the product

$$\left[\Gamma\left(\frac{1}{2}\right)\right]^2 = 2\int_0^\infty e^{-u^2}\,du \cdot 2\int_0^\infty e^{-v^2}\,dv$$
$$= 4\int_0^\infty\int_0^\infty e^{-(u^2+v^2)}\,du\,dv,$$

where we have expressed the product of two integrals as an iterated integral. The presence of the term $(u^2 + v^2)$ in the integrand suggests a change to polar coordinates. Thus, we set

$$u = r\cos\theta, \qquad v = r\sin\theta,$$

and obtain

$$\left[\Gamma\left(\frac{1}{2}\right)\right]^2 = 4\int_0^{\pi/2}\int_0^\infty e^{-r^2} r\,dr\,d\theta$$
$$= 2\int_0^{\pi/2} d\theta \cdot \underbrace{2\int_0^\infty e^{-r^2} r\,dr.}_{\Gamma(1)}$$

This last integral involving r is recognized as simply $\Gamma(1)$, and thus

$$\left[\Gamma\left(\frac{1}{2}\right)\right]^2 = 2\int_0^{\pi/2} d\theta = \pi.$$

Finally, taking the positive square root of this expression yields the intended result

$$\Gamma\left(\frac{1}{2}\right) = \sqrt{\pi}.$$

The following is a short list of identities involving the gamma function for easy reference.

Basic Identities for $\Gamma(x)$

(G1): $\quad \Gamma(x) = \int_0^\infty e^{-t}t^{x-1}\,dt, \qquad x > 0$

(G2): $\Gamma(x) = 2\int_0^\infty e^{-t^2} t^{2x-1}\,dt, \qquad x > 0$

(G3): $\Gamma(x + 1) = x\Gamma(x)$

(G4): $\Gamma(n + 1) = n!, \qquad n = 0, 1, 2, \ldots$

(G5): $\Gamma(1/2) = \sqrt{\pi}$

(G6): $\Gamma(x) = \dfrac{\Gamma(x + k)}{x(x + 1)(x + 2)\cdots(x + k - 1)}, \qquad k = 1, 2, 3, \ldots$

(G7): $\sqrt{\pi}\,\Gamma(2x) = 2^{2x-1}\Gamma(x)\Gamma\left(x + \frac{1}{2}\right)$

(G8): $\Gamma\left(n + \frac{1}{2}\right) = \dfrac{(2n)!}{2^{2n}n!}\sqrt{\pi}, \qquad n = 0, 1, 2, \ldots$

(G9): $\Gamma(x)\Gamma(1 - x) = \dfrac{\pi}{\sin \pi x}, \qquad x$ nonintegral

(G10): $\Gamma(n + 1) \sim \sqrt{2\pi n}\, n^n e^{-n}, \qquad n \to \infty$ (Stirling's formula)*

□ 9.3.2 DIGAMMA FUNCTION

(This section can be omitted if Section 9.4.2 is also omitted.)

Closely associated with the derivative of the gamma function is the *logarithmic derivative function,* or *digamma function*†

$$\psi(x) \equiv \frac{d}{dx}\log \Gamma(x) = \frac{\Gamma'(x)}{\Gamma(x)}, \qquad x > 0, \tag{31}$$

where log denotes the natural logarithm. The function $\psi(x)$ satisfies relations somewhat analogous to those of the gamma function. For example, starting with the logarithm of the recurrence formula (21), i.e.,

$$\log \Gamma(x + 1) = \log \Gamma(x) + \log x,$$

and differentiating, we get

$$\frac{d}{dx}\log \Gamma(x + 1) = \frac{d}{dx}\log \Gamma(x) + \frac{1}{x},$$

from which we deduce

$$\psi(x + 1) = \psi(x) + \frac{1}{x}, \qquad x > 0. \tag{32}$$

When x is a positive integer n, it follows from (32) that

---

* The symbol $\sim$ means "is asymptotic to."

† The digamma function is also commonly called the *psi function*.

$$\psi(n+1) = \psi(n) + \frac{1}{n}$$

$$= \psi(n-1) + \frac{1}{n-1} + \frac{1}{n}$$

$$= \psi(n-2) + \frac{1}{n-2} + \frac{1}{n-1} + \frac{1}{n},$$

and so forth. By repeated application of (32), we are finally led to the result

$$\psi(n+1) = \psi(1) + 1 + \frac{1}{2} + \frac{1}{3} + \cdots + \frac{1}{n}. \tag{33}$$

However, from (31)

$$\psi(1) = \frac{\Gamma'(1)}{\Gamma(1)} = \Gamma'(1), \tag{34}$$

where*

$$\Gamma'(1) = \int_0^\infty e^{-t} \log t \, dt = -\gamma. \tag{35}$$

The constant γ is called *Euler's constant,* which is also defined by

$$\gamma = \lim_{n\to\infty}\left(1 + \frac{1}{2} + \frac{1}{3} + \cdots + \frac{1}{n} - \log n\right) = 0.577215 \ldots . \tag{36}$$

In terms of γ, we can rewrite (33) in the form

$$\psi(n+1) = -\gamma + \sum_{k=1}^{n} \frac{1}{k} \tag{37}$$

$(n = 1, 2, 3, \ldots)$.

Additional properties of $\psi(x)$ are taken up in Exercises 9.3.

EXERCISES 9.3

1. Evaluate the following (use the result $\Gamma(1/2) = \sqrt{\pi}$ where necessary):
   (a) $\Gamma(6)$
   (b) $\Gamma(3/2)$
   (c) $\Gamma(7/2)$
   (d) $\Gamma(-1/2)$
   (e) $\Gamma(-9/2)$
   (f) $\Gamma(8/3)/\Gamma(2/3)$
2. The binomial coefficient is defined by $(a \neq 0)$

$$\binom{a}{0} = 1, \quad \binom{a}{k} = \frac{a(a-1)\cdots(a-k+1)}{k!}, \quad k = 1, 2, 3, \ldots .$$

---

* By differentiating (19) under the integral sign, we obtain $\Gamma'(x) = \int_0^\infty e^{-t}t^{x-1}(\log t)\, dt$.

Show that

(a) $\binom{n}{k} = \dfrac{n!}{k!(n-k)!}, \qquad n = 0, 1, 2, \ldots, \qquad k = 0, 1, 2, \ldots, n$

(b) $\binom{-1/2}{n} = \dfrac{(-1)^n(2n)!}{2^{2n}(n!)^2}, \qquad n = 0, 1, 2, \ldots.$

(c) $\binom{a}{k} = \dfrac{\Gamma(a+1)}{k!\,\Gamma(a-k+1)}, \qquad k = 0, 1, 2, \ldots.$

In Problems 3–5, verify the integral formula.

3. $\displaystyle\int_0^\infty e^{-pt}t^x\,dt = \frac{\Gamma(x+1)}{p^{x+1}}, \qquad x > -1, \qquad p > 0$

4. $\displaystyle\int_0^\infty x^a b^{-x}\,dx = \frac{\Gamma(a+1)}{(\log b)^{a+1}}, \qquad a > -1, \qquad b > 1$

Hint: Let $t = x \log b$.

5. $\displaystyle\int_a^\infty e^{2ax-x^2}\,dx = \frac{1}{2}\sqrt{\pi}\,e^{a^2}$

Hint: $2ax - x^2 = a^2 - (x-a)^2$.

★6. Show that

$$\int_0^{\pi/2} \cos^{2x-1}\theta \sin^{2y-1}\theta\,d\theta = \frac{\Gamma(x)\Gamma(y)}{2\Gamma(x+y)}, \qquad x > 0, \qquad y > 0.$$

Hint: Consider the product

$$\Gamma(x)\Gamma(y) = 2\int_0^\infty e^{-u^2}u^{2x-1}\,du \cdot 2\int_0^\infty e^{-v^2}v^{2y-1}\,dv$$

and change to polar coordinates (see Example 2).

In Problems 7–10, use the result of Problem 6 to evaluate the integral.

7. $\displaystyle\int_0^{\pi/2} \sin\theta\cos^2\theta\,d\theta$

8. $\displaystyle\int_0^{\pi/2} \sqrt{\sin 2x}\,dx$

9. $\displaystyle\int_0^{\pi/2} \tan^{1/2}\theta\,d\theta$

10. $\displaystyle\int_0^{\pi} \sin^5\theta\,d\theta$

Hint: Use property (G9) in Section 9.3.1.

11. The *beta function* is defined by the integral

$$B(x, y) = \int_0^1 t^{x-1}(1-t)^{y-1}\,dt, \qquad x > 0, \qquad y > 0.$$

Show that

$$B(x, y) = \frac{\Gamma(x)\Gamma(y)}{\Gamma(x+y)}.$$

Hint: Let $t = \cos^2\theta$ and use Problem 6.

12. Using the result of Problem 11, evaluate the following:
    (a) $B(2, 3)$ (b) $B(1/2, 1)$
    (c) $B(2/3, 1/3)$ (d) $B(3/4, 1/4)$

    Hint: Use property (G9) in Section 9.3.1, if needed.

★13. By setting $y = x$ in the result of Problem 6,
    (a) show that

$$\frac{\Gamma(x)\Gamma(x)}{2\Gamma(2x)} = 2^{1-2x}\int_0^{\pi/2} \sin^{2x-1}\phi \, d\phi.$$

    Hint: Use the identity $\sin\theta\cos\theta = \frac{1}{2}\sin 2\theta$.
    (b) Evaluating the integral in (a), deduce the *Legendre duplication formula*

$$\sqrt{\pi}\,\Gamma(2x) = 2^{2x-1}\Gamma(x)\Gamma\left(x + \frac{1}{2}\right).$$

    (c) From (b), derive the relation

$$\Gamma\left(n + \frac{1}{2}\right) = \frac{(2n)!}{2^{2n}n!}\sqrt{\pi}, \qquad n = 0, 1, 2, \ldots.$$

14. Find the area inside the curve $x^{2/3} + y^{2/3} = 1$.
    Hint: See Problem 11.

15. Show that the area enclosed by the curve $x^4 + y^4 = 1$ is $[\Gamma(1/4)]^2/2\sqrt{\pi}$.

16. Find the total arc length of the lemniscate $r^2 = a^2 \cos 2\theta$.

★17. A particle of mass m starts from rest at $r = 1$ and moves along a radial line toward the origin $r = 0$ under the reciprocal force law $f = -k/r$, where k is a positive constant. The energy equation of the particle is given by

$$\frac{1}{2}m\left(\frac{dr}{dt}\right)^2 + k\log r = 0.$$

    Show that the time required for the particle to reach the origin is $(m\pi/2k)^{1/2}$.

★18. Starting with $f(t) = \int_0^t u^{x-1}(t-u)^{y-1}\,du, \quad x > 0, \quad y > 0,$
    (a) use the Laplace convolution theorem to show that

$$F(p) = \frac{\Gamma(x)\Gamma(y)}{p^{x+y}}$$

    (see Section 5.3.2).
    (b) Take the inverse Laplace transform of $F(p)$ and establish the relation

$$\int_0^1 u^{x-1}(1-u)^{y-1}\,du = \frac{\Gamma(x)\Gamma(y)}{\Gamma(x+y)}, \qquad x > 0, \qquad y > 0.$$

19. Take the logarithmic derivative of $\Gamma(x)\Gamma(1-x) = \pi\csc\pi x$ to deduce that the digamma function satisfies the identity

$$\psi(1-x) - \psi(x) = \pi\cot\pi x, \qquad (x \text{ nonintegral}).$$

20. By taking the logarithmic derivative of the Legendre duplication formula (see Problem 13),
    (a) deduce that

$$\psi(x) + \psi\left(x + \frac{1}{2}\right) + 2 \log 2 = 2\psi(2x).$$

    (b) From (a), show that $\psi(1/2) = -\gamma - 2 \log 2$.
    (c) Show that (for $n = 1, 2, 3, \ldots$)

$$\psi\left(n + \frac{1}{2}\right) = -\gamma - 2 \log 2 + 2 \sum_{k=1}^{n} \frac{1}{2k - 1}.$$

21. Show that (for $n = 0, 1, 2, \ldots$)

$$\psi(n + 1) = -\gamma + \sum_{k=1}^{\infty} \frac{n}{k(k + n)}.$$

9.4 BESSEL FUNCTIONS

Solutions of *Bessel's equation*

$$x^2y'' + xy' + (x^2 - \nu^2)y = 0 \tag{38}$$

are called *Bessel functions* of order ν. We assume that ν is real and, since only ν^2 occurs in the equation, it is customary to take $\nu \geqslant 0$. The equation is named in honor of the German astronomer F. W. Bessel who studied its solutions in connection with his studies of planetary motion.* The equation is also closely associated with probems featuring circular or cylindrical symmetry, such as the study of free vibrations of a circular membrane or finding the temperature distribution in a solid cylinder. Bessel's equation, and hence Bessel functions, also occur in the study of electromagnetic theory. Other areas of application are too numerous to mention. In fact, Bessel functions occur so frequently in engineering and physics applications that they are undoubtedly the most important functions beyond the study of elementary functions.

The point $x = 0$ is a regular singular point of (38). Rather than avoid the singularity, however, it is precisely that point about which we seek a series solution of the form†

$$y = x^s \sum_{n=0}^{\infty} c_n x^n = \sum_{n=0}^{\infty} c_n x^{n+s}, \tag{39}$$

---

* FRIEDRICH WILHELM BESSEL (1784–1846) first achieved fame by computing the orbit of Halley's comet. He was also the first to measure the distance of a fixed star from the sun, and he made the first systematic study of the solutions (known as Bessel functions) of the equation that bears his name.

† This technique is called the *method of Frobenius*. See Section 9.3 in L.C. Andrews, *Ordinary Differential Equations with Applications*, (Glenview, Ill.: Scott, Foresman and Co., 1982).

since the nature of the solutions near $x = 0$ will then be most clearly revealed. In this setting we must determine the value of the parameter s in addition to the constants c_n in obtaining our solution.

The substitution of (39) into (38) followed by some algebraic manipulation leads to

$$\sum_{n=0}^{\infty} [(n+s)(n+s-1)c_n + (n+s)c_n - \nu^2 c_n]x^{n+s} + \sum_{n=0}^{\infty} c_n x^{n+s+2} = 0,$$

which can be expressed in the equivalent form

$$c_0(s^2 - \nu^2)x^s + c_1[(s+1)^2 - \nu^2]x^{s+1} + \sum_{n=2}^{\infty} \{[(n+s)^2 - \nu^2]c_n + c_{n-2}\}x^{n+s} = 0. \tag{40}$$

The values of s leading to solutions of the form (39) are found by setting the coefficient of c_0 to zero; thus we see that $s = \pm\nu$. This choice of s leaves c_0 arbitrary, and the remaining constants must satisfy

$$\begin{aligned} c_1[(s+1)^2 - \nu^2] &= 0, \\ [(n+s)^2 - \nu^2]c_n + c_{n-2} &= 0, \qquad n = 2, 3, 4, \ldots. \end{aligned} \tag{41}$$

For our choice of s, however, it follows that $c_1 = 0$ and

$$c_n = -\frac{c_{n-2}}{n^2 \pm 2n\nu}, \qquad n = 2, 3, 4, \ldots. \tag{42}$$

Because $c_1 = 0$, it is clear that all c_n with odd index also vanish by virtue of (42).

To be precise and produce at least one solution of Bessel's equation, we will select $s = +\nu$, and then (42) for even values of n can be written as

$$c_n = c_{2m} = -\frac{c_{2m-2}}{2^2 m(m+\nu)}, \qquad m = 1, 2, 3, \ldots. \tag{43}$$

This last expression can also be written in the form

$$c_{2m} = \frac{(-1)^m c_0}{2^{2m} m!(\nu+1)(\nu+2)\cdots(\nu+m)}$$

(see Problem 4 in Exercises 9.4), or, in terms of gamma functions

$$c_{2m} = \frac{(-1)^m \Gamma(\nu+1)c_0}{2^{2m} m!\,\Gamma(\nu+m+1)}, \qquad m = 1, 2, 3, \ldots \tag{44}$$

[recall that $x\Gamma(x) = \Gamma(x+1)$]. One solution of Bessel's equation is therefore given by

$$y_1 = c_0 x^\nu + c_0 x^\nu \sum_{m=1}^{\infty} \frac{(-1)^m \Gamma(\nu+1)}{2^{2m} m!\,\Gamma(\nu+m+1)} x^{2m}$$

$$= c_0 \sum_{m=0}^{\infty} \frac{(-1)^m \Gamma(\nu+1)}{2^{2m} m! \Gamma(\nu+m+1)} x^{2m+\nu}, \tag{45}$$

where c_0 is arbitrary. It is customary to set

$$c_0 = \frac{1}{2^\nu \Gamma(\nu+1)}$$

and denote the resulting infinite series in (45) by the symbol

$$J_\nu(x) = \sum_{m=0}^{\infty} \frac{(-1)^m (x/2)^{2m+\nu}}{m! \Gamma(m+\nu+1)}, \tag{46}$$

known as the *Bessel function of the first kind* and order ν. In particular, when $\nu = 0$ and $\nu = 1$ we obtain, respectively, the series

$$J_0(x) = 1 - \frac{x^2}{2^2} + \frac{x^4}{2^4(2!)^2} - \frac{x^6}{2^6(3!)^2} + \cdots \tag{47}$$

and

$$J_1(x) = \frac{x}{2} - \frac{x^3}{2^3 2!} + \frac{x^5}{2^5 2! 3!} - \frac{x^7}{2^7 3! 4!} + \cdots. \tag{48}$$

The graphs of $J_0(x)$, $J_1(x)$, and $J_2(x)$ are shown in Figure 9.3. These functions exhibit an oscillatory behavior somewhat like that of the sinusoidal functions, except that the amplitude of the Bessel functions diminishes as x increases and the (infinitely many) zeros of these functions are not evenly spaced. For reference purposes, the first few zeros of some of the Bessel functions are listed in Table 9.1.*

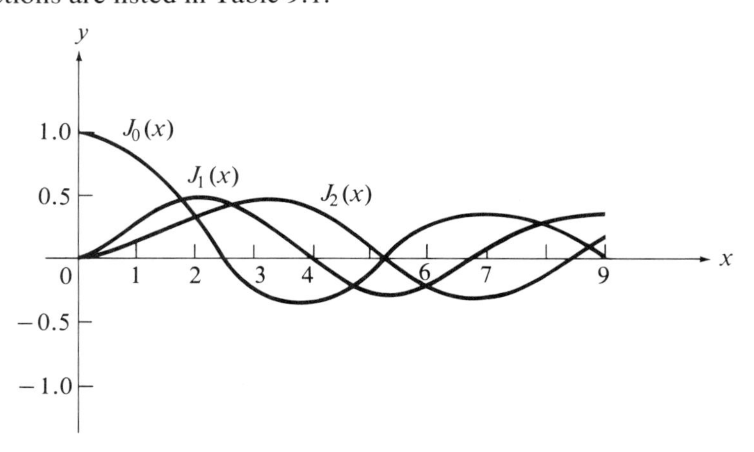

FIGURE 9.3

Graph of $J_n(x)$, $n = 0, 1, 2$

* The zeros of the Bessel functions are important in the study of eigenvalue problems (see Section 9.5).

TABLE 9.1 Zeros of Bessel Functions: $J_n(x_k) = 0$

n \ k	1	2	3	4	5
0	2.405	5.520	8.654	11.792	14.931
1	3.832	7.016	10.173	13.324	16.471
2	5.136	8.417	11.620	14.796	17.960
3	6.380	9.761	13.015	16.223	19.409
4	7.588	11.065	14.373	17.616	20.827

We have shown that $y_1 = J_\nu(x)$ is one solution of Bessel's equation (38). Because the equation is second-order, we expect to find a second linearly independent solution. Furthermore, it seems natural to assume that the other solution could be found by setting $s = -\nu$ in the recurrence formula for c_n. If we proceed in this fashion, we find the same calculations with ν replaced by $-\nu$ throughout, and so the formal replacement of ν with $-\nu$ in (46) gives us the function

$$J_{-\nu}(x) = \sum_{m=0}^{\infty} \frac{(-1)^m (x/2)^{2m-\nu}}{m!\,\Gamma(m-\nu+1)}. \tag{49}$$

It is a routine matter to verify that $y_2 = J_{-\nu}(x)$ is also a solution of (38). Moreover, for ν not an integer, $J_{-\nu}(x)$ is *linearly independent* of $J_\nu(x)$. This is so since $J_{-\nu}(x)$ has negative powers of x and hence becomes infinite at $x = 0$, while $J_\nu(x)$ remains finite at $x = 0$. Therefore, $J_{-\nu}(x)$ cannot be a constant multiple of $J_\nu(x)$, which means it is linearly independent of it. Under these conditions we write the general solution of (38) as

$$y = C_1 J_\nu(x) + C_2 J_{-\nu}(x), \qquad \nu \neq 0, 1, 2, \ldots, \tag{50}$$

where C_1 and C_2 are arbitrary constants.

For $\nu = n$ $(n = 0, 1, 2, \ldots)$, we find that (49) becomes

$$\begin{aligned} J_{-n}(x) &= \sum_{m=0}^{\infty} \frac{(-1)^m (x/2)^{2m-n}}{m!\,\Gamma(m-n+1)} \\ &= \sum_{m=n}^{\infty} \frac{(-1)^m (x/2)^{2m-n}}{m!\,\Gamma(m-n+1)}, \end{aligned}$$

where the first n terms vanish since $1/\Gamma(m-n+1) = 0$ for $m = 0, 1, \ldots, n-1$, by virtue of Theorem 9.1. With the change of index $m = k + n$, we can rewrite this last expression as

$$\begin{aligned} J_{-n}(x) &= \sum_{k=0}^{\infty} \frac{(-1)^{k+n} (x/2)^{2k+n}}{(k+n)!\,\Gamma(k+1)} \\ &= (-1)^n \sum_{k=0}^{\infty} \frac{(-1)^k (x/2)^{2k+n}}{k!\,\Gamma(k+n+1)}, \end{aligned} \tag{51}$$

from which we deduce

$$J_{-n}(x) = (-1)^n J_n(x), \qquad n = 0, 1, 2, \ldots. \tag{52}$$

Clearly, $J_n(x)$ and $J_{-n}(x)$ are *not* linearly independent functions, and consequently (50) does not represent the general solution of (38) when $\nu = n$ ($n = 0, 1, 2, \ldots$).

9.4.1 BESSEL FUNCTIONS OF THE SECOND KIND

In finding the general solution of Bessel's equation, it is preferable to introduce a second solution whose independence of $J_\nu(x)$ is not restricted to certain values of ν. This can be accomplished in several ways. First, it is easy to verify that*

$$y_2 = J_\nu(x) \int \frac{dx}{x[J_\nu(x)]^2} \tag{53}$$

is such a solution, although the integral is difficult to handle. Also, we could use the method of Frobenius to produce a second linearly independent solution of the form

$$y_2 = J_\nu(x) \log x + \sum_{n=0}^{\infty} b_n x^{n-\nu}. \tag{54}$$

However, it is customary to use a different approach from either of these and define a second linearly independent solution y_2 by the function

$$Y_\nu(x) = \frac{J_\nu(x) \cos \nu\pi - J_{-\nu}(x)}{\sin \nu\pi}, \tag{55}$$

called the *Bessel function of the second kind* and order ν.† It is clearly a solution of Bessel's equation since it is a linear combination of solutions, and for ν not an integer it is easy to show that it is linearly independent of the function $J_\nu(x)$ (see Problem 15 in Exercises 9.4). However, when $\nu = n$ ($n = 0, 1, 2, \ldots$), it requires further investigation since $Y_\nu(x)$ assumes the indeterminate form 0/0. The limit of $Y_\nu(x)$ as $\nu \to n$ does exist (see Section 9.4.2), and so we write

$$Y_n(x) = \lim_{\nu \to n} Y_\nu(x), \qquad n = 0, 1, 2, \ldots. \tag{56}$$

Moreover, for $n = 0, 1, 2, \ldots$, it can be shown that $Y_n(x)$ is linearly independent of $J_n(x)$.‡ We conclude, therefore, that the general solution of Bessel's equation (38) for arbitrary values of ν is given by

$$y = C_1 J_\nu(x) + C_2 Y_\nu(x). \tag{57}$$

REMARK: *Like $J_{-\nu}(x)$, the function $Y_\nu(x)$ is not bounded at $x = 0$.*

---

* See Problem 42 in Exercises 1.3.

† This function is also denoted by $N_\nu(x)$ in some of the literature and called the *Neumann function* after the German mathematician and physicist CARL NEUMANN (1839–1873).

‡ The linear independence of $J_n(x)$ and $Y_n(x)$ follows directly from the result of Problem 17(b) in Exercises 9.4.

□ 9.4.2 SERIES EXPANSIONS FOR THE FUNCTIONS $Y_n(x)$

We wish to derive an expression for the Bessel function of the second kind when ν takes on integer values. Because the limit (56) leads to the indeterminate form 0/0, we must apply L'Hôpital's rule, from which we deduce

$$Y_n(x) = \lim_{\nu \to n} \frac{1}{\pi}\left[\frac{\partial}{\partial \nu} J_\nu(x) - (-1)^n \frac{\partial}{\partial \nu} J_{-\nu}(x)\right], \qquad n = 0, 1, 2, \ldots \tag{58}$$

(see Problem 18 in Exercises 9.4). For $x > 0$, the derivative of the Bessel function with resepct to its order ν leads to

$$\begin{aligned}\frac{\partial}{\partial \nu} J_\nu(x) &= \sum_{m=0}^{\infty} \frac{(-1)^m}{m!}\left\{\frac{(x/2)^{2m+\nu}\log(x/2)}{\Gamma(m+\nu+1)} - \frac{(x/2)^{2k+\nu}\Gamma'(m+\nu+1)}{[\Gamma(m+\nu+1)]^2}\right\} \\ &= \sum_{m=0}^{\infty} \frac{(-1)^m (x/2)^{2m+\nu}}{m!\Gamma(m+\nu+1)}[\log(x/2) - \psi(m+\nu+1)],\end{aligned}$$

where $\psi(x)$ is the digamma function (see Section 9.3.2). We can further write this last expression as

$$\frac{\partial}{\partial \nu} J_\nu(x) = J_\nu(x)\log(x/2) - \sum_{m=0}^{\infty} \frac{(-1)^m (x/2)^{2m+\nu}}{m!\Gamma(m+\nu+1)} \psi(m+\nu+1). \tag{59}$$

By a similar analysis, it follows that

$$\frac{\partial}{\partial \nu} J_{-\nu}(x) = -J_{-\nu}(x)\log(x/2) + \sum_{m=0}^{\infty} \frac{(-1)^m (x/2)^{2m-\nu}}{m!\Gamma(m+\nu+1)} \psi(m-\nu+1). \tag{60}$$

At this point we wish to first consider the special case when $\nu \to 0$. Here we see that (58) reduces to

$$Y_0(x) = \frac{2}{\pi} \lim_{\nu \to 0} \frac{\partial}{\partial \nu} J_\nu(x),$$

and by using (59), we get

$$Y_0(x) = \frac{2}{\pi} J_0(x)\log(x/2) - \frac{2}{\pi}\sum_{m=0}^{\infty} \frac{(-1)^m (x/2)^{2m}}{(m!)^2} \psi(m+1) \tag{61}$$

where $x > 0$. Another form of (61) can be obtained by making the observation

$$\begin{aligned}&\sum_{m=0}^{\infty} \frac{(-1)^m (x/2)^{2m}}{(m!)^2} \psi(m+1) \\ &\quad = -\gamma + \sum_{m=1}^{\infty} \frac{(-1)^m (x/2)^{2m}}{(m!)^2}\left(-\gamma + 1 + \frac{1}{2} + \cdots + \frac{1}{m}\right) \\ &\quad = -\gamma \sum_{m=0}^{\infty} \frac{(-1)^m (x/2)^{2m}}{(m!)^2} + \sum_{m=1}^{\infty} \frac{(-1)^m (x/2)^{2m}}{(m!)^2}\left(1 + \frac{1}{2} + \cdots + \frac{1}{m}\right),\end{aligned} \tag{62}$$

from which we deduce

$$Y_0(x) = \frac{2}{\pi} J_0(x)[\log(x/2) + \gamma]$$

$$-\frac{2}{\pi}\sum_{m=1}^{\infty}\frac{(-1)^m(x/2)^{2m}}{(m!)^2}\left(1+\frac{1}{2}+\cdots+\frac{1}{m}\right) \tag{63}$$

where $x > 0$.

The derivation of the series for $Y_n(x)$, $n = 1, 2, 3, \ldots$, is a little more difficult to obtain, but it can be shown that*

$$Y_n(x) = \frac{2}{\pi} J_n(x)\log(x/2) - \frac{1}{\pi}\sum_{m=0}^{n-1}\frac{(n-m-1)!}{m!}(x/2)^{2m-n}$$

$$-\frac{1}{\pi}\sum_{m=0}^{\infty}\frac{(-1)^m(x/2)^{2m+n}}{m!(m+n)!}[\psi(m+n+1)+\psi(m+1)] \tag{64}$$

where $x > 0$.

Graphs of $Y_n(x)$ for various values of n are shown in Figure 9.4. Observe the logarithmic behavior as $x \to 0^+$. Also note that these functions have oscillatory characteristics similar to those of $J_n(x)$.

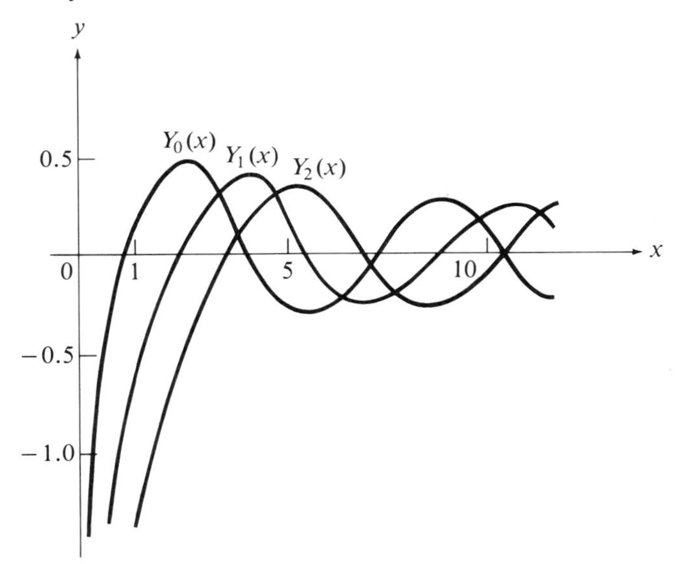

FIGURE 9.4

Graph of $Y_n(x)$, $n = 0, 1, 2$

9.4.3 BASIC IDENTITIES AND RECURRENCE FORMULAS

There are many recurrence relations connecting the Bessel functions, analogous to those of the Legendre polynomials. We will develop some of these here while others are taken up in the exercises.

---

* See Section 6.5.1 in L. C. Andrews, *Special Functions for Engineers and Applied Mathematicians*, (New York: Macmillan, 1985).

Suppose we multiply the series (46) for $J_\nu(x)$ by x^ν and then differentiate the result with respect to x. This action gives us

$$\frac{d}{dx}[x^\nu J_\nu(x)] = \frac{d}{dx}\left[\sum_{m=0}^{\infty} \frac{(-1)^m x^{2m+2\nu}}{2^{2m+\nu} m!\,\Gamma(m+\nu+1)}\right]$$

$$= \sum_{m=0}^{\infty} \frac{(-1)^m 2(m+\nu) x^{2m+2\nu-1}}{2^{2m+\nu} m!\,\Gamma(m+\nu+1)}$$

$$= x^\nu \sum_{m=0}^{\infty} \frac{(-1)^m (x/2)^{2m+(\nu-1)}}{m!\,\Gamma(m+\nu)},$$

or

$$\frac{d}{dx}[x^\nu J_\nu(x)] = x^\nu J_{\nu-1}(x). \tag{65}$$

Similarly, if we multiply $J_\nu(x)$ by $x^{-\nu}$, we will find that

$$\frac{d}{dx}[x^{-\nu} J_\nu(x)] = -x^{-\nu} J_{\nu+1}(x) \tag{66}$$

(see Problem 19 in Exercises 9.4).

If we perform the differentiation in (65) and (66) and divide the results by the factors x^ν and $x^{-\nu}$, respectively, we obtain

$$J_\nu'(x) + \frac{\nu}{x} J_\nu(x) = J_{\nu-1}(x) \tag{67}$$

and

$$J_\nu'(x) - \frac{\nu}{x} J_\nu(x) = -J_{\nu+1}(x). \tag{68}$$

The substitution of $\nu = 0$ in (68) leads to the special result

$$J_0'(x) = -J_1(x). \tag{69}$$

The identities (65) and (66) can also be expressed as integration formulas. By integrating each side once, we have that

$$\int x^\nu J_{\nu-1}(x)\,dx = x^\nu J_\nu(x) + C \tag{70}$$

and

$$\int x^{-\nu} J_{\nu+1}(x)\,dx = -x^{-\nu} J_\nu(x) + C, \tag{71}$$

where C is any constant. These integration formulas are frequently used in the evaluation of integrals involving Bessel functions.

EXAMPLE 3

Reduce $\int x^2 J_2(x)\, dx$ to an integral involving only $J_0(x)$.

SOLUTION To use (71), we first rewrite the integral in the form

$$\int x^2 J_2(x)\, dx = \int x^3[x^{-1}J_2(x)]\, dx$$

and use integration by parts with

$$u = x^3, \qquad dv = x^{-1}J_2(x)\, dx,$$
$$du = 3x^2 dx, \qquad v = -x^{-1}J_1(x).$$

Thus, we have

$$\int x^2 J_2(x)\, dx = -x^2 J_1(x) + 3\int xJ_1(x)\, dx,$$

and a second integration by parts finally gives

$$\int x^2 J_2(x)\, dx = -x^2 J_1(x) - 3xJ_0(x) + 3\int J_0(x)\, dx.$$

The last integral involving $J_0(x)$ cannot be evaluated in closed form, and so our integration is complete.

As a general rule, any integral of the form

$$\int x^m J_n(x)\, dx,$$

where m and n are integers such that $m + n > 0$, can be integrated in closed form when $(m + n)$ is odd, but will ultimately depend upon the residual integral $\int J_0(x)\, dx$ when $(m + n)$ is even. Since it cannot be evaluated in closed form, the integral $\int_0^x J_0(t)\, dt$ has been tabulated.*

EXAMPLE 4

Show that†

$$\int_0^\infty e^{-sx} J_0(x)\, dx = (s^2 + 1)^{-1/2}, \qquad s > 1.$$

* For example, see M. Abramowitz and I. Stegun (eds.), *Handbook of Mathematical Functions* (New York: Dover, 1965).

† We can interpret this integral as the *Laplace transform* of $J_0(x)$ (see Chapter 5).

SOLUTION For integrals of this type we often replace the Bessel function by its series representation and then integrate termwise. Doing so in this case, we get

$$\int_0^\infty e^{-sx} J_0(x)\, dx = \sum_{m=0}^\infty \frac{(-1)^m}{2^{2m}(m!)^2} \int_0^\infty e^{-sx} x^{2m}\, dx$$

$$= \sum_{m=0}^\infty \frac{(-1)^m (2m)!}{2^{2m}(m!)^2} s^{-(2m+1)}$$

where we are using the result of Problem 3 in Exercises 9.3. Next, recalling Problem 2(b) in Exercises 9.3, we can express this series in the form of a binomial series and deduce that*

$$\int_0^\infty e^{-sx} J_0(x)\, dx = \frac{1}{s} \sum_{m=0}^\infty \binom{-1/2}{m} s^{-2m}$$

$$= \frac{1}{s}(1 + s^{-2})^{-1/2}$$

$$= (s^2 + 1)^{-1/2}, \qquad s > 1,$$

which is our intended result.

For reference purposes, we provide the following lists of basic identities involving Bessel functions. Because $Y_\nu(x)$ is a linear combination of $J_\nu(x)$ and $J_{-\nu}(x)$, it follows that $Y_\nu(x)$ satisfies the same basic identities as $J_\nu(x)$.

Basic Identities for $J_\nu(x)$

(J1): $$J_\nu(x) = \sum_{m=0}^\infty \frac{(-1)^m (x/2)^{2m+\nu}}{m!\,\Gamma(m + \nu + 1)}$$

(J2): $$J_0(0) = 1; \qquad J_\nu(0) = 0, \qquad \nu > 0$$

(J3): $$J_{-n}(x) = (-1)^n J_n(x), \qquad n = 0, 1, 2, \ldots$$

(J4): $$\frac{d}{dx}[x^\nu J_\nu(x)] = x^\nu J_{\nu-1}(x)$$

(J5): $$\frac{d}{dx}[x^{-\nu} J_\nu(x)] = -x^{-\nu} J_{\nu+1}(x)$$

(J6): $$J_\nu'(x) + \frac{\nu}{x} J_\nu(x) = J_{\nu-1}(x)$$

---

*Recall the binomial series $(1 + x)^a = \sum_{n=0}^\infty \binom{a}{n} x^n$, $|x| < 1$.

(J7): $J'_\nu(x) - \frac{\nu}{x} J_\nu(x) = -J_{\nu+1}(x)$

(J8): $J_{\nu-1}(x) - J_{\nu+1}(x) = 2J'_\nu(x)$

(J9): $J_{\nu-1}(x) + J_{\nu+1}(x) = \frac{2\nu}{x} J_\nu(x)$

(J10): $\int x^\nu J_{\nu-1}(x)\,dx = x^\nu J_\nu(x) + C$

(J11): $\int x^{-\nu} J_{\nu+1}(x)\,dx = -x^{-\nu} J_\nu(x) + C$

(J12): $J_\nu(x) \sim \frac{(x/2)^\nu}{\Gamma(\nu+1)}, \qquad \nu \neq -1, -2, -3, \ldots, \qquad x \to 0^+$

(J13): $J_\nu(x) \sim \sqrt{\frac{2}{\pi x}} \cos\left[x - \left(\nu + \frac{1}{2}\right)\frac{\pi}{2}\right], \qquad x \to +\infty$

Basic Identities for $Y_\nu(x)$

(Y1): $Y_\nu(x) = \frac{(\cos \nu\pi) J_\nu(x) - J_{-\nu}(x)}{\sin \nu\pi}$

(Y2): $Y_{-n}(x) = (-1)^n Y_n(x), \qquad n = 0, 1, 2, \ldots$

(Y3): $\frac{d}{dx}[x^\nu Y_\nu(x)] = x^\nu Y_{\nu-1}(x)$

(Y4): $\frac{d}{dx}[x^{-\nu} Y_\nu(x)] = -x^{-\nu} Y_{\nu+1}(x)$

(Y5): $Y'_\nu(x) + \frac{\nu}{x} Y_\nu(x) = Y_{\nu-1}(x)$

(Y6): $Y'_\nu(x) - \frac{\nu}{x} Y_\nu(x) = -Y_{\nu+1}(x)$

(Y7): $Y_{\nu-1}(x) - Y_{\nu+1}(x) = 2Y'_\nu(x)$

(Y8): $Y_{\nu-1}(x) + Y_{\nu+1}(x) = \frac{2\nu}{x} Y_\nu(x)$

(Y9): $Y_0(x) \sim \frac{2}{\pi} \log x, \qquad x \to 0^+$

(Y10): $Y_\nu(x) \sim -\frac{\Gamma(\nu)}{\pi}\left(\frac{2}{x}\right)^\nu, \qquad \nu > 0, \qquad x \to 0^+$

(Y11): $Y_\nu(x) \sim \sqrt{\frac{2}{\pi x}} \sin\left[x - \left(\nu + \frac{1}{2}\right)\frac{\pi}{2}\right], \qquad x \to +\infty$

EXERCISES 9.4

1. Using the series representation (46), show that
   (a) $J_0(0) = 1$ and $J_\nu(0) = 0$ for $\nu > 0$.
   (b) $J_1'(0) = 1/2$ and $J_\nu'(0) = 0$ for $\nu > 1$.

2. Verify by direct substitution of the series that $y = J_{-\nu}(x)$ satisfies Bessel's equation (38).

3. Directly from the series (46), show that ($n = 0, 1, 2, \ldots$)
$$J_n(-x) = (-1)^n J_n(x).$$

4. Given the recurrence formula (43), show that it can be expressed in the equivalent form
$$c_{2m} = \frac{(-1)^m c_0}{2^{2m} m!(\nu + 1)(\nu + 2) \cdots (\nu + m)}, \qquad m = 1, 2, 3, \ldots .$$

5. Establish the following identities ($x > 0$):

   (a) $J_{1/2}(x) = \sqrt{\dfrac{2}{\pi x}} \sin x$ (b) $J_{-1/2}(x) = \sqrt{\dfrac{2}{\pi x}} \cos x$

   (c) $J_{1/2}(x) J_{-1/2}(x) = \dfrac{\sin 2x}{\pi x}$ (d) $[J_{1/2}(x)]^2 + [J_{-1/2}(x)]^2 = \dfrac{2}{\pi x}$

   Hint: For (a) and (b), compare series representations.

In Problems 6–9, express the general solution of the DE in terms of Bessel functions.

6. $x^2 y'' + xy' + (x^2 - 1)y = 0$

7. $xy'' + y' + xy = 0$

8. $4x^2 y'' + 4xy' + (4x^2 - 1)y = 0$

★9. $x^2 y'' + xy' + (4x^2 - 9)y = 0$
   Hint: Let $t = 2x$.

10. Show that for constant k,
$$\frac{d}{dx}[J_\nu(kx)] = -kJ_{\nu+1}(kx) + \frac{\nu}{x} J_\nu(kx).$$

11. Show that $y = J_0(kx)$, where k is a constant, is a solution of
$$xy'' + y' + k^2 xy = 0.$$

12. Show that the change of variable $y = u(x)/\sqrt{x}$ reduces Bessel's equation (38) to the form
$$u'' + \left[1 + \frac{\frac{1}{4} - \nu^2}{x^2}\right] u = 0.$$

13. Use the result of Problem 12 to find a general solution, in terms of trigonometric functions, of Bessel's equation when $\nu = 1/2$.

★14. Use Problem 12 to show that for large x, the solutions of Bessel's equation are described approximately by

$$y = C_1 \frac{\cos x}{\sqrt{x}} + C_2 \frac{\sin x}{\sqrt{x}}.$$

More precisely, it can be shown that for $x \gg 1$,

$$J_\nu(x) \sim \sqrt{\frac{2}{\pi x}} \cos\left[x - \left(\nu + \frac{1}{2}\right)\frac{\pi}{2}\right],$$

$$Y_\nu(x) \sim \sqrt{\frac{2}{\pi x}} \sin\left[x - \left(\nu + \frac{1}{2}\right)\frac{\pi}{2}\right].$$

★15. If y_1 and y_2 are linearly independent functions,
(a) show that y_1 and $y_3 = y_1 + y_2$ are linearly independent.
(b) Use (a) to show that $J_\nu(x)$ and $Y_\nu(x)$ are linearly independent for $\nu \neq$ integer.

16. If y_1 and y_2 are two solutions of Bessel's equation (38), show that for some constant C, their Wronskian is given by

$$W(y_1, y_2)(x) = \frac{C}{x}.$$

Hint: See Problem 41 in Exercises 1.3.

★17. From the result of Problem 16, show that

(a) $W(J_\nu, J_{-\nu})(x) = -\dfrac{2 \sin \nu\pi}{\pi x}.$

Hint: Use the relation $C = \lim_{x \to 0^+} xW(J_\nu, J_{-\nu})(x)$ and Problem 23(a).

(b) $W(J_\nu, Y_\nu)(x) = \dfrac{2}{\pi x}.$

Hint: Use part (a).

18. Using L'Hôpital's rule, show that

$$\lim_{\nu \to n} \frac{J_\nu(x) \cos \nu\pi - J_{-\nu}(x)}{\sin \nu\pi} = \lim_{\nu \to n} \frac{1}{\pi}\left[\frac{\partial}{\partial \nu} J_\nu(x) - (-1)^n \frac{\partial}{\partial \nu} J_{-\nu}(x)\right].$$

19. Show that

$$\frac{d}{dx}[x^{-\nu} J_\nu(x)] = -x^{-\nu} J_{\nu+1}(x).$$

20. By adding and subtracting (67) and (68), deduce that
(a) $2J_\nu'(x) = J_{\nu-1}(x) - J_{\nu+1}(x).$
(b) $\dfrac{2\nu}{x} J_\nu(x) = J_{\nu-1}(x) + J_{\nu+1}(x).$

★21. Derive the *generating function* relation

$$\exp\left[\frac{1}{2} x\left(t - \frac{1}{t}\right)\right] = \sum_{n=-\infty}^{\infty} J_n(x) t^n.$$

Hint: Write $\exp[\frac{1}{2}x(t - 1/t)] = \exp(\frac{1}{2}xt) \exp(-x/2t)$ and expand each exponential on the right as a power series.

★22. Setting $w(x, t) = \exp[\frac{1}{2}x(t - 1/t)]$, show that
(a) $w(x + y, t) = w(x, t)w(y, t)$.
(b) From (a), derive the *addition formula*

$$J_n(x + y) = \sum_{k=-\infty}^{\infty} J_k(x)J_{n-k}(y).$$

Hint: Use Problem 21.

23. Develop the following asymptotic formulas for $x \to 0^+$:

(a) $J_\nu(x) \sim \dfrac{(x/2)^\nu}{\Gamma(\nu + 1)}, \qquad \nu \neq -1, -2, -3, \ldots .$

(b) $Y_\nu(x) \sim -\dfrac{\Gamma(\nu)}{\pi}(2/x)^\nu, \qquad \nu \neq 0, -1, -2, \ldots .$

(c) $Y_0(x) \sim \dfrac{2}{\pi}\log x.$

24. Using (55), show that

(a) $\dfrac{d}{dx}[x^\nu Y_\nu(x)] = x^\nu Y_{\nu-1}(x).$

(b) $\dfrac{d}{dx}[x^{-\nu} Y_\nu(x)] = -x^{-\nu} Y_{\nu+1}(x).$

(c) $2Y_\nu'(x) = Y_{\nu-1}(x) - Y_{\nu+1}(x).$

(d) $\dfrac{2\nu}{x} Y_\nu(x) = Y_{\nu-1}(x) + Y_{\nu+1}(x).$

In Problems 25–29, use identities or integration by parts to verify the result.

25. $\displaystyle\int xJ_0(x)\,dx = xJ_1(x) + C$

26. $\displaystyle\int x^3J_0(x)\,dx = x^3J_1(x) - 2x^2J_2(x) + C$

27. $\displaystyle\int x^{-2}J_2(x)\,dx = -\frac{2}{3x^2}J_1(x) - \frac{1}{3}J_1(x) + \frac{1}{3x}J_0(x) + \frac{1}{3}\int J_0(x)\,dx$

Hint: Let $u = x^2J_2(x)$ and $dv = x^{-4}dx$, and integrate by parts.

28. $\displaystyle\int J_0(x)\cos x\,dx = xJ_0(x)\cos x + xJ_1(x)\sin x + C$

29. $\displaystyle\int J_0(x)\sin x\,dx = xJ_0(x)\sin x - xJ_1(x)\cos x + C$

★30. By expanding $\cos(x \sin\theta)$ in an infinite series and integrating the resulting series termwise, show that

$$J_0(x) = \frac{1}{\pi}\int_0^\pi \cos(x\sin\theta)\,d\theta.$$

In Problems 31–36, derive the given integral formula.

31. $\displaystyle\int_0^\infty e^{-ax}J_0(bx)\,dx = (a^2 + b^2)^{-1/2}, \qquad a > b$

Hint: Use Example 4 and let $t = bx$.

32. $\displaystyle\int_0^\infty x^2 e^{-ax} J_0(bx)\,dx = (2a^2 - b^2)(a^2 + b^2)^{-5/2}, \qquad a > b$

Hint: Differentiate both sides of Problem 31 with respect to a.

33. $\displaystyle\int_0^\infty e^{-a^2x^2} J_0(bx)x\,dx = \frac{1}{2a^2} e^{-b^2/4a^2}, \qquad a, b > 0$

★34. $\displaystyle\int_0^{\pi/2} J_0(x \cos \phi) \cos \phi\, d\phi = \frac{\sin x}{x}$

★35. $\displaystyle\int_0^{\pi/2} J_1(x \cos \phi)\, d\phi = \frac{1 - \cos x}{x}$

★36. $\displaystyle\int_0^a x(a^2 - x^2)^{-1/2} J_0(kx \sin \phi)\,dx = \frac{\sin(ka \sin \phi)}{k \sin \phi}$

★37. Given the nonhomogeneous boundary value problem, with ν not an integer,

$$\frac{d}{dx}(xy') + \left(x - \frac{\nu^2}{x}\right)y = f(x), \qquad y \text{ finite as } x \to 0^+, \qquad y(1) = 0,$$

(a) show that $z_1 = J_\nu(x)$ and $z_2 = J_{-\nu}(1)J_\nu(x) - J_\nu(1)J_{-\nu}(x)$ satisfy, respectively, the first and second prescribed boundary conditions.

(b) Show that the Green's function for this problem is

$$g(x, s) = \begin{cases} \dfrac{\pi J_\nu(s)}{2(\sin \pi\nu)J_\nu(1)} [J_{-\nu}(1)J_\nu(x) - J_\nu(1)J_{-\nu}(x)], & 0 < s \leq x \\ \dfrac{\pi J_\nu(x)}{2(\sin \pi\nu)J_\nu(1)} [J_{-\nu}(1)J_\nu(s) - J_\nu(1)J_{-\nu}(s)], & x < s < 1. \end{cases}$$

Hint: Use Problem 17.

★38. Find a Green's function for the boundary value problem in Problem 37 in terms of $J_\nu(x)$ and $Y_\nu(x)$, when $\nu = n$ $(n = 0, 1, 2, \ldots)$.

★39. Using the Green's function obtained in Problem 38, solve

$$xy'' + y' + xy = x, \qquad y \text{ finite as } x \to 0^+, \qquad y(1) = 0.$$

Hint: $J_0(x)Y_1(x) - J_1(x)Y_0(x) = -2/\pi x$.

9.5 EIGENVALUE PROBLEMS ASSOCIATED WITH BESSEL FUNCTIONS

In practice, Bessel's equation often appears in the form of an *eigenequation,* i.e.,

$$x^2y'' + xy' + (\lambda x^2 - \nu^2)y = 0, \qquad 0 < x < b. \tag{72}$$

Although (72) is more general than (38), it is still called *Bessel's equation*. In self-adjoint form, Equation (72) becomes

$$\frac{d}{dx}(xy') - \frac{\nu^2}{x}y + \lambda xy = 0, \qquad 0 < x < b, \tag{73}$$

(see Section 1.6) where we see that

$$p(x) = x, \qquad q(x) = -\frac{\nu^2}{x}, \qquad r(x) = x.$$

As previously noted, Bessel's equation has a singularity at $x = 0$. In particular, $p(0) = 0$, and for $\nu \neq 0$, $q(x)$ becomes infinite at $x = 0$.

If we are seeking solutions of (72) that satisfy the orthogonality property of Theorem 1.4, at the endpoint $x = 0$ we must impose the boundary condition

$$y(x), \quad y'(x) \quad \text{finite as} \quad x \to 0^+. \tag{74}$$

The endpoint $x = b$ is an ordinary point of the equation, and hence, any boundary condition of the form $a_{21}y(b) + a_{22}y'(b) = 0$ is suitable for our needs. For the sake of illustration, we take the simple case

$$y(b) = 0. \tag{75}$$

In searching for the eigenvalues and eigenfunctions of (72) subject to the boundary conditions (74) and (75), we begin by setting $\lambda = 0$, which reduces Bessel's equation to the *Cauchy-Euler equation*

$$x^2y'' + xy' - \nu^2 y = 0. \tag{76}$$

Depending on the value of ν, the general solution of (76) is

$$y = \begin{cases} C_1 + C_2 \log x, & \nu = 0 \\ C_3 x^{\nu} + C_4 x^{-\nu}, & \nu \neq 0. \end{cases} \tag{77}$$

The boundary condition (74) demands that we set $C_2 = C_4 = 0$. The remaining condition (75) further requires that $C_1 = C_3 = 0$, and thus we deduce that $\lambda = 0$ is not an eigenvalue of the problem.

Next we set $\lambda = k^2 > 0$, which yields

$$x^2y'' + xy' + (k^2x^2 - \nu^2)y = 0, \tag{78}$$

By making the change of variable $t = kx$ and applying the chain rule we find that

$$y' = \frac{dy}{dt}\frac{dt}{dx} = k\frac{dy}{dt},$$

$$y'' = k^2\frac{d^2y}{dt^2}.$$

These substitutions in (78) lead to the familiar form

$$t^2\frac{d^2y}{dt^2} + t\frac{dy}{dt} + (t^2 - \nu^2)y = 0. \tag{79}$$

The functional form of (79) is the same as (38), and thus its general solution is

$$y(t) = C_1 J_\nu(t) + C_2 Y_\nu(t). \tag{80}$$

Finally, letting $t = kx$ in this last expression, we obtain

$$y(x) = C_1J_\nu(kx) + C_2Y_\nu(kx) \tag{81}$$

as the general solution of (78).

The boundedness condition (74) necessitates that we set $C_2 = 0$, since $Y_\nu(kx)$ becomes infinite at $x = 0$. Imposing the second condition (75) then leads to

$$C_1J_\nu(kb) = 0, \tag{82}$$

which for $C_1 \neq 0$ can be satisfied only if the Bessel function itself is zero. It can be shown that the Bessel function $J_\nu(x)$ has infinitely many zeros on the interval $x > 0$,* so if we denote the nth such zero by μ_n, i.e., $J_\nu(\mu_n) = 0$, then k must assume one of the values μ_n/b. Hence, the positive eigenvalues are

$$\lambda_n = k_n^2 = \frac{\mu_n^2}{b^2}, \qquad n = 1, 2, 3, \ldots, \tag{83}$$

and the corresponding eigenfunctions are, therefore,

$$\phi_n(x) = J_\nu(k_nx), \qquad n = 1, 2, 3, \ldots. \tag{84}$$

For λ negative, the general solution of Bessel's equation (72) can be expressed in terms of *modified Bessel functions* (see Section 9.6). It can be shown that modified Bessel functions have no zeros on the positive axis, and this condition in turn implies there are no negative eigenvalues. Therefore, the only eigenvalues and eigenfunctions of (72) subject to conditions (74) and (75) are those given by (83) and (84).

□ 9.5.1 EQUATIONS REDUCIBLE TO BESSEL'S EQUATION

Elementary problems are regarded as solved when their solutions can be expressed in terms of tabulated functions such as trigonometric and exponential functions. The same can be said of more complicated problems when their solutions can be expressed in terms of Bessel functions, since extensive tables of Bessel functions have been formed for various values of x and ν.†

A fairly large number of DEs occurring in physics and engineering applications are specializations of the form

$$x^2y'' + (1 - 2a)xy' + [b^2c^2x^{2c} + (a^2 - c^2\nu^2)]y = 0, \qquad \nu \geq 0, \qquad b > 0, \tag{85}$$

the general solution of which, expressed in terms of Bessel functions, is

$$y = x^a[C_1J_\nu(bx^c) + C_2Y_\nu(bx^c)]. \tag{86}$$

---

* See Theorem 5.6 in L. C. Andrews, *Ordinary Differential Equations with Applications*, (Glenview, Ill.: Scott, Foresman and Co., 1982). Also recall Table 9.1 in Section 9.4 which lists some zeros of Bessel functions.

† For example, see Chapters 9 and 10 in M. Abramowitz and I. Stegun (eds.), *Handbook of Mathematical Functions*, (New York: Dover, 1965).

To derive this solution formula requires two transformations of variables. First, let us set

$$y = x^a z, \tag{87}$$

from which we obtain

$$y' = x^a z' + a x^{a-1} z,$$

$$y'' = x^a z'' + 2a x^{a-1} z' + a(a-1) x^{a-2} z,$$

and substituting these expressions into (85) and simplifying the results gives us

$$x^2 z'' + x z' + (b^2 c^2 x^{2c} - c^2 \nu^2) z = 0. \tag{88}$$

Next, we make the change of independent variable

$$t = x^c, \tag{89}$$

from which it follows (through application of the chain rule)

$$x z' = c x^c \frac{dz}{dt},$$

$$x^2 z'' = c(c-1) x^c \frac{dz}{dt} + c^2 x^{2c} \frac{d^2 z}{dt^2}.$$

Hence, Equation (88) becomes

$$t^2 \frac{d^2 z}{dt^2} + t \frac{dz}{dt} + (b^2 t^2 - \nu^2) z = 0, \tag{90}$$

which is of the form (78), and therefore has general solution

$$z(t) = C_1 J_\nu(bt) + C_2 Y_\nu(bt). \tag{91}$$

Transforming back to the original variables x and y leads us to the desired result (86). For those cases when ν is not an integer, we can express the general solution (86) in the alternate form

$$y = x^a [C_1 J_\nu(bx^c) + C_2 J_{-\nu}(bx^c)], \qquad \nu \neq \text{integer}. \tag{92}$$

EXAMPLE 5

Find the general solution of $x^2 y'' + 5xy' + (4x^2 + 3)y = 0$.

SOLUTION This DE is of the form (85) with

$$1 - 2a = 5, \qquad 2c = 2, \qquad b^2 c^2 = 4, \qquad a^2 - c^2 \nu^2 = 3.$$

These conditions are satisfied if $a = -2$, $c = 1$, $b = 2$, and $\nu = 1$. Thus,

$$y = x^{-2} [C_1 J_1(2x) + C_2 Y_1(2x)].$$

EXAMPLE 6

Find the general solution of Airy's equation $y'' + xy = 0$.

SOLUTION In order to compare this DE with (85), we must multiply through by x^2, putting it in the form

$$x^2y'' + x^3y = 0.$$

Here we find that $a = 1/2$, $b = 2/3$, $c = 3/2$, and $\nu = 1/3$. Hence,

$$y = x^{1/2}\left[C_1J_{1/3}\left(\frac{2}{3}x^{3/2}\right) + C_2Y_{1/3}\left(\frac{2}{3}x^{3/2}\right)\right],$$

or, since ν is not an integer, we can also write

$$y = x^{1/2}\left[C_1J_{1/3}\left(\frac{2}{3}x^{3/2}\right) + C_2J_{-1/3}\left(\frac{2}{3}x^{3/2}\right)\right].$$

EXAMPLE 7

Assuming $\lambda \geqslant 0$, determine the eigenvalues and eigenfunctions of

$$4x^2y'' + (\lambda x^2 - 3)y = 0, \qquad y, y' \text{ finite as } x \to 0^+, \qquad y'(4) = 0.$$

SOLUTION For $\lambda = 0$, the DE is a Cauchy-Euler equation with general solution

$$y = C_1x^{3/2} + C_2x^{-1/2}.$$

To satisfy the boundedness condition at $x = 0$ we must select $C_2 = 0$, and the second boundary condition leads to

$$y'(4) = 3C_1 = 0.$$

Thus, $\lambda = 0$ is not an eigenvalue.

Next, writing $\lambda = k^2 > 0$ and dividing the DE by 4, we get

$$x^2y'' + \left(\frac{k^2}{4}x^2 - \frac{3}{4}\right)y = 0,$$

where we can identify $a = 1/2$, $b = k/2$, $c = 1$, and $\nu = 1$ by comparison with (85). The general solution is, therefore,

$$y = x^{1/2}\left[C_1J_1\left(\frac{kx}{2}\right) + C_2Y_1\left(\frac{kx}{2}\right)\right].$$

To investigate the behavior of our solution for $x \to 0^+$, we use the result of Problem 23 in Exercises 9.4, which leads to

$$\lim_{x\to 0^+} x^{1/2}\left[C_1J_1\left(\frac{kx}{2}\right) + C_2Y_1\left(\frac{kx}{2}\right)\right] = \lim_{x\to 0^+} x^{1/2}\left[C_1\frac{kx}{4} - C_2\frac{4}{\pi x}\right].$$

Thus, to keep y finite as $x \to 0^+$ we must choose $C_2 = 0$, and so we are left with

$$y = C_1 x^{1/2} J_1\left(\frac{kx}{2}\right).^*$$

In order to satisfy the boundary condition at $x = 4$ we first compute

$$y' = C_1\left[\frac{1}{2}x^{-1/2}J_1\left(\frac{kx}{2}\right) + \frac{k}{2}x^{1/2}J_1'\left(\frac{kx}{2}\right)\right],$$

and hence,

$$y'(4) = C_1\left[\frac{1}{4}J_1(2k) + kJ_1'(2k)\right] = 0.$$

We conclude that the eigenvalues are $\lambda_n = k_n^2$ $(n = 1, 2, 3, \ldots)$, where $J_1(2k_n) + 4k_nJ_1'(2k_n) = 0$, and the corresponding eigenfunctions are

$$\phi_n(x) = x^{1/2}J_1(k_nx/2), \qquad n = 1, 2, 3, \ldots.$$

EXERCISES 9.5

In Problems 1–3, find the eigenvalues ($\lambda \geq 0$) and eigenfunctions. Give appropriate equations for calculating the eigenvalues.

1. $xy'' + y' + \lambda xy = 0, \quad y, y'$ finite as $x \to 0^+, \quad y(1) = 0$
2. $4x^2y'' + 4xy' + (\lambda x^2 - 1)y = 0, \quad y, y'$ finite as $x \to 0^+, \quad y(1) = 0$
3. $x^2y'' + xy' + (\lambda x^2 - 4)y = 0, \quad y, y'$ finite as $x \to 0^+, \quad y(3) = 0$

*4. Given that the Bessel function $J_\nu(x)$ has infinitely many zeros for $x > 0$, use Rolle's theorem from calculus to show that the function $J_\nu'(x)$ also has infinitely many positive zeros.

5. Given the eigenvalue problem

$$x^2y'' + xy' + (\lambda x^2 - \nu^2)y = 0, \qquad y, y' \text{ finite as } x \to 0^+, \qquad y'(1) = 0,$$

show that it possesses the eigenvalues $\lambda_n = k_n^2$, where k_n satisfies $J_\nu'(k_n) = 0$, $n = 1, 2, 3, \ldots$. What are the corresponding eigenfunctions?

6. Given the eigenvalue problem

$$x^2y'' + xy' + (\lambda x^2 - \nu^2)y = 0,$$

$$y, y' \text{ finite as } x \to 0^+, \qquad a_{21}y(1) + a_{22}y'(1) = 0,$$

show that it possesses the eigenvalues $\lambda_n = k_n^2$, where k_n satisfies

* This choice of y also satisfies y' finite as $x \to 0^+$.

$$hJ_\nu(k_n) + k_nJ'_\nu(k_n) = 0, \qquad (h = a_{21}/a_{22}), \qquad n = 1, 2, 3, \ldots.$$

What are the corresponding eigenfunctions?

In Problems 7–18, express the general solution of each equation in terms of Bessel functions.

7. $xy'' + y' + \frac{1}{4}y = 0$

8. $4x^2y'' + 4xy' + (x^2 - n^2)y = 0$

9. $x^2y'' + xy' + 4(x^4 - k^2)y = 0$

10. $xy'' - y' + xy = 0$

11. $xy'' + (1 + 2n)y' + xy = 0$

12. $x^2y'' - 5xy' + (64x^8 + 5)y = 0$

13. $x^2y'' - 7xy' + \left(36x^6 + \frac{175}{16}\right)y = 0$

14. $y'' + y = 0$

15. $y'' + k^2x^2y = 0$

16. $y'' + k^2x^4y = 0$

17. $4x^2y'' + (1 + 4x)y = 0$

18. $x^2y'' + 5xy' + (9x^2 - 12)y = 0$

19. Given the DE

$$y'' + ae^{mx}y = 0,$$

(a) show that the substitution $t = e^{mx}$ transforms it into

$$t\frac{d^2y}{dt^2} + \frac{dy}{dt} + \frac{a}{m^2}y = 0.$$

(b) Solve the DE in (a) in terms of Bessel functions.
(c) Find the general solution of the original DE.

20. Given the DE

$$x^2y'' + x(1 - 2x\tan x)y' - (x\tan x + n^2)y = 0,$$

(a) show that the transformation $y = u(x)\sec x$ leads to an equation in u solvable in terms of Bessel functions.
(b) Find the general solution of the original DE.

21. Show that the *spherical Bessel functions*

$$j_n(x) = \sqrt{\frac{\pi}{2x}}\,J_{n+\frac{1}{2}}(x), \qquad y_n(x) = \sqrt{\frac{\pi}{2x}}\,Y_{n+\frac{1}{2}}(x)$$

$(n = 0, 1, 2, \ldots)$, are solutions of

$$x^2y'' + 2xy' + [x^2 - n(n + 1)]y = 0.$$

22. Referring to Problem 21, show that

(a) $j_0(x) = \dfrac{\sin x}{x}$.

(b) $y_0(x) = -\dfrac{\cos x}{x}$.

★23. A particle of variable mass $m = (a + bt)^{-1}$, where a and b are constants, starts from rest at a distance r_0 from the origin 0 and is attracted to 0 by a force always directed toward 0 of magnitude k^2mr (k constant). The equation of motion is given by

$$\frac{d}{dt}\left(m\frac{dr}{dt}\right) = -k^2mr.$$

Solve this equation for r subject to the prescribed initial conditions.
Hint: Make the change of variable $bx = a + bt$, transforming the equation of motion to $x^2r'' - xr' + k^2x^2r = 0$.

In Problems 24–26, find the eigenvalues ($\lambda \geqslant 0$) and eigenfunctions. Give appropriate equations for calculating the eigenvalues. *Hint:* If necessary, use Problem 23 in Exercises 9.4.

24. $x^2y'' - xy' + (4\lambda x^4 - 3)y = 0, \qquad y, y' \text{ finite as } x \to 0^+, \qquad y(1) = 0$

25. $y'' + \lambda x^2 y = 0, \qquad y(0) = 0, \qquad y(1) = 0$

26. $xy'' + \lambda y = 0, \qquad y, y' \text{ finite as } x \to 0^+, \qquad y(1) = 0$

27. In a problem on the stability of a tapered strut, the displacement y satisfies the boundary value problem

$$4xy'' + K^2y = 0, \qquad y'(a) = 0, \qquad y'(b) = 0 \qquad (a < b).$$

Show that the determination of the positive constant K for which nontrivial solutions exist leads to the relation

$$J_0(K\sqrt{a})\,Y_0(K\sqrt{b}) - J_0(K\sqrt{b})\,Y_0(K\sqrt{a}) = 0.$$

28. The linear density of an elastic string of unit length varies according to $\rho = \rho_0(1 + x)^2$, where ρ_0 is a constant and x is measured over $0 \leqslant x \leqslant 1$. The ends of the string are fixed to an axis rotating with angular speed ω (see Section 1.5.2).
(a) Show that the governing equation can be written as

$$\frac{d^2y}{dz^2} + 4\lambda z^2 y = 0, \qquad y(1) = 0, \qquad y(2) = 0,$$

where $z = 1 + x$ and $\lambda = \rho_0\omega^2/4T$.
(b) Show that the nth critical speed is $\omega_n = 2\mu_n\sqrt{T/\rho_0}$, where μ_n denotes the roots of

$$J_{1/4}(\mu_n)\,Y_{1/4}(4\mu_n) - J_{1/4}(4\mu_n)\,Y_{1/4}(\mu_n) = 0, \qquad n = 1, 2, 3, \ldots.$$

★29. The small deflections of a uniform column of length b bending under its own weight are governed by

$$\theta'' + K^2x\theta = 0, \qquad \theta'(0) = 0, \qquad \theta(b) = 0,$$

where θ is the angle of deflection from the vertical and K is a positive constant.
(a) Show that the solution of the DE satisfying the first boundary condition (at $x = 0$) is

$$\theta(x) = Cx^{1/2}J_{-1/3}\left(\frac{2}{3}Kx^{3/2}\right),$$

where C is any constant.
(b) Show that the shortest column length for which buckling may occur (denoted by b_0) is $b_0 \simeq 1.99K^{-2/3}$.
Hint: The first zero of $J_{-1/3}(u)$ is $u \simeq 1.87$.

★30. An axial load P is applied to a column whose circular cross section is tapered so that the moment of inertia is $I(x) = (x/a)^4$. If the column is simply supported at the ends $x = 1$ and $x = a$, the buckling modes are described by solutions of

$$x^4y'' + \lambda y = 0, \qquad y(1) = 0, \qquad y(a) = 0 \qquad (a > 1),$$

where $\lambda = Pa^4/E$ and E is Young's modulus.

(a) Express the general solution of the DE in terms of Bessel functions, but do not apply the boundary conditions.

(b) By making the substitution $y = xu(x)$ followed by $x = 1/t$, show that the general solution of the DE can also be expressed in terms of sines and cosines.

(c) Apply the prescribed boundary conditions to the solution in (b), find the critical loads P_n, and show that the first buckling mode is described by

$$\phi_1(x) = x \sin\left[\frac{a}{a-1}\left(1 - \frac{1}{x}\right)\right].$$

In Problems 31–34, use the approximation methods of Section 1.7.

★31. Find an approximation to the first positive zero of $J_1(x)$ by approximating the smallest eigenvalue of

$$\frac{d}{dx}(xy') + \left(\lambda x - \frac{1}{x}\right)y = 0, \qquad y(0) = 0, \qquad y(1) = 0.$$

Use the trial solution $\psi = c_1x(1 - x)$ and

(a) the collocation method.

(b) the Galerkin method.

(c) Compare your answer in (a) and (b) with that found in Table 9.1 in Section 9.4.

★32. Use the trial solution $\psi = c_1(z - 1)(2 - z) + c_2(z - 1)^2(2 - z)$ and the method of collocation to approximate the first two critical speeds ω_1 and ω_2 of Problem 28.

★33. Using the trial solution $\psi = c_1(b - x^2)$ and Galerkin's method, show that the shortest column length b_0 in Problem 29(b) is a solution of (with $K = 1$)

$$b^4 - 3b^3 + 3b^2 + 4b - 12 = 0.$$

By synthetic division, or trial and error, determine the smallest positive root of this polynomial.

★34. Use the trial solution $\psi = c_1(x - 1)(2 - x)$ to approximate the smallest critical load P_1 in Problem 30 when $a = 2$.

(a) Use the collocation method and set $E_1 = 0$ at $x = 3/2$.

(b) Use the Galerkin method.

9.6 MODIFIED BESSEL FUNCTIONS

In the previous section we found that solutions of

$$x^2y'' + xy' + (k^2x^2 - \nu^2)y = 0 \tag{93}$$

can be expressed in the form

$$y = C_1 J_\nu(kx) + C_2 Y_\nu(kx). \tag{94}$$

The DE

$$x^2y'' + xy' - (x^2 + \nu^2)y = 0, \tag{95}$$

which bears a great resemblance to Bessel's equation, is *Bessel's modified equation*. It is of the form (93) with $k^2 = -1$, and so we can formally write the solution of (95) as

$$y = C_1 J_\nu(ix) + C_2 Y_\nu(ix), \tag{96}$$

where $i^2 = -1$.

The disadvantage of the general solution (96) is that it is expressed in terms of complex functions, and in most situations we prefer real solutions. The problem is similar to stating that

$$y = C_1 e^{ix} + C_2 e^{-ix}$$

is the general solution of $y'' + y = 0$. In order to avoid the imaginary arguments in (96), we introduce the *modified Bessel function of the first kind* and order ν,

$$I_\nu(x) = i^{-\nu} J_\nu(ix) = \sum_{m=0}^{\infty} \frac{(x/2)^{2m+\nu}}{m!\Gamma(m+\nu+1)}. \tag{97}$$

Thus, $y_1 = I_\nu(x)$ is one solution of (95), and when ν is not an integer, the function $I_{-\nu}(x)$, obtained by replacing ν with $-\nu$ in (97), is another solution. Moreover, $I_{-\nu}(x)$ is linearly independent of $I_\nu(x)$, because $J_\nu(ix)$ and $J_{-\nu}(ix)$ are linearly independent. However, when $\nu = n$ ($n = 0, 1, 2, \ldots$), we find that

$$\begin{aligned} I_{-n}(x) &= i^n J_{-n}(ix) \\ &= i^n(-1)^n J_n(ix) \\ &= (-1)^{2n} I_n(x), \end{aligned}$$

or

$$I_{-n}(x) = I_n(x), \qquad n = 0, 1, 2, \ldots. \tag{98}$$

Hence, $I_{-n}(x)$ is not independent of $I_n(x)$.

Rather than use $Y_\nu(ix)$ to define a second linearly independent solution of (95) which is not restricted to certain values of ν, it is customary in most applications to introduce the *modified Bessel function of the second kind* and order ν,

$$K_\nu(x) = \frac{\pi}{2} \frac{I_{-\nu}(x) - I_\nu(x)}{\sin \nu\pi}. \tag{99}$$

When ν takes on integer values, we define

$$K_n(x) = \lim_{\nu \to n} K_\nu(x), \qquad n = 0, 1, 2, \ldots \tag{100}$$

(see Problem 16 in Exercises 9.6). The functions $I_\nu(x)$ and $K_\nu(x)$ are linearly independent for all ν (see Problem 11 in Exercises 9.6) and thus we write the general solution of Bessel's modified equation as

$$y = C_1 I_\nu(x) + C_2 K_\nu(x) \tag{101}$$

for all ν.

The graphs of $I_n(x)$ and $K_n(x)$ for $n = 0, 1, 2$ are shown in Figures 9.5 and 9.6, respectively. Note that these functions do not have the oscillatory characteristics that we previously noted for the Bessel functions $J_n(x)$ and $Y_n(x)$. Fundamentally, this means that the modified Bessel functions do not have zeros on the open interval $(0, \infty)$.

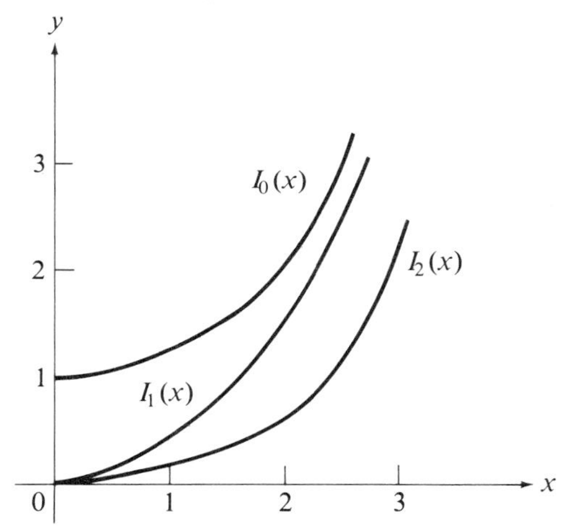

FIGURE 9.5

Graph of $I_n(x)$, $n = 0, 1, 2$

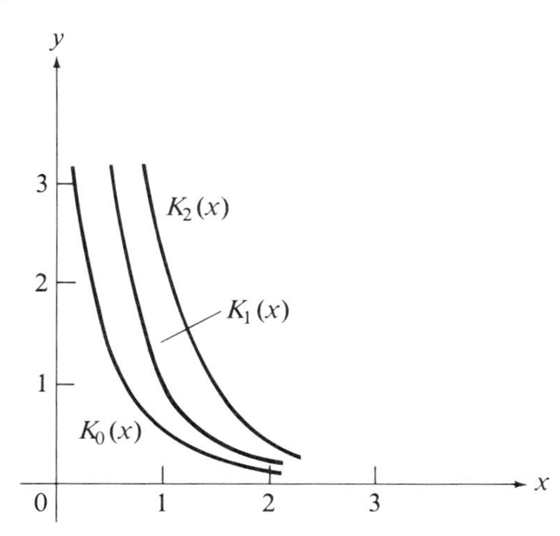

FIGURE 9.6

Graph of $K_n(x)$, $n = 0, 1, 2$

EXERCISES 9.6

1. Verify by direct substitution of the series that $y = I_\nu(x)$ is a solution of (95).

In Problems 2–6, express the general solution of the DE in terms of modified Bessel functions.

2. $x^2y'' + xy' - (x^2 + 1)y = 0$
3. $xy'' + y' - xy = 0$
4. $4x^2y'' + 4xy' - (4x^2 + 1)y = 0$
★5. $x^2y'' + xy' - (4x^2 + 1)y = 0$
6. $x^2y'' + xy' - (x^2 + 1/16)y = 0$
7. Show that

   (a) $I_{1/2}(x) = \sqrt{\dfrac{2}{\pi x}}\sinh x.$ (b) $I_{-1/2}(x) = \sqrt{\dfrac{2}{\pi x}}\cosh x.$

8. Show that the change of variable $y = u(x)/\sqrt{x}$ reduces Bessel's modified equation (95) to

$$u'' - \left[1 + \frac{\nu^2 - \frac{1}{4}}{x^2}\right]u = 0.$$

★9. Use Problem 8 to show that for large x, the solutions of Bessel's modified equation are described approximately by

$$y = C_1\frac{e^x}{\sqrt{x}} + C_2\frac{e^{-x}}{\sqrt{x}}.$$

More precisely, it can be shown that for $x \gg 1$,

$$I_\nu(x) \sim \frac{e^x}{\sqrt{2\pi x}} \quad \text{and} \quad K_\nu(x) \sim \sqrt{\frac{\pi}{2x}}\,e^{-x}.$$

10. If y_1 and y_2 are any two solutions of Bessel's modified equation (95), show that for some value of C their Wronskian is

$$W(y_1, y_2)(x) = \frac{C}{x}.$$

Hint: See Problem 41 in Exercises 1.3.

★11. From the result of Problem 10, show that

   (a) $W(I_\nu, I_{-\nu})(x) = -\dfrac{2\sin\nu\pi}{\pi x}.$ (b) $W(I_\nu, K_\nu)(x) = -\dfrac{1}{x}.$

12. Show that

   (a) $K_{-\nu}(x) = K_\nu(x).$ (b) $K_{1/2}(x) = \sqrt{\dfrac{\pi}{2x}}\,e^{-x}.$

   Hint: See Problem 7.

13. Show that

   (a) $\dfrac{d}{dx}[x^\nu I_\nu(x)] = x^\nu I_{\nu-1}(x).$ (b) $\dfrac{d}{dx}[x^{-\nu} I_\nu(x)] = x^{-\nu} I_{\nu+1}(x).$

14. Using the results of Problem 13, show that

(a) $I'_\nu(x) = I_{\nu-1}(x) - \dfrac{\nu}{x} I_\nu(x)$. (b) $I'_\nu(x) = \dfrac{\nu}{x} I_\nu(x) + I_{\nu+1}(x)$.

(c) $I'_\nu(x) = \dfrac{1}{2} [I_{\nu-1}(x) + I_{\nu+1}(x)]$. (d) $I_{\nu-1}(x) - I_{\nu+1}(x) = \dfrac{2\nu}{x} I_\nu(x)$.

15. Show that

(a) $\dfrac{d}{dx} [x^\nu K_\nu(x)] = -x^\nu K_{\nu-1}(x)$. (b) $\dfrac{d}{dx} [x^{-\nu} K_\nu(x)] = -x^{-\nu} K_{\nu+1}(x)$.

★16. Following the technique of Section 9.4.2, show that

$$K_0(x) = -I_0(x)[\log(x/2) + \gamma] + \sum_{m=1}^{\infty} \frac{(x/2)^{2m}}{(m!)^2} \left(1 + \frac{1}{2} + \cdots + \frac{1}{m}\right)$$

where $x > 0$.

★17. If b in the equation

$$x^2y'' + (1 - 2a)xy' + [b^2c^2x^{2c} + (a^2 - c^2\nu^2)]y = 0, \qquad \nu \geq 0,$$

is allowed to be purely imaginary, say $b = i\beta$, show that the general solution can be expressed as ($\beta > 0$)

$$y = x^a[C_1 I_\nu(\beta x^c) + C_2 K_\nu(\beta x^c)].$$

In Problems 18–23, use the result of Problem 17 to express the general solution of each DE in terms of modified Bessel functions.

18. $y'' - y = 0$

19. $y'' - xy = 0$

20. $x^2y'' + xy' - (4 + 36x^4)y = 0$

21. $xy'' - 3y' - 9x^5y = 0$

22. $y'' - k^2x^2y = 0$

23. $y'' - k^2x^4y = 0$

★24. Show that there are no negative eigenvalues of

$$x^2y'' + xy' + (\lambda x^2 - \nu^2)y = 0,$$

$$y, y' \text{ finite as } x \to 0^+, \qquad y(1) = 0.$$

9.7 GENERALIZED FOURIER SERIES

Generalized Fourier series were first introduced in Section 4.6, but have appeared occasionally throughout our discussion of solving PDEs by separating variables. These are series representations of a suitable function f, which have the general form

$$f(x) = \sum_{n=1}^{\infty} c_n \phi_n(x), \qquad x_1 < x < x_2, \tag{102}$$

where the Fourier coefficients are determined by

$$c_n = \| \phi_n(x) \|^{-2} \int_{x_1}^{x_2} r(x)f(x)\phi_n(x)\, dx, \qquad n = 1, 2, 3, \ldots . \tag{103}$$

The set of functions $\{\phi_n(x)\}$ is usually associated with an orthogonal set of eigenfunctions [with respect to a weight function $r(x)$] belonging to a Sturm-Liouville system. In this section we will discuss two important series of this type, known as *Fourier-Legendre series* and *Fourier-Bessel series*.

9.7.1 FOURIER-LEGENDRE SERIES

In Section 9.2, we found that the Legendre polynomials $P_n(x)$, $n = 0, 1, 2, \ldots$, are eigenfunctions of the singular Sturm-Liouville system

$$\begin{gathered} \frac{d}{dx}[(1 - x^2)y'] + \lambda y = 0, \qquad -1 < x < 1 \\ y(x),\ y'(x) \quad \text{finite as} \quad x \to -1^+, \qquad x \to 1^-. \end{gathered} \tag{104}$$

The Legendre operator $L = D[(1 - x^2)D]$ satisfies the identity

$$\int_{-1}^{1} (uL[v] - vL[u])\, dx = (1 - x^2)W(u, v)(x) \Big|_{-1}^{1} = 0 \tag{105}$$

[see Equation (68) in Section 1.6], where u and v are any continuous, twice differentiable functions fulfilling the boundary conditions in (104). It follows therefore that L is a *symmetric operator* whose eigenfunctions satisfy the orthogonality property

$$\int_{-1}^{1} P_n(x)P_k(x)\, dx = 0, \qquad n \neq k, \tag{106}$$

where the weighting function is $r(x) = 1$.

The generalized Fourier series associated with the Legendre polynomials is of the form*

$$f(x) = \sum_{n=0}^{\infty} c_n P_n(x), \qquad -1 < x < 1 \tag{107}$$

where the Fourier coefficients are given by

$$c_n = \| P_n(x) \|^{-2} \int_{-1}^{1} f(x)P_n(x)\, dx, \qquad n = 0, 1, 2, \ldots . \tag{108}$$

The normalization factor in (108) is

---

* Note that the series (107) begins with $n = 0$ since this value of n corresponds to the first eigenfunction $P_0(x)$.

$$\| P_n(x) \|^2 = \frac{2}{2n+1}, \qquad n = 0, 1, 2, \ldots \tag{109}$$

(see Problem 7 in Exercises 9.7), and the Fourier coefficients take the form

$$c_n = \left(n + \frac{1}{2}\right) \int_{-1}^{1} f(x)P_n(x)\,dx, \qquad n = 0, 1, 2, \ldots . \tag{110}$$

The series described by (107) and (110) is called a *Fourier-Legendre series*, or more simply, a *Legendre series*. Because the associated eigenvalue problem (104) is singular, Theorem 4.7 concerning the convergence of generalized Fourier series does not directly apply. We do have, however, the following theorem for Legendre series which we state without proof.†

THEOREM 9.2

> If f is a piecewise smooth function in the interval $[-1, 1]$, the Fourier-Legendre series (107) and (110) for f converges pointwise for all x in the interval. At points of continuity the series converges to $f(x)$, whereas at points of discontinuity of f in $(-1, 1)$, the series converges to the average value $\frac{1}{2}[f(x^+) + f(x^-)]$. At $x = -1$ the series converges to $f(-1^+)$ and at $x = 1$ the series converges to $f(1^-)$.

EXAMPLE 8

Find the Fourier-Legendre series for

$$f(x) = \begin{cases} -1, & -1 < x < 0 \\ 1, & 0 < x < 1. \end{cases}$$

SOLUTION In general, it is often difficult to obtain all the coefficients c_n simultaneously as we did in the case of trigonometric series. Thus, we ordinarily resort to separate computations for each value of n. That is the approach we will use here.

We first observe that f is an odd function, and since $P_n(x)$ is an even function or odd function, depending upon whether n is even or odd, it follows that

$$c_n = \left(n + \frac{1}{2}\right) \int_{-1}^{1} f(x)P_n(x)\,dx = 0, \qquad n = 0, 2, 4, \ldots .$$

For odd values of n we can write

---

† For a proof of Theorem 9.2, see Section 4.5 in L. C. Andrews, *Special Functions for Engineers and Applied Mathematicians*, (New York: Macmillan, 1985).

$$c_n = 2\left(n + \frac{1}{2}\right)\int_0^1 f(x)P_n(x)\,dx = (2n+1)\int_0^1 P_n(x)\,dx, \qquad n = 1, 3, 5, \ldots.$$

Recalling that $P_1(x) = x$, $P_3(x) = \frac{1}{2}(5x^3 - 3x)$, and $P_5(x) = \frac{1}{8}(63x^5 - 70x^3 + 15x)$, we find successively

$$c_1 = 3\int_0^1 P_1(x)\,dx = \frac{3}{2},$$

$$c_3 = 7\int_0^1 P_3(x)\,dx = -\frac{7}{8},$$

$$c_5 = 11\int_0^1 P_5(x)\,dx = \frac{11}{16},$$

and so forth, and

$$f(x) = \frac{3}{2}P_1(x) - \frac{7}{8}P_3(x) + \frac{11}{16}P_5(x) - \cdots, \qquad -1 < x < 1.$$

Of particular interest is the case when f is a polynomial. In such an instance the Fourier-Legendre series is necessarily finite since two polynomials can be equated only if they are of the same degree. In other words, if f is a polynomial of degree N, the series (107) will truncate after $n = N$ (see Problem 2 in Exercises 9.7).

REMARK: *Because the Fourier-Legendre series is finite when f is a polynomial, it follows that the series converges for* all *values of x in this case, and not just in the interval* $[-1, 1]$.

EXAMPLE 9

Find the Fourier-Legendre series for $f(x) = x^2$.

SOLUTION Only those coefficients for which $n \leq 2$ can be nonzero, so we write

$$\begin{aligned} x^2 &= c_0P_0(x) + c_1P_1(x) + c_2P_2(x) \\ &= c_0 + c_1x + c_2\frac{1}{2}(3x^2 - 1). \end{aligned}$$

Because there are only three Fourier coefficients to find this time, we simply equate like coefficients of both sides which yields the immediate result $c_0 = 1/3$, $c_1 = 0$, and $c_2 = 2/3$. [Of course, we could have computed the coefficients using (110).] Our series is, therefore, given by

$$x^2 = \frac{1}{3}P_0(x) + \frac{2}{3}P_2(x).$$

9.7.2 FOURIER-BESSEL SERIES

Another important set of orthogonal functions is the set of Bessel functions $J_\nu(k_n x)$, $n = 1, 2, 3, \ldots$, associated with the singular Sturm-Liouville system

$$\begin{gathered}\frac{d}{dx}(xy') - \frac{\nu^2}{x}y + \lambda xy = 0, \qquad 0 < x < b, \qquad \nu \geqslant 0,\\ y(x),\ y'(x) \quad \text{finite as} \quad x \to 0^+,\\ a_{21}y(b) + a_{22}y'(b) = 0,\end{gathered} \tag{111}$$

where the k_n ($n = 1, 2, 3, \ldots$) satisfy the relation

$$hJ_\nu(k_n b) + k_n J_\nu'(k_n b) = 0 \qquad (h = a_{21}/a_{22}). \tag{112}$$

We leave it to the reader to show that $L = D(xD) - \nu^2/x$ is symmetric. Noting from (111) that $r(x) = x$, the orthogonality relation of the Bessel functions is

$$\int_0^b xJ_\nu(k_n x)J_\nu(k_m x)\,dx = 0, \qquad n \neq m, \tag{113}$$

and the related *Fourier-Bessel series* is

$$f(x) = \sum_{n=1}^{\infty} c_n J_\nu(k_n x), \qquad 0 < x < b, \tag{114}$$

where

$$c_n = \|J_\nu(k_n x)\|^{-2}\int_0^b xf(x)J_\nu(k_n x)\,dx, \qquad n = 1, 2, 3, \ldots. \tag{115}$$

Series of this type arise frequently in solving certain partial DEs in circular or cylindrical shaped domains (see Section 10.3).

THEOREM 9.3

> If f is a piecewise smooth function on $[0, b]$, then the Fourier-Bessel series (114) and (115) converges pointwise to $f(x)$ at points of continuity of f and to the average value $\frac{1}{2}[f(x^+) + f(x^-)]$ at points of discontinuity of f on the interval $(0, b)$.*

The normalization factor $\|J_\nu(k_n x)\|^2$ for the Bessel functions may assume different forms depending upon the nature of the boundary condition prescribed at $x = b$. For example, the simple boundary condition

$$y(b) = 0 \tag{116}$$

---

*The series converges to zero at $x = 0$ if $\nu > 0$.

leads to the expression

$$\|J_\nu(k_n x)\|^2 = \frac{1}{2}b^2[J_{\nu+1}(k_n b)]^2, \qquad n = 1, 2, 3, \ldots \tag{117}$$

(see Problem 16 in Exercises 9.7) where $J_\nu(k_n b) = 0$. On the other hand, the more general boundary condition

$$hy(b) + y'(b) = 0 \qquad (h = a_{21}/a_{22}) \tag{118}$$

yields the result

$$\|J_\nu(k_n x)\|^2 = \frac{(k_n^2 + h^2)b^2 - \nu^2}{2k_n^2}[J_\nu(k_n b)]^2, \qquad n = 1, 2, 3, \ldots \tag{119}$$

(see Problem 15 in Exercises 9.7) where $hJ_\nu(k_n b) + k_n J'_\nu(k_n b) = 0$. When $h = 0$ in (118), we have the boundary condition

$$y'(b) = 0, \tag{120}$$

and by setting $h = 0$ in (119), we obtain the corresponding normalization factor

$$\|J_\nu(k_n x)\|^2 = \frac{k_n^2 b^2 - \nu^2}{2k_n^2}[J_\nu(k_n b)]^2, \qquad n = 1, 2, 3, \ldots. \tag{121}$$

EXAMPLE 10

Find the Fourier-Bessel series of

$$f(x) = \begin{cases} x, & 0 < x < 1 \\ 0, & 1 < x < 2, \end{cases}$$

corresponding to the set of functions $\{J_1(k_n x)\}$, where $J_1(2k_n) = 0$, $n = 1, 2, 3, \ldots$.

SOLUTION The series we seek has the form

$$f(x) = \sum_{n=1}^{\infty} c_n J_1(k_n x), \qquad 0 < x < 2,$$

where

$$c_n = \|J_1(k_n x)\|^{-2}\int_0^2 xf(x)J_1(k_n x)\,dx$$

$$= \|J_1(k_n x)\|^{-2}\int_0^1 x^2 J_1(k_n x)\,dx, \qquad n = 1, 2, 3, \ldots.$$

To evaluate the integral, we make the change of variable $t = k_n x$, which leads to

$$\int_0^1 x^2 J_1(k_n x)\,dx = \frac{1}{k_n^3}\int_0^{k_n} t^2 J_1(t)\,dt = \frac{J_2(k_n)}{k_n},$$

where we are recalling the integral formula

$$\int x^{\nu} J_{\nu-1}(x)\,dx = x^{\nu} J_{\nu}(x) + C$$

[see (70) in Section 9.4.3]. The condition $J_1(2k_n) = 0$ suggests that (117) is the correct normalization factor, i.e.,

$$\| J_1(k_n x) \|^2 = 2[J_2(2k_n)]^2, \qquad n = 1, 2, 3, \ldots .$$

Therefore, the Fourier coefficients are given by

$$c_n = \frac{J_2(k_n)}{2k_n[J_2(2k_n)]^2}, \qquad n = 1, 2, 3, \ldots ,$$

and

$$f(x) = \frac{1}{2}\sum_{n=1}^{\infty} \frac{J_2(k_n)}{k_n[J_2(2k_n)]^2} J_1(k_n x), \qquad 0 < x < 2.$$

EXERCISES 9.7

1. Find the first *three* nonzero coefficients in the Fourier-Legendre series of the following functions:

   (a) $f(x) = \begin{cases} 0, & -1 < x < 0 \\ x, & 0 < x < 1 \end{cases}$ (b) $f(x) = |x|, \quad -1 < x < 1$

★2. If $f(x)$ is a polynomial of degree N, show that its Fourier-Legendre series is the finite series

$$f(x) = c_0 P_0(x) + c_1 P_1(x) + \cdots + c_N P_N(x).$$

3. If $f(x)$ is a polynomial of degree less than m, show that

$$\int_{-1}^{1} f(x) P_m(x)\,dx = 0.$$

   Hint: Use Problem 2 and Equation (106).

4. Using the result of Problem 2, express the following polynomials in a Fourier-Legendre series.
   (a) $f(x) = x^3$ (b) $f(x) = 5x^3 + x^2$

5. Make the change of variable $x = \cos\phi$ in the Fourier-Legendre series (107) and (110) and show that the resulting expressions are

$$f(\phi) = \sum_{n=0}^{\infty} c_n P_n(\cos\phi), \qquad 0 < \phi < \pi,$$

   and

$$c_n = \left(n + \frac{1}{2}\right)\int_0^{\pi} f(\phi) P_n(\cos\phi)\sin\phi\,d\phi, \qquad n = 0, 1, 2, \ldots .$$

6. Using the result of Problem 5, find the first three nonzero terms of the Fourier-Legendre series for the function

$$f(\phi) = \begin{cases} 0, & 0 < \phi < \pi/2 \\ 1, & \pi/2 < \phi < \pi. \end{cases}$$

★7. Using the recurrence formula

$$P_n(x) = \left(\frac{2n-1}{n}\right) x P_{n-1}(x) - \left(\frac{n-1}{n}\right) P_{n-2}(x), \qquad n = 2, 3, 4, \ldots$$

and the orthogonality property of the Legendre polynomials, show that

(a) $\| P_n(x) \|^2 = \dfrac{2n-1}{n} \displaystyle\int_{-1}^{1} x P_n(x) P_{n-1}(x)\, dx, \qquad n = 2, 3, 4, \ldots.$

(b) By writing the above recurrence formula in the form

$$x P_n(x) = \frac{1}{2n+1} [(n+1) P_{n+1}(x) + n P_{n-1}(x)],$$

use (a) to obtain the recurrence relation

$$\| P_n(x) \|^2 = \left(\frac{2n-1}{2n+1}\right) \| P_{n-1}(x) \|^2, \qquad n = 2, 3, 4, \ldots.$$

(c) Calculate $\| P_0(x) \|^2$ and $\| P_1(x) \|^2$ directly and use successive applications of (b) to deduce that

$$\| P_n(x) \|^2 = \frac{2}{2n+1}, \qquad n = 0, 1, 2, \ldots.$$

In Problems 8–12, find a Fourier-Bessel series for $f(x)$ in terms of the orthogonal set $\{J_0(k_n x)\}$, where $J_0(k_n) = 0$, $n = 1, 2, 3, \ldots$.

8. $f(x) = 0.1 J_0(k_3 x), \quad 0 < x < 1$

9. $f(x) = 1, \quad 0 < x < 1$

10. $f(x) = \frac{1}{8}(1 - x^2), \quad 0 < x < 1$

11. $f(x) = \log x, \quad 0 < x < 1$

★12. $f(x) = x^4, \quad 0 < x < 1$

13. If $J_\nu(k_n) = 0$, $n = 1, 2, 3, \ldots$, for $\nu \geq 0$, show that

$$x^\nu = 2 \sum_{n=1}^{\infty} \frac{J_\nu(k_n x)}{k_n J_{\nu+1}(k_n)}, \qquad 0 < x < 1.$$

★14. Expand $f(x) = x^\nu$ in a series in terms of $\{J_\nu(k_n x)\}$, $0 < x < 1$, where $J_\nu'(k_n) = 0$, $n = 1, 2, 3, \ldots$.

★15. Multiply Bessel's DE

$$\frac{d}{dx}\left[x \frac{d}{dx} J_\nu(kx)\right] + \left(k^2 x - \frac{\nu^2}{x}\right) J_\nu(kx) = 0$$

by the factor $2x(d/dx)J_\nu(kx)$ and show that

(a) $$\frac{d}{dx}\left[x\frac{d}{dx}J_\nu(kx)\right]^2 + \left(k^2x^2 - \frac{\nu^2}{x}\right)\frac{d}{dx}[J_\nu(kx)]^2 = 0.$$

(b) Integrate the expression in (a) over the interval $(0, b)$ and deduce that

$$\|J_\nu(kx)\|^2 = \frac{1}{2k^2}\{[bkJ'_\nu(kb)]^2 + (kb^2 - \nu^2)[J_\nu(kb)]^2\}.$$

(c) Finally, set $k = k_n$ and use (112) to deduce that

$$\|J_\nu(k_nx)\|^2 = \frac{(k_n^2 + h^2)b^2 - \nu^2}{2k_n^2}[J_\nu(k_nb)]^2, \qquad n = 1, 2, 3, \ldots.$$

★16. Given that $J_\nu(k_nb) = 0$, $n = 1, 2, 3, \ldots$, use the result of Problem 15(b) and the identity

$$xJ'_\nu(x) = \nu J_\nu(x) - xJ_{\nu+1}(x)$$

to obtain the normalization factor

$$\|J_\nu(k_nx)\|^2 = \frac{1}{2}b^2[J_{\nu+1}(k_nb)]^2, \qquad n = 1, 2, 3, \ldots.$$

★17. Let $u(x, t)$ represent the temperature distribution in a nonhomogeneous insulated rod $-1 \leqslant x \leqslant 1$ along the x-axis, in which the thermal conductivity is proportional to $(1 - x^2)$. The heat equation in this case is replaced with the variable coefficient PDE

$$b\frac{\partial}{\partial x}[(1 - x^2)u_x] = u_t, \qquad b > 0.$$

The ends of the rod are insulated because the conductivity vanishes there, and thus the boundary conditions are

$$u(x, t),\ u_x(x, t) \ \text{ finite as } \ x \to -1^+, \qquad x \to 1^-.$$

(a) If the initial temperature is given by $f(x)$, show that the formal solution has the form

$$u(x, t) = \sum_{n=0}^{\infty} c_nP_n(x)e^{-n(n+1)bt},$$

where $P_n(x)$ is the nth Legendre polynomial.

(b) What is the temperature distribution for the special case when $f(x) = x^2$?

★18. If the tension in a taut string of unit length is $T(x) = ax$ and the mass density of the string is $\rho(x) = bx$, where a and b are constants, the governing PDE is

$$\frac{\partial}{\partial x}(xu_x) = \frac{b}{a}xu_{tt}, \qquad 0 < x < 1, \qquad t > 0$$

Determine the resulting motion of the string given that the boundary and initial conditions are described by

B.C.: $u(x, t)$ finite as $x \to 0^+$, $\quad u(1, t) = 0$
I.C.: $u(x, 0) = 1 - x^2$, $\quad u_t(x, 0) = 0$

CHAPTER 10

PROBLEMS IN SEVERAL DIMENSIONS

When problems of heat conduction or vibration are formulated for thin plates or membranes, two or more spatial coordinates in addition to time may be required in writing the governing PDE. Such problems, which generally involve three or more independent variables, are called *problems in several dimensions*. Fortunately, the theory for many of these problems is a natural extension of an already-developed theory for PDEs having two independent variables.

Using techniques analogous to the derivation of their one-dimensional equations, the *two-dimensional heat equation* and *two-dimensional wave equation* are derived in Section 10.1.1. Problems involving these equations in *rectangular domains* are discussed in Section 10.2. Here we find that *separation of variables* leads to an *eigenvalue problem* which features a PDE, but which can often be solved by another separation of variables. Finally, by using the *superposition principle* to satisfy the initial conditions, we are introduced to the notion of *double Fourier series*, the general theory of which is comparable to that of single Fourier series discussed in Chapter 4.

In using coordinate systems other than rectangular, the resulting PDEs will normally contain variable coefficients. The solutions of such equations cannot usually be expressed in terms of elementary functions, but often lead to *special functions*. In particular, the vibrating membrane problem in a *circular domain* (Section 10.3) and the steady-state temperature distribution in a cylindrical domain (Section 10.4) both have solutions that involve *Bessel functions*. On the other hand, the *Legendre polynomials* arise in the electrostatic potential problem in a *spherical domain* where the potential is independent of the azimuthal angle (Section 10.5).

In the final section we briefly discuss solutions of *Poisson's equation*, which is a nonhomogeneous potential equation.

10.1 INTRODUCTION

So far our considerations have been limited to problems involving no more than two independent variables. The solution technique of separating the variables, however, can also be applied in many instances when there are more than two independent variables. The primary distinction in these higher dimensional problems is that the associated eigenvalue problem is itself a PDE. Although we find that the eigenequation can often be solved through another separation of variables, the eigenfunctions that arise will contain two or more independent variables. As a result, we have to consider Fourier series of several variables at some point in the solution process. Fortunately, the generalization of the Fourier series method from one to several variables develops in a natural sort of way, and we can use our previous knowledge of the subject (Chapter 4) to help guide us in this development.

In their general forms, the equations of concern here are

$$\nabla^2 u = a^{-2} u_t \qquad \text{(heat equation)}, \tag{1}$$

$$\nabla^2 u = c^{-2} u_{tt} \qquad \text{(wave equation)}, \tag{2}$$

and

$$\nabla^2 u = 0 \qquad \text{(potential equation)}. \tag{3}$$

Since the Laplacian $\nabla^2 u$ is fundamental to all three equations, its form in various coordinate systems is of special importance. In particular, we have defined the Laplacian in rectangular coordinates by

$$\nabla^2 u = u_{xx} + u_{yy} + u_{zz}, \tag{4}$$

which has the advantge of constant coefficients.* In other coordinate systems the Laplacian will generally assume a form wherein the coefficients are not constant. This means that the DEs obtained by separating the variables will also have variable coefficients, and so their solutions often lead to *special functions* such as Bessel functions and Legendre polynomials.

REMARK: *The geometries for which separation of variables can be effective in solving the heat, potential, and wave equations in higher dimensions are quite limited. Even where solutions can be found by this method they often involve multiple infinite series that are cumbersome and are not generally practical for numerical computations. Nonetheless, we can often learn a great deal about the general characteristics of the solution by finding these infinite series representations where possible. Once this analysis has been completed, however, we may turn to numerical techniques such as finite-difference methods to produce actual numerical values of the solution.*

□ 10.1.1 TWO-DIMENSIONAL HEAT EQUATION

Consider a thin flat isotropic plate whose flat surfaces are insulated. We assume the temperature $u(x, y, t)$ of the plate depends only upon the time t

---

*Constant-coefficient equations are almost always the easiest to solve.

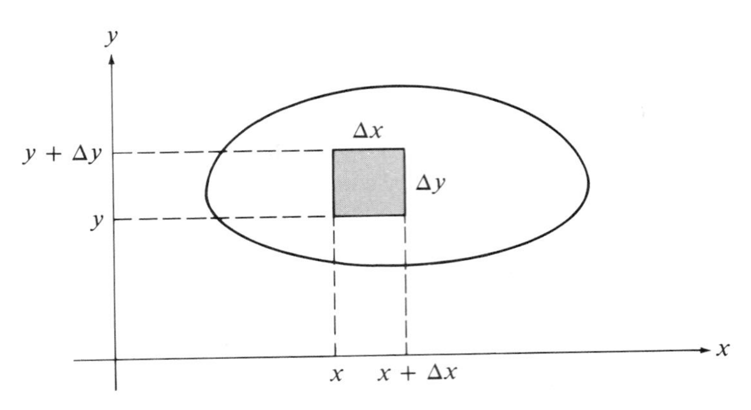

FIGURE 10.1

A flat plate

and position in the xy-plane (see Figure 10.1). As we did in Section 6.1 for heat flow in a slender rod, we will derive the governing DE by applying the law of conservation of energy to a small element of the plate.

Let $Q_1(x, y, t)$ and $Q_2(x, y, t)$ denote the rate of heat flow in the x- and y-directions, respectively, of a small volume element of dimensions Δx by Δy and unit thickness.* The rate of heat flow (or flux) into the element is, therefore, approximately

$$Q_1(x, y, t)\Delta y + Q_2(x, y, t)\Delta x,$$

whereas the rate at which heat leaves the element is approximately

$$Q_1(x + \Delta x, y, t)\Delta y + Q_2(x, y + \Delta y, t)\Delta x.$$

The rate of heat storage in the small element is proportional to the rate of change of temperature u_t and is approximately equal to $\Delta x \Delta y \rho c u_t$, where ρ is the mass density of the plate and c is the specific heat constant.

If we assume that no heat generation takes place within the area of the plate, then application of the *law of conservation of energy* leads to

$$Q_1(x, y, t)\Delta y + Q_2(x, y, t)\Delta x = Q_1(x + \Delta x, y, t)\Delta y + Q_2(x, y + \Delta y, t)\Delta x + \Delta x \Delta y \rho c u_t,$$

which can be rearranged as

$$-\left(\frac{Q_1(x + \Delta x, y, t) - Q_1(x, y, t)}{\Delta x}\right) - \left(\frac{Q_2(x, y + \Delta y, t) - Q_2(x, y, t)}{\Delta y}\right) = \rho c u_t. \tag{5}$$

---

*The assumption of unit thickness in the small element of the plate in no way influences the final form of the governing PDE, since the thickness parameter, if constant, is eventually eliminated no matter what dimension is assigned to it.

In the limit as both Δx and Δy tend to zero, Equation (5) reduces to

$$-\frac{\partial Q_1}{\partial x} - \frac{\partial Q_2}{\partial y} = \rho c u_t. \tag{6}$$

Fourier's law of heat conduction in two dimensions becomes

$$Q_1 = -ku_x, \qquad Q_2 = -ku_y, \tag{7}$$

where k is the heat conductivity constant. Finally, the substitution of (7) into (6) leads to the *two-dimensional heat equation*

$$u_{xx} + u_{yy} = a^{-2}u_t, \tag{8}$$

where $a^2 = k/\rho c$.

□ 10.1.2 TWO-DIMENSIONAL WAVE EQUATION

We wish to derive the equation governing the transverse deflections of a stretched membrane, e.g., a drumhead. The problem is essentially a two-dimensional version of a vibrating string. Thus, our present considerations will be similar to those in the case of the vibrating string (Section 7.1).

Let us assume the undeflected membrane is located in the xy-plane (see Figure 10.2). Other basic assumptions used in deriving the governing equation of motion are the following:

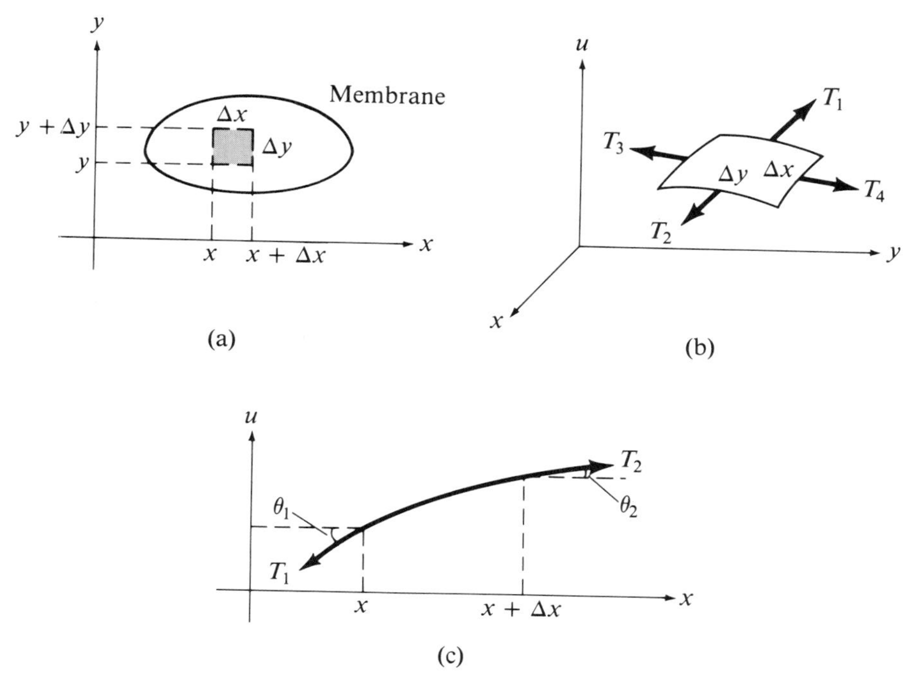

FIGURE 10.2

A two-dimensional membrane

1. The mass ρ per unit area of the membrane is constant, and the membrane is perfectly elastic.
2. The membrane is stretched and fixed along its entire boundary. The stretching tension T per unit area is constant and so large that gravitational forces can be neglected.
3. The transverse deflections $u(x, y, t)$ of the membrane during motion are "small" compared with the size of the membrane.

To begin, let us consider the forces T_1, T_2, T_3, and T_4 acting on a small element of the membrane as pictured in Figure 10.2(b). Because the deflections of the membrane are small, it is reasonable to approximate the sides of this small element by Δx and Δy, and to assume that the forces are tangent to the membrane. The forces T_1 and T_2 acting along the edges Δy of the small element are approximately

$$T_1 = T\Delta y, \qquad T_2 = T\Delta y.$$

If we assume the motion of the membrane in the xy-plane is negligibly small, we only need to sum forces in the transverse direction, denoted by the u-axis. The transverse components of the forces acting along the edges Δy sum to

$$T_2 \sin \theta_2 - T_1 \sin \theta_1 = T\Delta y \sin \theta_2 - T\Delta y \sin \theta_1 \tag{9}$$

[see Figure 10.2(c)]. Using small angle approximations, we may express this resultant component as

$$\begin{aligned} T\Delta y(\sin \theta_2 - \sin \theta_1) &\cong T\Delta y(\tan \theta_2 - \tan \theta_1) \\ &= T\Delta y[u_x(x + \Delta x, y, t) - u_x(x, y, t)]. \end{aligned} \tag{10}$$

Likewise, the resultant transverse component of the forces acting along the remaining two edges Δx is

$$T\Delta x[u_y(x, y + \Delta y, t) - u_y(x, y, t)].$$

By Newton's second law, the sum of the transverse force components is equal to the inertia force $\rho\Delta x\Delta y u_{tt}$, i.e.,

$$\begin{aligned} T\Delta y[u_x(x + \Delta x, y, t) - u_x(x, y, t)] + T\Delta x[u_y(x, y + \Delta y, t) - u_y(x, y, t)] \\ = \rho\Delta x\Delta y u_{tt}. \end{aligned} \tag{11}$$

If we divide this result by $\Delta x\Delta y$ and take the limit as both Δx and Δy tend to zero, it follows that

$$u_{xx} + u_{yy} = c^{-2}u_{tt}, \qquad c^2 = \frac{T}{\rho}, \tag{12}$$

which is called the *two-dimensional wave equation*.

10.2 RECTANGULAR DOMAINS

Problems of heat conduction and vibration phenomena are mathematically similar problems. Hence, for illustrative purposes it makes little difference whether we formulate our problems in terms of heat flow or wave motion. In this section we will consider problems where the spatial boundaries form a rectangular domain.

To start, we want to determine the small transverse displacements of a thin rectangular membrane occupying the region $0 \leq x \leq a, 0 \leq y \leq b$ of the xy-plane when at rest. If we assume all edges of the membrane are fixed, the initial displacement is described by the function $f(x, y)$, and the initial velocity is zero, the problem is characterized by

$$u_{xx} + u_{yy} = c^{-2}u_{tt}, \qquad 0 < x < a, \qquad 0 < y < b, \qquad t > 0$$

$$\text{B.C.:} \quad \begin{cases} u(0, y, t) = 0, & u(a, y, t) = 0 \\ u(x, 0, t) = 0, & u(x, b, t) = 0 \end{cases} \tag{13}$$

$$\text{I.C.:} \quad u(x, y, 0) = f(x, y), \qquad u_t(x, y) = 0$$

(see Figure 10.3).

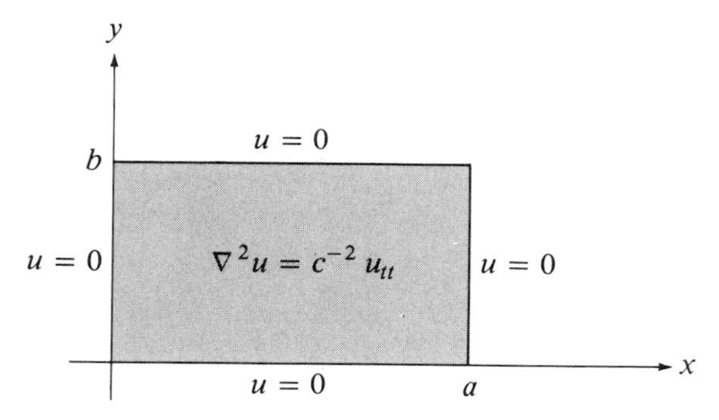

FIGURE 10.3

A rectangular membrane

To separate variables this time, we will start with the product form

$$u(x, y, t) = \psi(x, y)W(t), \tag{14}$$

which assumes a separation of spatial coordinates x and y from time t. The substitution of (14) into the wave equation in (13) leads to

$$\frac{\psi_{xx} + \psi_{yy}}{\psi} = \frac{W''}{c^2 W} = -\lambda,$$

from which we deduce

$$\psi_{xx} + \psi_{yy} + \lambda\psi = 0,$$

$$\text{B.C.:} \quad \psi(0, y) = 0, \qquad \psi(a, y) = 0, \qquad \psi(x, 0) = 0, \qquad \psi(x, b) = 0, \tag{15}$$

and

$$W'' + \lambda c^2 W = 0. \tag{16}$$

Equation (16) is the same time-dependent DE we found in solving the one-dimensional wave equation, and the DE in (15) is commonly called the *reduced wave equation,* or *Helmholtz equation.* Nontrivial solutions of (15) are possible only for particular values of λ, called *eigenvalues,* and the corresponding nontrivial solutions are called *eigenfunctions.*

In solving (15), we resort to the product solution form

$$\psi(x, y) = X(x)Y(y),$$

and find that (15) leads to the separated form

$$\frac{X''}{X} = -\frac{Y''}{Y} - \lambda = -\mu,$$

where μ is a new separation constant; thus

$$X'' + \mu X = 0, \qquad X(0) = 0, \qquad X(a) = 0, \tag{17}$$

$$Y'' + (\lambda - \mu)Y = 0, \qquad Y(0) = 0, \qquad Y(b) = 0. \tag{18}$$

The solution of (17) is familiar to us by now, and is

$$\mu = \mu_m = \frac{m^2\pi^2}{a^2}, \qquad X_m(x) = \sin\frac{m\pi x}{a}, \qquad m = 1, 2, 3, \ldots. \tag{19}$$

In the same fashion, we find the solution of (18) demands that

$$\lambda - \mu = \frac{n^2\pi^2}{b^2}, \qquad Y_n(y) = \sin\frac{n\pi y}{b}, \qquad n = 1, 2, 3, \ldots. \tag{20}$$

Based on (19) and (20), we see that the eigenvalues of (15) are given by

$$\lambda = \lambda_{mn} = \left(\frac{m^2}{a^2} + \frac{n^2}{b^2}\right)\pi^2; \qquad m, n = 1, 2, 3, \ldots \tag{21}$$

which form an (infinite) square matrix of values. The corresponding eigenfunctions $\phi_{mn}(x, y) = X_m(x)Y_n(y)$ have the explicit form

$$\phi_{mn}(x, y) = \sin\frac{m\pi x}{a}\sin\frac{n\pi y}{b}; \qquad m, n = 1, 2, 3, \ldots. \tag{22}$$

With λ restricted to those values occurring in (21), we find that the solutions of (16) are

$$W_{mn}(t) = A_{mn}\cos(\sqrt{\lambda_{mn}}\,ct) + B_{mn}\sin(\sqrt{\lambda_{mn}}\,ct); \qquad m, n = 1, 2, 3, \ldots. \tag{23}$$

It is interesting to observe that, depending upon the values of a and b, several eigenfunctions $\phi_{mn}(x, y)$ can correspond to the same eigenvalue λ_{mn}. Physically, this tells us that a vibrating membrane permits different vibration modes having the same angular frequency $\omega_{mn} = c\sqrt{\lambda_{mn}}$, but entirely different *nodal lines,* i.e., curves on the membrane defined by $\phi_{mn}(x, y) = 0$.*

Let us illustrate this multiplicity of eigenvalues by using the simplest example, which is a square membrane of unit side, i.e., $a = b = 1$. The frequencies of vibration are then given by

$$\omega_{mn} = \pi c\sqrt{m^2 + n^2}; \qquad m, n = 1, 2, 3, \ldots \tag{24}$$

and the modes of vibration by

$$\phi_{mn}(x, y) = \sin m\pi x \sin n\pi x; \qquad m, n = 1, 2, 3, \ldots. \tag{25}$$

For $m = n = 1$, we get the *fundamental mode* for which $\omega_{11} = \pi c\sqrt{2}$ and

$$\phi_{11}(x, y) = \sin \pi x \sin \pi y. \tag{26}$$

Hence, there are no nodal lines in this case. Next, we set $m = 1, n = 2$, or $m = 2, n = 1$, both cases of which yield the same frequency $\omega_{12} = \omega_{21} = \pi c\sqrt{5}$, corresponding to the two modes

$$\begin{aligned} \phi_{12}(x, y) &= \sin \pi x \sin 2\pi y, \\ \phi_{21}(x, y) &= \sin 2\pi x \sin \pi y. \end{aligned} \tag{27}$$

The mode ϕ_{12} has the nodal line $y = 1/2$, whereas the mode ϕ_{21} has the nodal line $x = 1/2$. Of course, the frequency ω_{12} (or ω_{21}) also corresponds to the compound mode

$$C_1\phi_{12}(x, y) + C_2\phi_{21}(x, y) = C_1 \sin \pi x \sin 2\pi y + C_2 \sin 2\pi x \sin \pi y \tag{28}$$

where C_1 and C_2 can be any constants. Depending upon the choice of C_1 and C_2, different nodal patterns can arise. Some of these and other nodal patterns are illustrated in Figure 10.4.

REMARK: *As the values of m and n increase, we find that more than two eigenfunctions can correspond to the same frequency. This is so because the sum of squares of two numbers does not uniquely determine the two numbers.*†

Finally, by applying the superposition principle to the solutions (22) and (23), we obtain

$$u(x, y, t) = \sum_{m=1}^{\infty} \sum_{n=1}^{\infty} W_{mn}(t)\phi_{mn}(x, y),$$

or explicitly,

---

* The nodal lines are curves along which the membrane remains at rest during a given vibration mode.

† For example, $1^2 + 8^2 = 4^2 + 7^2 = 65$.

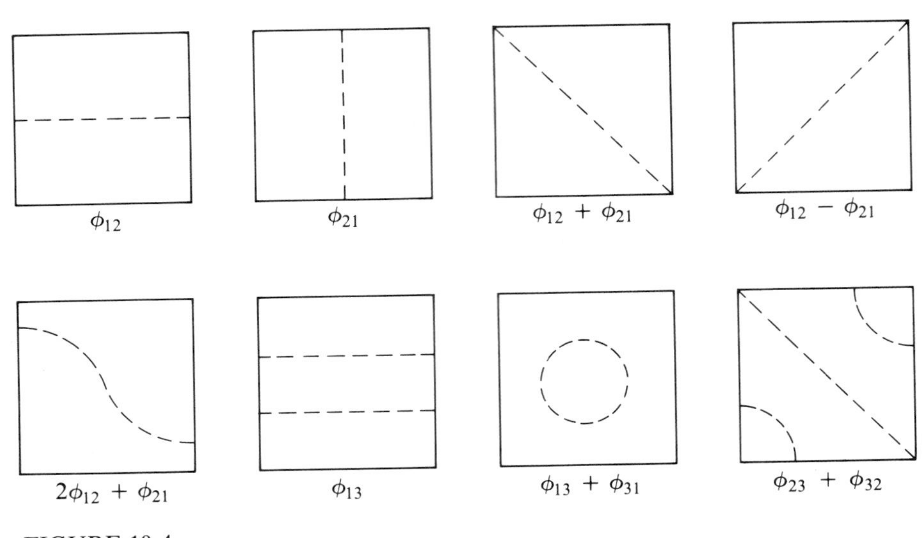

FIGURE 10.4

Nodal lines for a square membrane

$$u(x, y, t) = \sum_{m=1}^{\infty} \sum_{n=1}^{\infty} [A_{mn} \cos(\sqrt{\lambda_{mn}}\, ct) + B_{mn} \sin(\sqrt{\lambda_{mn}}\, ct)] \sin \frac{m\pi x}{a} \sin \frac{n\pi y}{b}. \tag{29}$$

Of course, we have yet to determine the constants A_{mn} and B_{mn}.

10.2.1 DOUBLE FOURIER SERIES

By imposing the initial conditions in (13) upon the solution function (29), we are led to

$$f(x, y) = \sum_{m=1}^{\infty} \sum_{n=1}^{\infty} A_{mn} \sin \frac{m\pi x}{a} \sin \frac{n\pi y}{b} \tag{30}$$

and

$$0 = \sum_{m=1}^{\infty} \sum_{n=1}^{\infty} \sqrt{\lambda_{mn}}\, cB_{mn} \sin \frac{m\pi x}{a} \sin \frac{n\pi y}{b}. \tag{31}$$

Equation (31) suggests that $B_{mn} = 0$ for all m and n, while (30) is what we call a *double Fourier sine series* of f. The theory of such series closely parallels that of single Fourier series, although we will not develop any of the theory here.* Rather, we will assume the validity of (30) and proceed formally to find the coefficients A_{mn}.

---

* For a discussion of double Fourier series, see Chapter 7 in G. P. Tolstov, *Fourier Series*, (New Jersey: Prentice-Hall, 1962).

Let us first introduce the notation

$$E_m(y) = \sum_{n=1}^{\infty} A_{mn} \sin \frac{n\pi y}{b}, \qquad m = 1, 2, 3, \ldots \tag{32}$$

so that (30) takes the form

$$f(x, y) = \sum_{m=1}^{\infty} E_m(y) \sin \frac{m\pi x}{a}. \tag{33}$$

Thus, if we think of y as being temporarily fixed, (33) is simply a Fourier sine series of $f(x, y)$ considered as a function of x alone. In this case, we have

$$E_m(y) = \frac{2}{a} \int_0^a f(x, y) \sin \frac{m\pi x}{a}\, dx, \qquad m = 1, 2, 3, \ldots. \tag{34}$$

Now that the function $E_m(y)$ has been determined, we can return to (32) and interpret this expression as a Fourier sine series of $E_m(y)$, for which

$$A_{mn} = \frac{2}{b} \int_0^b E_m(y) \sin \frac{n\pi y}{b}\, dy, \qquad n = 1, 2, 3, \ldots \tag{35}$$

where m is fixed. Finally, by combining (34) and (35), we deduce the *generalized Fourier formula*

$$A_{mn} = \frac{4}{ab} \int_0^a \int_0^b f(x, y) \sin \frac{m\pi x}{a} \sin \frac{n\pi y}{b}\, dy\, dx, \tag{36}$$

for $m, n = 1, 2, 3, \ldots.$

Although we haven't discussed the validity of such representations, it turns out that the series (30) will converge to $f(x, y)$ at every point of continuity of f in the rectangular domain $0 \leq x \leq a$, $0 \leq y \leq b$, provided f is at least piecewise smooth throughout this domain.*

While the particular problem (13) led to a *double Fourier sine series,* it is not the only type of double Fourier series to occur in practice. Another type of double series that arises is

$$\begin{aligned} f(x, y) = \frac{1}{4} A_{00} &+ \frac{1}{2} \sum_{m=1}^{\infty} A_{m0} \cos \frac{m\pi x}{a} + \frac{1}{2} \sum_{n=1}^{\infty} A_{0n} \cos \frac{n\pi y}{b} \\ &+ \sum_{m=1}^{\infty} \sum_{n=1}^{\infty} A_{mn} \cos \frac{m\pi x}{a} \cos \frac{n\pi y}{b}, \end{aligned} \tag{37}$$

called a *double Fourier cosine series.* This latter series might arise, for example, in solving a heat conduction problem in a rectangular region (e.g., a flat plate) where the edges of the plate are insulated. Naturally, other varieties of double Fourier series are also possible.

---

* By *piecewise smooth,* we mean that f, f_x, and f_y are all piecewise continuous throughout the domain.

EXAMPLE 1

Find a double Fourier sine series for $f(x, y) = xy$, $0 < x < \pi$, $0 < y < \pi$.

SOLUTION The series we seek is ($a = b = \pi$)

$$xy = \sum_{m=1}^{\infty} \sum_{n=1}^{\infty} A_{mn} \sin mx \sin ny,$$

where

$$A_{mn} = \frac{4}{\pi^2} \int_0^{\pi} \int_0^{\pi} xy \sin mx \sin ny \, dx \, dy$$

$$= (-1)^{m+n} \frac{4}{mn}.$$

Thus,

$$xy = 4 \sum_{m=1}^{\infty} \sum_{n=1}^{\infty} \frac{(-1)^{m+n}}{mn} \sin mx \sin ny.$$

REMARK: *In case the initial velocity $u_t(x, y, 0) = g(x, y)$ is not identically zero, the coefficients B_{mn} in (29) are determined by*

$$\sqrt{\lambda_{mn}}\, cB_{mn} = \frac{4}{ab} \int_0^{a} \int_0^{b} g(x, y) \sin \frac{m\pi x}{a} \sin \frac{n\pi y}{b} \, dy \, dx.$$

EXERCISES 10.2

1. Determine and graph the nodal lines of $\phi_{mn}(x, y) = \sin m\pi x \sin n\pi y$ ($0 < x < 1$, $0 < y < 1$) corresponding to
   (a) $\phi_{12} - \phi_{21}$.
   (b) $\phi_{12} + \phi_{21}$.
   (c) ϕ_{14}.
   (d) ϕ_{34}.

2. Show that the square membrane ($a = b$) is the one for which the frequency ω_{11} is the lowest among all the rectangular membranes with the same area $A = ab$ and the same value of c.

3. Find a similar result as in Problem 2 for the frequency ω_{mn} with m and n fixed.

★4. Find the eigenfunctions ϕ_{11}, ϕ_{12}, and ϕ_{21} associated with a rectangular membrane that is fixed at the boundary and occupies the domain $-a \leqslant x \leqslant a$, $-b \leqslant y \leqslant b$.

5. Find the eigenvalues and eigenfunctions of

$$x^2\psi_{xx} + x\psi_x + \psi_{yy} + \lambda\psi = 0, \qquad 1 < x \leqslant e, \qquad 0 < y < \pi$$

$$\text{B.C.:} \quad \psi(1, y) = 0, \qquad \psi(e, y) = 0, \qquad \psi_y(x, 0) = 0, \qquad \psi_y(x, \pi) = 0.$$

6. Verify that the eigenfunctions $\phi_{mn}(x, y) = \sin(m\pi x/a)\sin(n\pi y/b)$ satisfy the orthogonality relation

$$\int_0^a \int_0^b \phi_{mn}(x, y)\phi_{jk}(x, y)\,dy\,dx = \begin{cases} \dfrac{ab}{4}, & m = j \text{ and } n = k \\ 0, & \text{otherwise.} \end{cases}$$

In Problems 7–10, represent the given function on the domain $0 < x < a$, $0 < y < b$, by a double Fourier sine series of the form (30).

7. $f(x, y) = 1$

8. $f(x, y) = xy$

9. $f(x, y) = x$

10. $f(x, y) = xy(a - x)(b - y)$

11. Find the deflections of a rectangular membrane $0 \leqslant x \leqslant a$, $0 \leqslant y \leqslant b$, fixed on the boundary, given zero initial velocity and initially deflected in the shape

$$u(x, y, 0) = f(x, y) = A \sin\frac{3\pi x}{a}\sin\frac{5\pi y}{b}.$$

12. Find the small deflections of a square membrane of unit side (with $c = 1$), fixed on the boundary, given zero initial velocity, and initial deflection $f(x, y)$, where
(a) $f(x, y) = xy(1 - x)(1 - y)$.
(b) $f(x, y) = A\sin^2 \pi x \sin^2 \pi y$.

13. Find a formal solution of the heat conduction problem

$$u_{xx} + u_{yy} = a^{-2}u_t, \qquad 0 < x < \pi, \qquad 0 < y < b, \qquad t > 0$$

B.C.: $u(0, y, t) = 0$, $u(\pi, y, t) = 0$, $u(x, 0, t) = 0$, $u(x, b, t) = 0$
I.C.: $u(x, y, 0) = f(x, y)$.

14. By assuming the product form $u(x, y, z, t) = \psi(x, y, z)W(t)$, show that separation of variables applied to
(a) the *heat equation* $\nabla^2 u = a^{-2}u_t$ leads to the two equations

$$W' + \lambda a^2 W = 0,$$

$$\nabla^2\psi + \lambda\psi = 0.$$

(b) the *wave equation* $\nabla^2 u = c^{-2}u_{tt}$ leads to the two equations

$$W'' + \lambda c^2 W = 0,$$

$$\nabla^2\psi + \lambda\psi = 0.$$

(Note that the *Helmholtz equation* is common to both the heat and wave equations.)

*15. If R is any regular, closed region in the xy-plane with boundary curve C, show that the eigenfunctions of the Helmholtz problem

$$\nabla^2\psi + \lambda\psi = 0 \quad \text{in} \quad R$$

B.C.: $\psi = 0$ on C

are orthogonal if they correspond to different eigenvalues.
Hint: Recall Green's second identity

$$\iint_R (f\nabla^2 g - g\nabla^2 f)\, dx\, dy = \oint_C \left(f\frac{\partial g}{\partial n} - g\frac{\partial f}{\partial n}\right) ds.$$

16. Show that the Fourier coefficients for the double Fourier cosine series (37) are given by

$$A_{mn} = \frac{4}{ab}\int_0^a \int_0^b f(x, y) \cos\frac{m\pi x}{a} \cos\frac{n\pi y}{b}\, dy\, dx; \qquad m, n = 0, 1, 2, \ldots .$$

In Problems 17–20, represent the given function on the domain $0 < x < a, 0 < y < b$, by a double Fourier cosine series of the form (37) (see Problem 16).

17. $f(x, y) = 1$

18. $f(x, y) = xy$

19. $f(x, y) = x$

20. $f(x, y) = \sin^2\dfrac{\pi x}{a}\cos^2\dfrac{2\pi y}{b}$

In Problems 21–24, find a formal solution of the given problem.

21. $u_{xx} + u_{yy} = c^{-2}u_{tt}, \quad 0 < x < a, \quad 0 < y < b, \quad t > 0$

B.C.: $\begin{cases} u(0, y, t) = 0, & u(a, y, t) = 0 \\ u_y(x, 0, t) = 0, & u_y(x, b, t) = 0 \end{cases}$

I.C.: $u(x, y, 0) = f(x, y), \quad u_t(x, y, 0) = 0$

22. $u_{xx} + u_{yy} = a^{-2}u_t, \quad 0 < x < \pi, \quad 0 < y < \pi, \quad t > 0$

B.C.: $\begin{cases} u(0, y, t) = 0, & u_x(\pi, y, t) = 0 \\ u_y(x, 0, t) = 0, & u(x, \pi, t) = 0 \end{cases}$

I.C.: $u(x, y, 0) = f(x, y)$

23. $u_{xx} + u_{yy} = u_t, \quad 0 < x < a, \quad 0 < y < b, \quad t > 0$

B.C.: $\begin{cases} u(0, y, t) = 0, & u_x(a, y, t) = 0 \\ u(x, 0, t) = 0, & u_y(x, b, t) = 0 \end{cases}$

I.C.: $u(x, y, 0) = f(x, y)$

★24. $u_{xx} + u_{yy} + u_{zz} = a^{-2}u_t, \quad 0 < x < \pi, \quad 0 < y < \pi, \quad 0 < z < \pi, \quad t > 0$

B.C.: $\begin{cases} u(0, y, z, t) = 0, & u(\pi, y, z, t) = 0 \\ u(x, 0, z, t) = 0, & u(x, \pi, z, t) = 0 \\ u(x, y, 0, t) = 0, & u(x, y, \pi, t) = 0 \end{cases}$

I.C.: $u(x, y, z, 0) = f(x, y, z)$

★25. All four faces of an infinitely long rectangular prism bounded by the planes $x = 0, x = 1, y = 0, y = 1$, are kept at 0°C.

(a) If the initial temperature distribution is

$$u(x, y, 0) = A \sin \pi x \sin \pi y,$$

find the subsequent temperature distribution $u(x, y, t)$ for all later times.

(b) Show that very close to any corner of the prism, the lines of equal temperature and flow of heat are orthogonal systems of rectangular hyperbolas.

★26. Suppose that a square column is considered to be of infinite height on one side of its base $0 \leq x \leq 1, 0 \leq y \leq 1$, in the plane $z = 0$. If the lateral boundaries are maintained at 0°C and the temperature over the base is prescribed by

$u(x, y, 0) = f(x, y)$, show that the internal steady-state temperature distribution is of the form

$$u(x, y, z) = \sum_{m=1}^{\infty} \sum_{n=1}^{\infty} C_{mn} \sin(m\pi x) \sin(n\pi y) e^{-k_{mn}^2 z}.$$

Identify the constants k_{mn}^2 and the Fourier coefficients C_{mn}.

★27. Given the heat conduction problem

$$u_{xx} + u_{yy} = u_t, \qquad 0 < x < a, \qquad 0 < y < b, \qquad t > 0$$

$$\text{B.C.:} \quad \begin{cases} u(0, y, t) = 0, & u(a, y, t) = 0 \\ u(x, 0, t) = 0, & u(x, b, t) = g(x) \end{cases}$$

$$\text{I.C.:} \quad u(x, y, 0) = f(x, y),$$

assume that $u(x, y, t) = S(x, y) + v(x, y, t)$ and find appropriate boundary value problems for the *steady-state solution* $S(x, y)$ and *transient solution* $v(x, y, t)$. *Hint:* See Section 6.2.2.

★28. For the special case $a = b = 1$, $g(x) = \sin \pi x$ and $f(x, y) = 0$, find a solution for Problem 27.

10.3 CIRCULAR DOMAINS: BESSEL FUNCTIONS

In solving heat conduction problems or wave motion problems by the separation of variables method in circular domains, it is necessary to express the problem in polar coordinates. Recall from Section 8.3 that the Laplacian in polar coordinates (r, θ) is given by

$$\nabla^2 u = u_{rr} + \frac{1}{r} u_r + \frac{1}{r^2} u_{\theta\theta}. \tag{38}$$

For illustrative purposes, we will discuss problems in this section in terms of vibrating circular membranes, while similar problems involving heat flow in a circular plate will be taken up in the exercises.

10.3.1 RADIAL SYMMETRIC VIBRATING MEMBRANE

We wish to determine the small transverse displacements u of a thin circular membrane (such as a drumhead) of unit radius whose edge is rigidly fixed (see Figure 10.5). If the displacements depend only upon the radial distance r from the center of the membrane and on time t, then $u_{\theta\theta} = 0$ in (38) and the governing PDE is the *radial symmetric* form of the wave equation $\nabla^2 u = c^{-2} u_{tt}$ given by

$$u_{rr} + \frac{1}{r} u_r = c^{-2} u_{tt}, \qquad 0 < r < 1, \qquad t > 0. \tag{39}$$

Since we have assumed the membrane is rigidly fixed on the boundary, we impose the boundary condition

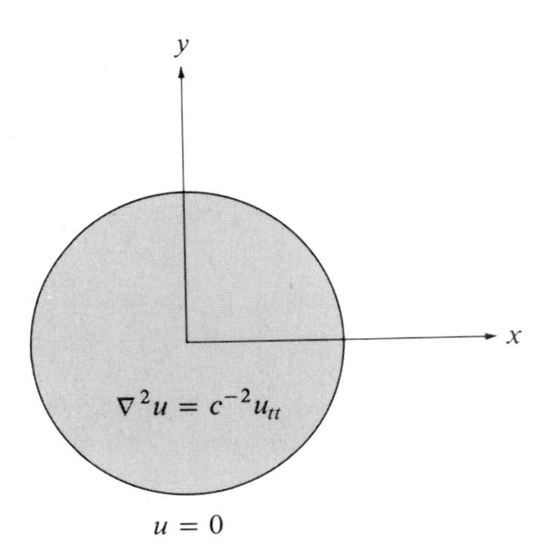

FIGURE 10.5

A circular membrane

B.C.: $$u(1, t) = 0. \tag{40}$$

Based on our discussion in Section 8.3.1, we see that $r = 0$ is a singular point of (39) and therefore we must impose the *implicit condition*

B.C.: $$u(r, t) \quad \text{finite as} \quad r \to 0^+. \tag{41}$$

Finally, if the membrane is initially deflected to the form $f(r)$ with velocity $g(r)$, we prescribe the initial conditions

I.C.: $$u(r, 0) = f(r), \qquad u_t(r, 0) = g(r), \qquad 0 < r < 1. \tag{42}$$

Expressing u in the product form $u(r, t) = R(r)W(t)$ and separating variables, we find

$$rR'' + R' + \lambda rR = 0,$$

B.C.: $$R(r) \quad \text{finite as} \quad r \to 0^+, R(1) = 0 \tag{43}$$

and

$$W'' + \lambda c^2 W = 0. \tag{44}$$

Equation (43) is *Bessel's equation* of order zero with general solution

$$R(r) = C_1 J_0(kr) + C_2 Y_0(kr) \tag{45}$$

($\lambda = k^2 > 0$).* In order to maintain finite displacements of the membrane at $r = 0$ we must set $C_2 = 0$, since Y_0 becomes unbounded when the argument is zero. The remaining solution, $R(r) = C_1 J_0(kr)$, must then satisfy the boundary condition

* Recall from Section 9.5 that Bessel's equation (43) has no eigenvalues for $\lambda \leq 0$.

$$R(1) = C_1 J_0(k) = 0. \tag{46}$$

It has previously been pointed out that the Bessel function J_0 has infinitely many zeros on the positive axis. Thus, for $C_1 \neq 0$, (46) is satisfied by selecting k as any one of the zeros of $J_0(k)$, which are denoted by $k_1, k_2, \ldots, k_n, \ldots$. Hence the eigenvalues of (43) are $\lambda_n = k_n^2$ with corresponding eigenfunctions

$$\phi(r) = J_0(k_n r), \qquad n = 1, 2, 3, \ldots. \tag{47}$$

With λ restricted as above, the solutions of (44) are

$$W_n(t) = a_n \cos k_n ct + b_n \sin k_n ct, \qquad n = 1, 2, 3, \ldots \tag{48}$$

and by combining results, we obtain the family of solutions

$$u_n(r, t) = (a_n \cos k_n ct + b_n \sin k_n ct) J_0(k_n r), \tag{49}$$

for $n = 1, 2, 3, \ldots$. These solutions are called *standing waves,* since each can be viewed as having fixed shape $J_0(k_n r)$ with varying amplitude $W_n(t)$. The situation, therefore, is similar to that in Section 7.2 where we discussed standing waves in connection with the vibrating string problem. However, because the zeros of the Bessel function J_0 are not regularly spaced (in contrast with the zeros of the sine functions appearing as eigenfunctions in the vibrating string problem), the sound emitted from a drum, for example, is quite different from that of, say, a violin. In musical tones the zeros are evenly spaced and the frequencies are integral multiples of the fundamental frequency. The zeros of our standing waves in this example appear as concentric circles, called *nodal lines.* When $n = 1$ there is no nodal line for $0 < r < 1$. When $n = 2$ there is one nodal line, and when $n = 3$ there are two nodal lines, and so forth (see Figure 10.6).

By forming a linear combination of the solutions (49) through the superposition principle, we get

$$u(r, t) = \sum_{n=1}^{\infty} (a_n \cos k_n ct + b_n \sin k_n ct) J_0(k_n r), \tag{50}$$

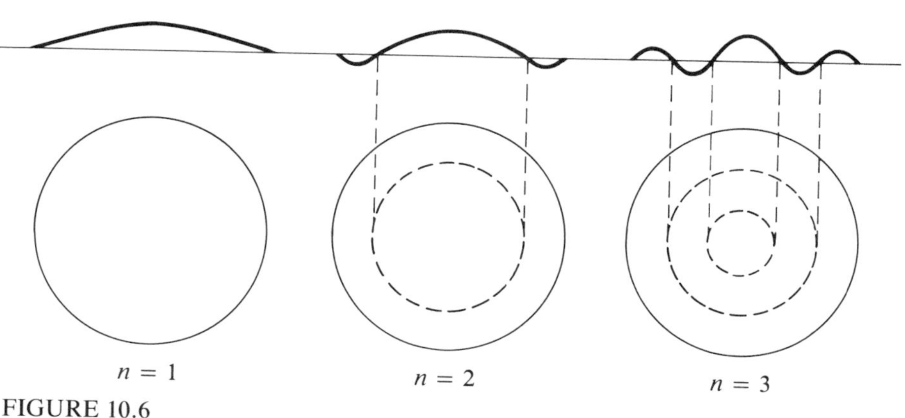

FIGURE 10.6

Nodal lines for a circular membrane

where the constants a_n and b_n are selected in such a way that the initial conditions (42) are satisfied. Hence,

$$u(r, 0) = f(r) = \sum_{n=1}^{\infty} a_n J_0(k_n r) \tag{51}$$

and

$$u_t(r, 0) = g(r) = \sum_{n=1}^{\infty} k_n c b_n J_0(k_n r), \tag{52}$$

which are recognized as *Fourier-Bessel series* (Section 9.7.2) for $f(r)$ and $g(r)$, respectively, where

$$a_n = \frac{2}{[J_1(k_n)]^2} \int_0^1 r f(r) J_0(k_n r)\, dr, \qquad n = 1, 2, 3, \ldots \tag{53}$$

and

$$k_n c b_n = \frac{2}{[J_1(k_n)]^2} \int_0^1 r g(r) J_0(k_n r)\, dr, \qquad n = 1, 2, 3, \ldots. \tag{54}$$

EXAMPLE 2

Find the displacements u of a thin circular membrane for which

$$u_{rr} + \frac{1}{r} u_r = c^{-2} u_{tt}, \qquad 0 < r < 1, \qquad t > 0$$

B.C.:
$$u(1, t) = 0, \qquad t > 0$$

I.C.:
$$u(r, 0) = \begin{cases} 1, & 0 < r < \dfrac{1}{2} \\ 0, & \dfrac{1}{2} < r < 1 \end{cases}, \qquad u_t(r, 0) = 0, \qquad 0 < r < 1.$$

SOLUTION The solution form is that given by (50), namely,

$$u(r, t) = \sum_{n=1}^{\infty} (a_n \cos k_n ct + b_n \sin k_n ct) J_0(k_n r),$$

where $J_0(k_n) = 0$, $n = 1, 2, 3, \ldots$. Since the initial velocity $g(r)$ is zero, it follows from (54) that $b_n = 0$, $n = 1, 2, 3, \ldots$, and the prescribed initial displacement $f(r)$ leads to

$$a_n = \frac{2}{[J_1(k_n)]^2} \int_0^{1/2} r J_0(k_n r)\, dr, \qquad n = 1, 2, 3, \ldots$$

[see (53)]. By making the change of variable $x = k_n r$, we get

$$a_n = \frac{2}{k_n^2 [J_1(k_n)]^2} \int_0^{k_n/2} x J_0(x)\, dx = \frac{J_1(k_n/2)}{k_n [J_1(k_n)]^2}, \qquad n = 1, 2, 3, \ldots,$$

where we have used the integral formula

$$\int x^{\nu}J_{\nu-1}(x) = x^{\nu}J_{\nu}(x) + C$$

[see (70) in Section 9.4.3]. Hence, our solution becomes

$$u(r, t) = \sum_{n=1}^{\infty} \frac{J_1(k_n/2)}{k_n[J_1(k_n)]^2} J_0(k_n r) \cos k_n ct.$$

□ 10.3.2 GENERAL VIBRATING MEMBRANE

If the displacement modes of the membrane discussed in the previous section also depend upon the polar angle θ, the problem is then characterized by

$$u_{rr} + \frac{1}{r} u_r + \frac{1}{r^2} u_{\theta\theta} = c^{-2}u_{tt}, \qquad 0 < r < 1, \qquad -\pi < \theta < \pi, \qquad t > 0$$

$$\text{B.C.:} \quad u(r, \theta, t) \text{ finite as } r \to 0^+, \quad u(1, \theta, t) = 0, \quad t > 0 \tag{55}$$

$$\text{I.C.:} \quad u(r, \theta, 0) = f(r, \theta), \quad u_t(r, \theta, 0) = g(r, \theta), \quad 0 < r < 1, \quad -\pi < \theta < \pi.$$

To keep the deflections single-valued, we must require the additional periodic conditions

$$\text{B.C.:} \quad u(r, -\pi, t) = u(r, \pi, t), \quad u_\theta(r, -\pi, t) = u_\theta(r, \pi, t) \tag{56}$$

(see also Section 8.3.1).

By starting with the product form $u(r, \theta, t) = \psi(r, \theta)W(t)$ and separating variables, we are led to the *Helmholtz eigenvalue problem*

$$r^2\psi_{rr} + r\psi_r + \psi_{\theta\theta} + \lambda r^2\psi = 0,$$

$$\text{B.C.:} \quad \begin{cases} \psi(r, \theta) \text{ finite as } r \to 0^+, & \psi(1, \theta) = 0 \\ \psi(r, -\pi) = \psi(r, \pi), & \psi_\theta(r, -\pi) = \psi_\theta(r, \pi), \end{cases} \tag{57}$$

and the time-dependent equation

$$W'' + \lambda c^2 W = 0. \tag{58}$$

To solve (57), we write $\psi(r, \theta) = R(r)H(\theta)$ and separate variables again to obtain

$$H'' + \mu H = 0, \qquad H(-\pi) = H(\pi), \qquad H'(-\pi) = H'(\pi) \tag{59}$$

and

$$\begin{aligned} &r^2R'' + rR' + (\lambda r^2 - \mu)R = 0, \\ \text{B.C.:} \quad &R(r) \text{ finite as } r \to 0^+, \quad R(1) = 0, \end{aligned} \tag{60}$$

where μ is the new separation constant. The solution of (59) requires that $\mu = \mu_n = n^2$ $(n = 0, 1, 2, \ldots)$, and in this case

$$H_n(\theta) = \begin{cases} a_0, & n = 0 \\ a_n \cos n\theta + b_n \sin n\theta, & n = 1, 2, 3, \ldots . \end{cases} \tag{61}$$

Knowing that $\mu = n^2$, we recognize (60) as *Bessel's equation* of order n, for which the general solution is ($\lambda = k^2 > 0$)

$$R(r) = C_1 J_n(kr) + C_2 Y_n(kr). \tag{62}$$

To keep our solution bounded at the origin ($r = 0$) we must set $C_2 = 0$ and then the remaining boundary condition in (60) yields

$$R(1) = C_1 J_n(k) = 0.$$

For each n, there exists a set of values k_{mn} ($m = 1, 2, 3, \ldots$) corresponding to the zeros of the Bessel function J_n. Thus we write $\lambda_{mn} = k_{mn}^2$ ($n = 0, 1, 2, \ldots, m = 1, 2, 3, \ldots$), and (62) becomes

$$R_{mn}(r) = J_n(k_{mn} r) \tag{63}$$

(set $C_1 = 1$). The solutions of (57) are $\psi_{mn}(r, \theta) = R_{mn}(r) H_n(\theta)$, which take the form (here we must redefine the constants to reflect both indices m and n)

$$\psi_{mn}(r, \theta) = \begin{cases} a_{m0} J_0(k_{m0} r); & n = 0, \quad m = 1, 2, 3, \ldots \\ (a_{mn} \cos n\theta + b_{mn} \sin n\theta) J_n(k_{mn} r); & \\ \qquad n = 1, 2, 3, \ldots, \quad m = 1, 2, 3, \ldots . \end{cases} \tag{64}$$

Returning now to (58), we obtain

$$W_{mn}(t) = \alpha_{mn} \cos k_{mn} ct + \beta_{mn} \sin k_{mn} ct \tag{65}$$

($n = 0, 1, 2, \ldots, m = 1, 2, 3, \ldots$). By combining all solutions by the superposition principle and redefining all constants, we are led to the result

$$\begin{aligned} u(r, \theta, t) = &\sum_{m=1}^{\infty} [A_{m0} \cos k_{m0} ct + C_{m0} \sin k_{m0} ct] J_0(k_{m0} r) \\ &+ \sum_{m=1}^{\infty} \sum_{n=1}^{\infty} [(A_{mn} \cos n\theta + B_{mn} \sin n\theta) \cos k_{mn} ct \\ &+ (C_{mn} \cos n\theta + D_{mn} \sin n\theta) \sin k_{mn} ct] J_n(k_{mn} r). \end{aligned} \tag{66}$$

Finally, the prescribed initial conditions yield the relations

$$\begin{aligned} f(r, \theta) = &\sum_{m=1}^{\infty} A_{m0} J_0(k_{m0} r) \\ &+ \sum_{m=1}^{\infty} \sum_{n=1}^{\infty} (A_{mn} \cos n\theta + B_{mn} \sin n\theta) J_n(k_{mn} r) \end{aligned} \tag{67}$$

and

$$g(r, \theta) = \sum_{m=1}^{\infty} k_{m0} c C_{m0} J_0(k_{m0} r)$$

$$+ \sum_{m=1}^{\infty} \sum_{n=1}^{\infty} k_{mn} c (C_{mn} \cos n\theta + D_{mn} \sin n\theta) J_n(k_{mn} r). \tag{68}$$

By following a procedure similar to that in Section 10.2.1, we can derive formulas for determining the coefficients appearing in the above series (see Problems 17–19 in Exercises 10.3).

REMARK: *Notice that the single series in (66) corresponding to $n = 0$ is the same solution obtained in Section 10.3.1 for the radial symmetric case.*

EXERCISES 10.3

1. Given that $J_0(k_1) = 0$, solve

$$u_{rr} + \frac{1}{r} u_r = c^{-2} u_{tt}, \qquad 0 < r < 1, \qquad t > 0$$

B.C.: $u(1, t) = 0$

I.C.: $u(r, 0) = 0.1 J_0(k_1 r), \qquad u_t(r, 0) = 0.$

2. Given the boundary value problem

$$u_{rr} + \frac{1}{r} u_r = c^{-2} u_{tt}, \qquad 0 < r < 1, \qquad t > 0$$

B.C.: $u(1, t) = 0$

I.C.: $u(r, 0) = 0, \qquad u_t(r, 0) = 1,$

show that

$$u(r, t) = \frac{2}{c} \sum_{n=1}^{\infty} \frac{\sin(k_n c t)}{k_n^2 J_1(k_n)} J_0(k_n r),$$

where $J_0(k_n) = 0$, $n = 1, 2, 3, \ldots$.

3. Determine the numerical values of the radii of the nodal lines shown in Figure 10.6 for the cases $n = 2$ and $n = 3$.
Hint: See Table 9.1 in Section 9.4.

4. Find a formal solution of

$$u_{rr} + \frac{1}{r} u_r = c^{-2} u_{tt}, \qquad 0 < r < 1, \qquad t > 0$$

B.C.: $u_r(1, t) = 0$

I.C.: $u(r, 0) = f(r), \qquad u_t(r, 0) = g(r).$

5. Solve the problem described by (39)–(42) for a circle of radius ρ.

6. If the temperature distribution in an insulated circular plate is independent of the polar angle θ, it is then described by solutions of

$$u_{rr} + \frac{1}{r} u_r = a^{-2} u_t, \qquad 0 < r < 1, \qquad t > 0$$

B.C.: $u_r(1, t) = 0$

I.C.: $u(r, 0) = f(r), \qquad 0 < r < 1.$

Show that the solutions are given by

$$u(r, t) = c_0 + \sum_{n=1}^{\infty} c_n J_0(k_n r) e^{-a^2 k_n^2 t},$$

where $J_0'(k_n) = 0$, $n = 1, 2, 3, \ldots$.

In Problems 7–9, solve the problem described by (39)–(42) when $g(r) = 0$ and

★7. $f(r) = r^4$.

8. $f(r) = T_0$ (constant).

9. $f(r) = 1 - r^2$.

10. Over a long, solid cylinder of unit radius and uniform temperature distribution T_1 is fitted a long, hollow cylinder $1 \leq r \leq 2$ of the same material at uniform temperature T_2. Assume that the outer cylinder surface is maintained at temperature T_2. Show that the temperature distribution throughout the two cylinders is given by

$$u(r, t) = T_2 + \frac{1}{2}(T_1 - T_2) \sum_{n=1}^{\infty} \frac{J_1(k_n)}{k_n [J_1(2k_n)]^2} J_0(k_n r) e^{-a^2 k_n^2 t},$$

where $J_0(2k_n) = 0$, $n = 1, 2, 3, \ldots$. What temperature exists throughout the two cylinders after a long time?

11. The temperature distribution $u(r, t)$ in a thin circular plate with heat exchanges from its faces into the surrounding medium at 0°C satisfies the boundary value problem

$$u_{rr} + \frac{1}{r} u_r - bu = u_t, \qquad 0 < r < 1, \qquad t > 0$$

B.C.: $u(1, t) = 0$

I.C.: $u(r, 0) = 1,$

where b is a positive constant. Show that

$$u(r, t) = 2e^{-bt} \sum_{n=1}^{\infty} \frac{J_0(k_n r)}{k_n J_1(k_n)} e^{-k_n^2 t},$$

where $J_0(k_n) = 0$, $n = 1, 2, 3, \ldots$.

★12. Given the boundary value problem

$$\frac{1}{r} \frac{\partial}{\partial r}(r u_r) = u_t, \qquad 0 < r < 1, \qquad t > 0$$

B.C.: $u_r(1, t) = 1$

I.C.: $u(r, 0) = 0,$

show that it has solution

$$u(r, t) = \frac{1}{4}\left[2r^2 - 1 + 8t - 8 \sum_{n=1}^{\infty} \frac{J_0(k_n r)}{k_n^2 J_0(k_n)} e^{-k_n^2 t}\right],$$

where $J_0'(k_n) = 0$, $n = 1, 2, 3, \ldots$.
Hint: Assume $u(r, t) = At + S(r) + v(r, t)$, where A is a constant to be determined, and $S(r)$ satisfies

$$\frac{1}{r}\frac{d}{dr}(rS') = A, \qquad S'(1) = 1.$$

Also, note that $\int_0^1 rJ_0(k_n r)\, dr = 0$.

13. The temperature distribution $u(r, t)$ in a thin circular plate, which is initially at 0°C, has its faces insulated and its boundary held at temperature T_0, is governed by the boundary value problem

$$u_{rr} + \frac{1}{r}u_r = u_t, \qquad 0 < r < 1, \qquad t > 0$$

B.C.: $u(1, t) = T_0$

I.C.: $u(r, 0) = 0$.

Find a bounded solution by assuming $u(r, t) = S(r) + v(r, t)$, where $S(r)$ is the steady-state solution and $v(r, t)$ the transient solution. After a long time, what is the temperature in the plate?

★14. Using the *Laplace transform,* find a bounded solution of Problem 13.

Hint: $\mathscr{L}^{-1}\left\{\frac{I_0(\sqrt{p}\,r)}{pI_0(\sqrt{p})};\ p \to t\right\} = 1 - 2\sum_{n=1}^{\infty} \frac{J_0(k_n r)}{k_n J_1(k_n)} e^{-k_n^2 t}$,

where $J_0(k_0) = 0$, $n = 1, 2, 3, \ldots$.

★15. Using the *Laplace transform,* find a bounded solution of

$$u_{rr} + \frac{1}{r}u_r = u_t, \qquad 0 < r < 1, \qquad t > 0$$

B.C.: $u(1, t) = 0$

I.C.: $u(r, 0) = T_0$ (constant).

Hint: See the hint in Problem 14.

★16. By assuming $u(r, t) = \sum_{n=1}^{\infty} E_n(t)J_0(k_n r)$, where $J_0(k_n) = 0$, $n = 1, 2, 3, \ldots$, follow the technique of Section 6.3.2 to to find a formal solution of

$$u_{rr} + \frac{1}{r}u_r = u_t - q(r, t), \qquad 0 < r < 1, \qquad t > 0$$

B.C.: $u(1, t) = 0$

I.C.: $u(r, 0) = 0$.

★17. By multiplying Equation (67) by $rJ_0(k_{q0} r)$ and integrating over the domain $-\pi \leqslant \theta \leqslant \pi$, $0 \leqslant r \leqslant 1$, deduce that ($m = 1, 2, 3, \ldots$)

$$A_{m0} = \frac{1}{\pi[J_1(k_{m0})]^2}\int_{-\pi}^{\pi}\int_0^1 rf(r, \theta)J_0(k_{m0} r)\, dr\, d\theta.$$

★18. By multiplying Equation (67) by $rJ_p(k_{qp}r)\cos p\theta$ and integrating over the domain $-\pi \leqslant \theta \leqslant \pi$, $0 \leqslant r \leqslant 1$, deduce that ($m, n = 1, 2, 3, \ldots$)

$$A_{mn} = \frac{2}{\pi[J_{n+1}(k_{mn})]^2} \int_{-\pi}^{\pi} \int_0^1 rf(r, \theta) J_n(k_{mn}r) \cos n\theta \, dr \, d\theta.$$

★19. By multiplying Equation (67) by $rJ_p(k_{qp}r)\sin p\theta$ and integrating over the domain $-\pi \leqslant \theta \leqslant \pi$, $0 \leqslant r \leqslant 1$, deduce that ($m, n = 1, 2, 3, \ldots$)

$$B_{mn} = \frac{2}{\pi[J_{n+1}(k_{mn})]^2} \int_{-\pi}^{\pi} \int_0^1 rf(r, \theta) J_n(k_{mn}r) \sin n\theta \, dr \, d\theta.$$

★20. The small displacements of a vibrating membrane in the shape of a quarter circle are governed by the boundary value problem

$$r^2 u_{rr} + ru_r + u_{\theta\theta} = c^{-2} r^2 u_{tt}, \qquad 0 < r < 1, \qquad 0 < \theta < \pi/2, \qquad t > 0$$

B.C.: $u(1, \theta, t) = 0, \qquad u(r, 0, t) = 0, \qquad u(r, \pi/2, t) = 0$

I.C.: $u(r, \theta, 0) = f(r, \theta), \qquad u_t(r, \theta, 0) = 0.$

Find a formal solution of this problem.

10.4 CYLINDRICAL DOMAINS: BESSEL FUNCTIONS

In cylindrical coordinates defined by $x = r\cos\theta$, $y = r\sin\theta$, $z = z$, the Laplacian takes the form

$$\nabla^2 u = u_{rr} + \frac{1}{r} u_r + \frac{1}{r^2} u_{\theta\theta} + u_{zz}. \tag{69}$$

As was the case in polar coordinates, the separation of variables method applied to PDEs involving (69) often leads to an ODE in the radial coordinate r that is of Bessel type. For illustrative purposes, we will discuss *potential problems* for which Laplace's equation $\nabla^2 u = 0$ is the governing PDE.

10.4.1 RADIAL SYMMETRIC POTENTIAL PROBLEM

Laplace's equation in cylindrical coordinates is associated with the problem of finding the steady-state temperature distribution in a solid cylinder made of homogeneous material (among other areas of application). If the temperature distribution u is independent of z, then the temperature distribution in every circular cross-section along the z-axis is the same. In this case the problem is mathematically equivalent to that discussed in Section 8.3.1. We will consider other problems here.

Let us consider a solid homogeneous cylinder with unit radius and height a units (see Figure 10.7). If the temperature on the surfaces of the cylinder are prescribed in such a way that they are a function of only the radial distance r and height z, the temperatures inside the cylinder will also depend on only these variables. The problem is one of steady-state, and is formulated by

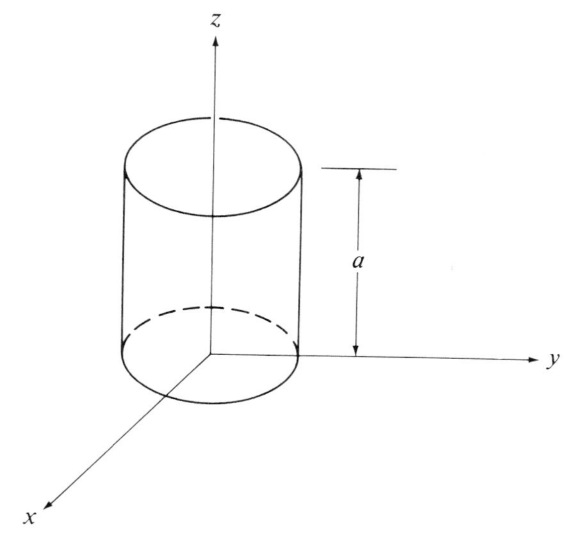

FIGURE 10.7

Cylindrical domain

$$u_{rr} + \frac{1}{r} u_r + u_{zz} = 0, \qquad 0 < r < 1, \qquad 0 < z < a$$

$$\text{B.C.:} \qquad u(1, z) = f(z), \qquad u(r, 0) = g(r), \qquad u(r, a) = h(r). \tag{70}$$

The problem described by (70) is a *Dirichlet problem*. To solve it we must first separate it into three problems, each of which has only *one* nonhomogeneous boundary condition. The three problems are*

$$u_{rr} + \frac{1}{r} u_r + u_{zz} = 0, \qquad 0 < r < 1, \qquad 0 < z < a$$

$$\text{B.C.:} \qquad u(1, z) = f(z), \qquad u(r, 0) = 0, \qquad u(r, a) = 0, \tag{71}$$

$$u_{rr} + \frac{1}{r} u_r + u_{zz} = 0, \qquad 0 < r < 1, \qquad 0 < z < a$$

$$\text{B.C.:} \qquad u(1, z) = 0, \qquad u(r, 0) = g(r), \qquad u(r, a) = 0, \tag{72}$$

and

$$u_{rr} + \frac{1}{r} u_r + u_{zz} = 0, \qquad 0 < r < 1, \qquad 0 < z < a$$

$$\text{B.C.:} \qquad u(1, z) = 0, \qquad u(r, 0) = 0, \qquad u(r, a) = h(r). \tag{73}$$

---

*The implicit boundary condition that u remain bounded as $r \to 0^+$ is usually taken for granted and not always explicitly stated.

We will solve (71) here, while (72) and (73) are left for the exercises.

If we set $u(r, z) = R(r)Z(z)$ in (71) and separate variables, we find

$$r^2R'' + rR' - \lambda r^2R = 0, \tag{74}$$

$$Z'' + \lambda Z = 0, \qquad Z(0) = 0, \qquad Z(a) = 0. \tag{75}$$

We recognize (75) as a familiar eigenvalue problem for which

$$\lambda_n = \frac{n^2\pi^2}{a^2}, \qquad Z_n(z) = \sin\frac{n\pi z}{a}, \qquad n = 1, 2, 3, \ldots. \tag{76}$$

For these restricted values of λ, (74) is *Bessel's modified equation* of order zero, with general solution

$$R_n(r) = c_nI_0\left(\frac{n\pi r}{a}\right) + d_nK_0\left(\frac{n\pi r}{a}\right) \tag{77}$$

(see Section 9.6). However, because K_0 is unbounded at $r = 0$, we must set $d_n = 0$ for all n. Then, combining solutions (76) and (77) through use of the superposition principle, we obtain

$$u(r, z) = \sum_{n=1}^{\infty} c_nI_0\left(\frac{n\pi r}{a}\right) \sin\frac{n\pi z}{a}. \tag{78}$$

The remaining task at this point is the determination of the constants c_n. By imposing the nonhomogeneous boundary condition in (71), we have

$$u(1, z) = f(z) = \sum_{n=1}^{\infty} c_nI_0\left(\frac{n\pi}{a}\right) \sin\frac{n\pi z}{a}, \tag{79}$$

which is a *Fourier sine series* for the function $f(z)$. Hence,

$$c_nI_0\left(\frac{n\pi}{a}\right) = \frac{2}{a}\int_0^a f(z) \sin\frac{n\pi z}{a}\,dz, \qquad n = 1, 2, 3, \ldots. \tag{80}$$

EXAMPLE 3

A solid homogeneous cylinder of unit radius and height π has its top and bottom insulated, and has temperature $f(z) = z$ prescribed on its lateral surface. Determine the steady-state temperature distribution throughout the cylinder.

SOLUTION The problem is characterized by

$$u_{rr} + \frac{1}{r}u_r + u_{zz} = 0, \qquad 0 < r < 1, \qquad 0 < z < \pi$$

$$\text{B.C.:} \qquad u(1, z) = z, \qquad u_z(r, 0) = 0, \qquad u_z(r, \pi) = 0.$$

By setting $u(r, z) = R(r)Z(z)$ and separating the variables, we obtain

$$Z'' + \lambda Z = 0, \qquad Z'(0) = 0, \qquad Z'(\pi) = 0$$

and

$$r^2R'' + rR' - \lambda r^2 R = 0.$$

The problem for $Z(z)$ leads to the solution

$$\lambda_0 = 0, \qquad Z_0(z) = 1,$$

$$\lambda_n = n^2, \qquad Z_n(z) = \cos nz, \qquad n = 1, 2, 3, \ldots .$$

For $\lambda = 0$, the equation in $R(r)$ becomes the Cauchy-Euler equation

$$r^2R'' + rR' = 0,$$

with general solution

$$R_0(r) = C_1 + C_2 \log r.$$

Here we must set $C_2 = 0$ to avoid infinite temperatures at $r = 0$. For $\lambda = n^2$, the equation in $R(r)$ is Bessel's modified equation, whose only bounded solutions are multiples of

$$R_n(r) = I_0(nr), \qquad n = 1, 2, 3, \ldots .$$

Combining solutions by the superposition principle, we have

$$u(r, z) = \frac{1}{2} c_0 + \sum_{n=1}^{\infty} c_n I_0(nr) \cos nz,$$

and by imposing the remaining boundary condition, we find

$$u(1, z) = z = \frac{1}{2} c_0 + \sum_{n=1}^{\infty} c_n I_0(n) \cos nz.$$

Hence, the Fourier coefficients are

$$c_0 = \frac{2}{\pi} \int_0^{\pi} z\, dz = \pi$$

(see Example 1 in Section 4.2) and

$$c_n I_0(n) = \frac{2}{\pi} \int_0^{\pi} z \cos nz\, dz = \frac{2}{\pi n^2} [(-1)^n - 1], \qquad n = 1, 2, 3, \ldots ,$$

and our solution takes the form

$$u(r, z) = \frac{\pi}{2} - \frac{4}{\pi} \sum_{\substack{n=1 \\ (\text{odd})}}^{\infty} \frac{I_0(nr)}{n^2 I_0(n)} \cos nz.$$

☐ 10.4.2 A MORE GENERAL POTENTIAL PROBLEM

If the temperatures inside the cylinder described in the last section are not independent of θ, then we must solve a more difficult problem.

To illustrate this, let us take a specific case where the cylinder height is π units; the problem is described by

$$u_{rr} + \frac{1}{r} u_r + \frac{1}{r^2} u_{\theta\theta} + u_{zz} = 0,$$
$$0 < r < 1, \quad -\pi < \theta < \pi, \quad 0 < z < \pi \tag{81}$$

B.C.: $u(1, \theta, z) = f(\theta, z), \quad u(r, \theta, 0) = 0, \quad u(r, \theta, \pi) = 0.$

By writing $u(r, \theta, z) = \psi(r, \theta)Z(z)$, we are led to the equations

$$\psi_{rr} + \frac{1}{r} \psi_r + \frac{1}{r^2} \psi_{\theta\theta} - \lambda\psi = 0, \tag{82}$$

$$Z'' + \lambda Z = 0, \quad Z(0) = 0, \quad Z(\pi) = 0. \tag{83}$$

The eigenvalue problem (83) has the well known solution

$$\lambda_m = m^2, \quad Z_m(z) = \sin mz, \quad m = 1, 2, 3, \ldots, \tag{84}$$

whereas (82) requires another separation of variables. Hence, we set $\psi(r, \theta) = R(r)H(\theta)$, from which we obtain

$$H'' + \mu H = 0, \tag{85}$$

$$r^2R'' + rR' - (m^2r^2 + \mu)R = 0, \tag{86}$$

where μ is a new separation constant.

Physical (and mathematical) considerations demand that $H(\theta)$ be periodic with period 2π so that the potential function $u(r, \theta, z)$ will remain single-valued. This requirement suggests that $\mu = n^2$ $(n = 0, 1, 2, \ldots)$, and accordingly

$$H_n(\theta) = \begin{cases} A_0, & n = 0 \\ A_n \cos n\theta + B_n \sin n\theta, & n = 1, 2, 3, \ldots. \end{cases} \tag{87}$$

Also, with $\mu = n^2$ we see that (86) is *Bessel's modified equation* of order n. The only bounded solutions of this equation are multiples of

$$R_{mn}(r) = I_n(mr), \tag{88}$$

which, when combined with (84) and (87) and summed over all possible values of m and n, leads to the double series

$$u(r, \theta, z) = \sum_{m=1}^{\infty} a_{m0} I_0(mr) \sin mz + \sum_{m=1}^{\infty} \sum_{n=1}^{\infty} (a_{mn} \cos n\theta + b_{mn} \sin n\theta) I_n(mr) \sin mz \tag{89}$$

(after redefining the constants).

The constants appearing in (89) can be determined by setting $r = 1$ and requiring that

$$f(\theta, z) = \sum_{n=0}^{\infty} \left[\sum_{m=1}^{\infty} a_{mn} I_n(m) \sin mz \right] \cos n\theta$$

$$+ \sum_{n=1}^{\infty} \left[\sum_{m=1}^{\infty} b_{mn} I_n(m) \sin mz \right] \sin n\theta \tag{90}$$

(for convenience we have rearranged the terms). Let us first introduce the functions

$$\alpha_n(z) = \sum_{m=1}^{\infty} a_{mn} I_n(m) \sin mz, \qquad n = 0, 1, 2, \ldots \tag{91}$$

and

$$\beta_n(z) = \sum_{m=1}^{\infty} b_{mn} I_n(m) \sin mz, \qquad n = 1, 2, 3, \ldots, \tag{92}$$

and then (90) becomes

$$f(\theta, z) = \alpha_0(z) + \sum_{n=1}^{\infty} [\alpha_n(z) \cos n\theta + \beta_n(z) \sin n\theta]. \tag{93}$$

Thus, for a fixed value of z, (93) is a Fourier series for which

$$\alpha_0(z) = \frac{1}{2\pi} \int_{-\pi}^{\pi} f(\theta, z)\, d\theta, \tag{94a}$$

$$\alpha_n(z) = \frac{1}{\pi} \int_{-\pi}^{\pi} f(\theta, z) \cos n\theta\, d\theta, \qquad n = 1, 2, 3, \ldots, \tag{94b}$$

$$\beta_n(z) = \frac{1}{\pi} \int_{-\pi}^{\pi} f(\theta, z) \sin n\theta\, d\theta, \qquad n = 1, 2, 3, \ldots. \tag{94c}$$

Once these functions of z have been determined, we can interpret (91) and (92) as Fourier sine series and obtain the results

$$a_{mn} I_n(m) = \frac{2}{\pi} \int_0^{\pi} \alpha_n(z) \sin mz\, dz, \qquad m = 1, 2, 3, \ldots \tag{95}$$

and

$$b_{mn} I_n(m) = \frac{2}{\pi} \int_0^{\pi} \beta_n(z) \sin mz, \qquad m = 1, 2, 3, \ldots. \tag{96}$$

EXERCISES 10.4

1. Solve the problem described by (71) when $f(z) = A \sin(3\pi z/a)$.
2. Find a formal solution of the problem described by (72).
3. Show that a formal solution of the problem described by (73) is

$$u(r, z) = \sum_{n=1}^{\infty} c_n J_0(k_n r) \frac{\sinh k_n z}{\sinh k_n a},$$

where $J_0(k_n) = 0$, $n = 1, 2, 3, \ldots$.

4. Show that

$$u_{rr} + \frac{1}{r} u_r + u_{zz} = 0, \qquad 0 < r < 1, \qquad 0 < z < 1$$

B.C.: $u(1, z) = 0, \quad u_z(r, 0) = 0, \quad u(r, 1) = 1,$

has the solution

$$u(r, z) = 2 \sum_{n=1}^{\infty} \frac{J_0(k_n r)}{k_n J_1(k_n)} \frac{\cosh k_n z}{\cosh k_n},$$

where $J_0(k_n) = 0$, $n = 1, 2, 3, \ldots$.

5. Solve the problem described by (71) for a cylinder of radius ρ.

In Problems 6–9, find a solution of the given potential problem

6. $u_{rr} + \frac{1}{r} u_r + u_{zz} = 0, \quad 0 < r < \rho, \quad 0 < z < a$

B.C.: $u(\rho, z) = z(a - z), \quad u(r, 0) = 0, \quad u(r, a) = 0$

7. $u_{rr} + \frac{1}{r} u_r + u_{zz} = 0, \quad 0 < r < 1, \quad 0 < z < \pi$

B.C.: $u(1, z) = T_0, \quad u(r, 0) = 0, \quad u_z(r, \pi) = 0 \quad$ (T_0 constant)

8. $u_{rr} + \frac{1}{r} u_r + u_{zz} = 0, \quad 0 < r < 1, \quad 0 < z < 1$

B.C.: $u(1, z) = 0, \quad u(r, 0) = 0, \quad u(r, 1) = J_0(k_3 r), \quad$ where $J_0(k_3) = 0$

9. $u_{rr} + \frac{1}{r} u_r + u_{zz} = 0, \quad 0 < r < 1, \quad 0 < z < 1$

B.C.: $u(1, z) = 0, \quad u(r, 0) = 0, \quad u(r, 1) = 1 - r^2$

★10. The steady-state temperature distribution in a semi-infinite cylindrical domain satisfies the conditions

$$u_{rr} + \frac{1}{r} u_r + u_{zz} = 0, \qquad 0 < r < 1, \qquad z > 0$$

B.C.: $u(1, z) = 0, \quad u(r, 0) = 1.$

Using the *Fourier sine transform,* deduce that

$$u(r, z) = 1 - \frac{2}{\pi} \int_0^{\infty} \frac{I_0(rs)}{I_0(s)} \frac{\sin sz}{s} \, ds.$$

11. The temperature distribution in a solid cylinder of unit height and unit radius is governed by

$$u_{rr} + \frac{1}{r} u_r + u_{zz} = a^{-2} u_t, \qquad 0 < r < 1, \qquad 0 < z < 1, \qquad t > 0$$

B.C.: $u(r, 0, t) = 0, \quad u(r, 1, t) = 0, \quad u(1, z, t) = 0$

I.C.: $u(r, z, 0) = f(r, z).$

Show that

$$u(r, z, t) = \sum_{m=1}^{\infty} \sum_{n=1}^{\infty} C_{mn} J_0(k_m r) \sin(n\pi z) e^{-a^2(k_m^2 + n^2\pi^2)t}.$$

12. Solve Problem 11 when $f(r, z) = T_0$ (constant).

13. Suppose a cylindrical column of unit radius is considered to be of infinite height extending along the z-axis. If the lateral boundary is maintained at zero temperature and the initial temperature distribution inside the column is prescribed as $f(r, \theta)$, show that the subsequent temperature distribution of the column is given by

$$u(r, \theta, t) = \sum_{m=1}^{\infty} \sum_{n=0}^{\infty} J_n(k_{mn} r)(a_{mn} \cos n\theta + b_{mn} \sin n\theta) e^{-a^2 k_{mn}^2 t},$$

where k_{mn} is the mth positive root of $J_n(k) = 0$, and where the constants a_{mn} and b_{mn} are selected in such a way that

$$f(r, \theta) = \sum_{m=1}^{\infty} \sum_{n=0}^{\infty} J_n(k_{mn} r)(a_{mn} \cos n\theta + b_{mn} \sin n\theta)$$

for $r < 1$, $-\pi < \theta < \pi$.

★14. Show that the Fourier coefficients appearing in Problem 13 are given by

$$a_{m0} = \frac{1}{\pi [J_1(k_{m0})]^2} \int_0^1 \int_{-\pi}^{\pi} r f(r, \theta) J_0(k_{m0} r)\, d\theta\, dr, \qquad m = 1, 2, 3, \ldots,$$

$$a_{mn} = \frac{1}{\pi [J_{n+1}(k_{mn})]^2} \int_0^1 \int_{-\pi}^{\pi} r f(r, \theta) J_n(k_{mn} r) \cos n\theta\, d\theta\, dr; \qquad m, n = 1, 2, 3, \ldots,$$

$$b_{mn} = \frac{1}{\pi [J_{n+1}(k_{mn})]^2} \int_0^1 \int_{-\pi}^{\pi} r f(r, \theta) J_n(k_{mn} r) \sin n\theta\, d\theta\, dr; \qquad m, n = 1, 2, 3, \ldots.$$

★15. Consider a cylindrical column of unit radius that extends infinitely far along the positive z-axis. The temperature of the lateral surface is maintained at zero while that of the base is prescribed by $u(r, \theta, 0) = f(r, \theta)$. Show that the internal steady-state temperature distribution is given by

$$u(r, \theta, z) = \sum_{m=1}^{\infty} \sum_{n=0}^{\infty} J_n(k_{mn} r)(a_{mn} \cos n\theta + b_{mn} \sin n\theta) e^{-k_{mn} z},$$

where $J_n(k_{mn}) = 0$ and the Fourier coefficients are the same as in Problem 14.

10.5 SPHERICAL DOMAINS: LEGENDRE POLYNOMIALS

The *Legendre polynomials* (Section 9.2) are closely associated with problems in spherical domains. In particular, we find Legendre's equation evolving out of the separation of variables method when applied to potential problems where the solution is independent of the azimuthal angle in spherical-shaped regions. The general form of the Laplacian in spherical coordinates is

$$\nabla^2 u = u_{rr} + \frac{2}{r} u_r + \frac{1}{r^2} u_{\phi\phi} + \frac{\cot \phi}{r^2} u_\phi + \frac{1}{r^2 \sin^2 \phi} u_{\theta\theta} \tag{97}$$

(see Problem 15 in Exercises 10.5) which can also be expressed as

$$\nabla^2 u = \frac{1}{r^2}\left[\frac{\partial}{\partial r}\left(r^2 \frac{\partial u}{\partial r}\right) + \frac{1}{\sin\phi}\frac{\partial}{\partial\phi}\left(\sin\phi \frac{\partial u}{\partial\phi}\right) + \frac{1}{\sin^2\phi}\frac{\partial^2 u}{\partial\theta^2}\right] \tag{98}$$

(see Problem 16 in Exercises 10.5). The potential equation $\nabla^2 u = 0$ can be formed by using either (97) or (98), although (98) is generally more convenient.

10.5.1 ELECTRIC POTENTIAL DUE TO A SPHERE

Suppose that on the surface of a hollow sphere of unit radius, a fixed distribution of electric potential is maintained in such a way that it is independent of the polar azimuthal angle θ (see Figure 10.8). In the absence of any further charges inside the sphere, we wish to find the potential distribution $u(r, \phi)$ within the spherical cavity. If f denotes the electric potential on the spherical shell, the problem is characterized by the *Dirichlet problem*

$$\frac{\partial}{\partial r}\left(r^2 \frac{\partial u}{\partial r}\right) + \frac{1}{\sin\phi}\frac{\partial}{\partial\phi}\left(\sin\phi \frac{\partial u}{\partial\phi}\right) = 0, \quad 0 < r < 1, \quad 0 < \phi < \pi \tag{99}$$

$$\text{B.C.:} \quad u(1, \phi) = f(\phi), \quad 0 < \phi < \pi,$$

where we have used (98) for the Laplacian.

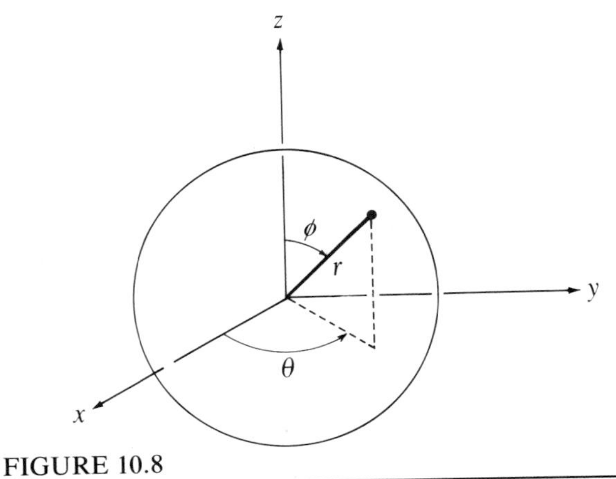

FIGURE 10.8

Spherical domain

Employing the separation of variables method with $u(r, \phi) = R(r)\Phi(\phi)$, we are led to

$$r^2 R'' + 2rR' - \lambda R = 0, \quad 0 < r < 1, \tag{100}$$

$$\frac{1}{\sin\phi}\frac{d}{d\phi}(\sin\phi\, \Phi') + \lambda\Phi = 0, \quad 0 < \phi < \pi. \tag{101}$$

By setting $x = \cos\phi$ in (101), we get the more familiar form

$$\frac{d}{dx}\left[(1-x^2)\frac{d\Phi}{dx}\right] + \lambda\Phi = 0, \qquad -1 < x < 1, \tag{102}$$

which is *Legendre's equation* (see Problem 5 in Exercises 9.2). Physical considerations demand that the potential u everywhere on and inside the sphere remain bounded. The only bounded solutions of (102) occur when λ is restricted to the set of values

$$\lambda_n = n(n+1), \qquad n = 0, 1, 2, \ldots, \tag{103}$$

and in this case the corresponding eigenfunctions are the *Legendre polynomials*

$$\Phi_n(\phi) \equiv P_n(x) = P_n(\cos\phi), \qquad n = 0, 1, 2, \ldots. \tag{104}$$

With the separation constant restricted to values given by (103), Equation (100) takes the form

$$r^2R'' + 2rR' - n(n+1)R = 0, \tag{105}$$

which is a Cauchy-Euler equation with general solutions

$$R_n(r) = a_nr^n + b_nr^{-(n+1)}, \qquad n = 0, 1, 2, \ldots. \tag{106}$$

To avoid infinite values of $R_n(r)$ at $r = 0$, we must select $b_n = 0$ for all n. Therefore, (106) reduces to

$$R_n(r) = a_nr^n, \qquad n = 0, 1, 2, \ldots, \tag{107}$$

and by summing over all possible solutions from (104) and (107), we obtain

$$u(r, \phi) = \sum_{n=0}^{\infty} a_nr^nP_n(\cos\phi). \tag{108}$$

Equation (108) represents a bounded solution of Laplace's equation for any choice of the constants a_n. To satisfy the boundary condition in (99), however, we must select the constants such that

$$u(1, \phi) = f(\phi) = \sum_{n=0}^{\infty} a_nP_n(\cos\phi), \qquad 0 < \phi < \pi. \tag{109}$$

This last expression is a *Fourier-Legendre series* for which

$$a_n = \left(n + \frac{1}{2}\right)\int_0^{\pi} f(\phi)P_n(\cos\phi)\sin\phi\, d\phi, \qquad n = 0, 1, 2, \ldots \tag{110}$$

(see Problem 5 in Exercises 10.5).

For problems involving electric potential it is also natural to inquire about the potential outside the sphere ($r > 1$). To determine the potential in this region we must again solve (99). In this case, however, our boundedness condition is not prescribed at $r = 0$ (which is outside the region of interest), but for $r \to \infty$. Hence, this time we set $a_n = 0$ ($n = 0, 1, 2, \ldots$) in Equation (106), and obtain

$$R_n(r) = \frac{b_n}{r^{n+1}}, \qquad n = 0, 1, 2, \ldots. \tag{111}$$

Combining (104) and (111) by the superposition principle leads to

$$u(r, \phi) = \sum_{n=0}^{\infty} \frac{b_n}{r^{n+1}} P_n(\cos\phi). \tag{112}$$

The determination of the constants b_n from the prescribed boundary condition yields the same integral as before [see (110)].

EXAMPLE 4

Find the electric potential inside a hollow unit sphere when the potential on the boundary is prescribed by

$$u(1, \phi) = \begin{cases} U_0, & 0 < \phi < \pi/2 \\ -U_0, & \pi/2 < \phi < \pi. \end{cases}$$

SOLUTION The solution is given by (108), where the Fourier coefficients are determined from (110), which yields

$$a_n = \left(n + \frac{1}{2}\right) U_0 \left[\int_0^{\pi/2} P_n(\cos\phi)\sin\phi\, d\phi - \int_{\pi/2}^{\pi} P_n(\cos\phi)\sin\phi\, d\phi\right]$$

$$= \left(n + \frac{1}{2}\right) U_0 \left[\int_0^1 P_n(x)\, dx - \int_{-1}^0 P_n(x)\, dx\right].$$

The last step follows from the change of variables $x = \cos\phi$. Owing to the even-odd character of the Legendre polynomials, we see that the replacement of x by $-x$ in the last integral leads to the conclusion

$$a_n = 0, \qquad n = 0, 2, 4, \ldots$$

and

$$a_n = (2n + 1) U_0 \int_0^1 P_n(x)\, dx, \qquad n = 1, 3, 5, \ldots.$$

Recalling Example 8 in Section 9.7.1, we have

$$a_1 = 3/2, \qquad a_3 = -7/8, \qquad a_5 = 11/16, \ldots.$$

Therefore, our solution becomes

$$u(r, \phi) = \frac{3}{2} r P_1(\cos\phi) - \frac{7}{8} r^3 P_3(\cos\phi) + \frac{11}{16} r^5 P_5(\cos\phi) - \cdots.$$

EXERCISES 10.5

1. Find the electric potential in the *interior* of a hollow unit sphere assuming the potential on the surface is
   (a) $f(\phi) = 1$
   (b) $f(\phi) = \cos\phi$
   (c) $f(\phi) = \cos^2\phi$
   (d) $f(\phi) = \cos 2\phi$

2. Find the electric potential in the *exterior* of the hollow unit sphere assuming the potential on the surface is prescribed as given in Problem 1.

3. Solve the potential problem (99) for a sphere of radius ρ.

4. Show that the orthogonality relation for the functions $\Phi_n(\phi) = P_n(\cos\phi)$ is

$$\int_0^\pi \Phi_m(\phi)\Phi_n(\phi)\sin\phi\, d\phi = 0, \qquad m \neq n.$$

5. Using the result of Problem 4, formally derive (110) for the Fourier coefficients.

6. Find the general solution of the PDE in (99) that depends only upon ϕ.

7. Find the general solution of the PDE in (99) that depends only upon r.

8. After a long period of time, the temperature $u(r, \phi)$ on the surface of a sphere of unit radius is maintained at

$$u(1, \phi) = T_0(1 - \cos^2\phi), \qquad T_0 \text{ constant}.$$

   Find the steady-state temperature everywhere within the solid homogeneous sphere.

9. If the potential on the surface of a sphere of unit radius is kept at constant potential U_0, show that at points far from the spherical surface the potential is (approximately) given by

$$u(r, \phi) \simeq \frac{3U_0}{2r^2}, \qquad r \gg 1.$$

10. A spherical shell has an inner radius of one unit and an outer radius of two units. The prescribed temperatures on the inner and outer surfaces are given respectively by

$$u(1, \phi) = 30 + 10\cos\phi, \qquad u(2, \phi) = 50 - 20\cos\phi.$$

   Determine the steady-state temperature everywhere within the spherical shell.

★11. The temperature on the surface of a solid homogeneous sphere of unit radius is prescribed by

$$u(1, \phi) = \begin{cases} T_0, & 0 < \phi < \pi/2 \\ 0, & \pi/2 < \phi < \pi. \end{cases}$$

   Show that the steady-state temperature distribution throughout the sphere is given by the series

$$u(r, \phi) = T_0 \sum_{n=0}^{\infty} (-1)^n \left(\frac{4n+3}{2n+2}\right) \frac{(2n)!}{2^{2n}(n!)^2} r^{2n+1} P_{2n+1}(\cos\phi).$$

   Hint: Use properties (L3), (L5), and (L10) in Section 9.2.2.

*12. Let $u(r, \phi, t)$ denote the temperature distribution in a solid homogeneous sphere of unit radius (independent of the angle θ) whose surface is maintained at 0°C, and whose initial temperature distribution is described by

$$u(r, \theta, 0) = f(r, \phi), \qquad 0 < r < 1, \qquad 0 < \phi < \pi.$$

(a) Show that the temperature distribution throughout the sphere has the formal solution

$$u(r, \phi, t) = \sum_{m=1}^{\infty} \sum_{n=0}^{\infty} C_{mn} \frac{J_{n+\frac{1}{2}}(k_{mn} r)}{\sqrt{r}} P_n(\cos \phi) e^{-a^2 k_{mn}^2 t},$$

where k_{mn} denotes the mth solution of $J_{n+\frac{1}{2}}(k) = 0$, $n = 0, 1, 2, \ldots$.

(b) If $f(r, \phi) = T_0$ (constant), show that the solution in (a) is independent of ϕ and that it reduces to

$$u(r, t) = \frac{2T_0}{\pi} \sum_{m=1}^{\infty} (-1)^{m+1} \left(\frac{\sin m\pi r}{mr} \right) e^{-a^2 m^2 \pi^2 t}.$$

Hint: See Problem 5 in Exercises 9.4.

*13. Using the *Laplace transform,* show that a bounded solution of the exterior temperature distribution problem

$$u_{rr} + \frac{2}{r} u_r = u_t, \qquad 1 < r < \infty, \qquad t > 0$$

B.C.: $u(1, t) = T_0$, $\quad u(r, t)$ finite as $r \to \infty$

I.C.: $u(r, 0) = 0$,

is given by

$$u(r, t) = \frac{T_0}{r} \operatorname{erfc}\left(\frac{r-1}{2\sqrt{t}} \right),$$

where erfc(x) is the complementary error function.
Hint: Show that the solution of the transformed problem is

$$U(r, p) = \frac{T_0}{rp} e^{-(r-1)\sqrt{p}},$$

and invert it by use of Problem 17 in Exercises 5.2.

*14. Solve Problem 13 when the initial condition is changed to

$$u(r, 0) = T_1 \text{ (constant)}.$$

*15. Show that the Laplacian in spherical coordinates (r, θ, ϕ) defined by

$$x = r \cos \theta \sin \phi, \qquad y = r \sin \theta \sin \phi, \qquad z = r \cos \phi,$$

is given by

$$\nabla^2 u = u_{rr} + \frac{2}{r} u_r + \frac{1}{r^2} u_{\phi\phi} + \frac{\cot \phi}{r^2} u_\phi + \frac{1}{r^2 \sin^2 \phi} u_{\theta\theta}.$$

16. Show that the result in Problem 15 can also be expressed as

$$\nabla^2 u = \frac{1}{r^2} \left[\frac{\partial}{\partial r}(r^2 u_r) + \frac{1}{\sin \phi} \frac{\partial}{\partial \phi}(\sin \phi \, u_\phi) + \frac{1}{\sin^2 \phi} u_{\theta\theta} \right].$$

★17. If $u(r, \theta, \phi)$ satisfies $\nabla^2 u = 0$, show that $v(r, \theta, \phi) = u\left(\frac{1}{r}, \theta, \phi\right) \Big/ r$ satisfies $\nabla^2 v = 0$.

10.6 POISSON'S EQUATION

The general form of the heat equation in a plane region R where there exists a known heat source is given by the *nonhomogeneous* equation

$$\nabla^2 u = a^{-2} u_t - q(x, y, t), \tag{113}$$

where $q(x, y, t)$ is proportional to the heat source.* When u and q are both independent of time, (113) reduces to the steady-state case

$$\nabla^2 u = -q(x, y), \tag{114}$$

called *Poisson's equation.*† This same equation arises also in solving for the gravitational potential in a region where a mass distribution exists, or in solving for the electric potential in the presence of a charge distribution. There are also other situations which lead to Poisson's equation.

A general treatment of Poisson's equation goes beyond the intended scope of this text. However, we will outline a general solution procedure and illustrate it for a simple example involving rectangular coordinates. Other geometries will be considered in the exercises.

In the xy-plane, the general problem we wish to address is characterized by

$$\begin{aligned} \nabla^2 u &= -q(x, y) \quad \text{in } R, \\ u &= f \quad \text{on } C, \end{aligned} \tag{115}$$

where R is a bounded region with boundary curve C. Our solution technique will closely parallel the eigenfunction expansion method illustrated in Section 4.6.1 for ordinary DEs. To begin, we divide (115) into two problems, one of which has a homogeneous DE and the other a homogeneous boundary condition. Thus, we seek solutions of

$$\begin{aligned} \nabla^2 u &= 0 \quad \text{in } R, \\ u &= f \quad \text{on } C, \end{aligned} \tag{116}$$

which we will denote by $u_H(x, y)$, and solutions of

$$\begin{aligned} \nabla^2 u &= -q(x, y) \quad \text{in } R, \\ u &= 0 \quad \text{on } C, \end{aligned} \tag{117}$$

which will be denoted by $u_P(x, y)$. The problem described by (116) is a Dirichlet problem for which solution techniques have already been developed

---

*The one-dimensional analog of (113) was derived as (5) in Section 6.1.1.

†Observe that Poisson's equation (114) reduces the potential equation when $q = 0$.

(Chapter 8). Therefore we will now address only (117). We leave it to the reader to show that $u(x\ y) = u_H(x, y) + u_P(x, y)$ is the solution of (115).

To solve (117), we assume there exists a solution of the form

$$u_P(x, y) = \sum_{m=1}^{\infty}\sum_{n=1}^{\infty} C_{mn}\phi_{mn}(x, y), \tag{118}$$

where $\phi_{mn}(x, y)$ denotes the eigenfunctions of the related Helmholtz equation

$$\begin{aligned} \nabla^2\phi + \lambda\phi &= 0 \quad \text{in} \quad R, \\ \phi &= 0 \quad \text{on} \quad C. \end{aligned} \tag{119}$$

The substitution of (118) into the PDE in (117) leads to

$$\sum_{m=1}^{\infty}\sum_{n=1}^{\infty} C_{mn}\nabla^2\phi_{mn}(x, y) = -q(x, y),$$

or upon rearranging the equation and using the relation $\nabla^2\phi_{mn} = -\lambda_{mn}\phi_{mn}$, we get

$$q(x, y) = \sum_{m=1}^{\infty}\sum_{n=1}^{\infty} C_{mn}\lambda_{mn}\phi_{mn}(x, y). \tag{120}$$

This last relation is simply a *generalized Fourier series in two variables* for the function $q(x, y)$. By multiplying (120) by $\phi_{jk}(x, y)$ and integrating the resulting expressions over R (using the orthogonality property of the eigenfunctions), we can derive the formal result*

$$C_{mn}\lambda_{mn} = \frac{1}{\|\phi_{mn}(x, y)\|^2}\iint_R q(x, y)\phi_{mn}(x, y)\,dx\,dy, \tag{121}$$

where

$$\|\phi_{mn}(x, y)\|^2 = \iint_R [\phi_{mn}(x, y)]^2\,dx\,dy. \tag{122}$$

10.6.1 RECTANGULAR DOMAINS

To be more specific, let us assume the region R is the rectangular domain $0 \le x \le a$, $0 \le y \le b$. In this case, (117) becomes

$$\begin{aligned} &u_{xx} + u_{yy} = -q(x, y), \quad 0 < x < a, \quad 0 < y < b \\ \text{B.C.:}\quad &u(0, y) = 0, \quad u(a, y) = 0, \quad u(x, 0) = 0, \quad u(x, b) = 0. \end{aligned} \tag{123}$$

---

* If there is more than one eigenfunction for the same eigenvalue, we assume that all eigenfunctions have been chosen to be mutually orthogonal. Also, in coordinate systems other than rectangular, the results (121) and (122) may involve a weighting function $r(x, y)$.

The related Helmholtz eigenvalue problem was solved in Section 10.2, which led to the solution

$$\lambda_{mn} = \left(\frac{m^2}{a^2} + \frac{n^2}{b^2}\right)\pi^2; \qquad m, n = 1, 2, 3, \ldots,$$
$$\phi_{mn}(x, y) = \sin\frac{m\pi x}{a}\sin\frac{n\pi y}{b}; \qquad m, n = 1, 2, 3, \ldots. \tag{124}$$

Therefore, (120) becomes a double Fourier sine series for which

$$C_{mn} = \frac{4}{ab\lambda_{mn}}\int_0^b\int_0^a q(x, y)\sin\frac{m\pi x}{a}\sin\frac{n\pi y}{b}\,dx\,dy;$$
$$m, n = 1, 2, 3, \ldots. \tag{125}$$

EXAMPLE 5

Solve the nonhomogeneous potential problem

$$u_{xx} + u_{yy} = -xy, \qquad 0 < x < \pi, \qquad 0 < y < \pi$$

B.C.: $$\begin{cases} u(0, y) = 0, & u(\pi, y) = 0 \\ u(x, 0) = 0, & u(x, \pi) = T_0 \text{ (constant)}. \end{cases}$$

SOLUTION The two problems we need to solve are the Dirichlet problem

$$u_{xx} + u_{yy} = 0, \qquad 0 < x < \pi, \qquad 0 < y < \pi$$

B.C.: $$\begin{cases} u(0, y) = 0, & u(\pi, y) = 0 \\ u(x, 0) = 0, & u(x, \pi) = T_0 \end{cases}$$

and

$$u_{xx} + u_{yy} = -xy, \qquad 0 < x < \pi, \qquad 0 < y < \pi$$

B.C.: $$\begin{cases} u(0, y) = 0, & u(\pi, y) = 0 \\ u(x, 0) = 0, & u(x, \pi) = 0. \end{cases}$$

The Dirichlet problem has the solution form [see (12) in Section 8.2]

$$u_H(x, y) = \sum_{n=1}^{\infty} A_n \sin nx \sinh ny,$$

where

$$A_n \sinh n\pi = \frac{2}{\pi}\int_0^{\pi} T_0 \sin nx\,dx$$
$$= \frac{2T_0}{n\pi}[1 - (-1)^n], \qquad n = 1, 2, 3, \ldots$$

Hence,

$$u_H(x, y) = \frac{4T_0}{\pi} \sum_{\substack{n=1 \\ (\text{odd})}}^{\infty} \frac{\sinh ny}{n \sinh n\pi} \sin nx.$$

To solve the second problem for $u_P(x, y)$, we simply substitute $q(x, y) = xy$, $a = b = \pi$, and $\lambda_{mn} = m^2 + n^2$ into (125) to obtain

$$C_{mn} = \frac{4}{(m^2 + n^2)\pi^2} \int_0^{\pi} \int_0^{\pi} xy \sin mx \sin ny \, dx \, dy$$

$$= \frac{(-1)^{m+n}4}{mn(m^2 + n^2)}; \qquad m, n = 1, 2, 3, \ldots,$$

where we are recalling the result of Example 1 in Section 10.2.1. It follows that

$$u_P(x, y) = 4 \sum_{m=1}^{\infty} \sum_{n=1}^{\infty} \frac{(-1)^{m+n}}{mn(m^2 + n^2)} \sin mx \sin ny,$$

and, therefore, the complete solution is

$$u(x, y) = u_H(x, y) + u_P(x, y)$$

$$= \frac{4T_0}{\pi} \sum_{\substack{n=1 \\ (\text{odd})}}^{\infty} \frac{\sinh ny}{n \sinh n\pi} \sin nx + 4 \sum_{m=1}^{\infty} \sum_{n=1}^{\infty} \frac{(-1)^{m+n}}{mn(m^2 + n^2)} \sin mx \sin ny.$$

EXERCISES 10.6

In Problems 1–4, find formal solutions by the eigenfunction expansion method.

1. $u_{xx} + u_{yy} = -1, \quad 0 < x < \pi, \quad 0 < y < 1$
   B.C.: $u(0, y) = 0, \quad u(\pi, y) = 0, \quad u(x, 0) = 0, \quad u(x, 1) = \sin x$

2. $u_{xx} + u_{yy} = -x, \quad 0 < x < \pi, \quad 0 < y < \pi$
   B.C.: $u(0, y) = 0, \quad u(\pi, y) = 1, \quad u(x, 0) = 0, \quad u(x, \pi) = 0$

3. $u_{xx} + u_{yy} = -\sin 3y, \quad 0 < x < \pi, \quad 0 < y < \pi$
   B.C.: $u(0, y) = 0, \quad u(\pi, y) = 0, \quad u(x, 0) = 1, \quad u(x, \pi) = 1$
   Hint: See Problem 7 in Exercises 8.2.

4. $u_{xx} + u_{yy} = -xy(a - x)(b - y), \quad 0 < x < a, \quad 0 < y < b$
   B.C.: $u(0, y) = 0, \quad u(a, y) = 0, \quad u(x, 0) = 0, \quad u(x, b) = 0$

★5. Find a formal solution of

$$u_{rr} + \frac{1}{r} u_r + u_{zz} = -q(r, z), \qquad 0 < r < 1, \qquad 0 < z < a$$

B.C.: $u(1, z) = f(z), \quad u(r, 0) = 0, \quad u(r, a) = 0.$

★6. Solve Problem 5 for the special case $q(r, z) = 1$ and $f(z) = z$.

★7. Find a formal solution of

$$\frac{1}{r^2}\left[\frac{\partial}{\partial r}\left(r^2\frac{\partial u}{\partial r}\right)+\frac{1}{\sin\phi}\frac{\partial}{\partial\phi}\left(\sin\phi\frac{\partial u}{\partial\phi}\right)\right]=-q(r,\phi),\qquad 0<r<1,\qquad 0<\phi<\pi$$

B.C.: $u(1, \phi) = f(\phi), \qquad 0 < \phi < \pi.$

★8. Given

$$u_{xx} + u_{yy} = -1, \qquad 0 < x < 1, \qquad 0 < y < 1$$

B.C.: $u(0, y) = 0, \qquad u(1, y) = 0, \qquad u(x, 0) = 0, \qquad u(x, 1) = 0,$

(a) approximate the value $u(\frac{1}{2}, \frac{1}{2})$ by using the first term of the series (118).

(b) Compare your answer in (a) with that for Problem 10 in Exercises 8.6 by the collocation and Galerkin methods.

BIBLIOGRAPHY

Abramowitz, M., and I. A. Stegun, eds. *Handbook of Mathematical Functions*. New York: Dover, 1965.

Andrews, L.C. *Special Functions for Engineers and Applied Mathematicians*. New York: Macmillan, 1985.

Hochstadt, H. *The Functions of Mathematical Physics*. New York: John Wiley & Sons, 1971.

Lebedev, N. N. *Special Functions and Their Applications*. Translated and edited by R. A. Silverman. New York: Dover, 1972.

Rainville, E. D. *Special Functions*. New York: Chelsea, 1960.

Watson, G. N. *A Treatise on the Theory of Bessel Functions*. 2nd ed. London: Cambridge University Press, 1952.

APPENDIX A

TABLE OF LAPLACE TRANSFORMS

The following is a short table of Laplace transforms and their inverses.

	$F(p) = \int_0^\infty e^{-pt} f(t)\, dt$	$f(t) = \mathcal{L}^{-1}\{F(p); t\}$
1	$\frac{1}{p}$	1
2	$\frac{1}{p^2}$	t
3	$\frac{1}{p^n} \quad (n = 1, 2, 3, \ldots)$	$\frac{t^{n-1}}{(n-1)!}$
4	$\frac{1}{\sqrt{p}}$	$\frac{1}{\sqrt{\pi t}}$
5	$\frac{1}{p^{3/2}}$	$2\sqrt{t/\pi}$
6	$\frac{1}{p^x}, \quad x > 0$	$\frac{t^{x-1}}{\Gamma(x)}$
7	$\frac{1}{p-a}$	e^{at}
8	$\frac{1}{(p-a)^2}$	te^{at}
9	$\frac{1}{(p-a)^n} \quad (n = 1, 2, 3, \ldots)$	$\frac{t^{n-1}e^{at}}{(n-1)!}$
10	$\frac{1}{(p-a)^x}, \quad x > 0$	$\frac{t^{x-1}e^{at}}{\Gamma(x)}$
11	$\frac{1}{(p-a)(p-b)}, \quad a \neq b$	$\frac{e^{at} - e^{bt}}{a - b}$
12	$\frac{p}{(p-a)(p-b)}, \quad a \neq b$	$\frac{ae^{at} - be^{bt}}{a - b}$

	$F(p) = \int_0^\infty e^{-pt} f(t)\, dt$	$f(t) = \mathcal{L}^{-1}\{F(p); t\}$
13	$\dfrac{1}{p^2 + a^2}$	$\dfrac{1}{a} \sin at$
14	$\dfrac{p}{p^2 + a^2}$	$\cos at$
15	$\dfrac{1}{p^2 - a^2}$	$\dfrac{1}{a} \sinh at$
16	$\dfrac{p}{p^2 - a^2}$	$\cosh at$
17	$\dfrac{1}{(p - a)^2 + b^2}$	$\dfrac{1}{b} e^{at} \sin bt$
18	$\dfrac{p - a}{(p - a)^2 + b^2}$	$e^{at} \cos bt$
19	$\dfrac{1}{p(p^2 + a^2)}$	$\dfrac{1}{a^2} (1 - \cos at)$
20	$\dfrac{1}{p^2(p^2 + a^2)}$	$\dfrac{1}{a^3} (at - \sin at)$
21	$\dfrac{1}{(p^2 + a^2)^2}$	$\dfrac{1}{2a^3} (\sin at - at \cos at)$
22	$\dfrac{p}{(p^2 + a^2)^2}$	$\dfrac{t}{2a} \sin at$
23	$\dfrac{p^2}{(p^2 + a^2)^2}$	$\dfrac{1}{2a} (\sin at + at \cos at)$
24	$\dfrac{p}{(p^2 + a^2)(p^2 + b^2)}, \qquad a^2 \neq b^2$	$\dfrac{1}{b^2 - a^2} (\cos at - \cos bt)$
25	$\dfrac{1}{p^4 + 4a^4}$	$\dfrac{1}{4a^3} (\sin at \cosh at - \cos at \sinh at)$
26	$\dfrac{p}{p^4 + 4a^4}$	$\dfrac{1}{2a^2} \sin at \sinh at$
27	$\dfrac{1}{p^4 - a^4}$	$\dfrac{1}{2a^3} (\sinh at - \sin at)$
28	$\dfrac{p}{p^4 - a^4}$	$\dfrac{1}{2a^2} (\cosh at - \cos at)$
29	$\dfrac{1}{\sqrt{p^2 + a^2}}$	$J_0(at)$
30	$\dfrac{1}{\sqrt{p}(p - a)}$	$\dfrac{1}{\sqrt{a}} e^{at} \operatorname{erf}(\sqrt{at})$
31	$\dfrac{1}{\sqrt{p}(\sqrt{p} + \sqrt{a})}$	$e^{at} \operatorname{erfc}(\sqrt{at})$

	$F(p) = \int_0^\infty e^{-pt} f(t)\, dt$	$f(t) = \mathcal{L}^{-1}\{F(p); t\}$
32	$e^{-ap}, \quad a > 0$	$\delta(t - a)$
33	$\frac{1}{p} e^{-ap}, \quad a > 0$	$h(t - a)$
34	$\frac{1}{p^2} e^{-ap}, \quad a > 0$	$(t - a)h(t - a)$
35	$\frac{1}{p} e^{-a/p}, \quad a > 0$	$J_0(2\sqrt{at})$
36	$\frac{1}{\sqrt{p}} e^{-a/p}, \quad a > 0$	$\frac{1}{\sqrt{\pi t}} \cos(2\sqrt{at})$
37	$e^{-a\sqrt{p}}, \quad a > 0$	$\frac{a}{2\sqrt{\pi t^3}} e^{-a^2/4t}$
38	$\frac{1}{\sqrt{p}} e^{-a\sqrt{p}}, \quad a \geqslant 0$	$\frac{1}{\sqrt{\pi t}} e^{-a^2/4t}$
39	$\frac{1}{p} e^{-a\sqrt{p}}, \quad a \geqslant 0$	$\operatorname{erfc}(a/2\sqrt{t})$
40	$\arctan(a/p)$	$\frac{\sin at}{t}$
41	$\frac{1}{p} \arctan(a/p)$	$\operatorname{Si}(at)$
42	$e^{a^2p^2} \operatorname{erfc}(as), \quad a > 0$	$\frac{1}{a\sqrt{\pi}} e^{-t^2/4a^2}$
43	$\frac{1}{p} e^{a^2p^2} \operatorname{erfc}(as), \quad a > 0$	$\operatorname{erf}(t/2a)$
44	$e^{ap} \operatorname{erfc}(\sqrt{ap}), \quad a > 0$	$\frac{\sqrt{a}}{\pi\sqrt{t}(t + a)}$
45	$\frac{1}{\sqrt{p}} e^{ap} \operatorname{erfc}(\sqrt{ap}), \quad a \geqslant 0$	$\frac{1}{\sqrt{\pi(t + a)}}$
46	$\operatorname{erf}(a/\sqrt{p})$	$\frac{1}{\pi t} \sin(2a\sqrt{t})$
47	$\frac{1}{\sqrt{p}} e^{a/p} \operatorname{erfc}(\sqrt{a/p}), \quad a \geqslant 0$	$\frac{1}{\sqrt{\pi t}} e^{-2\sqrt{at}}$

APPENDIX B

TABLE OF FOURIER TRANSFORMS

The following are short tables of Fourier transforms, cosine transforms, and sine transforms.

TABLE B-1 Fourier Transforms

	$f(t) = \frac{1}{\sqrt{2\pi}} \int_{-\infty}^{\infty} e^{-ist} F(s)\, ds$	$F(s) = \frac{1}{\sqrt{2\pi}} \int_{-\infty}^{\infty} e^{ist} f(t)\, dt$
1	1	$\sqrt{2\pi}\,\delta(s)$
2	$\frac{1}{t}$	$\sqrt{\pi/2}\, i \operatorname{sgn}(s)$
3	$\frac{1}{t^2 + a^2}, \quad a > 0$	$\sqrt{\pi/2}\, \frac{1}{a}\, e^{-a\|s\|}$
4	$\frac{t}{t^2 + a^2}, \quad a > 0$	$-\sqrt{\pi/2}\, \frac{is}{2a}\, e^{-a\|s\|}$
5	$\frac{t^2 - a^2}{t(t^2 + a^2)}, \quad a > 0$	$i\sqrt{2/\pi}\,(2e^{-a\|s\|} - 1) \operatorname{sgn}(s)$
6	e^{iat}	$\sqrt{2\pi}\,\delta(s + a)$
7	$e^{-a\|t\|}, \quad a > 0$	$\sqrt{2/\pi}\, \frac{a}{s^2 + a^2}$
8	$te^{-a\|t\|}, \quad a > 0$	$\sqrt{2/\pi}\, \frac{2ais}{(s^2 + a^2)^2}$
9	$\|t\| e^{-a\|t\|}, \quad a > 0$	$\sqrt{2/\pi}\, \frac{a^2 - s^2}{(s^2 + a^2)^2}$
10	$e^{-a^2 t^2}, \quad a > 0$	$\frac{1}{\sqrt{2}a}\, e^{-s^2/4a^2}$

(Continued)

TABLE B-1 Fourier Transforms *(Continued)*

	$f(t) = \frac{1}{\sqrt{2\pi}} \int_{-\infty}^{\infty} e^{-ist} F(s)\, ds$	$F(s) = \frac{1}{\sqrt{2\pi}} \int_{-\infty}^{\infty} e^{ist} f(t)\, dt$
11	$\cos\left(\frac{1}{2} t^2\right)$	$\frac{1}{\sqrt{2}} \left[\cos\left(\frac{1}{2} s^2\right) + \sin\left(\frac{1}{2} s^2\right) \right]$
12	$\sin\left(\frac{1}{2} t^2\right)$	$\frac{1}{\sqrt{2}} \left[\cos\left(\frac{1}{2} s^2\right) - \sin\left(\frac{1}{2} s^2\right) \right]$
13	$e^{-a\|t\|/\sqrt{2}}[\cos(at/\sqrt{2}) + \sin(a\|t\|/\sqrt{2})]$, $a > 0$	$\frac{2a^3}{\sqrt{\pi}} \frac{1}{s^4 + a^4}$
14	$e^{-\|t\|} \frac{\sin t}{t}$	$\frac{1}{\sqrt{2\pi}} \arctan(2/s^2)$
15	$\operatorname{sgn}(t)$	$\sqrt{\frac{2}{\pi}} \frac{i}{s}$
16	$t \operatorname{sgn}(t)$	$-\sqrt{\frac{2}{\pi}} \frac{1}{s^2}$
17	$h(1 - \|t\|)$	$\sqrt{\frac{2}{\pi}} \frac{\sin s}{s}$
18	$(1 - \|t\|)h(1 - \|t\|)$	$\frac{1}{\sqrt{2\pi}} \left(\frac{\sin s/2}{s/2}\right)^2$
19	$(a^2 - t^2)^{-1/2} h(a - \|t\|), \quad a > 0$	$\sqrt{\frac{\pi}{2}} J_0(as)$
20	$\delta(t - a)$	$\frac{1}{\sqrt{2\pi}} e^{ias}$

TABLE B-2 Cosine Transforms

	$f(t) = \sqrt{\frac{2}{\pi}} \int_0^\infty F_C(s) \cos st \, ds$	$F_C(s) = \sqrt{\frac{2}{\pi}} \int_0^\infty f(t) \cos st \, dt$
1	1	$\sqrt{2\pi}\,\delta(s)$
2	$\frac{1}{\sqrt{t}}$	$\frac{1}{\sqrt{s}}$
3	$\frac{1}{t^2 + a^2}, \quad a > 0$	$\sqrt{\frac{\pi}{2}} \frac{1}{a} e^{-as}$
4	$e^{-at}, \quad a > 0$	$\sqrt{\frac{2}{\pi}} \frac{a}{s^2 + a^2}$
5	$te^{-at}, \quad a > 0$	$\sqrt{\frac{2}{\pi}} \frac{a^2 - s^2}{(s^2 + a^2)^2}$
6	$e^{-a^2t^2}, \quad a > 0$	$\frac{1}{\sqrt{2}a} e^{-s^2/4a^2}$
7	$e^{-at} \cos at, \quad a > 0$	$\sqrt{\frac{2}{\pi}} \frac{as^2 + 2a^3}{s^4 + 4a^4}$
8	$e^{-at} \sin at, \quad a > 0$	$\sqrt{\frac{2}{\pi}} \frac{2a^3 - as^2}{s^4 + 4a^4}$
9	$\cos\left(\frac{1}{2} t^2\right)$	$\frac{1}{\sqrt{2}} \left[\cos\left(\frac{1}{2} s^2\right) + \sin\left(\frac{1}{2} s^2\right)\right]$
10	$\sin\left(\frac{1}{2} t^2\right)$	$\frac{1}{\sqrt{2}} \left[\cos\left(\frac{1}{2} s^2\right) - \sin\left(\frac{1}{2} s^2\right)\right]$

TABLE B-3 Sine Transforms

	$f(t) = \sqrt{\frac{2}{\pi}} \int_0^\infty F_S(s) \sin st \, ds$	$F_S(s) = \sqrt{\frac{2}{\pi}} \int_0^\infty f(t) \sin st \, dt$
1	$\frac{1}{t}$	$\sqrt{\frac{\pi}{2}} \operatorname{sgn}(s)$
2	$\frac{1}{\sqrt{t}}$	$\frac{1}{\sqrt{s}}$
3	$\frac{t}{t^2 + a^2}, \quad a > 0$	$\sqrt{\frac{\pi}{2}} e^{-as}$
4	$\frac{t}{(t^2 + a^2)^2}, \quad a > 0$	$\frac{1}{\sqrt{2\pi}} \frac{s}{a} e^{-as}$
5	$\frac{1}{t(t^2 + a^2)}, \quad a > 0$	$\sqrt{\frac{\pi}{2}} \frac{1}{a^2} (1 - e^{-as})$
6	$e^{-at}, \quad a > 0$	$\sqrt{\frac{2}{\pi}} \frac{s}{s^2 + a^2}$
7	$te^{-at}, \quad a > 0$	$\sqrt{\frac{2}{\pi}} \frac{2as}{(s^2 + a^2)^2}$
8	$te^{-a^2t^2}, \quad a > 0$	$\frac{s}{2\sqrt{2}a^3} e^{-s^2/4a^2}$
9	$\frac{1}{t} e^{-at}, \quad a > 0$	$\sqrt{\frac{2}{\pi}} \arctan(s/a)$
10	$e^{-at} \cos at, \quad a > 0$	$\sqrt{\frac{2}{\pi}} \frac{s^3}{s^4 + 4a^4}$
11	$e^{-at} \sin at, \quad a > 0$	$\sqrt{\frac{2}{\pi}} \frac{2a^2 s}{s^4 + 4a^4}$

CRAMER'S RULE

Determinants play a basic role in developing the theory of *linear systems of equations*. Such systems of equations arise in a wide variety of applications. In particular, we are frequently led to systems of linear equations in solving boundary value problems, and so a brief review of these equations may be helpful.

Suppose we wish to find values of x and y such that

$$\begin{aligned} a_1x + b_1y &= c_1, \\ a_2x + b_2y &= c_2. \end{aligned} \tag{1}$$

We assume that none of a_1, b_1, a_2, and b_2 are zero; otherwise, the system can be solved directly. To solve (1), we multiply the first equation by b_2 and the second equation by b_1, which gives us

$$\begin{aligned} a_1b_2x + b_1b_2y &= c_1b_2 \\ a_2b_1x + b_1b_2y &= c_2b_1. \end{aligned}$$

By subtracting the second equation from the first, we have

$$(a_1b_2 - a_2b_1)x = c_1b_2 - c_2b_1.$$

Thus we obtain the solution

$$x = \frac{c_1b_2 - c_2b_1}{a_1b_2 - a_2b_1} = \frac{\begin{vmatrix} c_1 & b_1 \\ c_2 & b_2 \end{vmatrix}}{\Delta}, \qquad \Delta \neq 0 \tag{2}$$

where Δ is the coefficient determinant defined by

$$\Delta = \begin{vmatrix} a_1 & b_1 \\ a_2 & b_2 \end{vmatrix}. \tag{3}$$

In a similar fashion, it can be shown that

$$y = \frac{\begin{vmatrix} a_1 & c_1 \\ a_2 & c_2 \end{vmatrix}}{\Delta}. \tag{4}$$

The solution formulas (2) and (4) constitute what is known as *Cramer's rule** for the solution of a system of two linear equations in two variables. Cramer's

* Named after GABRIEL CRAMER (1704–1752).

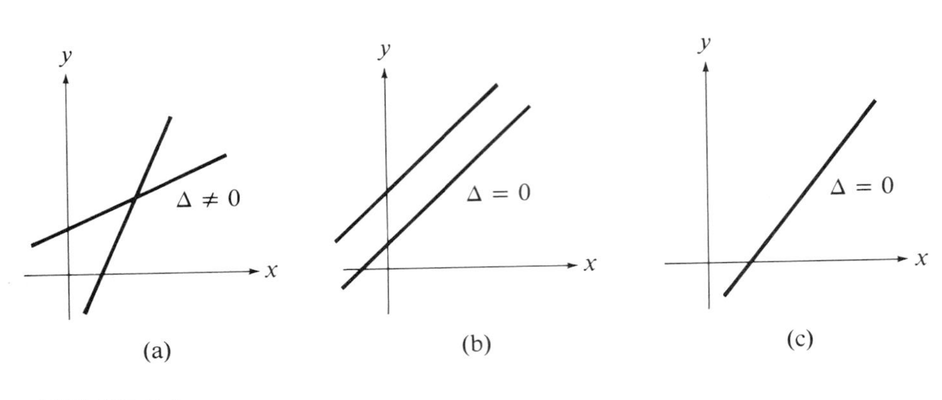

FIGURE C.1

(a) Unique solution; (b) no solution; (c) infinitely many solutions.

rule can easily be extended to n linear equations in n variables, although we will not do so here.

Geometrically, the system (1) represents two lines in the xy-plane. The solution is the point of intersection of the two lines. We say the system is *consistent* if it possesses at least one solution. If the two lines are distinct and nonparallel, exactly one solution exists [see Figure C.1(a)]. If the two lines are the same line, the solution set becomes infinite [see Figure C.1(c)]. If no solution of the system exists, the lines are parallel [see Figure C.1(b)] and the system is said to be *inconsistent*.

When $c_1 = c_2 = 0$, the system (1) is called *homogeneous*, and *nonhomogeneous* otherwise. For a homogeneous system it is clear that $x = y = 0$ is always a solution, called the *trivial solution*. Thus, *a homogeneous system is always consistent*. From a geometrical point of view, the homogeneous system represents two lines passing through the origin. Only if one line lies directly on top of the other will a nontrivial solution exist, and in this case there will be an infinite number of solutions. If the two lines are indeed the same, then the two equations in (1) must be proportional and this implies $\Delta = 0$.*

The nonhomogeneous system can be interpreted as a translation of the two lines defined by the homogeneous system. Hence, if only the trivial solution is possible for the homogeneous system ($c_1 = c_2 = 0$), the nonhomogeneous system will have a *unique solution*. If a nontrivial solution of the homogeneous system exists, however, the nonhomogeneous system will either have *no solution* or *infinitely many solutions* since the two lines must remain at least parallel. Let us summarize these conclusions.

* If two rows (or columns) of a determinant are proportional, the value of the determinant is zero.

Homogeneous System $(c_1 = c_2 = 0)$:

(a) If $\Delta \neq 0$, only the trivial solution exists.

(b) If $\Delta = 0$, infinitely many solutions exist.

Nonhomogeneous System $(c_1^2 + c_2^2 \neq 0)$:

(a) If $\Delta \neq 0$, a unique solution exists (i.e., the related homogeneous system has only the trivial solution).

(b) If $\Delta = 0$, either no solution or infinitely many solutions exist (i.e., the related homogeneous system has a nontrivial solution).

ANSWERS TO SELECTED ODD-NUMBERED PROBLEMS

CHAPTER 1

EXERCISES 1.2

1. $y = T_1 + (T_2 - T_1)\dfrac{x}{b}$

3. $y = \dfrac{1}{2}T_1(1 - x)$

5. $y = T_1\left[\cosh x + \dfrac{(1 - \cosh b)}{\sinh b}\sinh x\right]$

7. (a) $y = T_0 + (T_1 - T_0)\left[\cosh\sqrt{h}x - \left(\dfrac{h\cosh\sqrt{h}b + \sqrt{h}\sinh\sqrt{h}b}{h\sinh\sqrt{h}b + \sqrt{h}\cosh\sqrt{h}b}\right)\sinh\sqrt{h}x\right]$

   (b) $y = T_0 + (T_1 - T_0)(\cosh\sqrt{h}x - \sinh\sqrt{h}x)$

9. Only if $T_1 = T_2$.

EXERCISES 1.3

1. no nontrivial solutions

3. $y = C\cos 3x$, C arbitrary

5. $y = C_1\cos\pi x + C_2\sin\pi x$, C_1, C_2 arbitrary

7. no nontrivial solutions

9. no solution

11. $y = \dfrac{2\sin kx}{\sin k}$, unique if $k \neq n\pi$, $n = 0, 1, 2, \ldots$

13. $y = \dfrac{1}{2e}(3\sinh x - \cosh x)$, unique

15. $y = 3xe^{3(x-1)}$, unique

19. $k = n\pi$, $n = 1, 2, 3, \ldots$; $y = C\sin n\pi x$, C arbitrary

23. (a) $b - a = n\pi$, $n = 1, 3, 5, \ldots$
    (b) $b - a \neq n\pi$, $n = 1, 2, 3, \ldots$
    (c) $b - a = n\pi$, $n = 2, 4, 6, \ldots$

25. no solution

27. $y = -\dfrac{3}{\pi^2}\sin\pi x + \dfrac{4x}{\pi} - 2$, unique

29. $y = \dfrac{1}{5}e^{-x} - \dfrac{6}{5}\cos 2x$, unique

31. $y = (x - 1)e^{-x} + x - 2$, unique

33. $y = \dfrac{1}{6}\left[\left(\dfrac{e^3 - e - 6}{e^3 - e^{-2}}\right)e^{-2x} + \left(\dfrac{e - e^{-2} + 6}{e^3 - e^{-2}}\right)e^{3x} - e^x\right]$, unique

35. $y = Cx^3 \sin(4 \log x)$, C arbitrary

37. $y = \left(1 - \frac{1}{2} \log x\right) x$

39. $y = \frac{4}{5} x^{1/2}(x - 1)$

EXERCISES 1.4

1. $\lambda_n = \frac{1}{4}(2n - 1)^2\pi^2$, $\phi_n(x) = \sin \frac{1}{2}(2n - 1)\pi x$, $n = 1, 2, 3, \ldots$

3. $\lambda_n = n^2$, $\phi_n(x) = \cos nx$, $n = 0, 1, 2, \ldots$

5. $\lambda_n = -n^2\pi^2$, $\phi_n(x) = e^{-x} \sin n\pi x$, $n = 1, 2, 3, \ldots$

7. $\lambda_n = \frac{1}{4} + n^2\pi^2$, $\phi_n(x) = e^{-x/2} \sin n\pi x$, $n = 1, 2, 3, \ldots$

9. $\lambda_n = \frac{9}{8} + \frac{n^2\pi^2}{2}$, $\phi_n(x) = e^{3x/2} \sin n\pi x$, $n = 1, 2, 3, \ldots$

11. $\lambda_n = n^2\pi^2$, $\phi_n(x) = \sin(n\pi \log x)$, $n = 1, 2, 3, \ldots$

13. $\lambda_n = \frac{(2n - 1)^2\pi^2}{4(\log 2)^2}$, $\phi_n(x) = \cos\left[\frac{(2n - 1)\pi \log x}{2 \log 2}\right]$, $n = 1, 2, 3, \ldots$

15. (b) negative eigenvalue is $\lambda = -k^2$, where k is the only solution of $\tanh k = k/h$.

17. (a) $\lambda_2 \simeq 24.472$ (b) $\lambda_2 \simeq 24.139$

19. 1.875

21. $\lambda_1 \simeq 0.620$, $\phi_n(x) = \sin k_n x$, where $\tan k_n\pi = -k_n$, $n = 1, 2, 3, \ldots$

23. no real eigenvalues

25. $\lambda_0 = 0$, $\phi_0(x) = x - 1$

27. for $\lambda = -k^2 < 0$, $y = \frac{\sinh kx}{\sinh k}$;
for $\lambda = 0$, $y = x$;
for $\lambda = k^2 > 0$, $k \neq n\pi$, $y = \frac{\sin kx}{\sin k}$, $n = 1, 2, 3, \ldots$

29. for $\lambda = -k^2 < 0$, $y = e^{-2x}\left[\frac{\sinh 3k(2 - x)}{\sinh 6k}\right]$;
for $\lambda = 0$, $y = \left(1 - \frac{x}{2}\right)e^{-2x}$;
for $\lambda = k^2 > 0$, $k \neq \frac{n\pi}{6}$, $y = e^{-2x}\left[\frac{\sin 3k(2 - x)}{\sin 6k}\right]$, $n = 1, 2, 3, \ldots$

31. $\lambda_n = n^4$, $\phi_n(x) = \sin nx$, $n = 1, 2, 3, \ldots$

33. $\lambda_n = k_n^4$, where $\cosh k_n \cos k_n = 1$, $n = 1, 2, 3, \ldots$
$\phi_n(x) = \sinh k_n(1 - x) + \sin k_n(1 - x) + \cosh k_n \sin k_n x$
$- \sin k_n \cosh k_n x + \cos k_n \sinh k_n x - \sinh k_n \cos k_n x$

EXERCISES 1.5

1. (a) $y = \frac{x}{6}(\pi^2 + 1 - x^2)$, $\quad y_{max} = \frac{(\pi^2+1)^{3/2}}{9\sqrt{3}}$

   (b) $y = \frac{1}{6}(\pi^3 - x^3)$, $\quad y_{max} = \frac{\pi^3}{6}$

   (c) no solution

3. $y = \sin x$, $\quad y_{max} = 1$

5. $y = \frac{1}{12}\left[\frac{1}{16} - \left(x - \frac{1}{2}\right)^4\right]$, $\quad y_{max} = \frac{1}{192}$

7. (a) $y = \frac{mg}{2T}x(1-x) + (\beta - \alpha)x + \alpha$

   (c) $y\left(\frac{1}{2}\right) = \frac{mg}{8T}$ from (a)

   $y\left(\frac{1}{2}\right) = \frac{mg}{8T} + \frac{1}{48}\left(\frac{mg}{2T}\right)^3$ from (b)

   (d)

x	(a)	(b)
0	0	0
0.2	0.08	0.082
0.4	0.12	0.123
0.6	0.12	0.123
0.8	0.08	0.083
1.0	0	0

11. (a) $y = C_1 \sin \frac{(2n-1)\pi x}{2b}$, $\quad n = 1, 2, 3, \ldots$

    (b) $y = C_1 \cos \frac{(2n-1)\pi x}{2b}$, $\quad n = 1, 2, 3, \ldots$

    (c) $y = C_1 \cos \frac{n\pi x}{b}$, $\quad n = 0, 1, 2, \ldots$

15. $\omega_1 = \frac{1}{2}\sqrt{\frac{1+4\pi^2}{\rho}}$, $\quad \phi_1(x) = Cx^{-1/2}\sin(\pi \log x)$

17. (a) $y = \frac{x^4}{24} - \frac{x^3}{12} + \frac{x}{24}$ $\quad$ (b) $M(x) = EI\frac{x}{2}(x-1)$

    (c) $V(x) = EI\left(x - \frac{1}{2}\right)$

19. (a) $y = \frac{x^5}{120} - \frac{x^3}{60} + \frac{x}{120}$ $\quad$ (b) $M(x) = EI\frac{x}{2}\left(\frac{x^2}{3} - \frac{1}{5}\right)$

    (c) $V(x) = \frac{EI}{2}\left(x^2 - \frac{1}{5}\right)$

21. (a) $y = \sin x$ $\quad$ (b) $M(x) = -EI \sin x$ $\quad$ (c) $V(x) = -EI\cos x$

23. $x = 0.447$, $\quad y_{max} = 0.0024$

25. (b) $\phi_n(x) = 1 - \cos 2n\pi x$, $\quad n = 1, 2, 3, \ldots$

    $\phi_n(x) = (\sin k_n - k_n)(1 - \cos k_n x) - (1 - \cos k_n)(\sin k_n x - k_n)$,

    where $\tan(k_n/2) = k_n/2$, $\quad n = 1, 2, 3, \ldots$

27. $\omega_n = \frac{n^2\pi^2}{16}\sqrt{\frac{EI}{\rho}}$, $\quad n = 1, 2, 3, \ldots$

EXERCISES 1.6

1. $y'' + \frac{\lambda}{x} y = 0$

3. $\frac{d}{dx}(xe^{-x}y') + \lambda e^{-x}y = 0$

5. $\frac{d}{dx}(xy') + \left(\lambda x - \frac{n^2}{x}\right)y = 0$

9. (a) $C_n = \sqrt{\frac{2}{\pi}}, \quad n = 1, 2, 3, \ldots$

   (b) $C_0 = \frac{1}{\sqrt{\pi}}, \quad C_n = \sqrt{\frac{2}{\pi}}, \quad n = 1, 2, 3, \ldots$

11. $p_0(x) = 1, \quad p_1(x) = 1 - x, \quad p_2(x) = 1 - 2x + \frac{1}{2}x^2$

13. $\int_0^1 (1 + x)\phi_n(x)\phi_k(x)\, dx = 0, \quad n \neq k$

15. $\int_0^1 x\phi_n(x)\phi_k(x)\, dx = 0, \quad n \neq k$

17. $\lambda_0 = 0, \quad \phi_0(x) = 1$

    $\lambda_n = n^2 + \frac{9}{4}, \quad \phi_n(x) = e^{3x/2}\left(\sin nx - \frac{2x}{3}\cos nx\right), \quad n = 1, 2, 3, \ldots$

19. $\lambda_n = 16n^2, \quad \phi_n(x) = \sin[4n(\tan^{-1} x)], \quad n = 1, 2, 3, \ldots$

21. $\lambda_n = n^2\pi^2, \quad \phi_n(x) = \cos n\pi x, \quad \psi_n(x) = \sin n\pi x, \quad n = 0, 1, 2, \ldots$

23. $\lambda_n = 4n^2\pi^2, \quad \phi_n(x) = \cos 2n\pi x, \quad \psi_n(x) = \sin 2n\pi x, \quad n = 0, 1, 2, \ldots$

25. $\alpha = \int_{x_1}^{x_2} r(x)\phi_n(x)\psi_n(x)\, dx \Big/ \int_{x_1}^{x_2} r(x)[\phi_n(x)]^2\, dx$

27. $\lambda = k^2, \quad \phi(x) = \sin(k \log x),$ k any positive number

29. $y(x), \quad y'(x)$ finite as $x \to -1^+, \quad x \to 1^-$

EXERCISES 1.7

1. $c_1 = 0.286$

3. $c_1 = 0.247$

5. $c_1 = -1.535, \quad c_2 = 0.995$

7. $c_1 = 0.278$

9. $c_1 = 0.249$

11. $c_1 = -1.609, \quad c_2 = 0.999$

13. (a) $c_1 = bq_0/\sqrt{3}(K + EI\pi^4/b^4), \quad c_2 = bq_0/3\sqrt{3}(K + 16EI\pi^4/b^4)$

    (b) $c_1 = 2bq_0/\pi(K + EI\pi^4/b^4), \quad c_2 = bq_0/\pi(K + 16EI\pi^4/b^4)$

15. $\Lambda_1 = \frac{8}{3} \simeq 2.667$

17. $\Lambda_1 = \frac{16}{3} \simeq 5.333$

19. $\Lambda_1 = \frac{5}{2} = 2.5$

21. $\Lambda_1 = 6$

23. (a) $\Lambda_1 = 3.804$, $\psi_1 = x(1-x) - 0.634x^2(1-x)$
$\Lambda_2 = 14.196$, $\psi_2 = 0.423x(1-x) - x^2(1-x)$
(b) $\Lambda_1 = 4.237$, $\psi_1 = \sin \pi x + 0.106 \sin 2\pi x$
$\Lambda_2 = 20.437$, $\psi_2 = 0.424 \sin \pi x - \sin 2\pi x$

25. $\Lambda_0 = \dfrac{\omega H}{2} + \dfrac{3\epsilon H^2}{4m^2\omega^2}$

CHAPTER 2

EXERCISES 2.2

1. $g_1(t, \tau) = t - \tau$

3. $g_1(t, \tau) = \dfrac{1}{\sqrt{5}} \sin \sqrt{5}(t - \tau)$

5. $g_1(t, \tau) = 2e^{(t-\tau)} \sin \dfrac{1}{2}(t - \tau)$

7. $g_1(t, \tau) = \dfrac{\tau}{8}\left[\left(\dfrac{t}{\tau}\right)^4 - \left(\dfrac{\tau}{t}\right)^4\right]$

9. $g_1(t, \tau) = \dfrac{1}{2}(1 - \tau^2) \log \dfrac{(1+t)(1-\tau)}{(1-t)(1+\tau)}$

11. $y = e^t - 1$

13. $y = \dfrac{1}{25}[(60e^2 + 9 - 5t)e^{-t} + (15e^{-8} + e^{-10})e^{4t}]$

15. $y = 2 + 2e^{-t/2} \sinh \dfrac{3t}{2}$

17. $y = \dfrac{1}{6} + \dfrac{1}{3}e^{-t} + e^{-2t}\left[\dfrac{1}{2}\cos \sqrt{2}t - \dfrac{8}{3\sqrt{2}} \sin \sqrt{2}t\right]$

19. $y = e^{-t}\left(\dfrac{1}{2}t^3 + 6t + 4\right)$

21. $y = \dfrac{1}{24}t^{-5} - \dfrac{1}{8}t^{-1} + \dfrac{1}{12}t$

23. $y = \dfrac{1}{12}t^{-2} + \dfrac{1}{18}t^3 + \dfrac{1}{36} - \dfrac{1}{6}\log t$

25. $y = t^{-1/2}\left[\log t + \dfrac{1}{4}(\log t)^3\right]$

35. (b) $C_1 = \dfrac{\begin{vmatrix} k_0 & y_2(t_0) & y_3(t_0) \\ k_1 & y_2'(t_0) & y_3'(t_0) \\ k_2 & y_2''(t_0) & y_3''(t_0) \end{vmatrix}}{W(y_1, y_2, y_3)(t_0)}$, $C_2 = \dfrac{\begin{vmatrix} y_1(t_0) & k_0 & y_3(t_0) \\ y_1'(t_0) & k_1 & y_3'(t_0) \\ y_1''(t_0) & k_2 & y_3''(t_0) \end{vmatrix}}{W(y_1, y_2, y_3)(t_0)}$

$C_3 = \dfrac{\begin{vmatrix} y_1(t_0) & y_2(t_0) & k_0 \\ y_1'(t_0) & y_2'(t_0) & k_1 \\ y_1''(t_0) & y_2''(t_0) & k_2 \end{vmatrix}}{W(y_1, y_2, y_3)(t_0)}$

37. $g_1(t, \tau) = \dfrac{1}{2}(e^{t-\tau} - e^{\tau-t}) - (t - \tau)$

39. $$g_1(t, \tau) = \frac{(a-1)e^{b(t-\tau)} - (b-1)e^{a(t-\tau)} + (b-a)e^{t-\tau}}{(b-a)(a-1)(b-1)},$$

where $a = -\frac{1}{4}(3 + \sqrt{33})$ and $b = -\frac{1}{4}(3 - \sqrt{33})$

41. $$g_1(t, \tau) = \frac{1}{2}e^{t-\tau} - e^{2(t-\tau)} + \frac{1}{2}e^{3(t-\tau)}$$

43. $$y = \frac{1}{4}\cos 2t + \frac{1}{2}\sin 2t - \frac{1}{4}$$

EXERCISES 2.3

1. $$y = \begin{cases} e^{-t}(\cos t + \sin t), & 0 \leq t < \pi \\ e^{-t}(\cos t + \sin t) - e^{-(t-\pi)}\sin t, & t > \pi \end{cases}$$

3. $$y = \begin{cases} \cos t - \cos 2t, & 0 \leq t < \pi \\ \cos t - \cos 2t - \sin t, & t > \pi \end{cases}$$

5. $$y = \begin{cases} 2\sin t, & 0 \leq t < \pi/2 \\ 2\sin t - 5\cos t, & t > \pi/2 \end{cases}$$

15. $$g(t, \tau) = \begin{cases} e^{2(t-\tau)}\sin(t - \tau), & t_0 \leq \tau \leq t \\ 0, & t < \tau < \infty \end{cases}$$

EXERCISES 2.4

1. $y'' + x^{-1}y = x$

3. $$\frac{d}{dx}(x^{-1}y') + x^{-3}y = x^{-2}e^{-x}$$

5. $$\frac{d}{dx}(xy') + xy = 4x^2e^{-2x}$$

7. (a) $$g(x, s) = \begin{cases} \dfrac{(s-a)(x-b)}{a-b}, & a < s \leq x \\ \dfrac{(x-a)(s-b)}{a-b}, & x < s < b \end{cases}$$

(b) $$g(x, s) = \begin{cases} s, & 0 < s \leq x \\ x, & x < s < 1 \end{cases}$$

(c) $$g(x, s) = \begin{cases} 1 - x, & 0 < s \leq x \\ 1 - s, & x < s < 1 \end{cases}$$

(d) $g(x, s)$ does not exist

9. (a) $$g(x, s) = \begin{cases} \dfrac{\sinh s \sinh(1-x)}{\sinh 1}, & 0 < s \leq x \\ \dfrac{\sinh x \sinh(1-s)}{\sinh 1}, & x < s < 1 \end{cases}$$

(b) $$g(x, s) = \begin{cases} e^{1-s}\sinh(1-x), & 0 < s \leq x \\ e^{1-x}\sinh(1-s), & x < s < 1 \end{cases}$$

11. $g(x, s) = \begin{cases} -\dfrac{1}{7} s^6(x^6 - x^{-1}), & 0 < s \leq x \\ -\dfrac{1}{7} x^6(s^6 - s^{-1}), & x < s < 1 \end{cases}$

13. $g(x, s) = \begin{cases} \dfrac{(sx)^{1/2} \sinh\left(\dfrac{\sqrt{5}}{2} \log s\right) \sinh\left[\dfrac{\sqrt{5}}{2} (\log x - 1)\right]}{\dfrac{\sqrt{5}}{2} \sinh \dfrac{\sqrt{5}}{2}}, & 1 < s \leq x \\ \dfrac{(xs)^{1/2} \sinh\left(\dfrac{\sqrt{5}}{2} \log x\right) \sinh\left[\dfrac{\sqrt{5}}{2} (\log s - 1)\right]}{\dfrac{\sqrt{5}}{2} \sinh \dfrac{\sqrt{5}}{2}}, & x < s < e \end{cases}$

15. $g(x, s) = \begin{cases} \dfrac{2(sx)^{-1/2}}{\sqrt{3}} \sin\left(\dfrac{\sqrt{3}}{2} \log s\right) \cos\left(\dfrac{\sqrt{3}}{2} \log x\right), & 1 < s \leq x \\ \dfrac{2(xs)^{-1/2}}{\sqrt{3}} \sin\left(\dfrac{\sqrt{3}}{2} \log x\right) \cos\left(\dfrac{\sqrt{3}}{2} \log s\right), & x < s < e^{\pi/\sqrt{3}} \end{cases}$

17. $y = \dfrac{1}{2} x - \dfrac{1}{6} x^3$

19. $y = -\dfrac{1}{\pi} x - \dfrac{1}{\pi^2} \sin \pi x$

21. $y = 1$

23. $y = 2 \cosh x + (3 - 2 \cosh 1) \dfrac{\sinh x}{\sinh 1} - x^2 - 2$

25. $y = \dfrac{3 \sin 2(2 - x)}{4 \sin 4} + \dfrac{1}{4}\left[1 - \dfrac{\sin 2x}{\sin 4}\right] + g(x, 1)$

27. The Green's function does not exist, but the problem has infinitely many solutions given by (C arbitrary)

$$y = 2x(x - 1)e^x + C(x - x^3).$$

33. (a) no
(b) yes, infinitely many given by $y = C - x^2/2$ (C arbitrary)

35. no

37. (a) $g(x, s) = -\log x, \quad x > s, \quad p(x) = x$
(b) $g(x, s) = \dfrac{s}{2x} (1 - x^2), \quad x > s, \quad p(x) = x$

39. (b) $y = -\log(1 + x)$ (c) $y = -x/2$

43. (a) $y = \dfrac{x}{6T} (1 - x^2) + \dfrac{P}{T} g\left(x, \dfrac{1}{2}\right)$ (b) $y = \dfrac{x}{6T} (1 - x^2) + \dfrac{P}{T} g\left(x, \dfrac{1}{4}\right)$

CHAPTER 3

EXERCISES 3.2

1. hyperbolic: $u(x, y) = F(x + y) + G(2x + y)$
3. elliptic: $u(x, y) = F(y + iax) + G(y - iax)$

5. parabolic: $u(x, t) = F(2x + 3t) + tG(2x + 3t)$

7. elliptic: $u(x, y) = F[y - (1 - 2i)x] + G[y - (1 + 2i)x]$

9. hyperbolic for $x > 0$, elliptic for $x < 0$, and parabolic for $x = 0$

11. hyperbolic for $x^2 + y^2 > 1$, elliptic for $x^2 + y^2 < 1$, and parabolic for $x^2 + y^2 = 1$

13. elliptic for all x and y

19. (a) $u(x, t) = \frac{1}{2}\left[\frac{1}{1 + 2(x + ct)^2} + \frac{1}{1 + 2(x - ct)^2}\right]$

EXERCISES 3.3

3. $\lambda = k^2 > 0$: $u(x, y) = (C_1 \cos kx + C_2 \sin kx)(C_3 \cosh ky + C_4 \sinh ky)$
   $\lambda = 0$: $u(x, y) = (C_1 + C_2x)(C_3 + C_4y)$
   $\lambda = -k^2 < 0$: $u(x, y) = (C_1 \cosh kx + C_2 \sinh kx)(C_3 \cos ky + C_4 \sin ky)$

5. $v'' + n^2v = 0$; $v(r) = C_1 \cos nr + C_2 \sin nr$

7. $u(x, y) = 2e^{3(x+y)}$

9. $u(x, t) = \sin 3t \cos 3t$

11. $u(x, t) = \frac{b \sin(1 - x)}{\cos 1} e^{-kt}$

13. $u(x, y) = x \sin y$

15. $u(x, y) = x^2y^2 + ye^x + \frac{1}{2}y^2 - y + 2$

17. $u(x, y) = \left(2x + \frac{1}{3}x^3\right)e^y - 2x$

CHAPTER 4

EXERCISES 4.2

1. period can be any real (nonzero) number

5. $f(x) = (x - 2\pi)^2, \quad \pi \leqslant x < 3\pi$

7. $f(x) = \frac{1}{2} + \frac{6}{\pi}\sum_{n=1}^{\infty} \frac{\sin(2n - 1)x}{2n - 1}$

9. $f(x) = 1 - \cos 2x$

11. $f(x) = \frac{2 \sinh \pi}{\pi}\left\{\frac{1}{2} + \sum_{n=1}^{\infty}\left[\frac{(-1)^n}{n^2 + 1} \cos nx + \frac{(-1)^{n-1}n}{n^2 + 1} \sin nx\right]\right\}$

13. $f(x) = \frac{\pi^2}{6} + \sum_{n=1}^{\infty}\left\{\frac{2(-1)^n}{n^2} \cos nx + \left(\frac{(-1)^{n-1}\pi}{n} + \frac{2}{\pi n^3}[(-1)^n - 1]\right) \sin nx\right\}$

15. $f(x) = 2\sum_{\substack{n=1 \\ \text{(odd)}}}^{\infty}\left(\frac{2 \cos nx}{\pi n^2} + \frac{\sin nx}{n}\right)$

17. odd

19. odd

21. neither

23. even

25. even

27. $f(x) = 2\sum_{n=1}^{\infty} \frac{(-1)^{n-1}}{n} \sin nx$

29. $f(x) = -2\sum_{n=1}^{\infty} \frac{1}{n} \sin nx$

31. $f(x) = \frac{1}{\pi}(1 - e^{-\pi}) + \frac{2}{\pi}\sum_{n=1}^{\infty}\left(\frac{1-(-1)^n e^{-\pi}}{n^2+1}\right)\cos nx$

37. (a) $f(x) = 3 + (x^5 - 7x)$ (b) $f(x) = \cosh x + \sinh x$

(c) $f(x) = \frac{x^2}{x^2-1} + \frac{x}{x^2-1}$

EXERCISES 4.3

1. smooth
3. none of these
5. smooth
7. piecewise smooth
11. (a) 1/2 (b) 5/2 (c) 2 (d) 1/2 (e) 1
15. problems 9, 11, 13
17. $f(x) = \frac{\pi^2}{3} + 4\sum_{n=1}^{\infty}\frac{(-1)^n}{n^2}\cos nx$

EXERCISES 4.5

1. $f(x) = \frac{1}{3} + \frac{4}{\pi^2}\sum_{n=1}^{\infty}\frac{(-1)^n}{n^2}\cos n\pi x$

3. $f(x) = \frac{3}{4} - \frac{1}{\pi}\sum_{n=1}^{\infty}\frac{\sin(n\pi/2)}{n}\cos\frac{n\pi x}{2} + \frac{1}{\pi}\sum_{n=1}^{\infty}\frac{1}{n}\left[\cos\frac{n\pi}{2} + 1 - 2(-1)^n\right]\sin\frac{n\pi x}{2}$

5. $f(x) = \frac{3}{2} - \frac{1}{\pi}\sum_{n=1}^{\infty}\frac{\sin 2n\pi x}{n}$

9. $f(t) = \frac{E}{\pi} + \frac{2E}{\pi}\sum_{n=1}^{\infty}\frac{(-1)^n}{1-4n^2}\cos 200n\pi x$

11. $f(t) = \frac{2E}{\pi} + \frac{4E}{\pi}\sum_{n=1}^{\infty}\frac{\cos 2n\omega t}{1-4n^2}$

13. (a) $f(x) = \frac{\pi}{2} + \frac{4}{\pi}\sum_{n=1}^{\infty}\frac{\cos(2n-1)x}{(2n-1)^2}$ (b) $f(x) = 2\sum_{n=1}^{\infty}\frac{\sin nx}{n}$

(c) $f(x) = \frac{\pi}{2} + \sum_{n=1}^{\infty}\frac{\sin 2nx}{n}$

15. (a) $f(x) = 1$ (b) $f(x) = \frac{4}{\pi}\sum_{n=1}^{\infty}\frac{\sin(2n-1)\pi x}{2n-1}$

17. (a) $f(x) = 1 + \frac{8}{\pi^2}\sum_{n=1}^{\infty}\frac{\cos(n-\frac{1}{2})\pi x}{(2n-1)^2}$ (b) $f(x) = \frac{4}{\pi}\sum_{n=1}^{\infty}\frac{(-1)^{n-1}}{n}\sin\frac{n\pi x}{2}$

19. (a) $f(x) = \frac{2}{\pi} + \frac{4}{\pi}\sum_{n=1}^{\infty}\frac{\cos 2nx}{1-4n^2}$ (b) $f(x) = \sin x$

21. (a) $f(x) = \frac{e^2-1}{2} + 4\sum_{n=1}^{\infty}\left(\frac{e^2(-1)^n - 1}{n^2\pi^2+4}\right)\cos n\pi x$

(b) $f(x) = 2\pi\sum_{n=1}^{\infty}\left(\frac{n[1-e^2(-1)^n]}{n^2\pi^2+4}\right)\sin n\pi x$

EXERCISES 4.6

3. $\phi_0(x) = \dfrac{1}{\sqrt{\pi}}, \qquad \phi_n(x) = \sqrt{\dfrac{2}{\pi}} \cos n\pi x, \qquad n = 1, 2, 3, \ldots$

5. $\phi_n(x) = \dfrac{\sqrt{2} \sin k_n x}{\sqrt{1 + \cos^2 k_n}}, \qquad n = 1, 2, 3, \ldots,$

   where each k_n satisfies $\sin k_n + k_n \cos k_n = 0$

7. $f(x) = \dfrac{4}{\pi} \sum_{n=1}^{\infty} \dfrac{\sin(n - \frac{1}{2})\pi x}{(2n - 1)}$

9. $f(x) = 1 + 2 \sum_{n=1}^{\infty} \dfrac{(-1)^{n-1}}{n} \sin nx$

11. $f(x) = 2\pi e^{3x/2} \sum_{n=1}^{\infty} \left(\dfrac{n}{n^2\pi^2 + \frac{9}{4}}\right)[1 + (-1)^{n-1}e^{-3/2}] \sin n\pi x$

13. $f(x) = \dfrac{2}{\pi} \sum_{n=1}^{\infty} \dfrac{(-1)^{n-1}}{n} \sin(n\pi \log x)$

15. $y = \dfrac{2}{\pi^3} \sum_{n=1}^{\infty} \dfrac{(-1)^{n-1}}{n^3} \sin n\pi x$

17. $y = 1 + \sin x + \left(\dfrac{\cos 1 - 1}{\sin 1}\right) \cos x$

19. $y = 2 \cosh(x - 1) + 2 \sum_{n=1}^{\infty} \left(\dfrac{(-1)^{n-1}e + (2n - 1)\pi/2}{[1 + (2n - 1)^2\pi^2/4]^2}\right) \sin\left[\dfrac{(2n - 1)\pi x}{2}\right]$

21. $y = 3(1 - x)e^{-x} - \dfrac{2}{\pi} e^{-x} \sum_{n=1}^{\infty} \dfrac{1 - (-1)^n e^2}{n(n^2\pi^2 + 4)} \sin n\pi x$

23. $y = \dfrac{1}{2} x(3 - x^2) - \dfrac{8x^2}{\pi} \sum_{n=1}^{\infty} \dfrac{\sin[(2n - 1)\pi(\log x)/(\log 2)]}{(2n - 1)[1 + (2n - 1)^2\pi^2/(\log 2)^2]}$

25. $y = -\dfrac{\cos(\log x)}{\sin 2} + 1 + \dfrac{8}{\pi^2} \sum_{\substack{n=1 \\ (\text{odd})}}^{\infty} \left[\dfrac{1}{n^2[(n^2\pi^2/4) - 1]}\right] \cos\left(\dfrac{n\pi \log x}{2}\right)$

27. $y = C - \dfrac{1}{\pi^2} \cos \pi x, \qquad C$ arbitrary

29. no solution

31. $y = C \sin(4 \log x) + \dfrac{1}{7} \sin(3 \log x), \qquad C$ arbitrary

33. $g(x, s) = \dfrac{2(b - a)}{\pi^2} \sum_{n=1}^{\infty} \dfrac{\sin[n\pi(x - a)/(b - a)] \sin[n\pi(s - a)/b - a)]}{n^2}$

35. $g(x, s) = 8 \sum_{n=1}^{\infty} \dfrac{\cos(n - \frac{1}{2})\pi x \cos(n - \frac{1}{2})\pi s}{(2n - 1)^2\pi^2 - 4}$

37. $g(x, s) = \dfrac{1}{\pi} + \dfrac{2}{\pi} \sum_{n=1}^{\infty} \dfrac{\cos nx \cos ns}{n^2 + 1}$

39. $g(x, s) = 2 \sum_{n=1}^{\infty} \dfrac{\sin(n\pi \log x) \sin(n\pi \log s)}{n^2\pi^2 - 1}$

EXERCISES 4.7

5. $f(x) = \frac{4}{\pi} \int_0^\infty \left(\frac{\sin s - s \cos s}{s^3} \right) \cos sx \, ds, \qquad I = \frac{3\pi}{16}$

11. $\frac{1}{2} [\mathrm{Si}(b^2) - \mathrm{Si}(a^2)]$

13. $\mathrm{Si}(\mu x) - \frac{1}{2} \{\mathrm{Si}[\mu(x + 1)] + \mathrm{Si}[\mu(x - 1)]\}$

15. $\frac{1}{2} \cos 2[\mathrm{Ci}(8) - \mathrm{Ci}(6) - \mathrm{Ci}(4) + \mathrm{Ci}(2)] + \frac{1}{2} \sin 2[\mathrm{Si}(8) - \mathrm{Si}(6) + \mathrm{Si}(4) - \mathrm{Si}(2)]$

CHAPTER 5

EXERCISES 5.2

1. $\frac{2}{p^3}$

3. $\frac{p}{p^2 + k^2}$

5. $\frac{15}{8p^3} \sqrt{\frac{\pi}{p}}$

7. $\frac{1}{(p - 2)^2}$

9. $\frac{p^2 + 2k^2}{p(p^2 + 4k^2)}$

11. $\frac{e^{-ap}}{p}$

EXERCISES 5.3

1. $\frac{3}{(p - 2)^2}$

3. $\frac{6}{(p - 1)^2 + 9}$

5. $\frac{p^3}{p^4 + 4k^4}$

11. $\frac{p}{p^2 + 16}$

13. $\frac{2p}{(p^2 + 1)^2}$

15. $\frac{(p + 2)^2 - 1}{[(p + 2)^2 + 1]^2}$

17. $\frac{3(p + 4)}{(p + 4)^2 + 16} - \frac{24(p + 4)}{[(p + 4)^2 + 16]^2}$

21. $\log \frac{p + 1}{p - 1}$

23. $\frac{1}{p} \arctan \frac{1}{p}$

25. $3 \cos \sqrt{5}t + \frac{7}{\sqrt{5}} \sin \sqrt{5}t$

27. $e^{3t} \sin t$

29. $\frac{1}{8} + \frac{3}{4} t - \frac{1}{4} t^2 - \frac{1}{8} \cos 2t - \frac{3}{8} \sin 2t$

31. $\frac{5}{6} \sinh 2t - \frac{2}{3} \sinh t$

33. $\frac{1}{3} e^{-2t/3} \left(5 \cos \frac{2\sqrt{5}}{3} t - \frac{8\sqrt{5}}{5} \sin \frac{2\sqrt{5}}{3} t \right)$

35. $\frac{3}{10} e^{2t} - \frac{1}{6} - \frac{2}{15} e^{-3t}$

37. $\frac{1}{\sqrt{\pi t}} \cos 2\sqrt{at}$

41. $\frac{p}{(p + 1)(p^2 + 1)}$

43. $\frac{1}{2} (\sin t - t \cos t)$

45. $\dfrac{\frac{1}{a} \sin at - \frac{1}{b} \sin bt}{b^2 - a^2}$

EXERCISES 5.4

1. $y = \sinh t$

3. $y = 1 - \cos t$

5. $y = e^{-t} \sin t - \dfrac{1}{2} \sin 2t$

7. $y = \dfrac{1}{6} + \dfrac{1}{3} e^{-t} + e^{-2t}\left(\dfrac{1}{2} \cos \sqrt{2}t - \dfrac{8}{3\sqrt{2}} \sin \sqrt{2}t\right)$

9. $y = \dfrac{1}{4}(1 + t) + \dfrac{1}{4}(3 - 7t)e^{2t}$

11. (a) $y = e^{-at}\left(y_0 \cosh \alpha t + \dfrac{v_0 + ay_0}{\alpha} \sinh \alpha t\right)$

    (b) $y = e^{-at}[y_0 + (v_0 + ay_0)t]$

    (c) $y = e^{-at}\left(y_0 \cos \mu t + \dfrac{v_0 + ay_0}{\mu} \sin \mu t\right)$

13. $g_1(t, \tau) = t - \tau$

15. $g_1(t, \tau) = \dfrac{1}{\sqrt{5}} \sin \sqrt{5}(t - \tau)$

17. $g_1(t, \tau) = 2e^{(t-\tau)} \sin \dfrac{1}{2} (t - \tau)$

19. $g_1(t, \tau) = \dfrac{1}{3} e^{-(t-\tau)} \sin \dfrac{1}{3} (t - \tau)$

23. $y = C(t - 1 + e^{-2t} + te^{-2t})$, where C is arbitrary

EXERCISES 5.5

5. $\dfrac{1}{\sqrt{2\pi}(a - is)}$

7. $\sqrt{\dfrac{2}{\pi}} \dfrac{\cos(\pi s/2)}{1 - s^2} e^{i\pi s/2}$

9. (a) $\sqrt{\dfrac{2}{\pi}}\, 2ias(s^2 + a^2)^{-2}$ (b) $\sqrt{\dfrac{2}{\pi}}\, (a^2 - s^2)(s^2 + a^2)^{-2}$

19. (a) $ise^{-s^2/2}$ (b) $(1 - s^2)e^{-s^2/2}$ (c) $\dfrac{1}{\sqrt{2}} e^{-(s-ib)^2/4}$

27. (a) $\dfrac{1}{\sqrt{2\pi}} 2s^3(s^4 + 4a^4)^{-1}$ (b) $\dfrac{1}{\sqrt{2\pi}} 4a^2 s(s^4 + 4a^4)^{-1}$

EXERCISES 5.6

3. $y = -\dfrac{1}{2} xe^{-x}$

5. $y = \begin{cases} 1 - e^{-1} \cosh x, & 0 < x < 1 \\ \sinh(1)e^{-x}, & 1 < x < \infty \end{cases}$

7. $y = e^{-kx} - \sqrt{\dfrac{2}{\pi}} \displaystyle\int_0^\infty \dfrac{\sin sx}{s^2 + k^2} F(s)\, ds$

9. $y = -\dfrac{1}{k} e^{-kx} - \sqrt{\dfrac{2}{\pi}} \displaystyle\int_0^\infty \dfrac{\cos sx}{s^2 + k^2} F(s)\, ds$

CHAPTER 6

EXERCISES 6.2

1. $u(x, t) = 3 \sin(\pi x)e^{-a^2\pi^2 t} - 5 \sin(4\pi x)e^{-16a^2\pi^2 t}$

3. $u(x, t) = \dfrac{8p^2}{\pi^3} \sum_{\substack{n=1 \\ (\text{odd})}}^{\infty} \dfrac{1}{n^3} \sin\left(\dfrac{n\pi x}{p}\right) e^{-a^2n^2\pi^2 t/p^2}$

5. $S(x) = T_1 + (T_2 - T_1)x/(1 + p)$

7. $u(x, t) = 1 - x + \dfrac{4}{\pi} \sum_{n=1}^{\infty} \dfrac{(-1)^{n-1}}{n} \sin(n\pi x)e^{-a^2n^2\pi^2 t}$

9. $u(x, t) = \dfrac{1}{2} T_0\left[1 - \cos\left(\dfrac{2\pi x}{p}\right)e^{-4a^2\pi^2 t/p^2}\right]$

11. $u(x, t) = T_0 + \dfrac{8}{\pi^3} \sum_{\substack{n=1 \\ (\text{odd})}}^{\infty} \dfrac{1}{n^3} \sin(n\pi x)e^{-a^2n^2\pi^2 t}$

13. $u(x, t) = \dfrac{\pi}{2} - \dfrac{4}{\pi} \sum_{\substack{n=1 \\ (\text{odd})}}^{\infty} \dfrac{1}{n^2} \cos(nx)e^{-a^2n^2 t}$

15. $u(x, t) = 100 + \sum_{n=1}^{\infty} c_n(\sin k_n x + k_n \cos k_n x)e^{-a^2k_n^2 t}$, where

    $\cos k_n\pi - k_n \sin k_n\pi = 0$ and $c_n = 400/\{k_n[(1 + k_n^2)\pi + 1]\}$,

    $n = 1, 2, 3, \ldots.$

17. $u(x, t) = 100 - x - \dfrac{200}{\pi} \sum_{n=1}^{\infty} \dfrac{1}{n} \sin\left(\dfrac{n\pi x}{50}\right) e^{-a^2n^2\pi^2 t/2500}$

19. $t = \dfrac{4}{a^2\pi^2} \log \dfrac{40}{3\pi^2}$

21. (a) $u(x, t) = 3x + \dfrac{10}{\pi} \sum_{n=1}^{\infty} \dfrac{1}{n} [5 + 7(-1)^n] \sin\left(\dfrac{n\pi x}{20}\right) e^{-0.0022n^2\pi^2 t}$

    (b) $u(5, 30) = 12.6°$ (c) $u(5, 30) = 14.11°$, no

    (d) approximately 2 min, 40 s

23. (a) $u(x, t) = T_0$

    (b) $u(x, t) = \dfrac{p^2}{6} - \dfrac{p^2}{\pi^2} \sum_{n=1}^{\infty} \dfrac{1}{n^2} \cos\left(\dfrac{2n\pi x}{p}\right) e^{-4a^2n^2\pi^2 t/p^2}$

25. $u(x, t) = \dfrac{4T_0}{\pi} \sum_{n=1}^{\infty} \dfrac{1}{n} \sin(n\pi x)e^{-(1+n^2\pi^2)t}$

27. $u(x, t) = \dfrac{\pi^2}{3} e^{-t} + 4 \sum_{n=1}^{\infty} \dfrac{(-1)^n}{n} \cos(nx)e^{-(1+3n^2)t}$

29. $u(x, t) = \dfrac{4}{\pi} e^{-x} \sum_{\substack{n=1 \\ (\text{odd})}}^{\infty} \dfrac{1}{n} \sin(nx)e^{-n^2 t}$

EXERCISES 6.3

1. $u(x, t) = \dfrac{1}{2}x(1-x) + \dfrac{1}{\pi}\sum_{\substack{n=1\\(\text{odd})}}^{\infty} \dfrac{1}{n}\left\{\dfrac{2}{n^2\pi^2}[(-1)^n - 1] - (-1)^n\right\} \sin(n\pi x)e^{-a^2n^2\pi^2 t}$

3. $u(x, t) = x^3$

5. $u(x, t) = \dfrac{a^2 A}{\omega}\sin \omega t$

7. $$u(x, t) = \sum_{n=1}^{\infty}\left\{\frac{4}{n^3\pi^3}(-1)^{n+1}(1 - e^{-n^2\pi^2 t}) - \frac{10}{n\pi}\frac{[1-(-1)^n]}{1+n^4\pi^4}(n^2\pi^2 \sin t - \cos t + e^{-n^2\pi^2 t})\right\}\sin n\pi x$$

9. $$u(x, t) = (1-x)\sin t + \sin(2\pi x)e^{-4\pi^2 t} + \frac{2}{\pi}\sum_{n=1}^{\infty}\frac{\sin nx}{n(1+n^4\pi^4)}[n^2\pi^2(e^{-n^2\pi^2 t} - \cos t) - \sin t]$$

11. $u(x, t) = t + (t^2 - t)x + \dfrac{x}{2}(x-1) + \dfrac{1}{\pi}\sum_{n=1}^{\infty}\dfrac{(-1)^{n-1}}{n}\sin(n\pi x)e^{-n^2\pi^2 t}$

EXERCISES 6.4

1. $u(x, t) = \dfrac{2}{\pi}\displaystyle\int_0^{\infty}\left(\frac{1-\cos cs}{s}\right)\sin(sx)e^{-a^2s^2t}\,ds$

3. $u(x, t) = \dfrac{2}{\pi}\displaystyle\int_0^{\infty}\left(\frac{s \sin sx}{s^2+1}\right)e^{-a^2s^2t}\,ds$

7. (a) $S(x) = T_1$

   (b) $v_{xx} = a^{-2}v_t$
   $v(0, t) = 0, \quad v, v_x$ finite as $x \to \infty$
   $v(x, 0) = -T_1$

   (c) No. The function $f(x) = T_1$ is not absolutely integrable and $B(s)$ does not exist.

9. $u(x, t) = \dfrac{2}{\pi}\displaystyle\int_0^{\infty}\left(\frac{\cos sx}{s^2+1}\right)e^{-a^2s^2t}\,ds$

11. $u(x, t) = \dfrac{T_0}{\pi}\displaystyle\int_0^{\infty}\left[\left(\frac{\sin s}{s}\right)\cos sx + \left(\frac{1-\cos s}{s}\right)\sin sx\right]e^{-a^2s^2t}\,ds$

EXERCISES 6.5

7. (b) $u(0, t) = 2aK\sqrt{\dfrac{t}{\pi}}$

9. (a) $t \simeq 1.12$ s (b) $t \simeq 4.47$ s (c) $t \simeq 46.54$ min

11. (b) 0

13. (b) $u(x, t) = \frac{1}{2} e^{|x| + a^2 t}\left[1 - \operatorname{erf}\left(a\sqrt{t} + \frac{|x|}{2a\sqrt{t}}\right)\right]$

$+ \frac{1}{2} e^{-|x| + a^2 t}\left[1 + \operatorname{erf}\left(a\sqrt{t} + \frac{|x|}{2a\sqrt{t}}\right)\right]$

15. $u(x, t) = \frac{T_0}{2}\left[\operatorname{erf}\left(\frac{c + x}{2a\sqrt{t}}\right) - \operatorname{erf}\left(\frac{c - x}{2a\sqrt{t}}\right)\right]$

17. $u(x, t) = \frac{T_0}{2}\left[\operatorname{erfc}\left(\frac{c + x}{2a\sqrt{t}}\right) + \operatorname{erfc}\left(\frac{c - x}{2a\sqrt{t}}\right)\right]$

19. $v(x, t) = \frac{1}{2a\sqrt{\pi t}} \int_0^\infty g(\xi)\left\{\exp\left[-\frac{(x - \xi)^2}{4a^2 t}\right] - \exp\left[-\frac{(x + \xi)^2}{4a^2 t}\right]\right\} d\xi,$

where $g(\xi) = f'(\xi) - kf(\xi)$

23. $u(x, t) = \int_0^\infty f(\xi)[k(x - \xi, t) + k(x + \xi, t)]\, d\xi$

$+ a^2 \int_0^\infty \int_0^t q(\xi, \tau)[k(x - \xi, t - \tau) + k(x + \xi, t - \tau)]\, d\tau\, d\xi,$

where

$k(x, t) = (2a\sqrt{\pi t})^{-1} e^{-x^2/4a^2 t}; \qquad g(x, \xi; t, \tau) = k(x - \xi, t - \tau) + k(x + \xi, t - \tau)$

EXERCISES 6.6

1. $S_1 = 0.3438, \quad S_2 = 0.6250, \quad S_3 = 0.8438$

3. $S_1 = 0.4375, \quad S_2 = 0.8125, \quad S_3 = 1.125$

9.

t \ x	0	1/4	1/2	3/4	1
0	0	0.2500	0.5000	0.7500	1
1/32	0	0.2500	0.5000	0.7500	0
1/16	0	0.2500	0.5000	0.2500	0
3/32	0	0.2500	0.2500	0.2500	0
1/8	0	0.1250	0.2500	0.1250	0

11.

t \ x	0	1/4	1/2	3/4	1
0	0	0	0	0	0
1/32	0	1.00	1.00	1.00	0
1/16	0	1.50	2.00	1.50	0
3/32	0	2.00	2.50	2.00	0
1/8	0	2.25	3.00	2.25	0

13.

t \ x	0	1/4	1/2	3/4	1
0	0.0000	0.2500	0.5000	0.7500	1.0000
1/32	0.2500	0.2500	0.5000	0.7500	1.0000
1/16	0.3750	0.3750	0.5000	0.7500	1.0000
3/32	0.4375	0.4375	0.5625	0.7500	1.0000
1/8	0.5000	0.5000	0.5938	0.7813	1.0000

17.

x	CENTRAL DIFFERENCE	EXACT
0	1.000	1.000
0.25	0.875	0.875
0.50	0.750	0.750
0.75	0.625	0.625
1.00	0.500	0.500

CHAPTER 7

EXERCISES 7.2

1. $u(x, t) = \dfrac{4v_0}{c\pi^2} \sum_{\substack{n=1 \\ \text{(odd)}}}^{\infty} \dfrac{1}{n^2} \sin nc\pi t \sin n\pi x$

3. $u(x, t) = \dfrac{8}{\pi^3} \sum_{\substack{n=1 \\ \text{(odd)}}}^{\infty} \dfrac{1}{n^3} \cos nc\pi t \sin n\pi x$

5. $\omega = \dfrac{\pi}{p} \sqrt{\dfrac{T}{m}}$

7. Maximum displacement is 10 cm at $x = \pi/2$ m and $t = \pi/80$ s.

9. $u(x, t) = \dfrac{2p^2}{b(p-b)\pi^2} \sum_{n=1}^{\infty} \dfrac{1}{n^2} \sin \dfrac{n\pi b}{p} \cos \dfrac{nc\pi t}{p} \sin \dfrac{n\pi x}{p}$

11. $u(x, t) = Ae^{-kt}\left(\cos \mu t + \dfrac{k}{\mu} \sin \mu t\right) \sin \pi x$, where $\mu = \sqrt{\pi^2 c^2 - k^2}$

EXERCISES 7.3

1. $$u(x, t) = \frac{4Pc}{\pi} \sum_{\substack{n=1 \\ \text{(odd)}}}^{\infty} \frac{\cos nct \sin nx}{n^2(\omega^2 - n^2c^2)} + \frac{Pc^2 \cos \omega t}{\omega^2}\left[\cos \frac{\omega x}{c} - 1\right] + \frac{Pc^2 \sin \dfrac{\omega x}{c} \cos \omega t}{\omega^2 \sin \dfrac{\omega\pi}{c}}\left[1 - \cos \frac{\omega\pi}{c}\right]$$

3. $u(x, t) = \frac{1}{2} x \cos x \cos t + \frac{8}{\pi} \sum_{n=1}^{\infty} \frac{(-1)^n n}{(4n^2 - 1)^2} \cos 2nt \sin 2nx$

5. $u(x, t) = \frac{4Pc^2}{\pi} \sum_{\substack{n=1 \\ (\text{odd})}}^{\infty} \frac{\sin nx}{n(\omega^2 - n^2c^2)} (\cos nct - \cos \omega t)$

7. $u(x, t) = \frac{4V_0}{\pi^2} \sum_{\substack{n=1 \\ (\text{odd})}}^{\infty} \frac{\sin n\pi x}{n^2} \sin n\pi t + \frac{2k}{\pi^3} \sum_{n=1}^{\infty} \frac{(-1)^{n-1}}{n^3} (1 - \cos n\pi t) \sin n\pi x$

9. $u(x, t) = 2 + \frac{2}{3\pi} \sin \frac{3\pi t}{2} \sin \frac{3\pi x}{2}$

$$+ \frac{160}{\pi^3} \sum_{n=1}^{\infty} \frac{1}{(2n-1)^3} \left[1 - \cos \frac{(2n-1)\pi t}{2}\right] \sin \frac{(2n-1)\pi x}{2}$$

13. $u(x, t) = \frac{1}{2} t^2(1 - x) - x \cos t$

$$+ \frac{2}{\pi} \sum_{n=1}^{\infty} \left[\frac{1 - \cos n\pi t}{n^2\pi^2} + \frac{2(\cos t - \cos n\pi t)}{n^2\pi^2 - 1}\right] \frac{(-1)^{n-1}}{n} \sin n\pi x$$

EXERCISES 7.4

1. $u(x, t) = \frac{2}{\pi c} \int_0^{\infty} \frac{\sin sx}{s^2 + 1} \sin sct \, ds$

3. $u(x, t) = \int_0^{\infty} [A(s) \cos sct + B(s) \sin sct] \cos sx \, ds$, where

$$A(s) = \frac{2}{\pi} \int_0^{\infty} f(x) \cos sx \, dx, \qquad B(s) = \frac{2}{\pi sc} \int_0^{\infty} g(x) \cos sx \, dx$$

7. $u(x, t) = \frac{2}{\pi c} \int_0^{\infty} \frac{\sin sct \cos sx}{s(s^2 + 1)} \, ds$

EXERCISES 7.5

3. (a) $u(x, t) = -\frac{1}{2} (t - x)^2 h(t - x)$

   (b) $u(x, t) = \frac{A}{\omega} [\cos \omega(t - x) - 1] h(t - x)$

7. $u(x, t) = x + \sqrt{\frac{2}{\pi}} \sum_{n=1}^{\infty} \frac{(-1)^n}{n} \sin n\pi x \cos n\pi ct$

11. $u(x, t) = \sqrt{\frac{2}{\pi}} \int_0^{\infty} \left[F(s) \cos cst + \frac{G(s)}{cs} \sin cst\right] \cos sx \, ds$

13. $u(x, t) = \frac{1}{\pi c} \int_{-\infty}^{\infty} e^{-isx} \frac{\sin cst}{s(s^2 + 1)} \, ds$

EXERCISES 7.6

1.

t \ x	0	1/4	1/2	3/4	1
0	0	0.7071	1.0000	0.7071	0
1/4	0	0.7071	1.0000	0.7071	0
1/2	0	0.2929	0.4142	0.2929	0
3/4	0	−0.2929	−0.4142	−0.2929	0
1	0	−0.7071	−1.0000	−0.7071	0

3.

t \ x	0	1/4	1/2	3/4	1
0	0	0	0	0	0
1/4	0	0.2500	0.2500	0.2500	0
1/2	0	0.2500	0.5000	0.2500	0
3/4	0	0.2500	0.2500	0.2500	0
1	0	0.0000	0.0000	0.0000	0

CHAPTER 8

EXERCISES 8.2

1. $u(x, y) = \dfrac{\sinh y}{\sinh 1} \sin x$

3. $u(x, y) = \dfrac{1}{3} + \dfrac{4}{\pi^2} \displaystyle\sum_{n=1}^{\infty} \frac{(-1)^n}{n^2 \cosh n\pi} \cosh[n\pi(1 - y)] \cos n\pi x$

7. $\alpha_n = \dfrac{2}{a \sinh \dfrac{n\pi b}{a}} \displaystyle\int_0^a f_2(x) \sin \frac{n\pi x}{a}\, dx,$

$\beta_n = \dfrac{2}{a \sinh \dfrac{n\pi b}{a}} \displaystyle\int_0^a f_1(x) \sin \frac{n\pi x}{a}\, dx, \qquad n = 1, 2, 3, \ldots$

9. $u(x, y) = 100 + \dfrac{400}{\pi} \displaystyle\sum_{\substack{n=1 \\ (\text{odd})}}^{\infty} \frac{\sinh \frac{n\pi y}{2}}{n \sinh n\pi} \sin \frac{n\pi x}{2}$

11. $u(x, y) = 2 \displaystyle\sum_{n=1}^{\infty} \frac{\sin k_n \cos k_n x \sinh k_n y}{k_n \sinh 2k_n(1 + \sin^2 k_n)},$

where $\tan k_n = 1/k_n, \qquad n = 1, 2, 3, \ldots$

EXERCISES 8.3

1. $u(r, \theta) = \frac{1}{2}\left[1 + \left(\frac{r}{\rho}\right)^2 \cos 2\theta\right]$

3. $u(r, \theta) = \frac{1}{2} a_0 + \frac{4}{\pi} \sum_{\substack{n=1 \\ (\text{odd})}}^{\infty} \frac{r^n}{n^2} \sin n\theta$

5. (a) $u(r, \theta) = r \sin \theta$ (b) $u(r, \theta) = \frac{4T_0}{\pi} \sum_{\substack{n=1 \\ (\text{odd})}}^{\infty} \frac{r^n}{n} \sin n\theta$

9. $u(r, \theta) = \dfrac{T_1 \log b - T_2 \log a + (T_2 - T_1) \log r}{\log(b/a)}$

EXERCISES 8.4

11. The compatibility condition (Theorem 8.4) leads to the value $\pi - \pi^2/2$ instead of zero.

13. $\int_0^1 f(x)\, dx = 0$

EXERCISES 8.5

1. $u(x, y) = \frac{4}{\pi} \int_0^\infty e^{-sy} \left(\frac{\sin s - s \cos s}{s^3}\right) \cos sx\, ds$

3. $u(x, y) = \frac{1}{2} + \frac{1}{\pi} \tan^{-1} \frac{x}{y}$

5. $u(x, y) = T_0$

7. $u(x, y) = \frac{y}{\pi} \int_{-\infty}^\infty \frac{f(t)}{(t - x)^2 + y^2}\, dt$

11. $u(x, y) = \frac{1}{\pi} \tan^{-1}\left(\frac{2cy}{x^2 + y^2 - b^2}\right) + \frac{2}{\pi} \tan^{-1} \frac{x}{y}$

EXERCISES 8.6

1. $\psi(x, y) = \frac{\sinh y}{\sinh 1} \sin x$

3. $\psi(x, y) = \frac{\sinh(1 - y)}{\sinh 1} \cos x$

5. $\psi(x, y) = \frac{\sinh y}{\sinh 1} \sin x$

7. $\psi(x, y) = \frac{\sinh(1 - y)}{\sinh 1} \cos x$

9. (a) $\psi(x, y) = x(1 - x)e^{-2\sqrt{2}y}$ (b) $\psi(x, y) = x(1 - x)e^{-\sqrt{10}y}$

11.

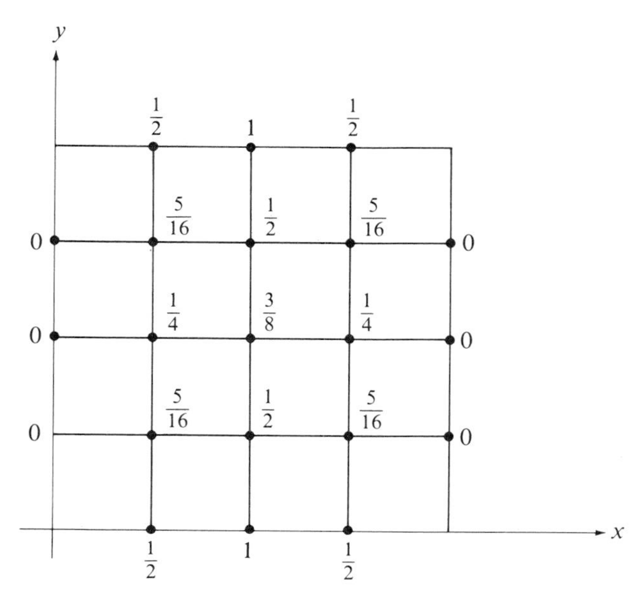

13.

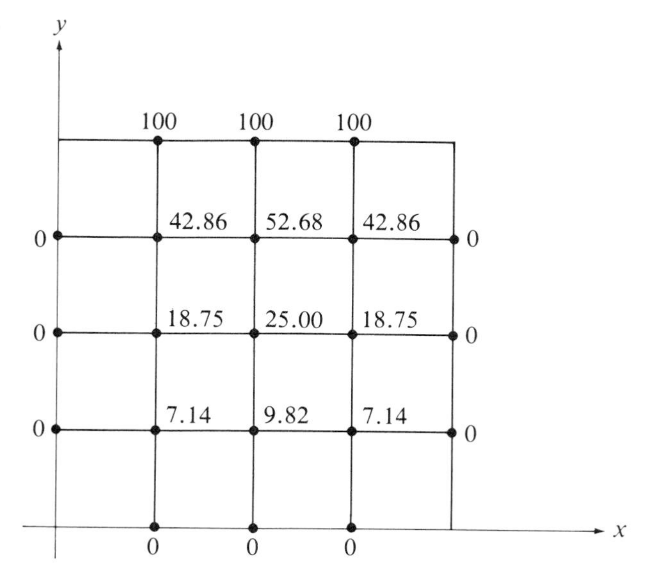

15. $4u_{i,j} - [u_{i+1,j} + u_{i-1,j} + u_{i,j+1} + u_{i,j-1}] = 0$

$u_{1,j} - u_{0,j} = 0, \qquad u_{n,j} - u_{n-1,j} = 0$

$u_{i,0} = \cos \dfrac{i\pi}{n}, \qquad u_{i,n} = \cos^2 \dfrac{i\pi}{n}$

$(i, j = 1, 2, \ldots, n - 1)$

CHAPTER 9

EXERCISES 9.2

1. (a), (b) $P_1(x) = x,\quad P_2(x) = \frac{1}{2}(3x^2 - 1),\quad P_3(x) = \frac{1}{2}(5x^3 - x),$

$P_4(x) = \frac{1}{8}(35x^4 - 30x^2 + 3)$

3. (a) $\lambda_n = n(n+1),\quad \phi_n(x) = P_n(x);\quad n = 1, 3, 5, \ldots$
   (b) $\lambda_n = n(n+1),\quad \phi_n(x) = P_n(x);\quad n = 0, 2, 4, \ldots$

7. (a) $y = CP_2(x) + \frac{1}{6}P_0(x),$ C arbitrary

   (b) $y = CP_2(x) + \frac{1}{4}P_1(x),$ C arbitrary

11. $y = C_1P_1(x) + C_2Q_1(x)$

13. $y = C_1P_5(x) + C_2Q_5(x)$

17. $g(x, s) = \begin{cases} Q_2(s)P_2(x), & 0 < s \leq x \\ Q_2(x)P_2(s), & x < s < 1 \end{cases}$

EXERCISES 9.3

1. (a) 120 (b) $\frac{1}{2}\sqrt{\pi}$ (c) $\frac{15}{8}\sqrt{\pi}$

   (d) $-2\sqrt{\pi}$ (e) $-\frac{32\sqrt{\pi}}{945}$ (f) $\frac{10}{9}$

7. $\frac{1}{3}$

9. $\frac{\pi}{\sqrt{2}}$

EXERCISES 9.4

7. $y = C_1J_0(x) + C_2Y_0(x)$

9. $y = C_1J_3(2x) + C_2Y_3(2x)$

13. $y = C_1\frac{\cos x}{\sqrt{x}} + C_2\frac{\sin x}{\sqrt{x}}$

39. $y = 1 - \frac{J_0(x)}{J_0(1)}$

EXERCISES 9.5

1. $\lambda_n = k_n^2,\quad \phi_n(x) = J_0(k_nx),\quad$ where $J_0(k_n) = 0,\quad n = 1, 2, 3, \ldots$

3. $\lambda_n = k_n^2,\quad \phi_n(x) = J_2(k_nx),\quad$ where $J_2(3k_n) = 0,\quad n = 1, 2, 3, \ldots$

5. $\phi_n(x) = J_\nu(k_nx),\quad n = 1, 2, 3, \ldots$

7. $y = C_1 J_0(\sqrt{x}) + C_2 Y_0(\sqrt{x})$

9. $y = C_1 J_k(x^2) + C_2 Y_k(x^2)$

11. $y = x^{-n}[C_1 J_n(x) + C_2 Y_n(x)]$

13. $y = x^4[C_1 J_{3/4}(2x^3) + C_2 J_{-3/4}(2x^3)]$

15. $y = \sqrt{x}\left[C_1 J_{1/4}\left(\frac{kx^2}{2}\right) + C_2 J_{-1/4}\left(\frac{kx^2}{2}\right)\right]$

17. $y = \sqrt{x}\,[J_0(2\sqrt{x}) + Y_0(2\sqrt{x})]$

19. (b) $y(t) = C_1 J_0\left(\frac{2}{m}\sqrt{at}\right) + C_2 Y_0\left(\frac{2}{m}\sqrt{at}\right)$

(c) $y(x) = C_1 J_0\left(\frac{2}{m}\sqrt{a}\,e^{mx/2}\right) + C_2 Y_0\left(\frac{2}{m}\sqrt{a}\,e^{mx/2}\right)$

23. $r(t) = \left(t + \frac{a}{b}\right)\left[AJ_1\left(kt + \frac{ka}{b}\right) + BY_1\left(kt + \frac{ka}{b}\right)\right]$, where

$A = \frac{\pi b r_0}{2a}\left[Y_1\left(\frac{ka}{b}\right) + \frac{ka}{b}Y_1'\left(\frac{ka}{b}\right)\right]$, $\quad B = -\frac{\pi b r_0}{2a}\left[J_1\left(\frac{ka}{b}\right) + \frac{ka}{b}J_1'\left(\frac{ka}{b}\right)\right]$

25. $\lambda_n = k_n^2$, $\quad \phi_n(x) = x^{1/2}J_{1/4}\left(\frac{1}{2}k_n x^2\right)$, where

$J_{1/4}\left(\frac{1}{2}k_n\right) = 0, \quad n = 1, 2, 3, \ldots$

31. (a) $k_1 = \sqrt{12} \simeq 3.464$ (b) $k_1 = \sqrt{15} \simeq 3.873$ (c) $k_1 = 3.832$

33. $b_0 = 2$

EXERCISES 9.6

3. $y = C_1 I_0(x) + C_2 K_0(x)$

5. $y = C_1 I_1(2x) + C_2 K_1(2x)$

19. $y = x^{1/2}\left[C_1 I_{1/3}\left(\frac{2}{3}x^{3/2}\right) + C_2 I_{-1/3}\left(\frac{2}{3}x^{3/2}\right)\right]$

21. $y = x^2[C_1 I_{2/3}(x^3) + C_2 I_{-2/3}(x^3)]$

23. $y = x^{1/2}\left[C_1 I_{1/6}\left(\frac{kx^3}{3}\right) + C_2 I_{-1/6}\left(\frac{kx^3}{3}\right)\right]$

EXERCISES 9.7

1. (a) $f(x) = \frac{1}{4}P_0(x) + \frac{1}{2}P_1(x) + \frac{5}{16}P_2(x) + \cdots$

(b) $f(x) = \frac{1}{2}P_0(x) + \frac{5}{8}P_2(x) - \frac{3}{16}P_4(x) + \cdots$

9. $f(x) = 2\sum_{n=1}^{\infty}\frac{J_0(k_n x)}{k_n J_1(k_n)}$

11. $f(x) = -2\sum_{n=1}^{\infty}\frac{J_0(k_n x)}{[k_n J_1(k_n)]^2}$

17. (b) $u(x, t) = \frac{1}{3} + \left(x^2 - \frac{1}{3}\right)e^{-6bt}$

CHAPTER 10

EXERCISES 10.2

1. (a) $y = x$ (b) $y = 1 - x$

   (c) $y = \frac{1}{4}, \frac{1}{2}, \frac{3}{4}$ (d) $x = \frac{1}{3}, \frac{2}{3}$; $y = \frac{1}{4}, \frac{1}{2}, \frac{3}{4}$

3. $a = \dfrac{mb}{n}$

5. $\lambda_{mn} = m^2\pi^2 + n^2$; $m = 1, 2, 3, \ldots,$ $n = 0, 1, 2, \ldots$

   $\phi_{mn}(x, y) = \sin(m\pi \log x) \cos(ny)$

7. $$f(x, y) = \frac{16}{\pi^2} \sum_{\substack{m=1 \\ (\text{odd})}}^{\infty} \sum_{\substack{n=1 \\ (\text{odd})}}^{\infty} \frac{1}{mn} \sin \frac{m\pi x}{a} \sin \frac{n\pi y}{b}$$

9. $$f(x, y) = \frac{8a}{\pi^2} \sum_{m=1}^{\infty} \sum_{\substack{n=1 \\ (\text{odd})}}^{\infty} \frac{(-1)^{m-1}}{mn} \sin \frac{m\pi x}{a} \sin \frac{n\pi y}{b}$$

11. $u(x, y, t) = A \cos \alpha ct \sin \dfrac{3\pi x}{a} \sin \dfrac{5\pi y}{b}$, where $\alpha^2 = \left(\dfrac{9}{a^2} + \dfrac{25}{b^2}\right)\pi^2$

13. $u(x, y, t) = \displaystyle\sum_{m=1}^{\infty} \sum_{n=1}^{\infty} A_{mn} \sin mx \sin \frac{n\pi y}{b} e^{-\lambda_{mn} a^2 t}$, where

    $\lambda_{mn} = m^2 + \dfrac{n^2\pi^2}{b^2}$; $m, n = 1, 2, 3, \ldots$

17. $f(x, y) = 1$

19. $$f(x, y) = \frac{a}{2} - \frac{4a}{\pi^2} \sum_{\substack{m=1 \\ (\text{odd})}}^{\infty} \frac{1}{m^2} \cos \frac{m\pi x}{a}$$

21. $$u(x, y, t) = \sum_{m=1}^{\infty} [A_{m0} \cos(\sqrt{\lambda_{m0}}\, ct) + B_{m0} \sin(\sqrt{\lambda_{m0}}\, ct)] \sin \frac{m\pi x}{a}$$
    $$+ \sum_{m=1}^{\infty} \sum_{n=1}^{\infty} [A_{mn} \cos(\sqrt{\lambda_{mn}}\, ct) + B_{mn} \sin(\sqrt{\lambda_{mn}}\, ct)] \sin \frac{m\pi x}{a} \cos \frac{n\pi y}{b},$$

    where $\lambda_{mn} = \left(\dfrac{m^2}{a^2} + \dfrac{n^2}{b^2}\right)\pi^2$; $m = 1, 2, 3, \ldots,$ $n = 0, 1, 2, \ldots.$

23. $u(x, y, t) = \displaystyle\sum_{m=1}^{\infty} \sum_{n=1}^{\infty} A_{mn} \sin \frac{(2m-1)\pi x}{2a} \sin \frac{(2n-1)\pi y}{2b} e^{-\lambda_{mn} t}$, where

    $\lambda_{mn} = \left[\dfrac{(2m-1)^2}{a^2} + \dfrac{(2n-1)^2}{b^2}\right] \dfrac{\pi^2}{4}$; $m, n = 1, 2, 3, \ldots$

25. (a) $u(x, y, t) = A \sin \pi x \sin \pi y e^{-2\pi^2 a^2 t}$

27. $S_{xx} + S_{yy} = 0$

    B.C.: $\begin{cases} S(0, y) = 0, & S(a, y) = 0 \\ S(x, 0) = 0, & S(x, b) = g(x) \end{cases}$

    $v_{xx} + v_{yy} = v_t$

B.C.: $\begin{cases} v(0, y, t) = 0, & v(a, y, t) = 0 \\ v(x, 0, t) = 0, & v(x, b, t) = 0 \end{cases}$

I.C.: $v(x, y, 0) = f(x, y) - S(x, y)$

EXERCISES 10.3

1. $u(r, t) = 0.1J_0(k_1 r) \cos k_1 ct$

3. $n = 2;\quad r = 0.4357$
   $n = 3;\quad r = 0.2779, \qquad r = 0.6379$

5. $u(r, t) = \sum_{n=1}^{\infty} (a_n \cos k_n ct + b_n \sin k_n ct) J_0(k_n r),$
   where $J_0(k_n \rho) = 0, \qquad n = 1, 2, 3, \ldots,$
   $$a_n = \frac{2}{\rho^2 [J_1(k_n \rho)]^2} \int_0^{\rho} r f(r) J_0(k_n r)\, dr, \qquad n = 1, 2, 3, \ldots$$
   $$k_n c b_n = \frac{2}{\rho^2 [J_1(k_n \rho)]^2} \int_0^{\rho} r g(r) J_0(k_n r)\, dr, \qquad n = 1, 2, 3, \ldots$$

7. $u(r, t) = 2 \sum_{n=1}^{\infty} \left(1 - \frac{16}{k_n^2} + \frac{64}{k_n^4}\right) \frac{J_0(k_n r)}{k_n J_1(k_n)} \cos k_n ct$

9. $u(r, t) = 8 \sum_{n=1}^{\infty} \frac{J_0(k_n r)}{k_n^3 J_1(k_n)} \cos k_n ct$

13. $u(r, t) = T_0 \left[1 - 2 \sum_{n=1}^{\infty} \frac{J_0(k_n r)}{k_n J_1(k_n)} e^{-k_n^2 t} \right],$ where $J_0(k_n) = 0, \quad n = 1, 2, 3, \ldots$
    $u(r, t) \to T_0$

15. $u(r, t) = 2T_0 \sum_{n=1}^{\infty} \frac{J_0(k_n r)}{k_n J_1(k_n)} e^{-k_n^2 t},$ where $J_0(k_n) = 0, \qquad n = 1, 2, 3, \ldots$

EXERCISES 10.4

1. $u(r, z) = AI_0\left(\frac{3\pi r}{a}\right) \sin \frac{3\pi z}{a} \Big/ I_0\left(\frac{3\pi}{a}\right)$

5. $u(r, z) = \sum_{n=1}^{\infty} c_n I_0\left(\frac{n\pi r}{a}\right) \sin \frac{n\pi z}{a},$ where
   $$c_n I_0\left(\frac{n\pi\rho}{a}\right) = \frac{2}{a} \int_0^a f(z) \sin \frac{n\pi z}{a}\, dz, \qquad n = 1, 2, 3, \ldots$$

7. $u(r, z) = \frac{4T_0}{\pi} \sum_{n=1}^{\infty} \frac{1}{2n - 1} \frac{I_0\left[\frac{(2n-1)r}{2}\right]}{I_0\left(\frac{2n-1}{2}\right)} \sin \frac{(2n-1)z}{2}$

9. $u(r, z) = 8 \sum_{n=1}^{\infty} \frac{\sinh k_n z}{\sinh k_n} \frac{J_0(k_n r)}{k_n^3 J_1(k_n)}$

EXERCISES 10.5

1. (a) $u(r, \phi) = 1$ (b) $u(r, \phi) = r \cos \phi$

   (c) $u(r, \phi) = \frac{2}{3} r^2 P_2(\cos \phi) + \frac{1}{3}$ (d) $u(r, \phi) = \frac{4}{3} r^2 P_2(\cos \phi) - \frac{1}{3}$

3. $u(r, \phi) = \sum_{n=0}^{\infty} a_n \left(\frac{r}{\rho}\right)^n P_n(\cos \phi)$, where

$$a_n = \left(n + \frac{1}{2}\right) \int_0^{\pi} f(\phi) P_n(\cos \phi) \sin \phi \, d\phi, \qquad n = 0, 1, 2, \ldots$$

7. $u(r) = C_1 + \frac{C_2}{r}$

EXERCISES 10.6

1. $u(x, y) = \frac{\sinh y}{\sinh 1} \sin x + \frac{16}{\pi^2} \sum_{\substack{m=1 \\ (\text{odd})}}^{\infty} \sum_{\substack{n=1 \\ (\text{odd})}}^{\infty} \frac{\sin mx \sin n\pi y}{mn(m^2 + n^2\pi^2)}$

3. $u(x, y) = \frac{4}{\pi} \sum_{\substack{n=1 \\ (\text{odd})}}^{\infty} \frac{\sinh ny + \sinh n(\pi - y)}{n \sinh n\pi} \sin nx + \frac{4 \sin 3y}{\pi} \sum_{\substack{m=1 \\ (\text{odd})}}^{\infty} \frac{\sin mx}{m(m^2 + 9)}$

5. $u(r, z) = \sum_{n=1}^{\infty} A_n I_0\left(\frac{n\pi r}{a}\right) \sin \frac{n\pi z}{a} + \sum_{m=1}^{\infty} \sum_{n=1}^{\infty} B_{mn} J_0(k_n r) \sin \frac{m\pi z}{a}$, where

$J_0(k_n) = 0, \qquad n = 1, 2, 3, \ldots,$

$$A_n I_0\left(\frac{n\pi}{a}\right) = \frac{2}{a} \int_0^a f(z) \sin \frac{n\pi z}{a} \, dz, \qquad n = 1, 2, 3, \ldots,$$

$$B_{mn} = \frac{4}{a[J_1(k_n)]^2 \left(\frac{m^2\pi^2}{a^2} + k_n^2\right)} \int_0^1 \int_0^a q(r, z) r J_0(k_n r) \sin \frac{m\pi z}{a} \, dz \, dr;$$

$m, n = 1, 2, 3, \ldots$

7. $u(r, \phi) = \sum_{n=0}^{\infty} a_n r^n P_n(\cos \phi) + r^{-1/2} \sum_{m=0}^{\infty} \sum_{n=1}^{\infty} C_{mn} P_m(\cos \phi) J_{m+(1/2)}(k_{mn} r)$,

where $J_{m+(1/2)}(k_{mn}) = 0; \qquad m = 0, 1, 2, \ldots, \; n = 1, 2, 3, \ldots,$

$$a_n = \left(n + \frac{1}{2}\right) \int_0^{\pi} f(\phi) P_n(\cos \phi) \sin \phi \, d\phi, \qquad n = 0, 1, 2, \ldots,$$

$$C_{mn} = \frac{1}{k_{mn}^2 \| \phi_{mn}(r, \phi) \|^2} \int_0^1 \int_0^{\pi} q(r, \phi) r^{3/2} J_{m+(1/2)}(k_{mn} r) P_m(\cos \phi) \sin \phi \, d\phi \, dr;$$

$m = 0, 1, 2, \ldots, \qquad n = 1, 2, 3, \ldots$

INDEX

A6
B7
C8
D9
E0
F1
G2
H3